SMALL ENGINE MECHANICS

Third Edition

William H. Crouse
Donald L. Anglin

Gregg Division ■ McGraw-Hill Book Company

New York Atlanta Dallas St. Louis
San Francisco Auckland Bogotá Guatemala
Hamburg Johannesburg Lisbon London
Madrid Mexico Montreal New Delhi
Panama Paris San Juan São Paulo
Singapore Sydney Tokyo Toronto

Sponsoring Editor: D. Eugene Gilmore
Editing Supervisor: Larry Goldberg
Design and Art Supervisor/
Cover Designer: Patricia F. Lowy
Production Supervisor: Al Rihner

Text Designer: Levavi & Levavi
Cover Illustration: Courtesy of Lawn Boy

ABOUT THE AUTHORS

William H. Crouse

Behind William H. Crouse's clear technical writing is a background of sound mechanical engineering training as well as a variety of practical industrial experience. After finishing high school, he spent a year working in a tinplate mill. Summers, while still in school, he worked in General Motors plants, and for three years he worked in the Delco-Remy Division shops. Later he became Director of Field Education in the Delco-Remy Division of General Motors Corporation, for which he prepared service bulletins and educational literature.

Mr. Crouse has contributed numerous articles to automotive and engineering magazines and has written several books about science and technology. He was the first Editor-in-Chief of the 15-volume *McGraw-Hill Encyclopedia of Science and Technology*. In addition, he has authored more than 50 technical books, including *Automotive Mechanics*, which has sold over a million copies. His books have been widely translated and used in automotive mechanics training throughout the world.

William H. Crouse's outstanding work in the automotive field has earned for him membership in the Society of Automotive Engineers and in the American Society of Engineering Education.

Donald L. Anglin

Trained in the automotive and diesel service field, Donald L. Anglin has worked both as a mechanic and as a service manager. He has taught automotive courses in high school, trade schools, community colleges, and universities. He has also worked as curriculum supervisor and school administrator for an automotive trade school. Interested in all types of vehicle performance, he has served as a racing-car mechanic and as a consultant to truck fleets on maintenance problems.

Currently, he devotes full time to technical writing, teaching, and visiting automotive instructors and service shops. Together with William H. Crouse, he has coauthored a number of magazine articles on automotive education, as well as many books in the McGraw-Hill Automotive Technology Series.

Donald L. Anglin is a Certified General Automotive Mechanic, a Certified General Truck Mechanic, and he holds many other licenses and certificates in automotive education, service, and related areas. Mr. Anglin's extensive work in the automotive service field has been recognized by membership in the American Society of Mechanical Engineers and the Society of Automotive Engineers.

Library of Congress Cataloging-in-Publication Data

Crouse, William Harry, (date)
Small engine mechanics.

Includes index.
1. Internal combustion engines, Spark ignition.
I. Anglin, Donald L. II. Title.
TJ790.C76 1986 621.43'4 85-25630
ISBN 0-07-014803-1

Small Engine Mechanics, Third Edition

1 2 3 4 5 6 7 8 9 0 SEMSEM 8 9 3 2 1 0 9 8 7 6

ISBN 0-07-014803-1

Contents

Preface

This is the third edition of *Small Engine Mechanics*. In this edition, many changes have been made to reflect the growing popularity of small-engine-powered equipment and the increasing variety of applications in which small engines are used.

The book has been shortened but strengthened, as suggested by users of the previous edition. New emphasis on liquid-cooled engines and electronic ignition is now included. Two new chapters on small diesel engines provide needed coverage in this rapidly developing area. New material on mower controls and mower blade brakes teach the student how to safely service this equipment. These and many other changes in content reflect the continuing significant technological advances that are now found on engines today.

Metric equivalents of all United States Customary measurements continue to be featured in this edition. For example, the dimensions of this book would be shown as 8½ × 11 inches [216 × 279 mm (millimeters)]. In addition to all the new developments covered in this edition, the book has been almost completely rewritten to simplify explanations, shorten sentences, and improve readability. There are objectives at the beginning of each chapter so that the student and instructor both know the expected learning results for that chapter.

A new edition of the *Workbook for Small Engine Mechanics* has also been prepared. It includes the basic engine-service jobs as proposed by small-engine instructors and the small-engine manufacturers. Taken together, *Small Engine Mechanics* and the workbook provide the user with the background information and hands-on experience needed to become a qualified small-engine mechanic and technician.

To assist the instructor, the *Instructor's Planning Guide for Small Engine Mechanics* is available. It was prepared to help the instructor do the best possible job of teaching by most effectively using the textbook, workbook, and other related instructional materials.

Together, the textbook, workbook, and instructor's guide make up an instructional program that will fit any teaching situation. The program is flexible enough to fit classroom instruction needs, shop activities, individual instruction, and do-it-yourself courses for hobbyists and consumers.

The authors are grateful to the many people, both in industry and in education, whose contributions and comments helped shape this book. They share, with the authors, a hope that this program will achieve the aims of all who work in the field of small-engine mechanics instruction: to train their students to become highly competent and qualified small-engine mechanics.

William H. Crouse
Donald L. Anglin

Acknowledgments

While preparing this book, the authors were given valuable aid and inspiration by many people in the small-engine field and in education. The authors gratefully acknowledge their indebtedness and offer their sincere thanks. All cooperated with the aim of providing accurate and complete information on how small engines are constructed, how they operate, and how to maintain and service them.

Special thanks are due to the following organizations for information and illustrations they supplied: American Honda Motor Company, Inc.; Ariens Company; ATW; Robert Bosch Corporation; Briggs & Stratton Corporation; The J. I. Case Company; Champion Spark Plug Company; Clinton Engines Corporation; Delco-Remy Division of General Motors Corporation; Deutz Corporation; Evinrude Motors Division of Outboard Marine Corporation; Federal-Mogul Corporation; Ford Motor Company; Fram Corporation; Husqvarna; Jacobsen Manufacturing Company; Kohler Company; Lawn Boy Division of Outboard Marine Corporation; Lister Diesels Division of Hawker Siddeley, Inc.; Neway Manufacturing, Inc.; Onan Corporation; Outboard Marine Corporation; Phillips Temro, Inc.; Selastomer Division of Microdot, Inc.; Sioux Tools, Inc.; Snap-on Tools Corporation; Suzuki Motor Company; Tecumseh Products Company; Teledyne Continental Motors; Teledyne Wisconsin Motor; TRW, Inc.; Yamaha Motor Company, Ltd.; and Yanmar Diesel Engine Company, Ltd.

To all these organizations and the people who represent them, sincere thanks.

William H. Crouse
Donald L. Anglin

SHOP SAFETY

Part 1 in* Small-Engine Mechanics *describes safety in the shop.
There is one chapter in Part 1.

CHAPTER

1

Safety in the Shop

After studying this chapter, you should be able to:

1. ***Describe shop layouts.***
2. ***Explain shop safety and what safety practice means.***
3. ***List shop hazards due to faulty working habits, equipment defects, and incorrect use of hand tools.***
4. ***Describe fire prevention in the shop and how to use the various types of fire extinguishers.***
5. ***List the shop safety rules.***

Δ1-1 SAFETY IS YOUR JOB Safety in the shop means protecting yourself and those around you from danger or injury. Safety is everybody's job. It is *your* job! When you are working in the shop, you are being "safe" if you are protecting your eyes, your fingers, your hands—your whole body—from danger at all times. And, just as important, safety also means looking out for those around you.

Δ1-2 LAYOUT OF THE SHOP The first thing you should do when you go into the shop is to find out where everything is located. A typical layout for a small lawn-mower repair shop is shown in Fig. 1-1. See what the layout of your shop is. You should do this regardless of whether it is a school shop or a shop where you are going to work.

You should note where the machine tools—the power tools—are located. Read the warning and caution signs posted on the walls. They are posted to warn you against potential danger. Notice where the fire extinguishers are located. You might need a fire extinguisher quickly some day.

Δ1-3 SHOP HAZARDS Federal laws have been enacted which are designed to ensure healthful and safe working conditions. The federally established National Institute for Occupational Safety and Health (NIOSH) makes studies of shop working conditions and reports on potential hazards that should be corrected. Further, the law requires that the shops with such hazards must eliminate them. Hazards found are sometimes the fault of management, and sometimes the fault of the workers. The following three sections discuss hazards that might be due to working conditions or habits, faulty or improperly used shop equipment, and faulty or improperly used hand tools.

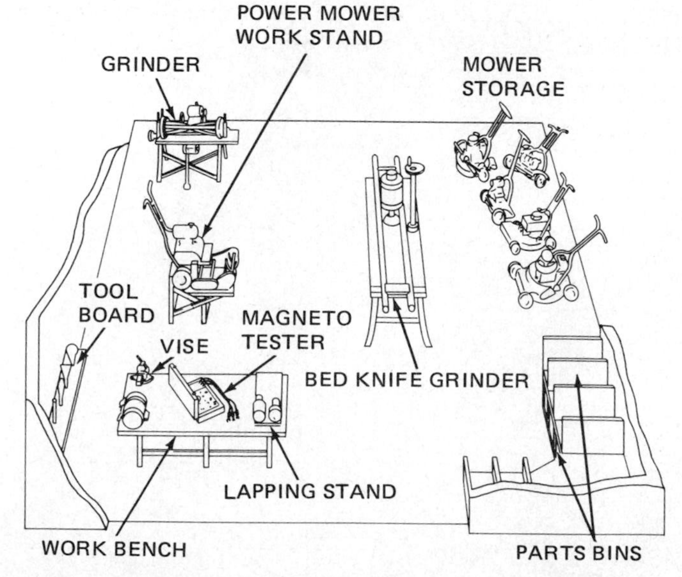

FIG. 1-1 Typical layout for a lawn-mower repair shop. *(Lawn Boy Division of Outboard Marine Corporation)*

Δ1-4 HAZARDS DUE TO FAULTY WORKING CONDITIONS OR HABITS

Here are some of the major hazards that might be due to working habits of the employees or to the general working conditions:

1. Smoking while handling dangerous materials such as gasoline or solvents (Fig. 1-2). This can result in a major fire or explosion.
2. Carelessly or incorrectly handling paint, thinners, solvents, or other flammable fluids. Figure 1-3 shows the correct arrangement for pumping a flammable liquid from a large container into a small one. Note the bond and ground wires. Without these, a spark might jump from the nozzle to the small container. This could cause an explosion and fire.
3. Blocking exits. Areas around exit doors and passageways leading to exits must be kept free of all obstructions. If you wanted to get out in an emergency—as, for example, when a fire or explosion occurred—a blocked exit could mean serious injury or even death.

FIG. 1-2 Do not smoke or have open flames around combustibles such as gasoline or solvents.

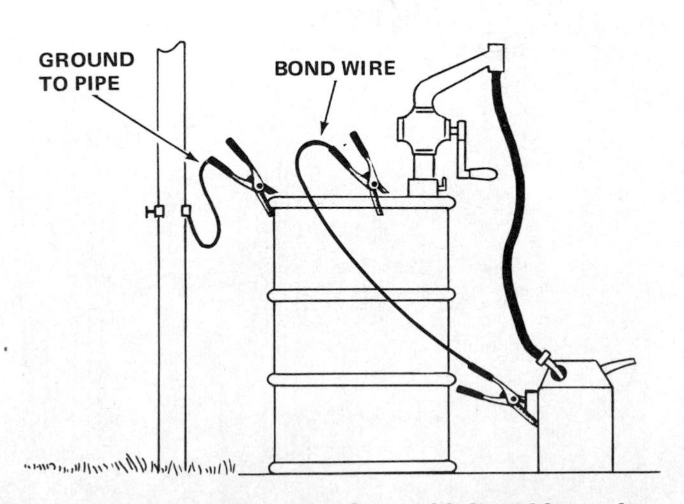

FIG. 1-3 Setup for pumping a flammable liquid from a large container into a small one.

Δ1-5 HAZARDS DUE TO EQUIPMENT DEFECTS OR MISUSE

Here are some common shop hazards due to equipment that is faulty or improperly used:

1. Using moving machinery with incorrect safety guarding. For example, fans should have adequate guards (Fig. 1-4). Air compressors should have proper guards over the belt and pulleys (Fig. 1-5).

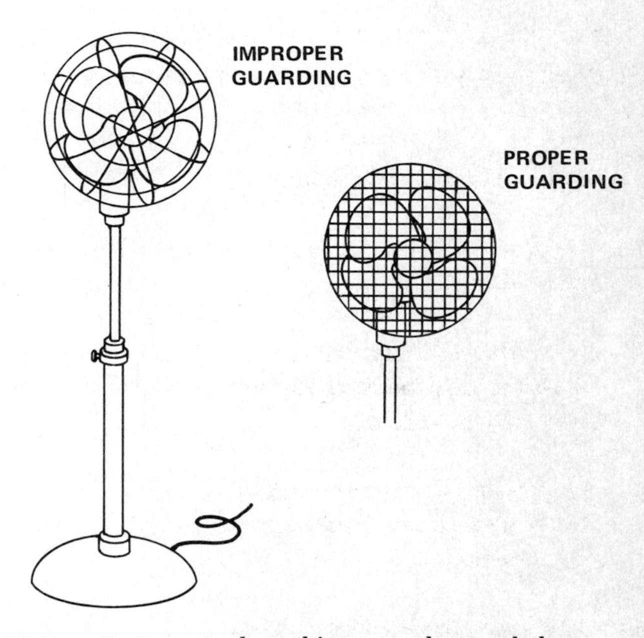

FIG. 1-4 Fans properly and improperly guarded.

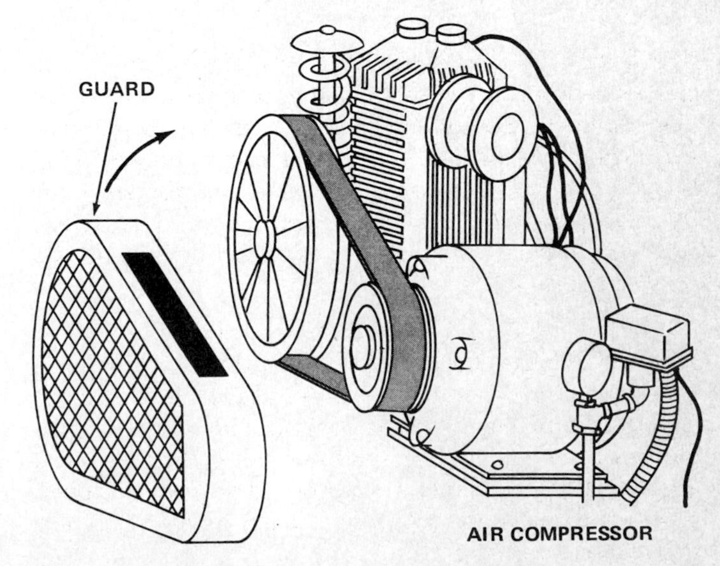

FIG. 1-5 Belts and pulleys on shop equipment should always be protected with guards.

2. Misusing flexible electric cords or using cords that are worn or frayed or cords that have been improperly spliced. Flexible cord should not be run through holes in the wall or tacked up on a wall. Any of these uses could cause a fire or someone to be electrocuted.
3. Using hand-held electric tools that are not properly grounded. All such tools should have a separate ground lead (Fig. 1-6) or be double-insulated to guard against shock.
4. Leaving a running machine unattended. Whenever you are using a power tool and have to leave it for a moment, turn it off! If you leave it running, someone might come along, not realize it is running, and be injured by it.
5. Playing with fire extinguishers. There have been cases where someone thought it would be fun to play with the fire extinguishers. But this presents two dangerous situations. One is that someone can get hurt from slipping and falling on the foam or liquid from the extinguisher or get a serious eye injury from the chemicals in the spray. The other is that the extinguisher's contents are used up, so the extinguisher is useless if a fire should break out.

Δ1-6 HAND-TOOL HAZARDS Hand tools should be kept clean and in good condition. Greasy and oily tools are hard to hold and use. Always wipe them before trying to use them. Do not use a hardened hammer or punch on a hardened surface. Hardened steel is brittle and can shatter from heavy blows. Slivers may fly out and enter the hand, or worse, the eye. Hammers with broken or cracked handles, chisels and punches with mushroomed heads, and broken or bent wrenches are other tool hazards that should be avoided.

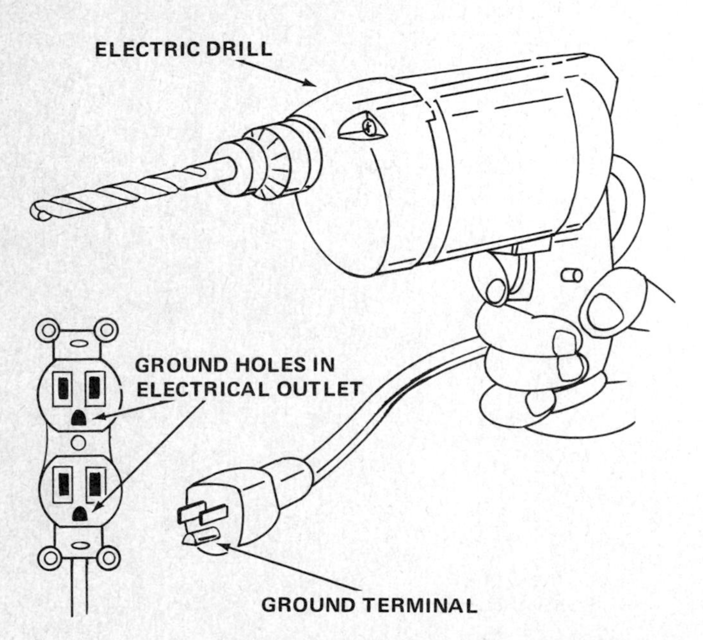

FIG. 1-6 Electric drill with three-wire cord. The third wire and terminal are to ground the electric drill motor.

Never use a tool that is in poor condition or that does not fit the job. Avoid hand-tool hazards so you won't get hurt.

Δ1-7 FIRE PREVENTION Gasoline is used so much in the shop that people forget it is very dangerous if not handled properly. A spark or lighted match in a closed place filled with gasoline vapor can cause an explosion. Even the spark from a light switch can set off an explosion. So you must always be careful with gasoline. Here are some tips.

There will be gasoline vapors around if gasoline is spilled or a fuel line is leaking. You should keep the shop doors open and the ventilating system running. Wipe up the spilled gasoline at once, and put the rags outside to dry. Never smoke or light cigarettes around gasoline. When you work on a leaky fuel line, carburetor, or fuel pump, catch the leaking gasoline in a container or with rags. Put the soaked rags outside to dry. Fix the leak as quickly as possible.

FIG. 1-7 Always store gasoline and all flammable liquids in approved safety containers. *(ATW)*

FIG. 1-8 Safety container for storage of oily rags.

Store gasoline in an approved safety container (Fig. 1-7). Never, *never* store gasoline in a glass jug. The jug could break and could cause an explosion and fire.

Oily rags can also be a source of fire. They can catch fire without a spark or flame. Oily rags and waste should be put into a special safety container where they can do no harm (Fig. 1-8).

Δ1-8 FIRE EXTINGUISHERS

Note the location of the fire extinguishers in the shop. Make sure you know how to use them. Figure 1-9 is a chart showing different types of fires and the kinds of fire extinguisher to use for each type. The quicker you begin to fight a fire, the easier it is to control. But you have to use the right kind of fire extinguisher, and use it correctly. The chart explains this. Talk over any questions with your instructor.

Δ1-9 SHOP SAFETY RULES

Some people say, "Accidents will happen!" But safety experts do not agree. They say, "Accidents are caused; they are caused by careless actions, by inattention to the job, by using

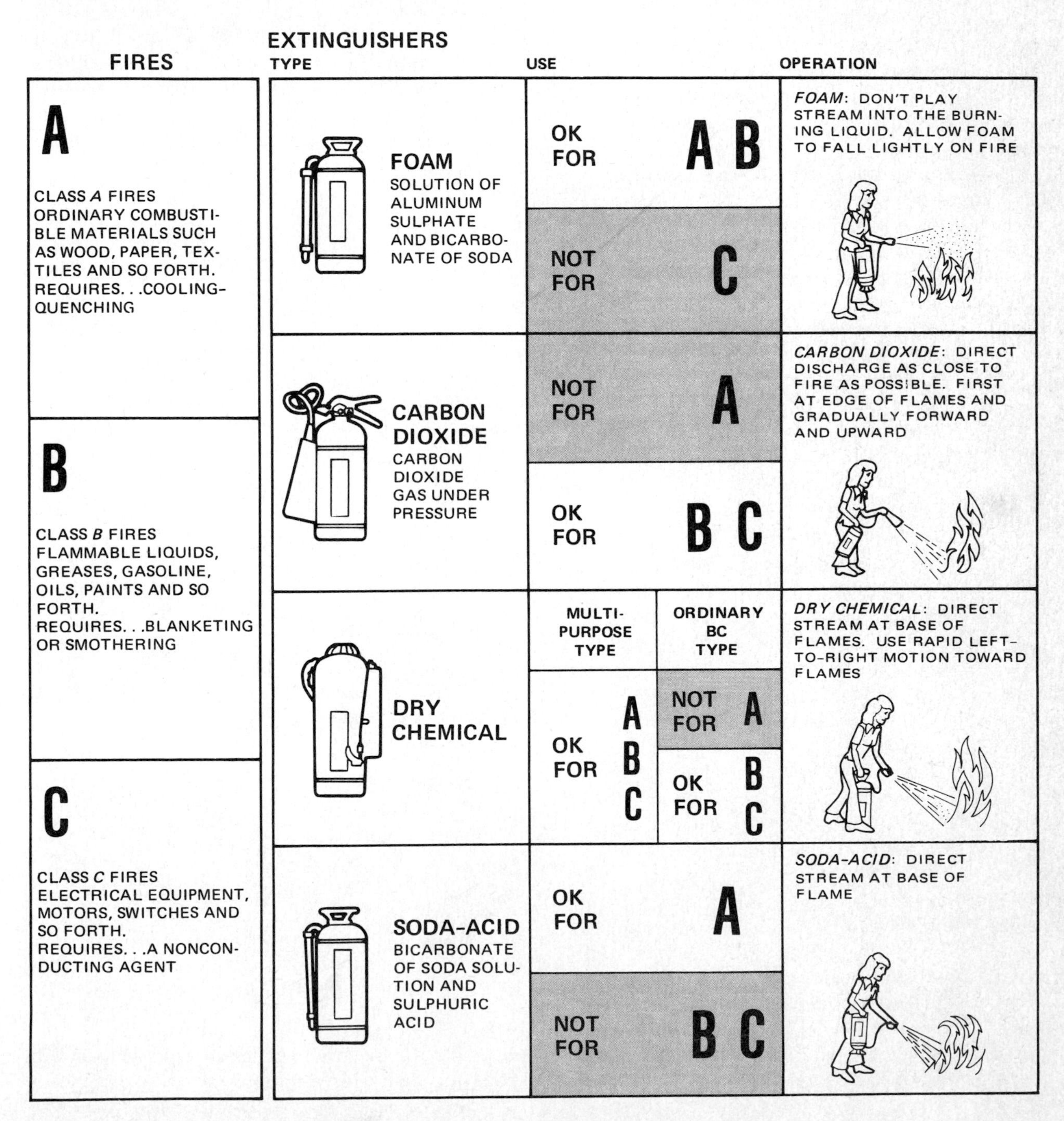

FIG. 1-9 Chart showing types of fire extinguishers and the classification of fires. *(Ford Motor Company)*

damaged or incorrect tools." Fewer accidents occur in a shop that is neat and clean.

To help prevent accidents from happening, follow these safety rules:

1. Work quietly and give the job you are doing your full attention.
2. Keep your tools and equipment under control.
3. Keep jack handles out of the way.
4. Never indulge in horseplay or other foolish activities. You could cause someone to get seriously hurt.
5. Don't put sharp objects, such as screwdrivers, in your pocket. You could cut yourself or get stabbed. Or you could damage the upholstery in a car.
6. Make sure your clothes are right for the job. Dangling sleeves or ties can get caught in machinery and cause serious injuries. Do not wear sandals or open-toe shoes. Wear full leather shoes with nonskid rubber heels and soles. Steel-toe safety shoes are best for shopwork. Keep long hair out of machinery by wearing a cap.
7. Do not wear rings, bracelets, or watches when working around moving machinery or electric equipment. Jewelry can catch in moving machinery with very serious results. Also, if a ring or bracelet should accidentally create a short circuit of the battery, the metal of the ring or bracelet could become white hot, almost instantly producing serious burns.
8. Wipe excess oil and grease off your hands and tools so that you can get a good grip on tools or parts.
9. If you spill oil, grease, or any liquid on the floor, clean it up so that no one will slip and fall.
10. Never use compressed air to blow dirt from your clothes. Never point a compressed-air hose at another person. Flying particles could put out an eye.
11. Always wear safety glasses, safety goggles, or a face shield when there are liquids or particles flying about. Always wear eye protection when using a grinding wheel (Fig. 1-10).
12. Watch out for sparks flying from a grinding wheel or welding equipment. The sparks can set your clothes on fire.
13. To protect your eyes, wear goggles when using chemicals, such as solvents. If you get a chemical in your eyes, wash them with water at once (Fig. 1-11). Then see the school nurse or a doctor as soon as possible.
14. Always use the right tool for the job. The wrong tool could damage the part being worked on and could cause you to get hurt.
15. If you have to lift a heavy object, do it right. You can strain your back and injure yourself if you try to lift too much or lift improperly (Fig. 1-12). When you must lift or move a heavy object, get help.
16. Never use your mouth and a piece of hose to siphon gasoline from a tank. Swallowing even a small amount of gasoline can cause serious respiratory infection and pneumonia. Also, the lead in gasoline is poisonous. If you should get gasoline in your mouth, spit it out. Then rinse out your mouth several times. Avoid taking deep breaths. If you swallow some gasoline, do not try to vomit. Instead, get medical help at once.

FIG. 1-10 Always wear safety glasses, safety goggles, or a face shield when using a grinder or machine that can throw chips or sparks.

FIG. 1-11 If solvent or some other chemical splashes in your eye, immediately wash your eye with water.

FIG. 1-12 When you must lift or move a heavy object, get help.

CAUTION Never run an engine in a closed room or shop that does not have a ventilating system. The exhaust gases contain carbon monoxide. Carbon monoxide is a colorless, odorless, tasteless, poisonous gas that can kill you! Enough carbon monoxide to kill you can accumulate in a closed one-car garage in only 3 minutes.

Δ1-10 USING POWER-DRIVEN SHOP EQUIPMENT

Several types of power-driven equipment are used in the shop. The instructions for using any equipment should be studied carefully before you attempt to operate it. Hands and clothes should be kept away from moving machinery and rotating parts. Do not attempt to feel the finish while the machine is in operation. There may be slivers of metal that will cut your hands. Sometimes you will work on a device with compressed springs, such as a windup starter or valves. Use great care to prevent the springs from slipping and jumping loose. If this happens, a spring may take off at high speed and hurt someone.

Never attempt to adjust or oil moving machinery unless the instructions tell you that this should be done.

Δ1-11 WHAT TO DO IN EMERGENCIES

If there is an accident and someone gets hurt, notify your instructor at once. The instructor will know what to do —give first aid, phone for the school nurse, a doctor, or an ambulance. Be very careful in giving first aid. You must know what you are doing. Trying first aid on an injured person can do more harm than good if it is done wrong. For example, a serious back injury could be made worse if the injured person is moved improperly. However, quick mouth-to-mouth resuscitation may save the life of a person who has suffered an electric shock. Talk to your instructor if you have any questions about this.

Δ1-12 OPERATING CAUTIONS FOR EQUIPMENT POWERED BY SMALL ENGINES

In the shop, you must also know how to safely start, stop, and operate a variety of equipment and machines powered by a small engine. Improper operation of power equipment creates hazards that can lead to personal injury and property damage. To prevent accidents, become thoroughly familiar with a machine before attempting to operate it. If anything appears unsafe, don't try to start the engine. Have your instructor's approval first.

Always wear eye protection, such as safety goggles, if appropriate. Read the operating instructions, know how to make emergency stops, practice operating or driving the machine before putting it to work, and always use your best judgment.

Several important operating cautions are listed below. (Chapter 18 covers more about how to safely and properly operate and maintain a small engine.)

1. Never allow children or other inexperienced persons to operate power equipment.
2. Never wear loose clothing such as scarves that could become entangled in the machine, choking you or pulling you into moving parts.
3. Make sure all guards and shields are in place and secure before starting.
4. To prevent unintentional starting when working on the equipment, always disconnect the spark-plug wire first.
5. Make sure hands, feet, and clothing are safely away from movable parts when starting.
6. Never attempt to start with the drive engaged. Make sure it is shifted into neutral and the brakes are set.
7. Never tamper with the governor setting to gain more power. The governor establishes safe operating limits. Overspeed is extremely hazardous and shortens the life of the equipment.
8. Modern horizontal-blade lawn mowers have an automatic braking device which stops the spinning blade when the handle is released. Some operators, who do not like the brake, bypass it by taping the control to the handle. If you service a lawn mover which has the control taped down, remove the tape. Explain to the operator how important it is for the operator's safety to allow the brake system to work (Δ12-15).
9. Keep people safely away from the operating area. Be especially watchful for children.
10. Watch for and avoid items such as stones and metal objects that could be picked up and thrown by blades. Clear the area of debris before operating.
11. Never attempt to unclog discharge chutes or to free stuck blades or any moving parts while the unit is operating. Stop the engine and disconnect the spark-plug wire first.
12. Never let a machine idle unattended even for a brief moment. Stop the engine whenever you leave the machine.
13. Watch out for and avoid steep inclines that could cause the machine to tip over.

Δ1-13 SERVICE CAUTIONS FOR EQUIPMENT POWERED BY SMALL ENGINES

Power mowers, garden tractors, snowblowers, and other machines powered by small engines have become so commonplace that it is easy to forget the potential dangers involved in servicing and operating such equipment. Some general precautions are listed below. But the best safeguard against accidents is to try to prevent them.

1. *Batteries* Be careful when handling storage batteries (Chap. 11). They contain acid that can eat through clothing and burn skin and cause blindness. A battery gives off highly flammable hydrogen gas while being charged. Avoid starting an engine until this gas is cleared from the area.

2. *Heat* Keep away from hot exhaust parts and allow time to cool before placing equipment in storage (Chap. 18). Never cover a hot engine with flammable materials, such as rags, plastic sheeting, or the cover. They may catch fire.
3. *Electric Shock* While electric energy from the ignition system (Chaps. 14 and 15) may not be strong enough to injure you, reaction to shock could cause you to pull away and come into contact with hot or moving parts. Keep hands away from the ignition system.
4. *Noise* Keep the sound level as low as possible. Do not operate the engine without a muffler or with a faulty exhaust system. Exposure to excessive noise is tiring and may cause hearing loss.

Δ1-14 DEADLY EXHAUST GASES When operating, internal-combustion engines discharge carbon monoxide as part of the exhaust gases. Carbon monoxide is very dangerous, because it is colorless, odorless, and hard to detect. Exhaust gases can cause death if inhaled for a short time. Always observe the following precautions:

1. Never operate an engine inside a closed building or in any area where exhaust gases can accumulate.
2. Be careful not to breathe exhaust fumes when working in the vicinity of an engine.
3. Keep the exhaust system tight and components in good condition at all times. Noise also can be harmful.
4. If the engine must be operated inside a shop for test purposes, make sure the exhaust gases are piped safely outside.
5. Exhaust-system parts get very hot. Keep your hands, feet, and clothing away from these parts while the engine is running and for a long time afterward.
6. Never operate an engine near a building where exhaust gases could seep inside — for example, through an open window or door.

Δ1-15 FUEL HAZARDS Gasoline is a highly volatile, extremely flammable fuel that is explosive as a vapor. The cautions listed below must be followed when storing, handling, and using gasoline:

1. Store gasoline only in an approved red container on which "GASOLINE" is clearly marked in large letters. Never store gasoline in a glass or household type of plastic container. The container could break and a disastrous explosion and fire result.
2. Store gasoline only in well-ventilated areas where escaping vapors can be safely dissipated.
3. Store gasoline containers safely out of reach of children.
4. Never store gasoline inside a home or in any area occupied by people. This can be extremely hazardous.
5. Do not store or pour gasoline near potential spark- or flame-producing equipment. Upon starting, appliances such as refrigerators and freezers can produce electric sparks that will ignite gasoline vapors. Even a spark from a light switch can ignite gasoline vapors.
6. Never use gasoline as a cleaning fluid. Observe "no smoking" rules whenever in the vicinity of a gasoline storage area or gasoline-fueled equipment.
7. Never add gasoline to the fuel tank of a small engine while it is running. Stop the engine and allow it to cool first, to prevent spilled gasoline from igniting on contact with hot parts or ignition sparks.
8. Make sure fuel lines and connections are tight and in good condition. This will prevent gasoline leakage and the resulting possibility of fire.
9. Avoid overfilling the fuel tank. To prevent the fuel from spilling and igniting on contact with the hot engine or an ignition spark, do not fill the tank completely.

CHAPTER 1: REVIEW QUESTIONS

Select the *one* correct, best, or most probable answer to each question. You can find the answers in the section indicated at the right of each question.

1. Safety in the shop means protecting from danger or harm yourself and (**Δ***1-1*)
   a. no one else
   b. those around you
   c. your boss
   d. none of the above
2. One of the most common causes of accidents in the shop is (**Δ***1-9*)
   a. failure to follow instructions
   b. following instructions
   c. following the wrong instructions
   d. none of the above
3. One liquid that is used so much in the shop that people forget it is very dangerous if not handled properly is (**Δ***1-7*)
   a. engine oil
   b. lubricating oil
   c. gasoline
   d. brake fluid
4. Never store gasoline in (**Δ***1-7*)
   a. an approved safety container
   b. a glass jug
   c. a metal tank
   d. none of the above
5. Oily rags must be stored in (**Δ***1-7*)
   a. a closed metal container
   b. a drawer in your workbench
   c. a plastic bag
   d. a corner of the shop

SMALL-ENGINE CONSTRUCTION AND OPERATION

This part of **Small-Engine Mechanics** *describes the construction and operation of small engines. This includes small two-cycle, four-cycle, and diesel engines. There are two basic types of small engines: spark ignition and compression ignition (diesel).*

Spark-ignition engines use spark plugs to ignite the air-fuel mixture in the engine cylinders. Diesel engines ignite the fuel by the heat of compression. This is the heat produced by compressing the air in the engine cylinders.

There are four chapters in Part 2:

CHAPTER

2 Engine Fundamentals

After studying this chapter, you should be able to:

1. ***Explain the difference between external-combustion engines and internal-combustion engines.***
2. ***Describe the combustion process in engine cylinders.***
3. ***Discuss the expansion of solids, liquids, and gases as they are heated.***
4. ***Define* atmospheric pressure *and explain what causes it.***
5. ***Explain how high pressure in the engine cylinder causes a piston to move, and how this straight-line motion is changed to rotary motion.***

Δ2-1 ENGINE TYPES The two basic types of engines are external-combustion and internal-combustion. The external-combustion engine operates because of combustion *outside* of the engine. The steam engine is an example. Combustion—a fire—outside of the engine boils water to produce steam. The steam then enters the engine and causes it to run.

The internal-combustion engine operates because the fuel is burned *inside* the engine. All small engines are internal-combustion engines.

Δ2-2 THE COMBUSTION PROCESS Combustion, or fire, results when any substance that will burn is ignited. The substance unites with oxygen in the air and the result is a high temperature, or heat.

To understand the combustion process, we first look at *elements*. Everything in the world is made up of elements. There are about 100 different elements, with names such as copper, iron, aluminum, gold, oxygen, and silver (Fig. 2-1). These elements unite in more than a million ways to make up all gases, liquids, and solids. Some of these materials and substances we know as air, water, ice, wood, paper, glass, cloth, and leather.

When elements unite, the action is called a *chemical reaction*. Combustion, or fire, is one common chemical reaction.

Most small engines are *spark-ignition engines*. During the combustion process, a mixture of air and gasoline vapor is compressed. Then the compressed air-fuel mixture is ignited by a spark at the spark plug. The air is about 20 percent oxygen (O). Gasoline is mostly hydrogen and carbon (and is called a *hydrocarbon*, or HC). The chemical reaction during combustion is between the three elements—hydrogen (H), oxygen (O), and carbon

Element	Symbol
Aluminum	Al
Calcium	Ca
Carbon	C
Chlorine	Cl
Copper	Cu
Hydrogen	H
Iron	Fe
Magnesium	Mg
Mercury	Hg
Nitrogen	N
Oxygen	O
Phosphorus	P
Potassium	K
Silver	Ag
Sodium	Na
Sulfur	S
Zinc	Zn

FIG. 2-1 **A list of a few common elements.**

(C). The equation for perfect combustion can be written as:

$$HC + O = H_2O + CO_2 + \text{heat}$$

So the result of perfect combustion is H_2O (water), CO_2 (carbon dioxide, a gas), and heat. However, perfect combustion never occurs in the engine. Some unburned gasoline (HC) and partly burned gasoline (CO, or carbon *monoxide*) remain in the exhaust gas.

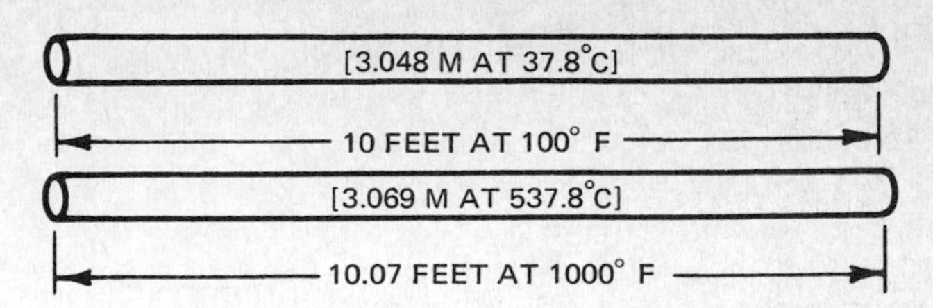

FIG. 2-2 A steel rod that measures 10 feet (3.048 m) at 100°F (37.8°C) will measure 10.07 feet (3.069 m) at 1000°F (537.8°C).

Δ2-3 EFFECT OF HEAT

The purpose of causing combustion in the engine is to produce heat that can be made to do work. Heat causes most substances to expand. When a piece of iron is heated, it expands. A steel rod that measures 10 feet in length [3.048 meters (m)] at 100 degrees Fahrenheit (°F) [37.8 degrees Celsius (°C)] will measure 10.07 feet [3.069 m] in length at 1000°F [357.8°C] (Fig. 2-2). As the rod is heated, it gets longer, due to the expansion of the steel.

Liquids and gases also expand when heated. One cubic foot of water at 39°F [3.89°C] will expand to 1.01 cubic feet when heated to 100°F [37.8°C]. A cubic foot is a cube measuring 1 foot on each side.

A cubic foot of air at 32°F [0°C], if heated to 100°F [37.8°C] *without a change of pressure,* will expand to 1.14 cubic feet.

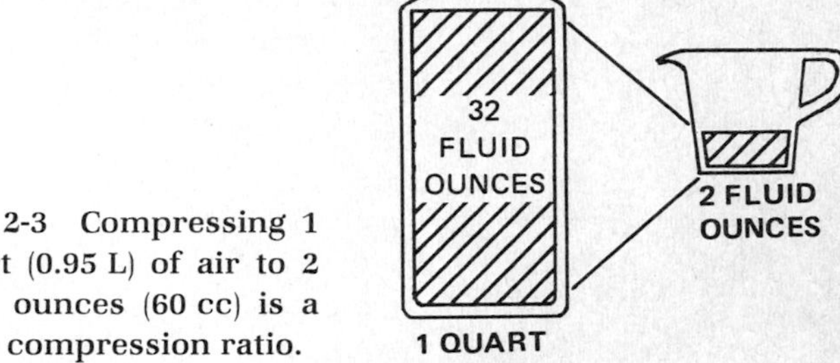

FIG. 2-3 Compressing 1 quart (0.95 L) of air to 2 fluid ounces (60 cc) is a 16:1 compression ratio.

Δ2-4 INCREASE OF PRESSURE

If the volume—the cubic foot—containing the air is held constant, heating the air causes the pressure to increase. Pressure is measured in force (push or pull) per unit area. For example, in the USCS, pressure is measured in pounds per square inch (psi). In the metric system, it is measured in kiloPascals [kPa].

If we start with a pressure of 15 psi [103 kPa] at 32°F [0°C], the pressure will increase to about 17 psi [117 kPa] if the air is heated to 100°F [37.8°C]. This assumes that the air is held in a closed container that prevents any expansion of the air.

Another way to increase the pressure is to compress the air into a smaller volume. This is what happens in the cylinders of a spark-ignition engine. The air-fuel mixture is compressed into about one-sixth to one-eighth of its original volume. This raises the temperature of the air.

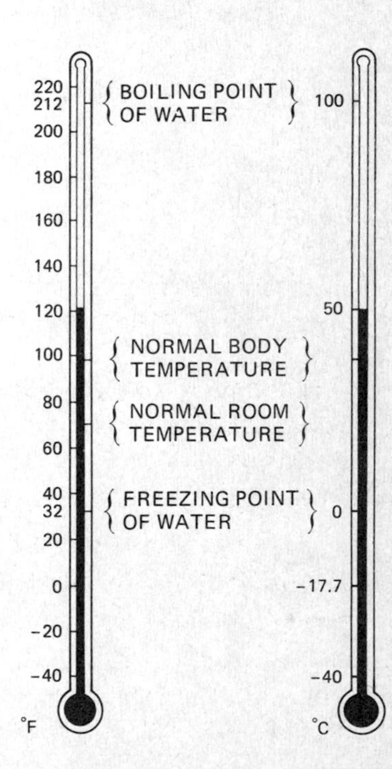

FIG. 2-4 Thermometers comparing Fahrenheit and Celsius (also called centigrade) readings.

Δ2-5 INCREASE OF TEMPERATURE

Pressure and temperature increase when a gas is compressed. In the spark-ignition engine, when the air-fuel mixture is compressed, its temperature increases several hundred degrees. In the diesel engine, air is compressed to one-sixteenth or less (sometimes as little as one twenty-second) of its original volume (Fig. 2-3). This is like compressing a quart [0.95 L (liter)] of air down to 2 fluid ounces [60 cc (cubic centimeters)]. Compressing air this much raises its temperature to at least 1000°F [538°C]. This temperature provides the "heat of compression" that causes the diesel engine to run (Chap. 5).

Δ2-6 THERMOMETER

The thermometer (Fig. 2-4) shows a familiar use of the expansion of liquids as temperature goes up. The liquid, usually mercury (a metal that is liquid at ordinary temperatures), is largely contained in the glass bulb at the bottom of the glass tube. As temperature increases, the mercury expands. Part of it is forced up through the hollow glass tube. The higher the temperature, the more the mercury expands and the higher it is forced up through the tube. The tube is marked to indicate the temperature in degrees.

Δ2-7 THERMOSTAT

Different metals expand at different rates with increasing temperatures. Aluminum expands about twice as much as iron as their temperatures go up. This difference in expansion rates is used in thermostats. Thermostats do numerous jobs in small engines. One type consists of a coil made up of two strips of different metals, such as brass and steel, welded together. When the coil is heated, one metal expands faster than the other, causing the coil to wind up or unwind.

Δ2-8 GRAVITY

Gravity is the attractive force between all objects. When we release a stone from our hand, the stone falls to earth. When a car is driven up a hill, part of the engine power is being used to lift the car against gravity. Likewise, a car can coast down a hill with the engine turned off, because gravity pulls downward on the car.

Gravitational attraction is usually measured in terms of weight. We put an object on a scale and see that it weights 10 pounds [4.5 kg]. What we mean is that the object has sufficient mass for the earth to register this much pull on it. Gravitational attraction gives any object its weight.

Δ2-9 ATMOSPHERIC PRESSURE

The air is also an "object" that is pulled toward the earth by gravity. At sea level and average temperature, 1 cubic foot of air weighs about 0.08 pound, or about 1.25 ounce. The blanket of air—our atmosphere—surrounding the earth is many miles thick. This means that there are, in effect, many thousands of cubic feet of air piled on top of one another, all adding their weight. The total weight, or downward push, of this air amounts to about 15 psi [103 kPa] at sea level. The pressure of all this air pushing downward is about 2160 pounds [9608 newtons (N)] on every square foot.

Δ2-10 VACUUM

A vacuum is the absence of air or any other matter. Astronauts, working in a space laboratory or traveling to the moon, are outside the blanket of air surrounding the earth. In space, there are only a few scattered particles of air, and so space is a vacuum.

Δ2-11 PRODUCING A VACUUM

There are many ways to produce a vacuum on earth. The engine, as it operates, produces a partial vacuum in the engine cylinders. This causes atmospheric pressure to push air, or air-fuel mixture, into the cylinders. Some small engines have a fuel pump that works by producing a partial vacuum.

Δ2-12 PISTON-ENGINE OPERATION

Now, let's discuss how the principles described above affect engine construction and operation. First, imagine a can with one end cut out (Fig. 2-5*a*). Imagine a second can slightly smaller in size which will fit snugly into the first can (Fig. 2-5*b*). Now suppose you push the smaller can rapidly up into the larger can, trapping air above it. This compresses the air, as it is pushed into a smaller space than it had previously occupied. If the air contains a small amount of gasoline vapor, and if an electric spark occurs in the compressed air-fuel mixture, a rapid burning (or combustion) occurs. Heat from the combustion causes the compressed air to expand, pushing the smaller can down inside the larger can (Fig. 2-5*c*).

This is about what happens in the internal-combustion engine. Actually, the larger can is called the *cylinder*. The smaller can is called the *piston*. To make the engine run, the piston must repeatedly slide up and down (or *reciprocate*) in the cylinder (Fig. 2-6).

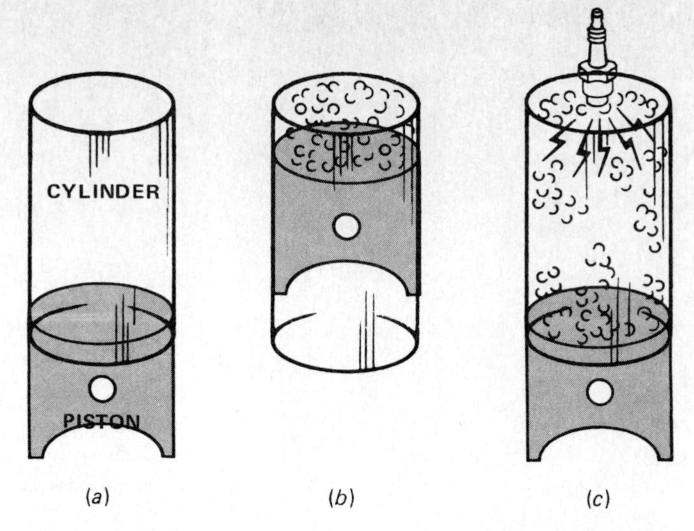

FIG. 2-5 Three views showing the actions in an engine cylinder. *(a)* The piston is a metal plug that fits snugly into the engine cylinder. *(b)* When the piston is pushed up into the cylinder, air is trapped and compressed. The cylinder is drawn as though it were transparent so the piston can be seen. *(c)* The increase of pressure as the gasoline vapor and air mixture is ignited pushes the piston down in the cylinder.

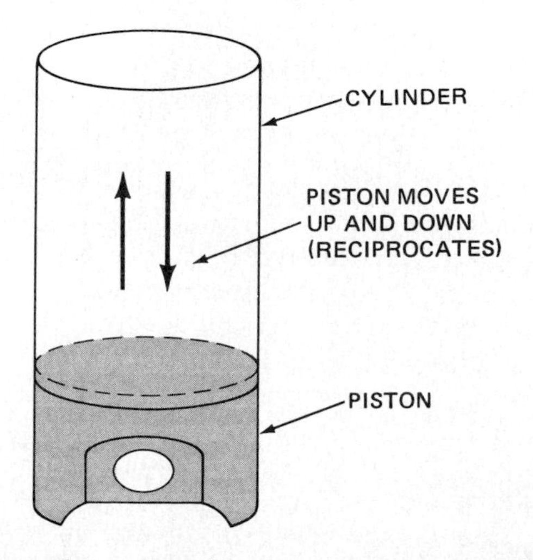

FIG. 2-6 In the engine, the piston moves up and down, or *reciprocates.*

Δ2-13 ENGINE CYLINDER

Most small engines have only one or two cylinders. However, some small engines have three or four cylinders. There is no set rule about how many cylinders a small engine should have, as opposed to a "big" engine. Regardless of how many cylinders an engine has, the same actions go on in each cylinder.

Basically, the engine cylinder is a round hole in a block of metal. The piston repeatedly slides up and down in the cylinder as long as the engine runs.

Δ2-14 PISTONS AND PISTON RINGS

The power that the engine produces depends on the pressure developed in the cylinder when the fuel burns. The piston must be a fairly loose fit in the cylinder. If the piston fits too tightly, it will expand as it gets hot and could stick in the cylinder. If the piston sticks, it could ruin the engine. But too much clearance allows some of the combustion pressure to leak past (or *blow by*) the piston (Fig. 2-7). The remaining pressure is lower, which provides a weaker push on the piston. Therefore, blowby results in a loss of engine power.

To control blowby, piston rings are used to provide a leak-tight sliding seal between the piston and cylinder wall (Fig. 2-8). So that the piston rings can be fitted to the piston, a groove for each ring is cut into the upper part of the piston. These grooves are called *piston-ring grooves.* The rings are made of cast iron or other metal. Each ring is split at one point. This allows the ring to be expanded and slipped over the head of the piston and into the ring groove.

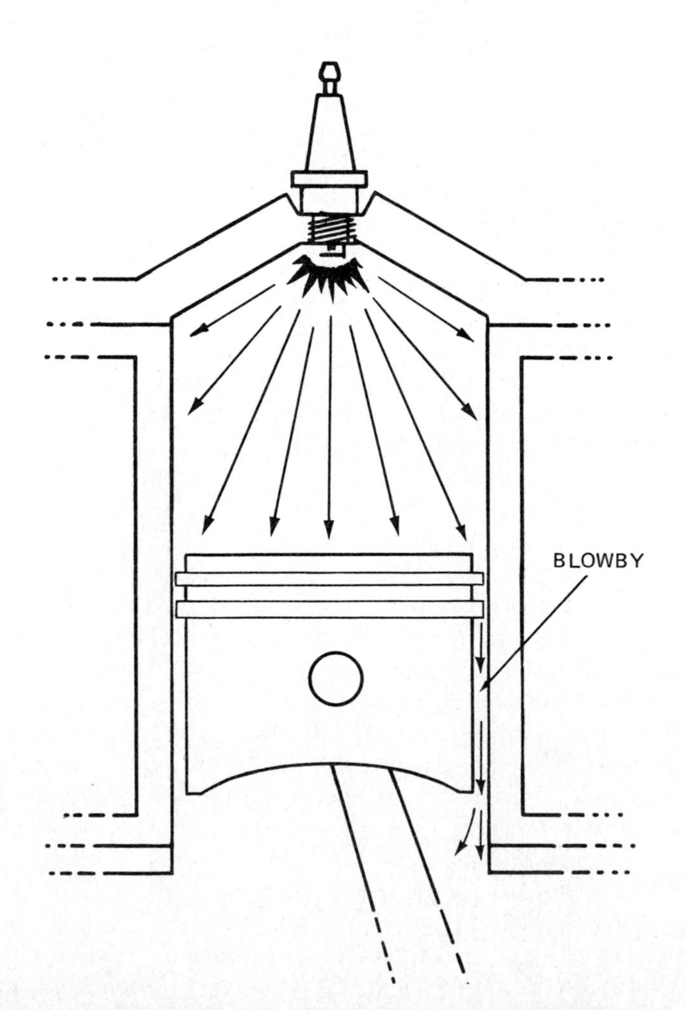

FIG. 2-7 If the piston fits too loosely in the cylinder, much of the pressure will be lost as some of the mixture leaks past, or "blows by," the piston.

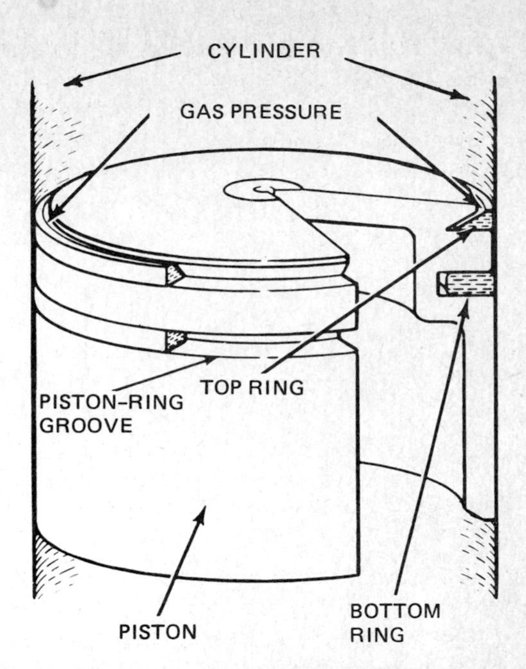

FIG. 2-8 Typical piston with piston rings in place. When the piston is installed in the cylinder, the rings are compressed into the grooves in the piston. *(Lawn Boy Division of Outboard Marine Corporation)*

Pistons for two-cycle engines have one or two rings, which are pinned or locked in place. This prevents the end of the ring from catching in a port, or hole, in the cylinder wall and breaking (**Δ**3-3). The split ends of each ring should almost touch. The gap formed by the split in the ring is called the piston-ring *end gap.* The lubricating system (Chap. 6) coats the piston, piston rings, and cylinder wall with oil. This allows the rings and piston to move easily up and down the cylinder.

Figure 2-9 shows how the piston ring works to hold the compression and combustion pressures within the combustion chamber. The arrows show the pressure from above the piston passing through the clearance between the piston and the cylinder wall. The pressure presses down against the top and back of the piston ring, as shown by the arrows in Fig. 2-9. This pushes the piston ring firmly against the cylinder wall and against the bottom of the piston-ring groove. As a result, there are good seals at both of these points. The higher the pressure in the combustion chamber, the greater the sealing force.

Small two-cycle engines have one or two rings on the piston (Fig. 2-8). Both are compression rings. Two rings divide up the job of holding in the compression and combustion pressures. This produces better sealing with less ring force against the cylinder wall.

The pistons in four-cycle engines usually have three rings. These are free to move around in their ring

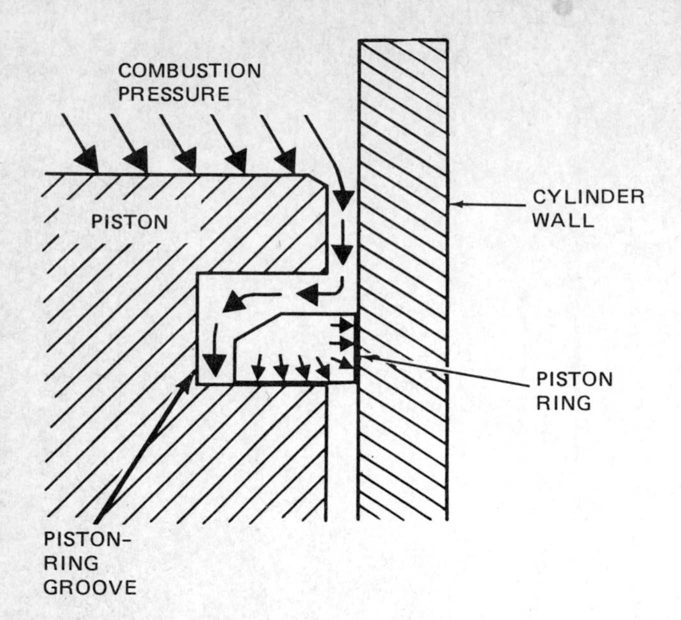

FIG. 2-9 Pressure in the combustion chamber above the piston, either from compression of the air-fuel mixture or from its combustion, presses the ring against the cylinder wall and the lower side of the piston-ring groove.

grooves. The third ring is called the *oil-control ring* (Fig. 2-10). It is needed because four-cycle engines get much more oil on the cylinder wall than do two-cycle engines. This additional oil must be scraped off to prevent it from getting into the combustion chamber. Once in the combustion chamber, excess oil will burn and cause trouble.

In the two-cycle engine, oil is mixed with the gasoline and this mixture enters the crankcase, as explained later. The gasoline goes on into the combuston chamber, where the fuel is burned. Part of the oil covers the cylinder walls. This allows the piston and rings to slide up and down with little friction.

Some pistons used in four-cycle engines have four rings. The bottom two rings are both oil-control rings.

Δ2-15 THE CRANK The piston moves up and down in the cylinder. This up-and-down motion is called *reciprocating motion* (Fig. 2-6). The piston moves in a straight line. This straight-line motion must be changed

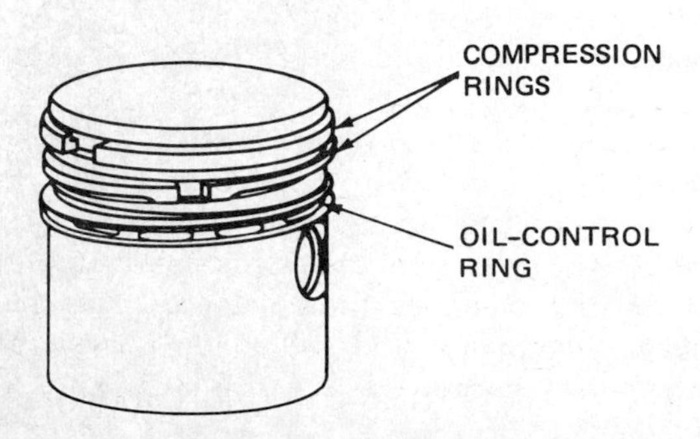

FIG. 2-10 The piston of the four-cycle engine has three rings. The upper two are compression rings, and the lower is an oil-control ring.

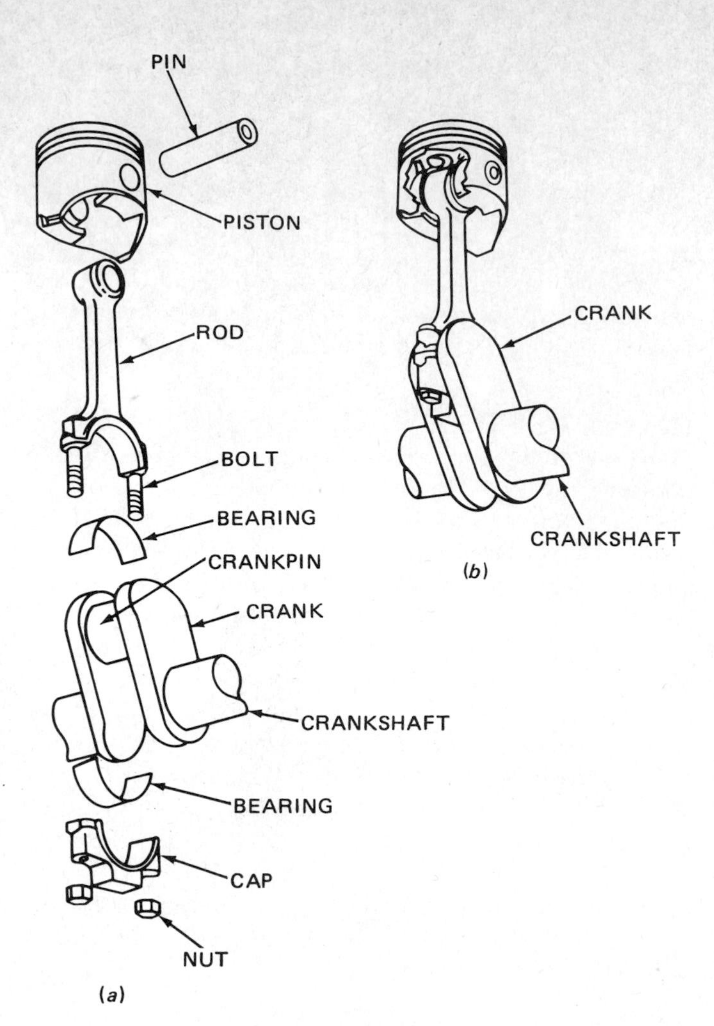

FIG. 2-11 *(a)* Piston, connecting rod, piston pin, and crankpin on an engine crankshaft shown disassembled. *(b)* Piston and connecting-rod assembly attached to the crankpin on a crankshaft.

to *rotary motion* before it can be used. Rotary motion is required to make shafts and wheels turn. To change the reciprocating motion to rotary motion, a crank and connecting rod are used (Fig. 2-11). The connecting rod connects the piston to the crank.

The crank is a simple device used in many machines. It is an offset part of a shaft. When the shaft rotates, the crank and crankpin swing in a circle. When the piston is pushed down in the cylinder by the pressure, the push on the piston, carried through the connecting rod to the crank, causes the crankshaft to turn. Figure 2-12 shows the motions that the piston, connecting rod, crank, and crankshaft go through. As the piston moves down and up, the top end of the connecting rod moves down and up with it. The bottom end of the connecting rod swings in a circle along with the crank.

The piston end of the connecting rod is attached to the piston by a piston pin, or wrist pin. The other end of the connecting rod is attached to the crankpin of the crank by a rod-bearing cap (Fig. 2-11). There are bearings at both ends of the connecting rod so that the rod can move with relative freedom. Bearings are described in Δ2-17.

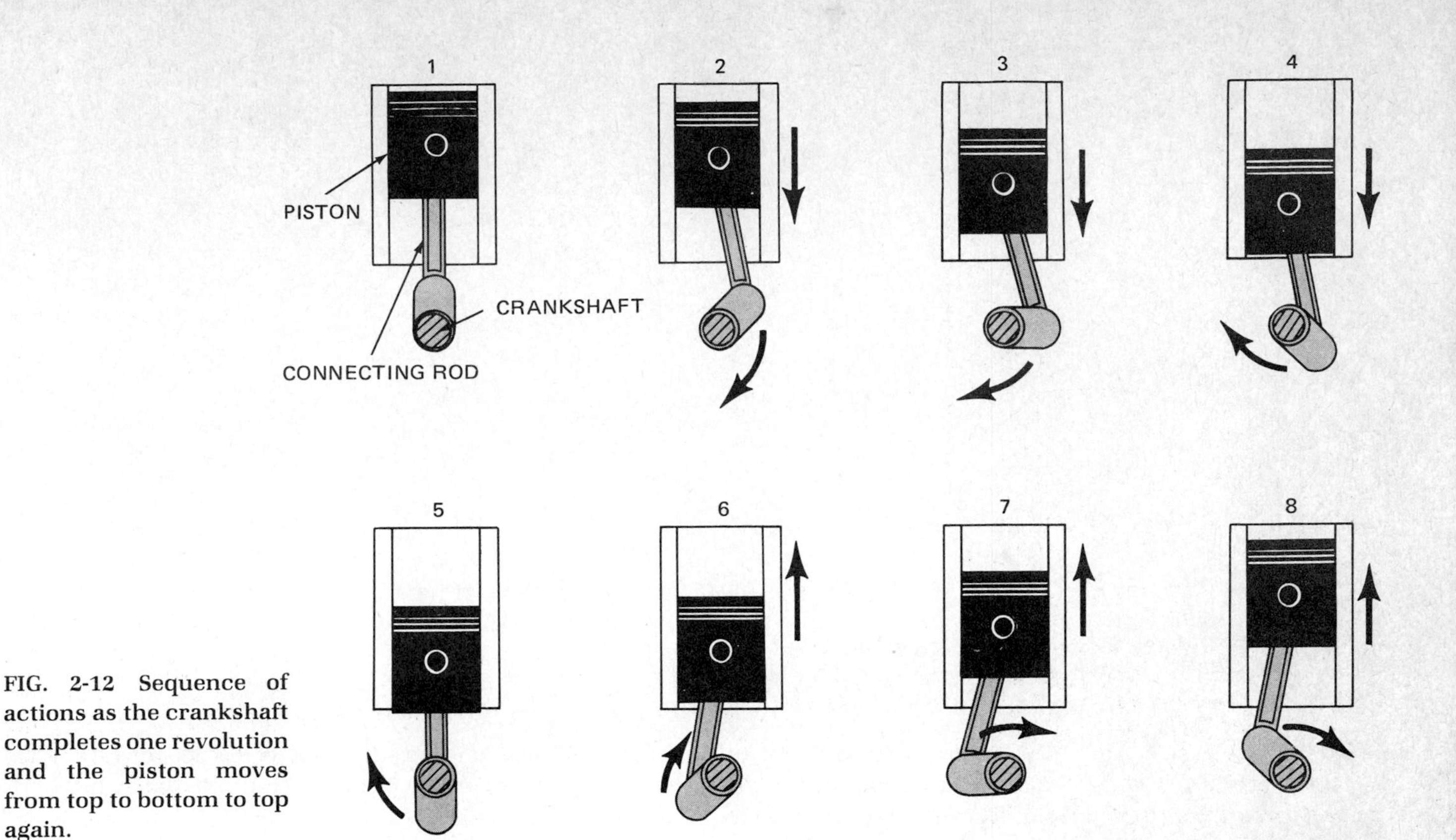

FIG. 2-12 Sequence of actions as the crankshaft completes one revolution and the piston moves from top to bottom to top again.

NOTE The crank end of the connecting rod is called the rod *big end.* The piston end of the connecting rod is called the rod *small end.*

△2-16 ENGINE CRANKSHAFT

The crank is an offset section of the crankshaft to which the connecting-rod big end is attached by a bearing. The crankshaft is mounted in the engine on bearings which allow the crankshaft to rotate. As the crankshaft rotates, the crank swings in a full circle (Fig. 2-12).

The crankshaft is assembled with flywheels, or counterweights (Fig. 2-13). These counterweights balance the weight of the crankpin and the piston-and-connecting-rod assembly. This reduces the tendency of the crankshaft to go out-of-round while it is rotating. The result is a smoother-running engine and less wear on the crankshaft bearings.

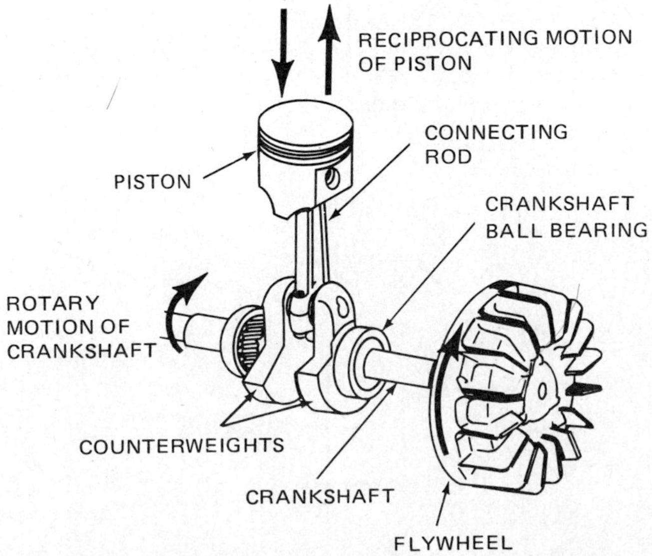

FIG. 2-13 The crankshaft converts the reciprocating motion to rotary motion. The crankshaft mounts in bearings which encircle the journals so it can rotate freely.

△2-17 ENGINE BEARINGS

The crankshaft is supported by bearings. In fact, everywhere there is rotary motion in the engine, bearings are used to support the moving parts. The purpose of the bearings is to reduce friction and allow the parts to move easily. Bearings are lubricated with oil to make the relative motion easier. Friction, engine oil, and lubricating systems are described in Chap. 6.

Bearings used in engines are of two types, sliding and rolling (Fig. 2-14). The sliding type of bearing is called a *plain bushing,* or *sleeve bearing.* It forms the shape of a sleeve that fits around the rotating journal or shaft. Connecting-rod big-end bearings (rod bearings) and some crankshaft supporting bearings (main bearings) are *split-sleeve bearings.* They must be split so they can be assembled in the engine. For this reason, split-sleeve bearings are also called *insert-type bearings,* or simply *bearing inserts.*

The construction of a typical insert-bearing half is shown in Fig. 2-15. It is made of a steel or bronze back to which a lining of relatively soft bearing material is ap-

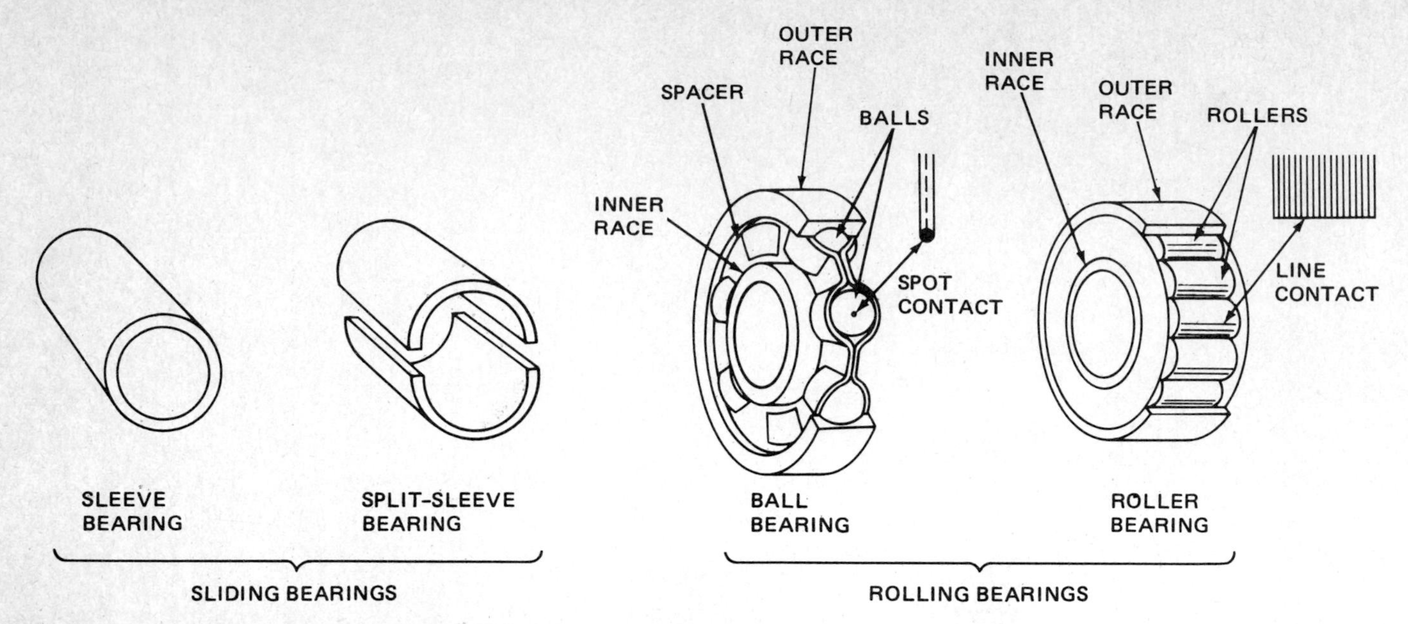

FIG. 2-14 Sleeve, ball, and roller bearings.

plied. This relatively soft bearing material is made of several materials, such as copper, lead, tin, and other metals. Each of these metals has the ability to conform to slight irregularities of the shaft rotating against it. If wear does take place, it is the bearing that wears. Then, the bearing can be replaced instead of the much more expensive crankshaft or other engine part.

The rolling type of bearing uses balls or rollers between the stationary support and the rotating shaft, as shown in Fig. 2-16. Since the balls or rollers provide rolling contact, the frictional resistance to movement is much less. In some roller bearings, the rollers are so small that they are hardly bigger than needles. These bearings are called *needle bearings.* Also, some roller bearings have the rollers set at an angle, so the races the rollers roll in are tapered. These bearings are called *tapered roller bearings.* A crankshaft supported by tapered roller bearings is shown in Fig. 2-16.

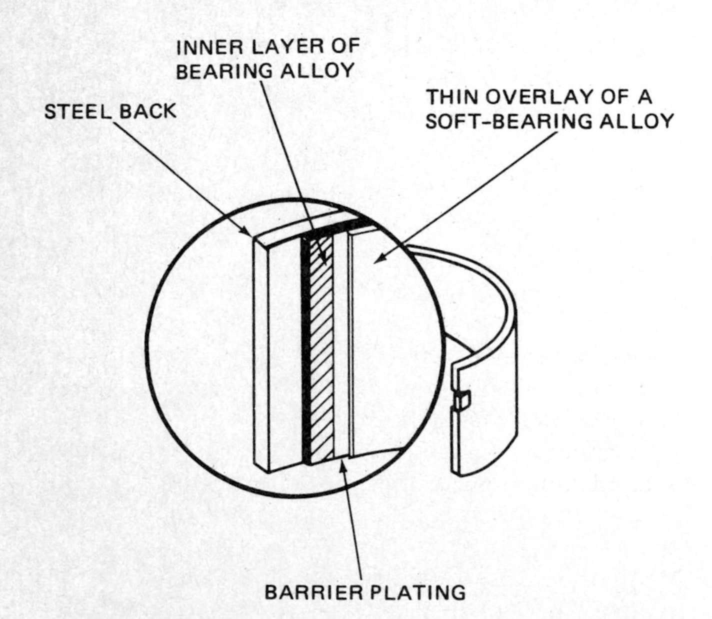

FIG. 2-15 Construction of an insert-type, or sleeve, bearing. *(Federal-Mogul Corporation)*

Some ball and roller bearings are sealed with their lubricant already in place. These bearings require no other lubrication. Other bearings require lubrication from the oil in the gasoline (two-cycle engines) or from the engine lubricating system (four-cycle engines).

Δ2-18 MAKING THE ENGINE RUN

For an engine to run, each cylinder must continuously receive a mixture of air and fuel. In the spark-ignition engine, this mixture is delivered to the cylinder from a mixing device called the *carburetor.* (The diesel engine uses a fuel-injection system, which is described in Chap. 5.)

The air-fuel mixture enters the engine cylinder and is compressed by the piston as it moves up. Then the com-

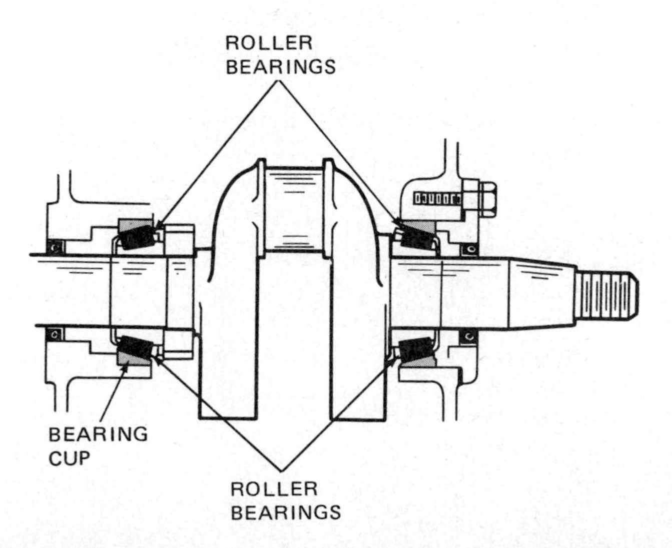

FIG. 2-16 Crankshaft mounted on tapered roller bearings. *(Kohler Company)*

pressed mixture is ignited by a spark from the spark plug. The mixture burns and produces a high temperature and high pressure. This pushes the piston down so the crankshaft rotates and the engine produces power (Fig. 2-5).

Next, the burned gases must be removed from the cylinder so that a fresh charge of air-fuel mixture can enter the cylinder. These actions go on all the time that the engine is running.

Δ2-19 THE PISTON STROKE

In the piston engine, the movement of the piston from one limiting position to the other is called a *piston stroke.* The upper limiting position of the piston is called *top dead center* (TDC), and the lower limiting position is called *bottom dead center* (BDC).

A piston stroke takes place when the piston moves from TDC to BDC or from BDC to TDC (Fig. 2-17). When the piston moves from TDC to BDC, after combustion has taken place, the stroke is called the *power stroke.* The high combustion pressure forces the piston to move down the cylinder during the power stroke. The result is engine power that is available to do work.

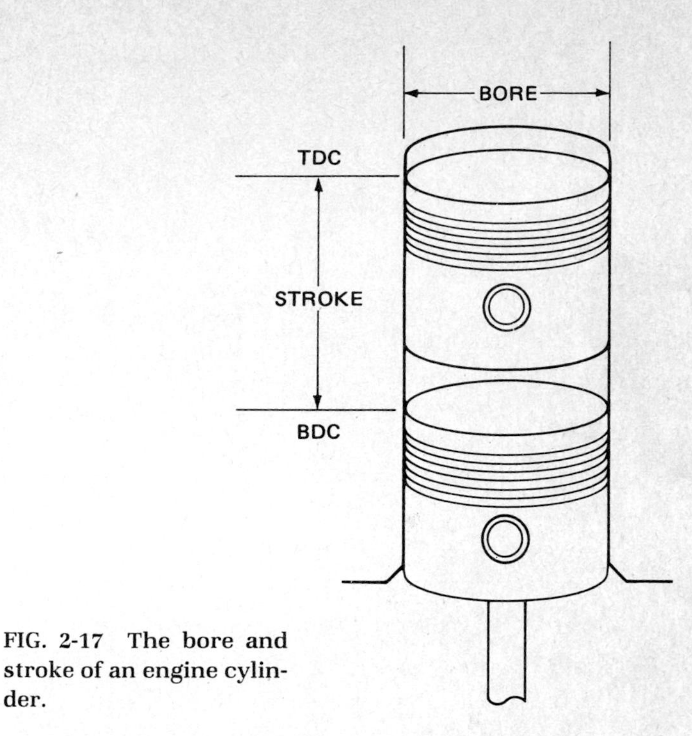

FIG. 2-17 The bore and stroke of an engine cylinder.

CHAPTER 2: REVIEW QUESTIONS

Select the *one* correct, best, or most probable answer to each question. You can find the answers in the section indicated at the right of each question.

1. When elements unite, the action is called (Δ2-2)
   a. a chemical action
   b. an element action
   c. a chemical reaction
   d. fire
2. In the internal-combustion engine, the chemical reaction (Δ2-2)
   a. is called combustion
   b. is between the fuel and oxygen
   c. results in water and carbon dioxide
   d. all of the above
3. When heated, solids and liquids will (Δ2-3)
   a. expand
   b. ignite
   c. produce a chemical reaction
   d. make the engine run
4. If a gas is heated in a closed container, the (Δ2-4)
   a. gas pressure will decrease
   b. gas will expand
   c. gas pressure will increase
   d. none of the above
5. If a gas is compressed, (Δ2-4)
   a. its temperature will rise
   b. its pressure will go up
   c. its volume gets smaller
   d. all of the above
6. If air is compressed to one-sixteenth of its original volume, its temperature will go up to about (Δ2-5)
   a. 100°F
   b. 1000°F
   c. 10°F
   d. 10,000°F
7. Mechanic A says that air pressure is due to gravity. Mechanic B says that air pressure is due to the weight of the air. Who is right? (Δ2-9)
   a. Mechanic A
   b. Mechanic B
   c. both A and B
   d. neither A nor B
8. In the piston engine, combustion causes the (Δ2-12)
   a. pressure in the cylinder to increase
   b. temperature of the gas to increase
   c. piston to be pushed down
   d. all of the above
9. The purpose of piston rings is to (Δ2-15)
   a. prevent blowby
   b. provide a sliding seal between the piston and cylinder wall
   c. prevent pressure loss during combustion
   d. all of the above
10. A piston stroke is movement of the piston from (Δ2-19)
    a. TDC to BDC
    b. BDC to TDC
    c. either a or b
    d. neither a nor b

CHAPTER

3

Two-Cycle Spark-Ignition Engines

After studying this chapter, you should be able to:

1. ***Explain the basic difference between two-cycle and four-cycle engines.***
2. ***Describe the construction and operation of the two-cycle engine.***
3. ***Explain why two-cycle engines use crank-case compression.***
4. ***Define* scavenging.**
5. ***List the various types of valves used in two-cycle engines and describe their operation.***

Δ3-1 TWO-CYCLE AND FOUR-CYCLE ENGINES Piston engines can be divided into two groups—two-cycle and four-cycle. In the two-cycle engine, it takes two piston movements, or strokes, to complete one cycle of engine operation. A stroke is a movement of the piston in the cylinder from top to bottom or from bottom to top (Fig. 2-17). The two limiting positions of the piston, the top and bottom, are TDC (top dead center) and BDC (bottom dead center). These terms are used often in working with engines.

The piston in the two-cycle engine takes two strokes to complete one cycle. Every downward movement of this piston from TDC to BDC is a power stroke. Every upward movement of the piston from BDC to TDC is a compression stroke. When the piston moves up, it compresses a charge of the air-fuel mixture. Then, as the piston nears TDC, the charge is ignited by a spark from the ignition system. The ignited mixture burns, and the pressure pushes the piston down. In this way, every other stroke produces power (Fig. 3-1*a*).

NOTE The word *cycle* means a series of events that repeat themselves. For example, the cycle of the seasons—spring, summer, fall, winter—is repeated every year. In a similar way, the two piston strokes in the two-cycle engine form a cycle that is repeated continuously as long as the engine runs.

The full name of the two-cycle engine is the *two-stroke-cycle engine.* Two piston strokes are required to complete one cycle. This engine is usually called a *two-cycle engine,* a *two-stroke engine,* or a *two-stroker.*

In the four-cycle engine (or the *four-stroke-cycle engine*) it takes four piston strokes to produce a single power stroke (Fig. 3-1*b*). Three of the piston strokes, or movements between TDC and BDC, are required to get the cylinder ready for the fourth stroke, which produces the power. When comparing the two-cycle with the four-cycle engine, the four-cycle engine has several advantages in automobiles, trucks, and buses. However, the two-cycle engine is simpler in construction, easier to maintain, lighter in weight, and generally less expensive to manufacture. For these reasons, many air-cooled small engines are two-cycle engines.

The two-cycle engine has only three major moving parts: the piston assembly, connecting rod, and crankshaft. The major difference between the one-cylinder engine and the multiple-cylinder engine is that the multicylinder engine has more parts. In the two-cycle multicylinder engine, each cylinder must have its own sealed, or airtight, crankcase.

Δ3-2 OPERATION OF TWO-CYCLE ENGINES The spark-ignition engine runs because air-fuel mixture is burned in the engine cylinder. This produces heat and high pressure that push the piston down the cylinder, forcing the crankshaft to rotate.

During each upward stroke of the piston in a two-cycle engine, three separate events must take place. First, the burned gases remaining from the previous power stroke must be cleared from the cylinder. Second, a fresh charge of the air-fuel mixture must be brought into the cylinder. Third, this charge of the air-fuel mixture must

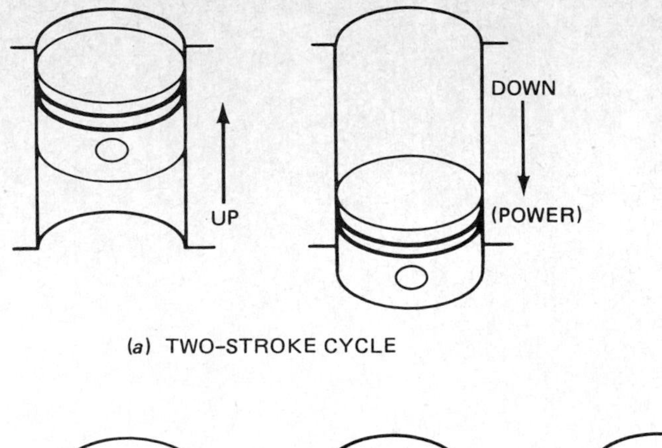

(a) TWO-STROKE CYCLE

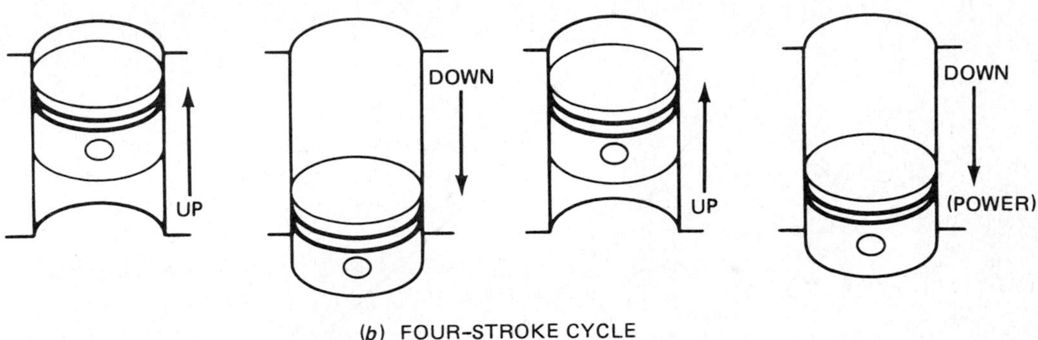

(b) FOUR-STROKE CYCLE

FIG. 3-1 *(a)* Two-stroke cycle. *(b)* Four-stroke cycle. In the two-cycle engine, every other piston stroke is a power stroke.

be compressed in readiness for ignition. To accomplish all this, the engine has openings, or *ports,* in the side of the cylinder wall (Fig. 3-2). The piston, as it moves up and down, opens or closes these ports. The air-fuel mixture enters the cylinder through one of the ports. The burned gases leave through another port.

Δ3-3 CYLINDER PORTS The three ports in the cylinder of a small air-cooled two-cycle engine are shown in Fig. 3-2. They are the intake port, the transfer port, and the exhaust port. The intake port is the port through which the fresh air-fuel mixture flows from the carburetor into the crankcase. The port through which the air-fuel mixture enters the cylinder is the transfer port. The port through which the burned gases leave the cylinder is the exhaust port.

Δ3-4 PORT ACTIONS Let's follow the sequence of actions during one complete cycle of operation in the two-cycle engine. The action starts when the fresh air-fuel mixture enters the crankcase. The piston, as it moves up and down in the cylinder, acts as a valve to block off and then open the ports. In Fig. 3-3*a*, the piston has moved up to TDC, and the compressed air-fuel mixture has been ignited. The power stroke begins, pushing the piston down.

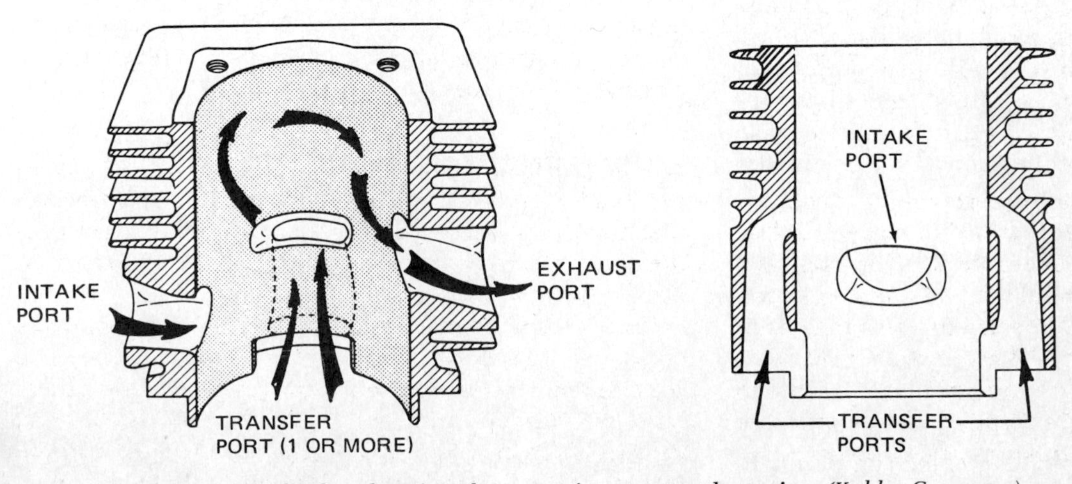

FIG. 3-2 A cutaway cylinder showing the ports in a two-cycle engine. *(Kohler Company)*

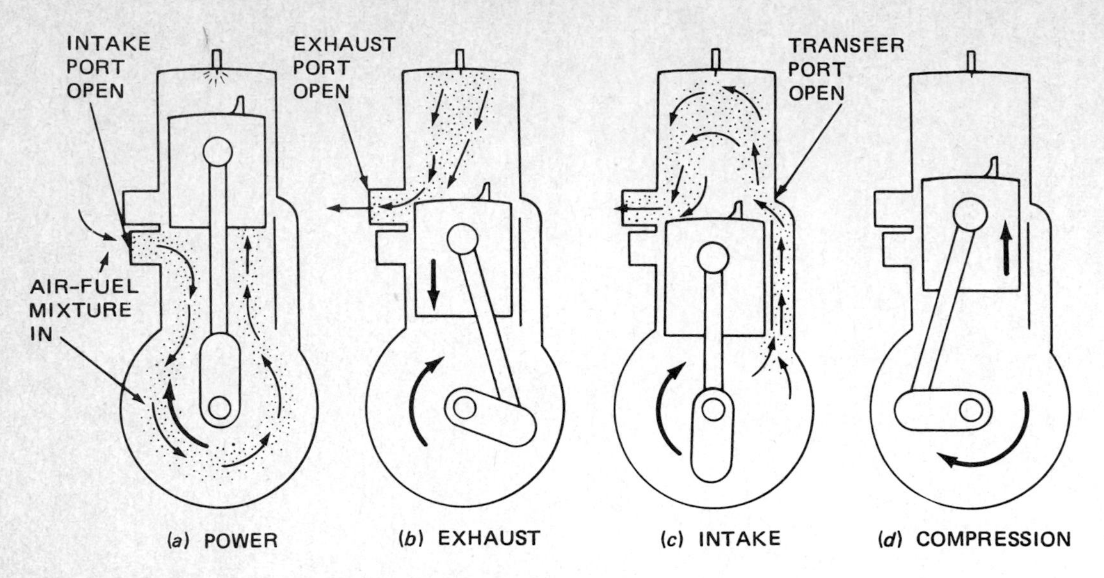

FIG. 3-3 The events in a two-cycle three-port engine.

Notice in Fig. 3-3*a* that the air-fuel mixture is flowing into the crankcase. This occurs because a vacuum (Δ2-10) is created in the crankcase as the piston moves up. Now atmospheric pressure causes air to flow through the carburetor, pick up a charge of fuel, and then flow into the crankcase. The bottom of the piston has moved above the intake port, opening it. This allows the air-fuel mixture to enter the crankcase.

At *b* in Fig. 3-3, the piston has been pushed down almost to BDC. This has opened the exhaust port so that the burned gases can begin to flow out, or exhaust, from the cylinder. When the piston has moved down to BDC, as shown at *c* in Fig. 3-3, the piston has cleared the transfer port. Now the air-fuel mixture in the crankcase flows through the transfer port into the cylinder, as shown by the arrows in Fig. 3-3*c*. The reason for the flow is that the piston, in moving down, slightly compresses the air-fuel mixture trapped below it in the crankcase.

Notice in Fig. 3-3*b* that the piston has closed off the intake port so that the crankcase is sealed. Therefore, as the piston moves down, it compresses the trapped air-fuel mixture in the crankcase.

Compression is shown in Fig. 3-3*d*. The piston, in moving up, closes off the transfer and exhaust ports. Therefore, the air-fuel mixture above the piston is compressed by the upward-moving piston. As the piston nears TDC (Fig. 3-3*a*), ignition takes place. Then the complete cycle (Fig. 3-3*a* to *d*) is repeated continuously as long as the engine runs.

A port can be either one large opening or two or more smaller openings. If the single port is too large, the piston rings could catch on the side of the opening and break. To prevent this, a series of smaller openings is used in some two-cycle engines.

Δ3-5 EVENTS DURING OPERATING CYCLE

The four actions of intake, compression, power, and exhaust take place during two piston strokes in the two-cycle engine. As a result, there is a power impulse every time the piston reaches TDC. To diagnose troubles in a two-cycle engine, the technician must have a thorough knowledge of the two-stroke cycle, and know how the actions during each event relate to each other. Now, let's cover these actions using a single-cylinder two-cycle engine.

As the piston moves up the cylinder, the lower edge of the piston moves above the intake port (Fig. 3-4). This clears or opens the port and allows air-fuel mixture to

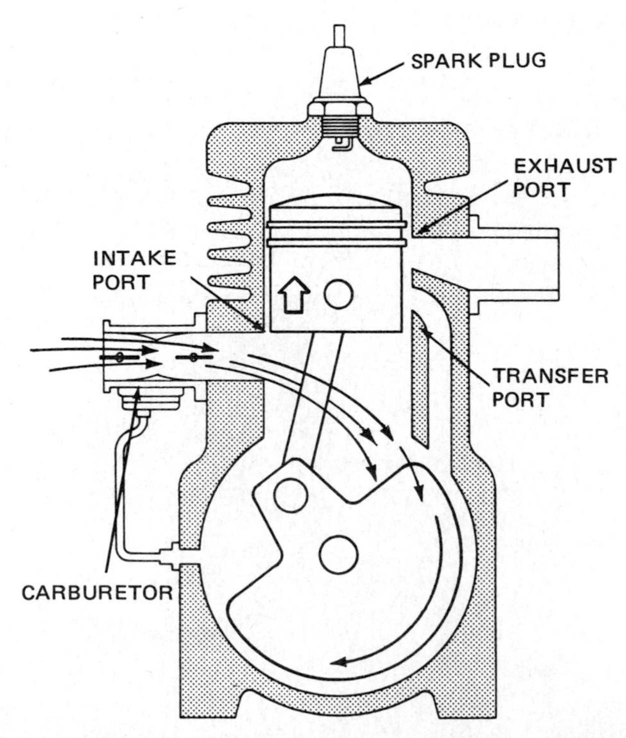

FIG. 3-4 As the piston moves up in the cylinder, the lower edge of the piston moves above the intake port. This allows fresh air-fuel mixture to enter the crankcase. *(Kohler Company)*

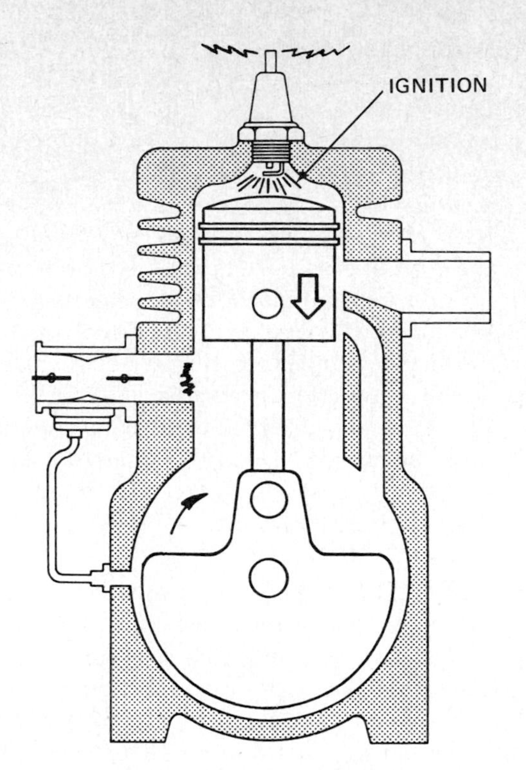

FIG. 3-5 As the piston nears TDC, ignition occurs. The resulting high combustion pressure forces the piston down. *(Kohler Company)*

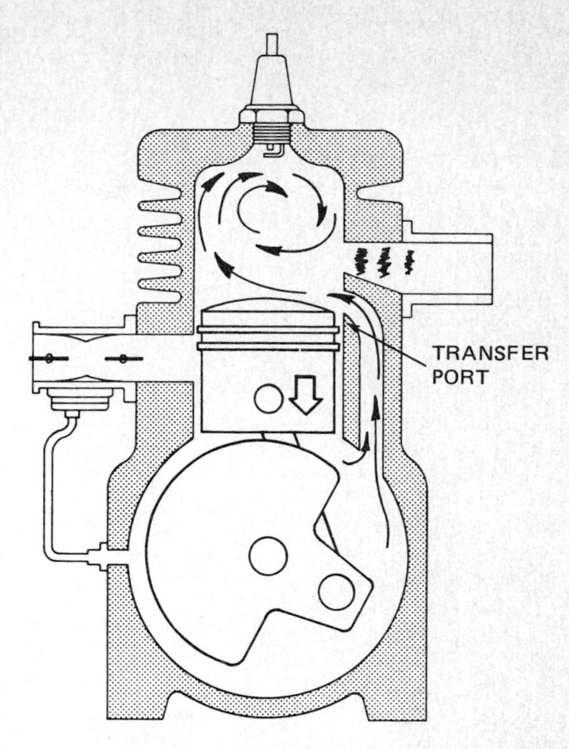

FIG. 3-7 As the piston nears BDC, it opens the transfer port, allowing fresh air-fuel mixture to flow from the crankcase into the cylinder. *(Kohler Company)*

flow from the carburetor through the intake port into the crankcase. At the same time, the piston has closed off the exhaust port and the transfer port (Fig. 3-4). This traps the air-fuel mixture that already is in the combustion

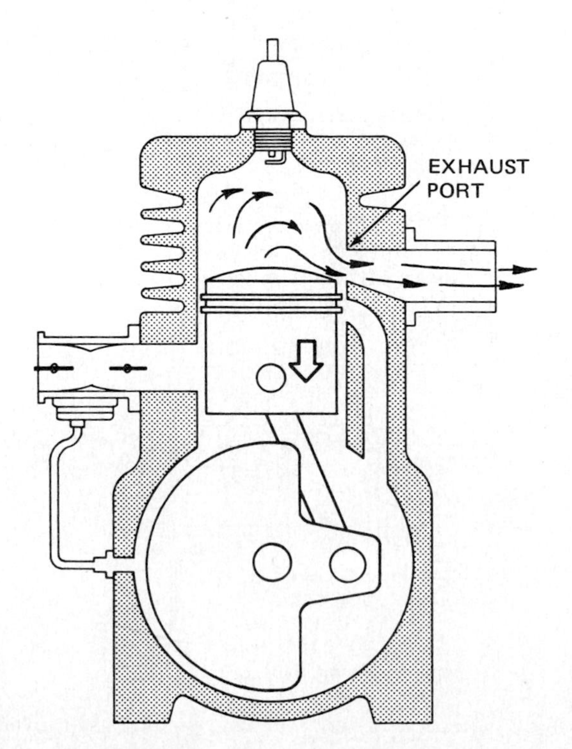

FIG. 3-6 As the piston moves down the cylinder, the piston opens the exhaust port, allowing the burned exhaust gases to flow from the cylinder. *(Kohler Company)*

chamber. As the piston moves up, this air-fuel mixture is compressed.

As the piston nears TDC, ignition takes place (Fig. 3-5). The high combustion pressure drives the piston down. The force applied through the connecting rod against the crankpin turns the crankshaft.

As the piston moves down the cylinder, the top of the piston moves below the exhaust port in the cylinder wall (Fig. 3-6). This opens the exhaust port. Burned gases, still under some pressure, begin to flow out of the cylinder through the exhaust port. In Fig. 3-6, notice that the piston also has closed the intake port into the crankcase.

As the piston nears BDC, the top edge of the piston moves below the transfer port, opening it (Fig. 3-7). Now the air-fuel mixture in the crankcase transfers to the cylinder by flowing through the open transfer port.

After the piston has passed through BDC and starts up again, it closes the transfer port and the exhaust port (Fig. 3-4). Now the fresh air-fuel charge above the piston is compressed and ignited. This same sequence of events takes place again and continues as long as the engine runs.

Δ3-6 CRANKCASE COMPRESSION In a two-cycle engine, the air-fuel mixture is delivered to the cylinder under pressure. This pressure is applied to the air-fuel mixture while it is trapped in the crankcase. To trap the air-fuel mixture, all two-cycle engines must have some type of crankcase-valve arrangement to open and close the ports. Some engines use the piston as a valve

(Fig. 3-3). In other engines, a reed valve or a rotary valve is used. The different types of crankcase valves are described in the following sections.

Δ3-7 REED VALVE Many two-cycle engines have a reed valve, or leaf valve, installed between the carburetor and the crankcase (Fig. 3-8). In a running engine, the reed valve acts like an automatically controlled flap. Changes in crankcase pressure cause the flap to open and close the intake port. When the piston moves up, a partial vacuum is created in the sealed crankcase (Fig. 3-8, left). Atmospheric pressure forces the reed valve off its seat and pushes the air-fuel mixture from the carburetor through the open reed valve into the crankcase.

After the piston passes TDC and starts down again, pressure begins to build up in the crankcase. This pressure closes the reed valve. Further downward movement of the piston compresses the air-fuel mixture that is trapped in the crankcase. The pressure which builds up on the air-fuel mixture then causes it to flow quickly through the transfer port into the engine cylinder. This occurs as soon as the piston moves down enough to open the transfer port (Fig. 3-8, right).

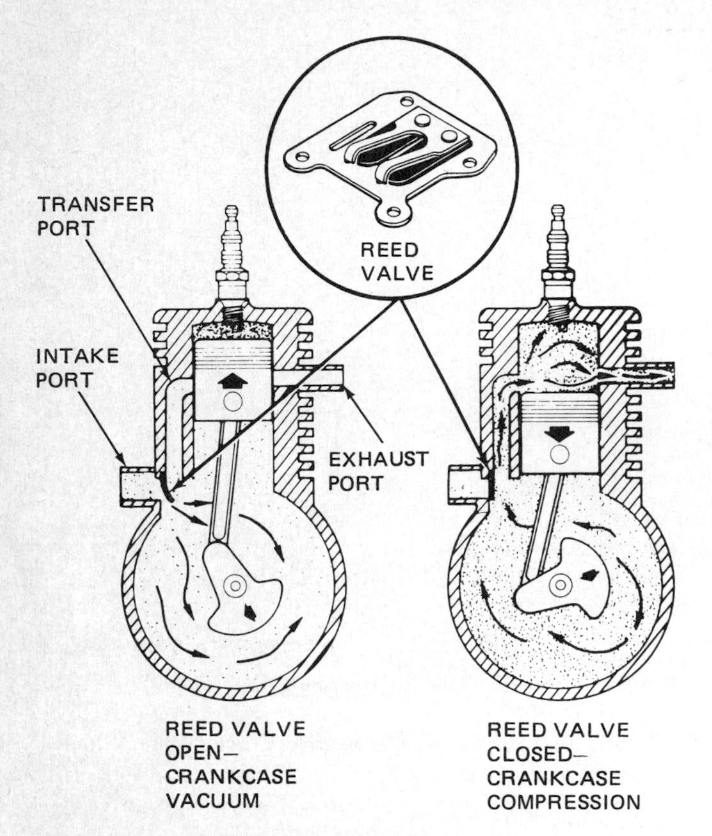

FIG. 3-8 A reed valve is used in the intake port of many two-cycle engines.

Δ3-8 ROTARY VALVE Some two-cycle engines have a rotating disk valve, or rotary valve, that opens and closes the intake port (Fig. 3-9). The positions of the rotary valve when the intake port is open and when it has just been closed are shown in Fig. 3-10 and 3-11. Notice that the rotary valve has rotated almost 180° from the time the port starts to open until it closes. This allows the air-fuel mixture to flow through the intake port into the

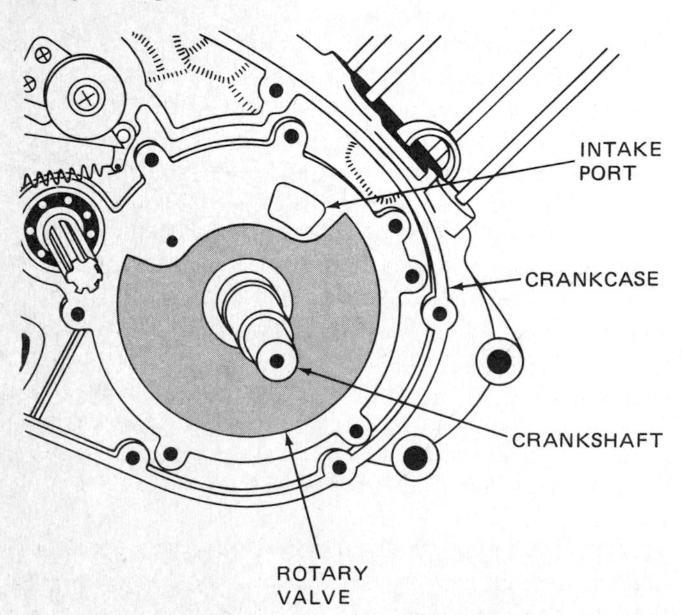

FIG. 3-9 The carburetor and side cover removed from a two-cycle engine to show the rotary disk valve, or rotary valve.

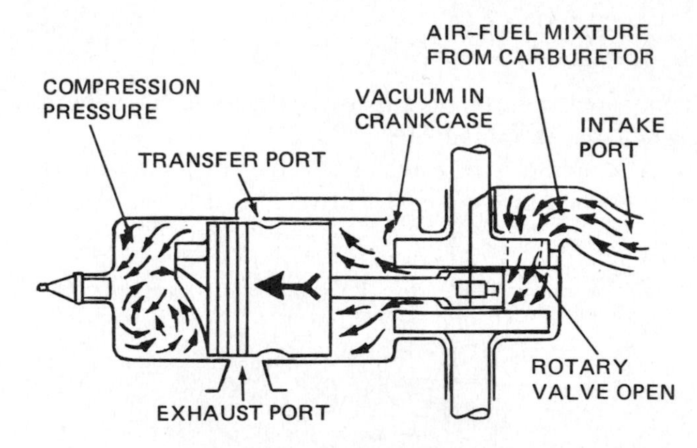

FIG. 3-10 As the crankshaft rotates, the valve port in the crankshaft lines up with the inlet port in the crankcase to admit the air-fuel mixture to the crankcase.

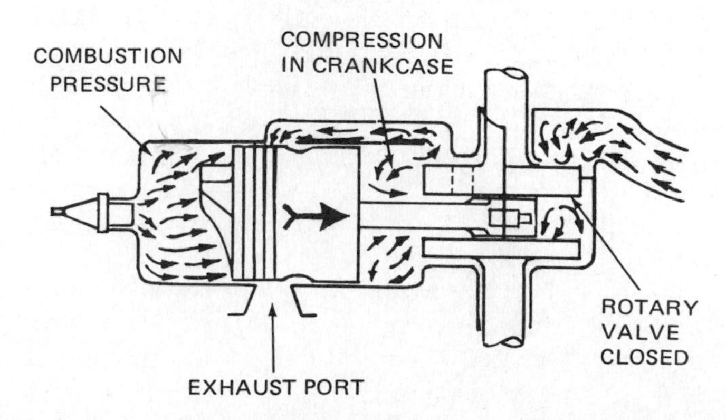

FIG. 3-11 Further rotation of the crankshaft moves the space in the rotary valve past the intake port, so that no air-fuel mixture can enter the crankcase.

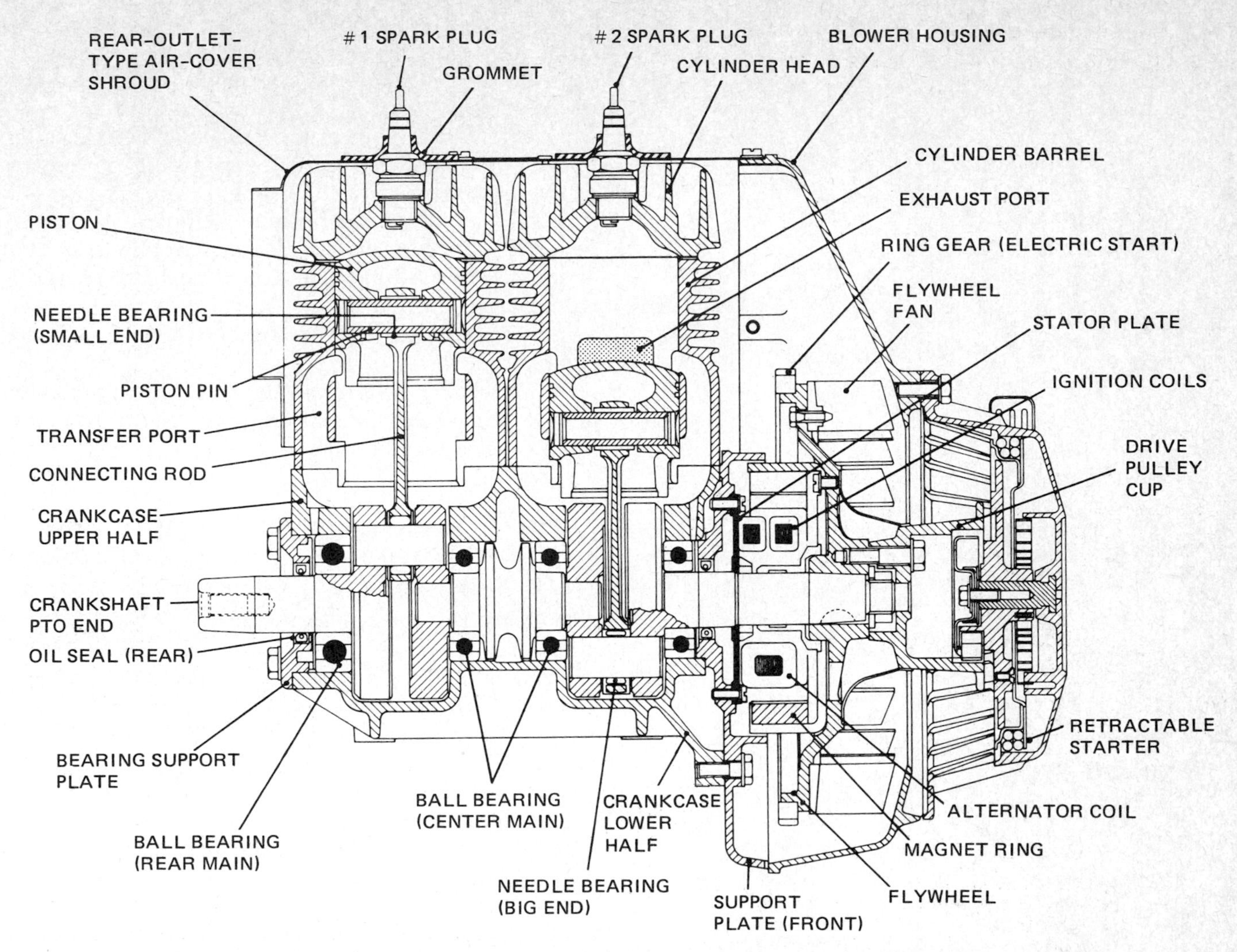

FIG. 3-12 In a two-cycle engine with more than one cylinder, such as this two-cycle engine, each of the cylinders must have its own sealed crankcase. The sealed ball bearings placed between the cranks of the crankshaft in this engine provide the seal. *(Kohler Company)*

crankcase for almost half of each crankshaft revolution. This is a much longer time than a piston-controlled port, or piston-port engine, can provide.

Δ3-9 CRANKCASE PRESSURE IN MULTIPLE-CYLINDER ENGINES

In multiple-cylinder two-cycle engines, each cylinder must have its own sealed crankcase. Therefore, in a two-cylinder two-cycle engine, the crankcase is divided into two sealed compartments. Each compartment provides primary compression of the air-fuel mixture for the cylinder directly above it.

Figure 3-12 shows a two-cylinder two cycle engine. Notice how the crankcase for each cylinder is sealed from the other.

Δ3-10 CYLINDER SCAVENGING

Most two-cycle engines produce less power than four-cycle engines of the same size. This is because less air-fuel mixture is burned in the cylinder of the two-cycle engine. For example, at full throttle in a two-cycle engine, the amount of air-fuel mixture burned may be only about half of the amount burned in a four-cycle engine. One cause of this difference is cylinder scavenging.

To scavenge means to clean away dirt or refuse. In the engine, this means to use part of the fresh air-fuel mixture entering the cylinder to help remove the burned exhaust gases. Cylinder scavenging greatly affects the operation of the two-cycle engine. For example, at full throttle in some two-cycle engines, as much as 25 percent of the air-fuel mixture that enters the cylinder escapes through the exhaust ports before they are closed.

The two basic methods of cylinder scavenging used in two-cycle engines are *cross-scavenging* and *loop-scavenging* (Fig. 3-13). Most two-cycle small engines and motorcycle engines are loop-scavenged. With this method of scavenging, the transfer port and exhaust port are 90° apart. This causes the incoming cooler air-fuel mixture to swirl through the cylinder, cleaning out the burned gases. As a result, the top of the piston runs cooler.

In a cross-scavenged engine, the transfer and exhaust ports are opposite each other (Fig. 3-13*a* and *b*). Therefore, this system requires a deflector, which faces the transfer port, on the top of the piston. The deflector causes the incoming air-fuel mixture to flow up and across the top of the combustion chamber. Without the deflector, the air-fuel mixture would flow almost straight across the top of the piston and out the exhaust port. This would leave burned air-fuel mixture in the cylinder. Cross-scavenging is widely used in two-cycle outboard engines.

Notice that Fig. 3-13 shows the various ways of cylinder scavenging and the various types of crankcase valving used in two-cycle engines.

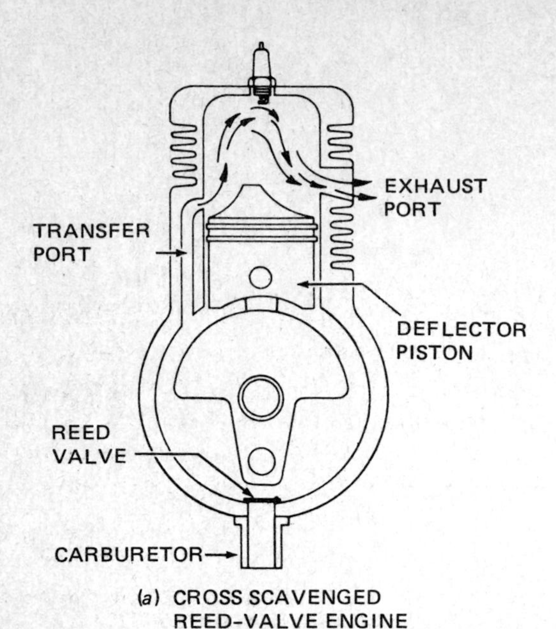

(*a*) CROSS SCAVENGED REED-VALVE ENGINE

(*b*) CROSS SCAVENGED ROTARY-VALVE ENGINE

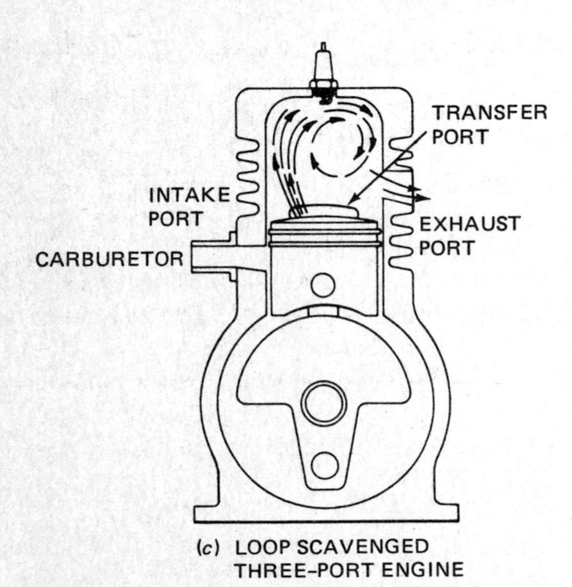

(*c*) LOOP SCAVENGED THREE-PORT ENGINE

FIG. 3-13 Various ways of scavenging and valving a two-cycle engine. *(Kohler Company)*

Δ3-11 CONSTRUCTION OF TWO-CYCLE-ENGINES

Figure 3-14 shows an air-cooled single-cylinder two-cycle engine that is completely disassembled. The construction is typical of two-cycle engines used in rotary lawn mowers and other power equipment. The engine is started by operation of either a mechanical or electric starter. A carburetor is attached to the crankcase, on the side opposite the cylinder. Air-fuel mixture flows from the carburetor, through a reed valve, and into the crankcase. The spark plug is located in the center of the combustion chamber. The plug provides a spark to ignite the air-fuel mixture every time the piston reaches TDC.

In an air-cooled engine, the cylinder and head are finned (Fig. 3-14). The fins help carry away excess heat which could cause overheating and damage to the engine. The operation of small-engine cooling systems is covered in Chap. 9.

Note that the cylinder is placed horizontally in Fig. 3-14 so the crankshaft is positioned up and down (vertically). This type of engine is called a *vertical-crankshaft engine.* Engines with a horizontal crankshaft are called *horizontal-crankshaft engines.* The engine classification is based on the position of the crankshaft (*not* the position of the cylinder). Other two-cycle small engines, such as those used in chain saws, must be capable of operation in any position. Therefore, these engines are classified as having a *multiposition* crankshaft.

Figure 3-15 shows a similar engine completely assembled. The engine assembly includes a rope-rewind starter and a shroud for directing cooling air between the fins on the cylinder and head. All small engines do not have a removable cylinder head, such as shown in Fig. 3-14. Some two-cycle engines are made with a one-piece head-and-cylinder assembly. To disassemble these engines, the cylinder is unbolted from the crankcase. Then the piston is pulled out from the bottom of the cylinder.

A cutaway single-cylinder two-cycle engine is shown in Fig. 3-16. This engine has a vertical crankshaft, and is the type used on most rotary lawn mowers (Fig. 3-17).

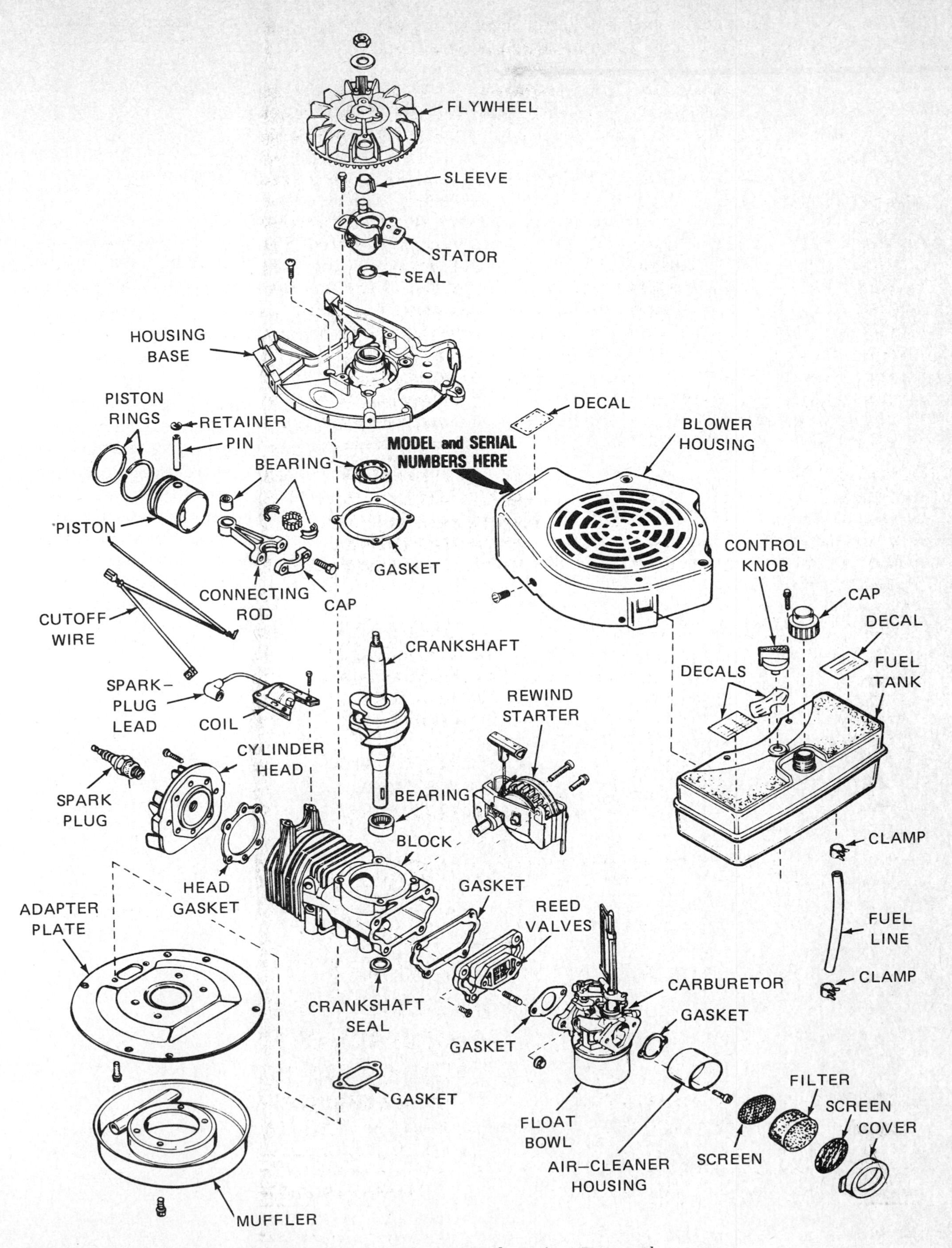

FIG. 3-14 A disassembled air-cooled single-cylinder two-cycle engine. *(Tecumseh Products Company)*

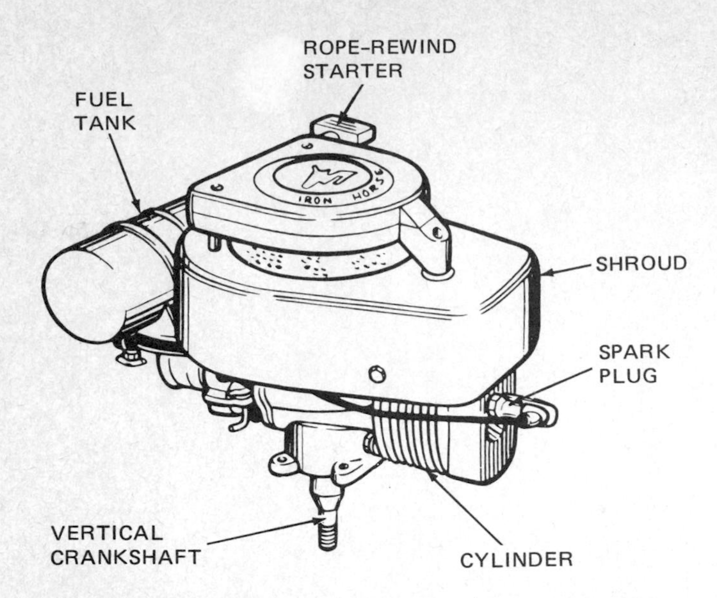

FIG. 3-15 **A single-cylinder two-cycle engine removed from a lawn mower.** *(Lawn Boy Division of Outboard Marine Corporation)*

The blade is installed on the lower end of the crankshaft.

Figure 3-18 shows a single-cylinder two-cycle engine with the starter, flywheel housing, flywheel, ignition magneto, and carburetor removed. These parts must be taken off before the piston and crankshaft can be removed from the crankcase. Figure 3-19 shows the engine with the cylinder head removed. An engine that has a removable cylinder head must have a gasket installed between the cylinder and the head. After the head bolts are tightened, the gasket compresses to seal the compression and combustion pressures in the cylinder.

The same engine is shown completely disassembled in Fig. 3-20. In this engine, the crankshaft turns in ball bearings (Δ2-17). A seal is installed on the outside of each crankshaft bearing to make the crankcase airtight. The connecting-rod big-end bearing is a roller bearing. The rollers are held in position by a split cage. Although many other small two-cycle engines have only one or two compression rings, three are shown for the piston in Fig. 3-20.

Δ3-12 ENGINE CRANKSHAFTS

Most small engines use a single-piece crankshaft (Figs. 3-20 and 3-21). This allows the connecting-rod big end to be split. Then a sliding bearing can be assembled to the crankpin by installing the connecting-rod cap (Fig. 3-21). However, some small engines (and many motorcycle engines) use a crankshaft that is assembled from several pieces (Fig. 3-22). This kind of crankshaft is sometimes called a *built-up crankshaft.* The connecting rod may not be split on the big end. Therefore, to install it on the crankshaft, the crankpin is assembled through the big-end roller bearing. Then the crankpin is pressed into holes in the two crankshaft halves (Fig. 3-22).

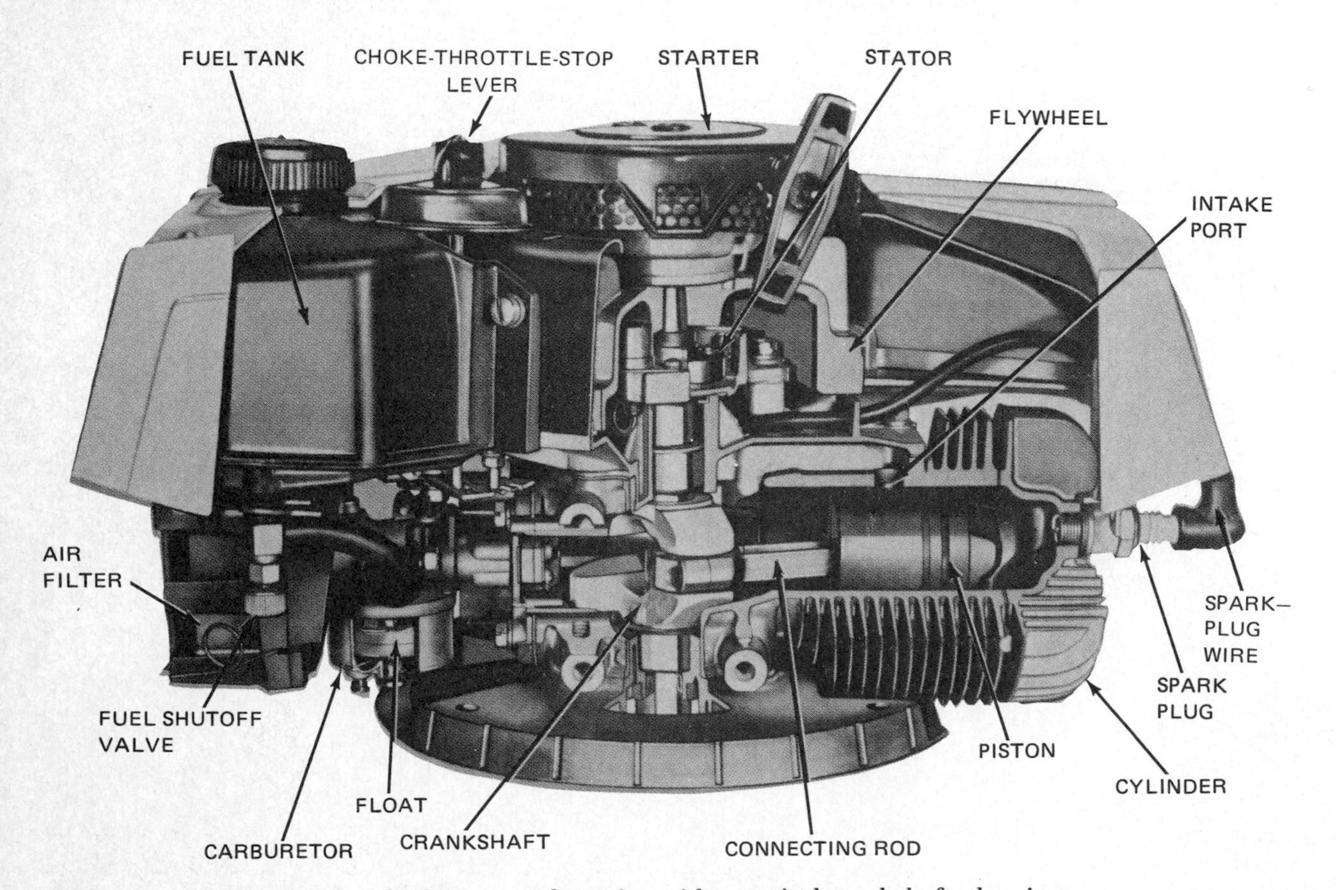

FIG. 3-16 **A cutaway single-cylinder two-cycle engine with a vertical crankshaft, showing the carburetor float and magneto stator.** *(Jacobsen Manufacturing Company)*

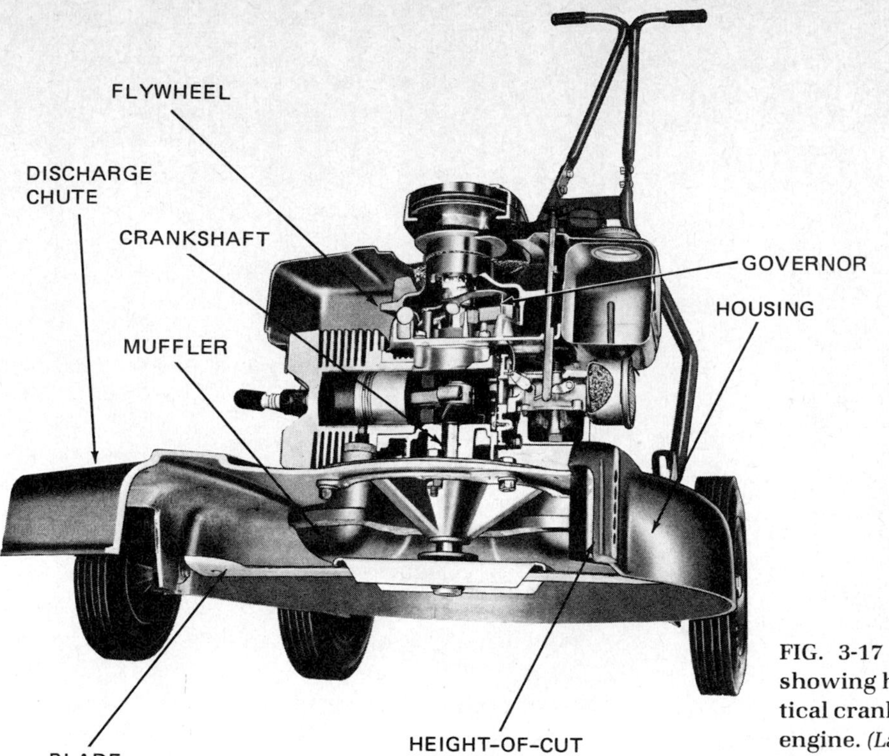

FIG. 3-17 A cutaway rotary lawn mower, showing how the blade is attached to the vertical crankshaft of a single-cylinder two-cycle engine. *(Lawn Boy Division of Outboard Marine Corporation)*

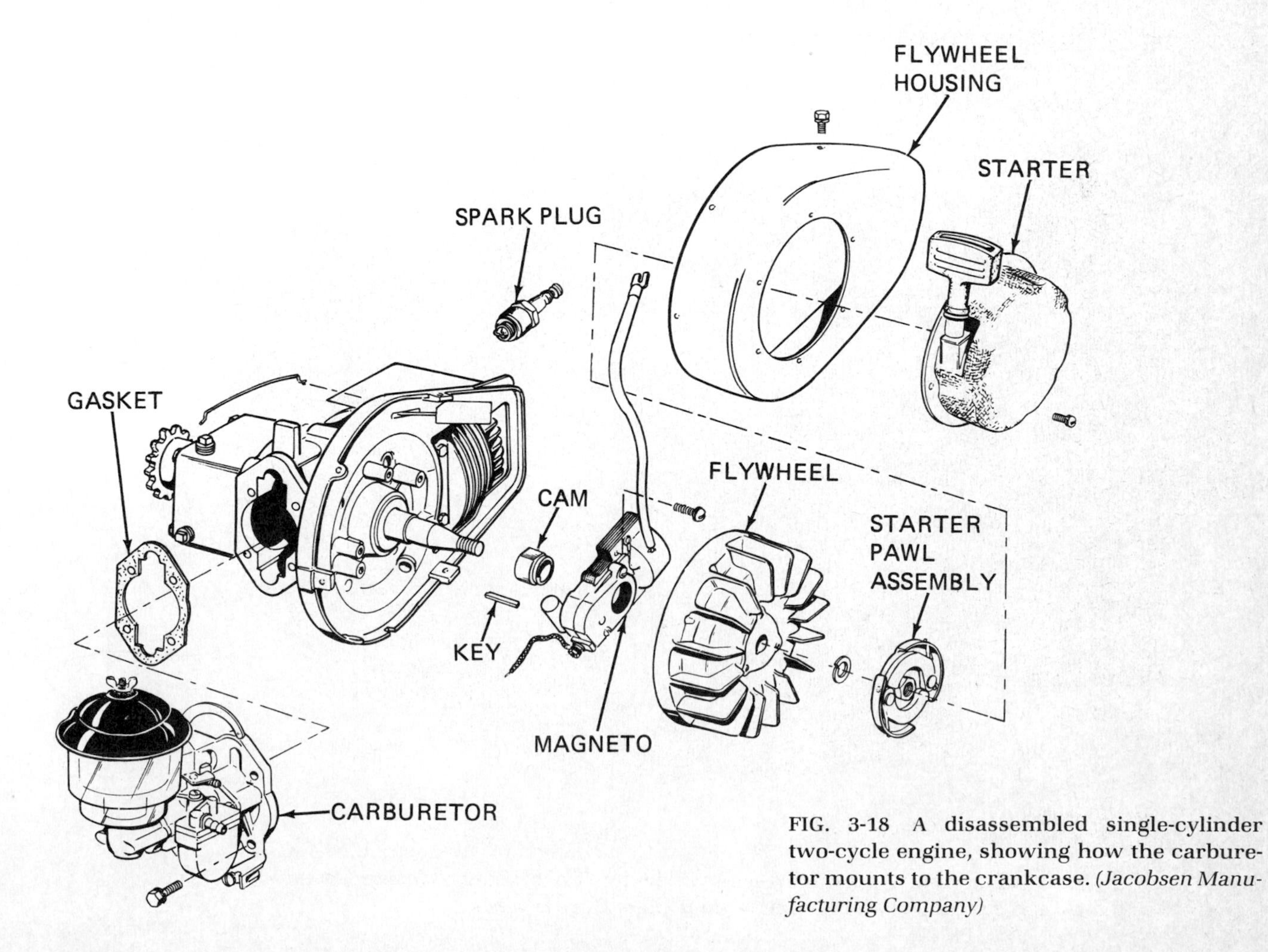

FIG. 3-18 A disassembled single-cylinder two-cycle engine, showing how the carburetor mounts to the crankcase. *(Jacobsen Manufacturing Company)*

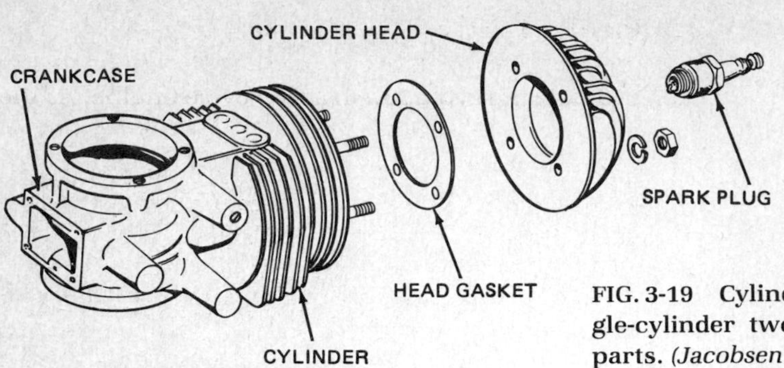

FIG. 3-19 Cylinder head and cylinder of a single-cylinder two-cycle engine, with related parts. *(Jacobsen Manufacturing Company)*

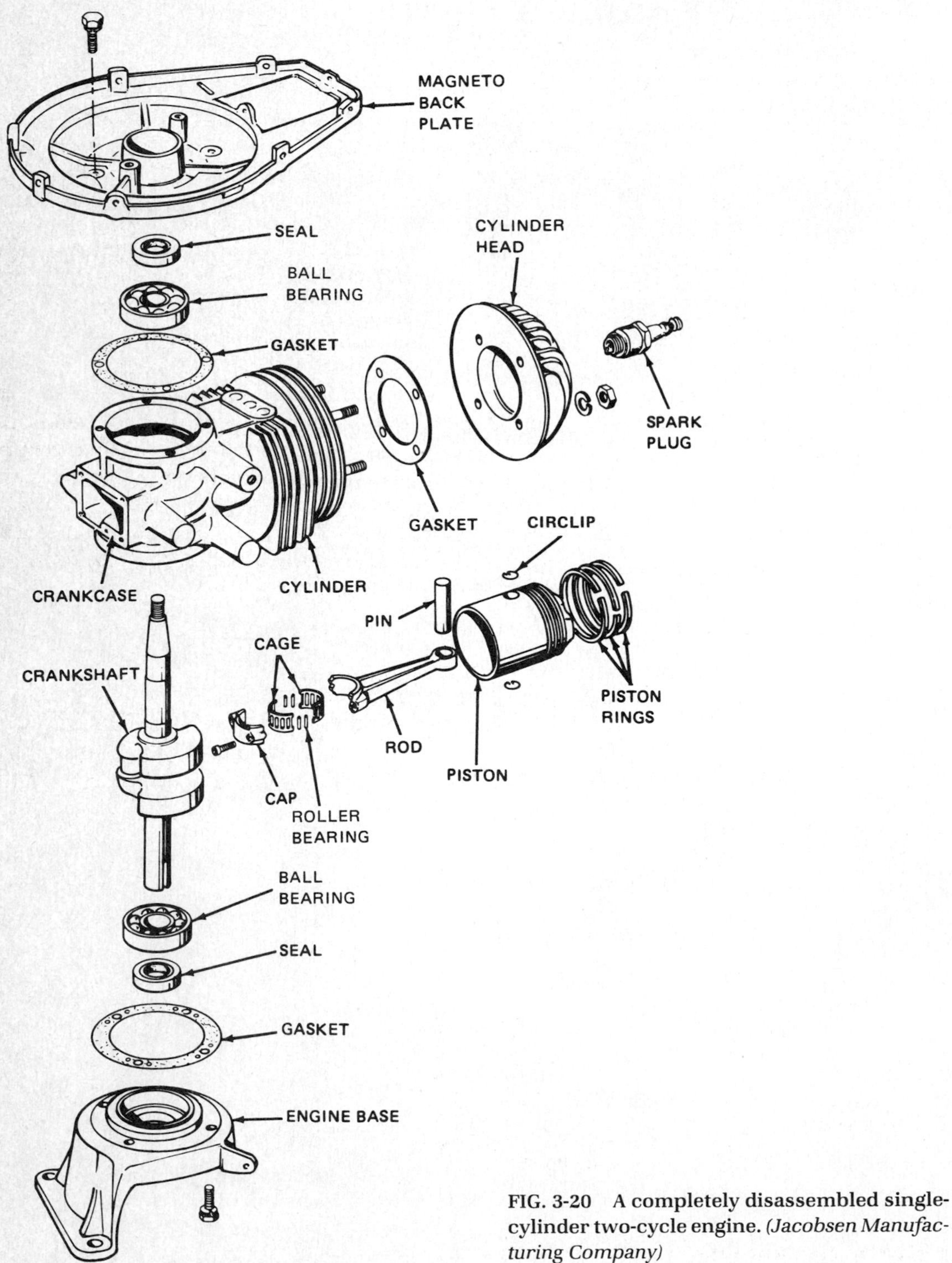

FIG. 3-20 A completely disassembled single-cylinder two-cycle engine. *(Jacobsen Manufacturing Company)*

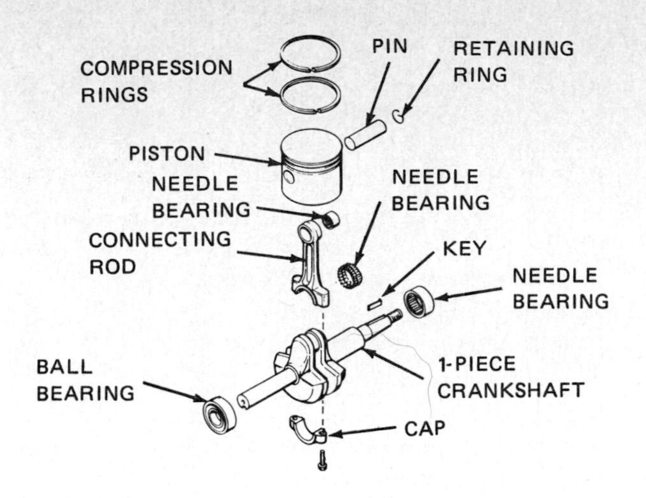

FIG. 3-21 A single-piece crankshaft, with piston-and-connecting-rod assembly. *(Tecumseh Products Company)*

Lubrication for the bearings in a two-cycle engine usually is provided by mixing a small amount of lubricating oil with the fuel. Another method of getting oil into the gasoline is to spray the oil into the air-fuel mixture before it enters the crankcase. Then, as the air-fuel mixture passes through the crankcase, the oil provides the required lubrication. Roller bearings need less lubrication than sliding bearings (Chap. 6). However, to be sure that the roller bearings receive adequate lubrication, some connecting-rod big ends have slits to allow the oil mist to pass through.

Some multiple-cylinder two-cycle engines have built-up crankshafts. Most "larger" small engines and almost all automotive engines use a one-piece crankshaft.

Δ3-13 FLYWHEELS The power impulses resulting from the power strokes occur for less than half of each crankshaft revolution in a two-cycle engine. As the piston passes TDC, the high combustion pressure pushes the piston down the cylinder. However, the push does not last long. As soon as the top of the piston moves below the top of the exhaust port, the gases escape and the pressure drops (Fig. 3-6).

During the rest of the cycle (as the piston passes BDC and moves back up the cylinder), no power is produced. Only the momentum of the moving parts carries the piston up to TDC so another power stroke can take place. This causes the crankshaft to speed up slightly during the power stroke and then to slow down during the rest of the cycle.

To smooth out the speed-up and slow-down action, most small engines have a *flywheel* mounted on one end of the crankshaft (Fig. 2-13). Other engines (especially motorcycle engines) have the flywheel inside the crankcase (Fig. 3-23). This allows the flywheel to be made in two halves. Each half is drilled so that one end of the crankpin can be pressed in during the assembly of a built-up crankshaft.

When the engine is running, the flywheel makes use of the property of *inertia* that all material things have. An object in motion tends to stay in motion. An object at rest (stationary) tends to remain at rest. These are the rules of inertia. So the flywheel, once in motion, tries to continue spinning at the same speed. When the power stroke occurs and the crankshaft tries to speed up, the inertia of the flywheel helps prevent a sudden increase in speed. When the crankshaft tries to slow down during the nonpower part of the cycle, the flywheel helps maintain the

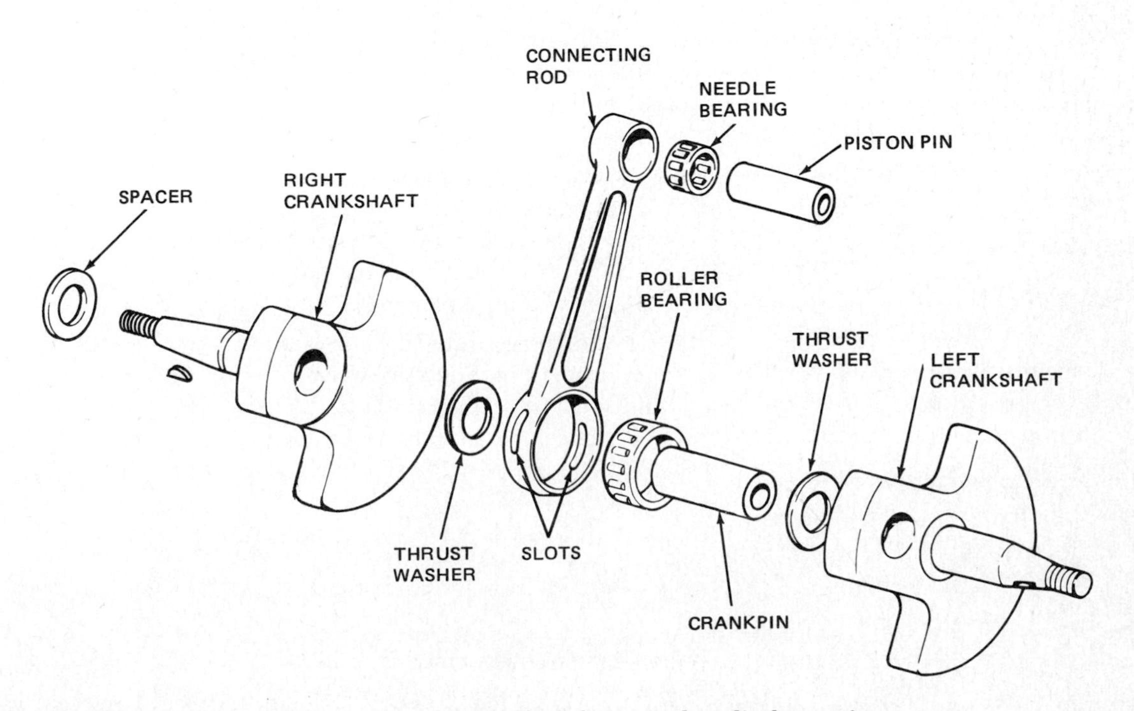

FIG. 3-22 A disassembled built-up crankshaft for a single-cylinder engine.

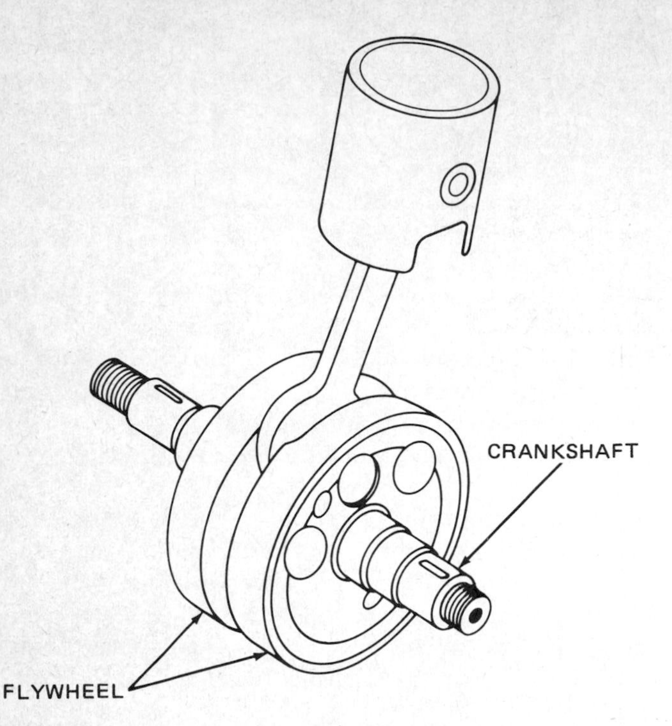

FIG. 3-23 An assembled two-cycle engine crankshaft, showing the flywheels.

crankshaft speed. The flywheel gives up energy to keep the crankshaft moving during the nonpower time. Then the flywheel takes in and stores energy when the crankshaft tries to speed up.

Multiple-cylinder engines also use a flywheel. The power strokes in these engines are arranged to follow one another in sequence (overlap). The overlap of the power impulses makes a smoother-running engine. However, a flywheel is still needed to further smooth out the speed of the crankshaft.

In small engines, the external flywheel may carry one or more small magnets (Fig. 3-18). These magnets are part of the magneto-ignition system and also part of the alternator that produces an electric current to charge the battery. Ignition and charging systems are described in later chapters.

CHAPTER 3: REVIEW QUESTIONS

Select the *one* correct, best, or most probable answer to each question. You can find the answers in the section indicated at the right of each question.

1. In the two-cycle engine, the burned gases leave the cylinder, and fresh air-fuel mixture enters, through (**Δ3**-2)
   a. valves in the cylinder head
   b. ports in the cylinder head
   c. ports in the cylinder wall
   d. reed valves in the cylinder head

2. The basic actions required to run an engine are to (**Δ3**-2)
   a. get burned gases out of the cylinder
   b. get fresh air-fuel mixture into the cylinder
   c. make the piston move up and down
   d. all of the above

3. In the two-cycle engine, the air-fuel mixture is ignited (**Δ3**-4)
   a. as the piston nears TDC
   b. in the crankcase
   c. as the piston nears BDC
   d. at the transfer point

4. In a two-cycle engine, air-fuel mixture is delivered to the cylinder under pressure because (**Δ3**-6)
   a. the valves are closed
   b. the piston is stationary at TDC
   c. pressure builds up in the crankcase
   d. of atmospheric pressure

5. The reed valve (**Δ3**-7)
   a. opens to admit air-fuel mixture to the crankcase
   b. closes to hold pressure in the crankcase
   c. works because of pressure and vacuum
   d. all of the above

6. In multiple-cylinder two-cycle engines, each cylinder (**Δ3**-9)
   a. has its own crankshaft
   b. has its own sealed compartment in the crankcase
   c. uses overhead valves
   d. none of the above

7. The rotation of the rotary valve (**Δ3**-8)
   a. closes and opens the intake port
   b. closes and opens the transfer port
   c. closes and opens the exhaust port
   d. all of the above

8. Some two-cycle small engines are (**Δ3**-10)
   a. loop-scavenged
   b. cross-scavenged
   c. either a or b
   d. neither a nor b

9. In most small engines, the crankshaft is (**Δ3**-12)
   a. a one-piece crankshaft
   b. an assembled, or built-up, crankshaft
   c. made of hollow castings
   d. designed to flex easily

10. The purpose of the flywheel is to (**Δ3**-13)
    a. prevent engine runaway
    b. prevent engine stalling
    c. smooth out engine operation
    d. keep the engine running after the ignition is turned off

CHAPTER

4

Four-Cycle Spark-Ignition Engines

After studying this chapter, you should be able to:

1. ***Describe the construction of a four-cycle engine and explain how it differs from the construction of a two-cycle engine.***
2. ***Discuss the four-stroke cycle and explain the actions during each piston stroke.***
3. ***Describe the differences between an L-head engine and an overhead-valve engine.***
4. ***Explain how the valves are operated in an L-head engine.***
5. ***Discuss the need for automatic compression release, and explain how it operates.***
6. ***Explain why the four-cycle engine requires at least one oil-control ring.***
7. ***Describe how to identify two-cycle and four-cycle engines by visual inspection.***

Δ4-1 CONSTRUCTION OF FOUR-CYCLE ENGINES All automotive engines operate on the four-cycle principle. Many small engines also are four-cycle engines. The basic difference between two-cycle and four-cycle engines is in the way the air-fuel mixture gets into the engine cylinders and in the way burned gases are removed from the cylinders.

The previous chapter explained how the two-cycle engine does all this. There are intake, transfer, and exhaust ports in the cylinder. The piston, as it moves down and up, opens and closes ports. This allows the burned gases to escape and a fresh charge of the air-fuel mixture to enter.

The four-cycle engine also has intake and exhaust ports. However, they are opened and closed in a different manner, as explained in later sections.

There are many similarities between two-cycle and four-cycle engines. The pistons, cylinder, connecting rod, and crankshaft are very similar (Fig. 4-1). However, there are several important differences, both in con-

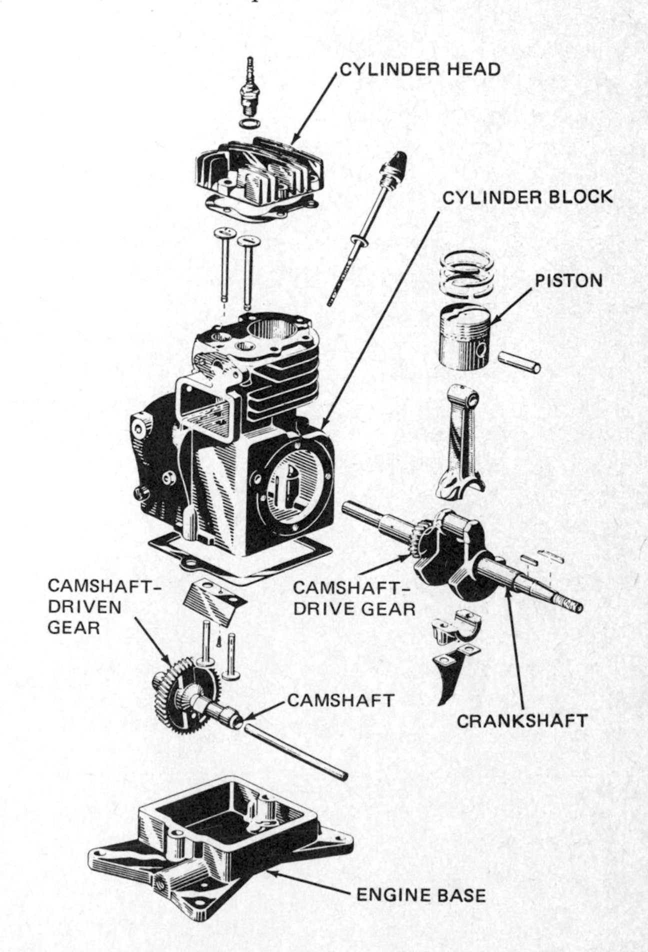

FIG. 4-1 **A disassembled one-cylinder four-cycle engine. The head and cylinder block are separate parts.** *(Clinton Engines Corporation)*

struction and in operation. Notice the much more complex construction of the four-cycle engine (Fig. 4-2). Then compare Fig. 3-20, which shows a disassembled two-cycle engine, with Fig. 4-2. There are many more parts in a four-cycle engine than in a two-cycle engine.

Δ4-2 THE FOUR-STROKE CYCLE In the four-stroke-cycle engine, one out of every four piston strokes is a power stroke. The four-stroke-cycle engine is usually called a *four-cycle engine*, or a *four-stroker*. This engine requires two valves that open and close to permit

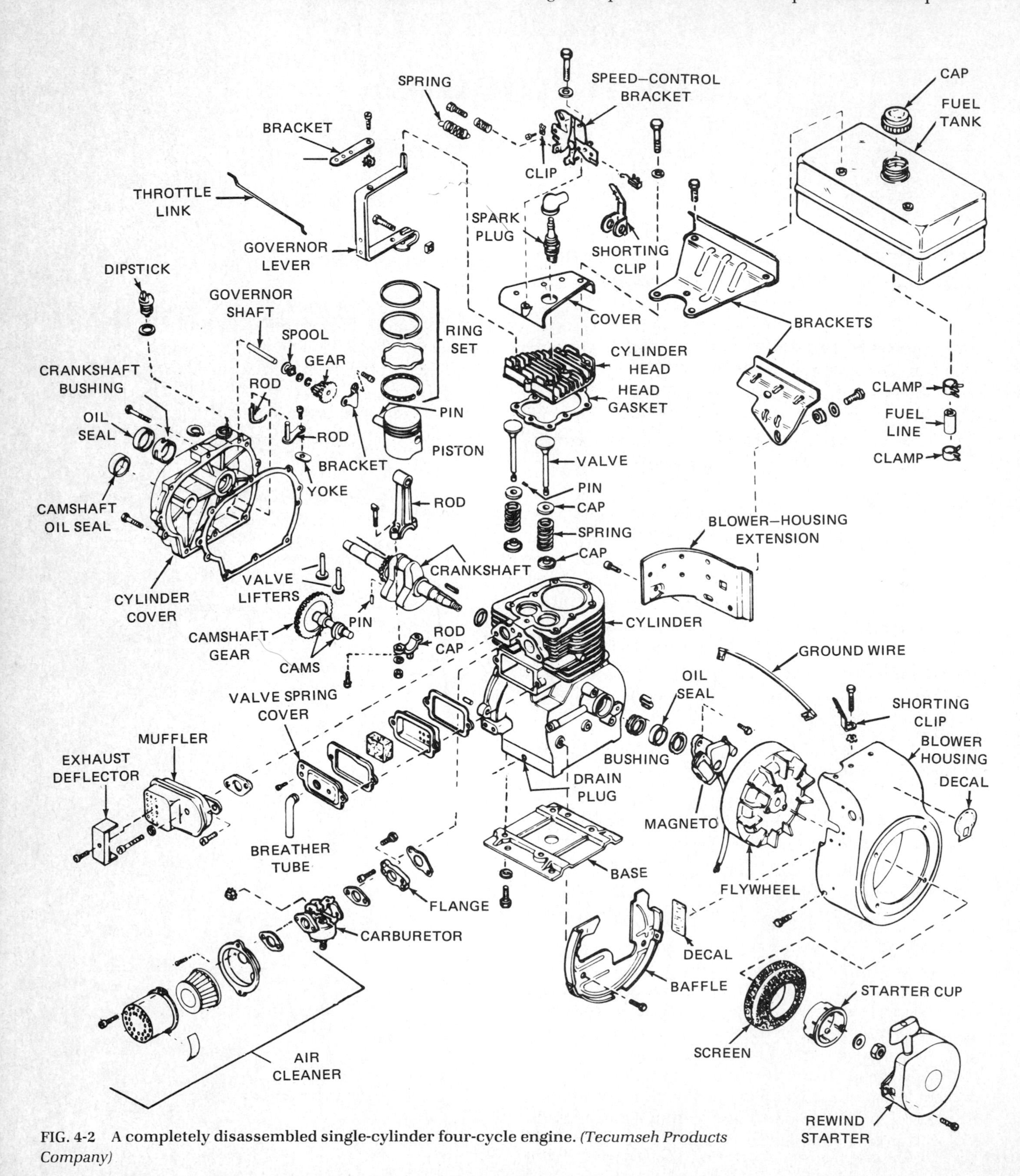

FIG. 4-2 A completely disassembled single-cylinder four-cycle engine. *(Tecumseh Products Company)*

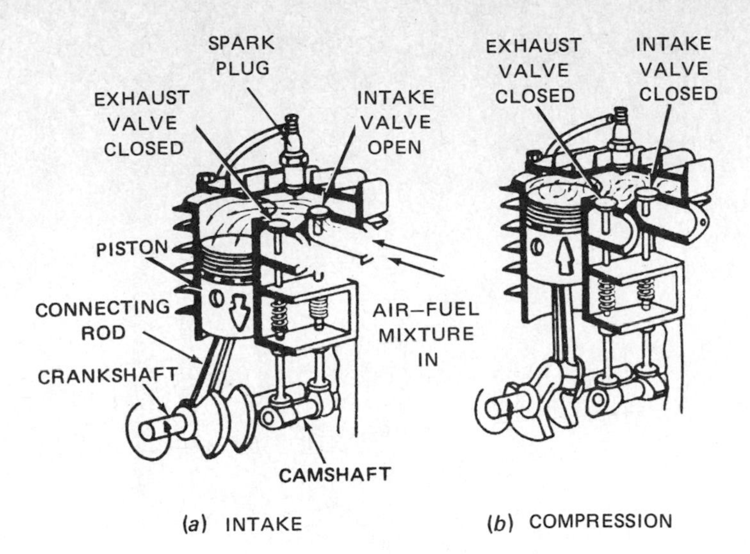

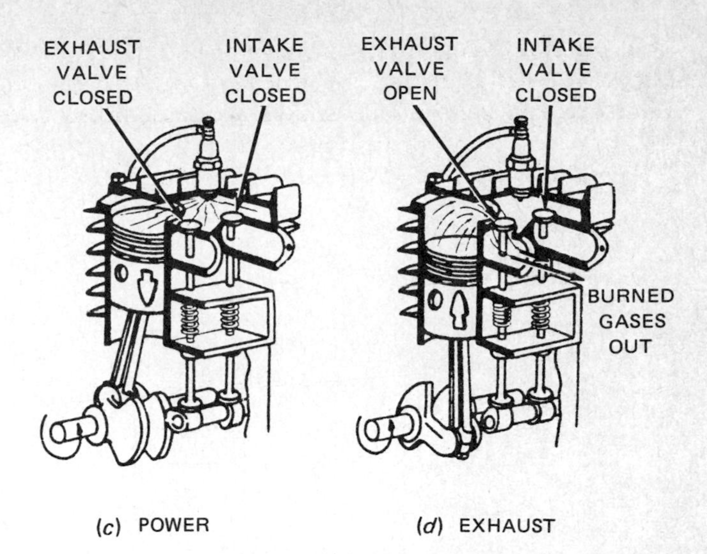

FIG. 4-3 The four strokes in a four-stroke-cycle engine.

intake and exhaust. Figure 4-3 shows the four strokes. How the valves are opened and closed is described later.

In Fig. 4-3*a*, the piston is moving down on the intake stroke. The intake valve is open and the carburetor is delivering an air-fuel mixture to the cylinder. After the intake stroke ends, the intake valve closes. Now the piston moves up on the compression stroke (Fig. 4-3*b*). The air-fuel mixture is trapped and therefore compressed in the top of the cylinder (the *combustion chamber*).

As the piston nears TDC on the compression stroke, the ignition system delivers a spark to the spark plug. This ignites the compressed air-fuel mixture, and the resulting high pressure forces the piston to move down on the power stroke (Fig. 4-3*c*). The downward push on the piston may total as much as 4000 pounds [17,792 N] in a modern small engine. This force is carried through the connecting rod to a crank on the engine crankshaft (Fig. 4-3*c*). The ignition system, which produces the spark at the spark plug, is explained in a later chapter.

The fourth stroke in the four-stroke cycle is the exhaust stroke. As the piston nears BDC on the power stroke, the exhaust valve opens. Now, as the piston passes BDC and moves up on the exhaust stroke, the burned gases in the cylinder are forced out (Fig. 4-3*d*).

As the piston nears TDC on the exhaust stroke, the intake valve opens. Then, after TDC, the exhaust valve closes, and the whole cycle of events is repeated once again. The cycle is repeated continuously as long as the engine runs.

The valves are opened and closed at the proper times by the *valve train*. The valve train includes a camshaft which is driven from the engine crankshaft. In small four-cycle engines, the camshaft is located in the cylinder block, near the crankshaft (Fig. 4-3). Valve trains and how they operate the valves are described later.

Δ4-3 INTAKE AND EXHAUST VALVES

In small four-cycle engines, there are two openings, or ports, that open into the combustion chamber. These openings are either to one side of the piston in the cylinder block or above the piston in the cylinder head. One opening is the intake port. The air-fuel mixture flows in through this port when the intake valve is open (Fig. 4-3*a*). The other opening is the exhaust port. The burned gases flow out, or exit, through this port when the exhaust valve is open (Fig. 4-3*d*).

An engine valve is a long metal stem which is flared out at one end (Fig. 4-4). When a valve moves into a valve port, the valve face seats on, or makes contact with, the

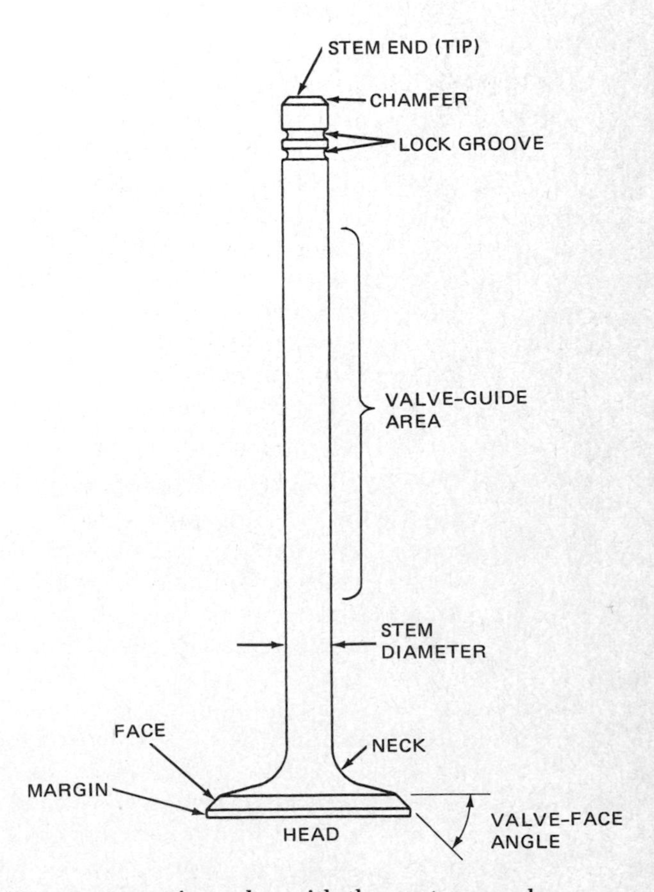

FIG. 4-4 An engine valve with the parts named.

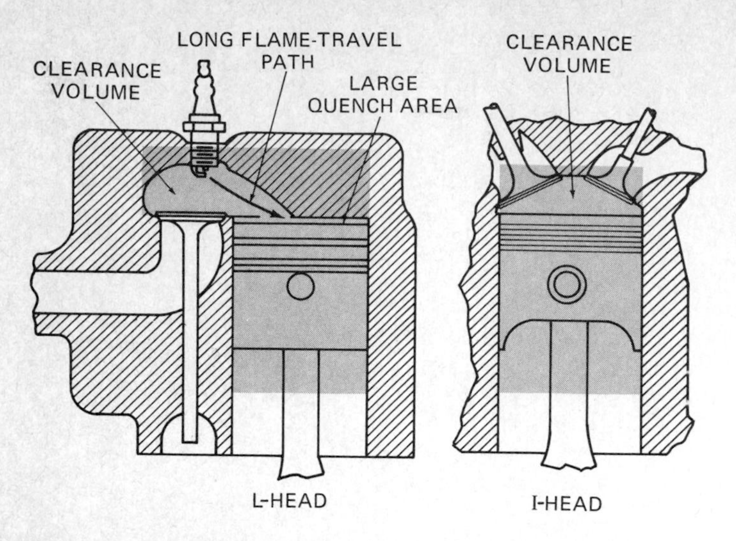

FIG. 4-5 Comparison of L-head (side-valve) and I-head (overhead-valve) combustion chambers and valve arrangements.

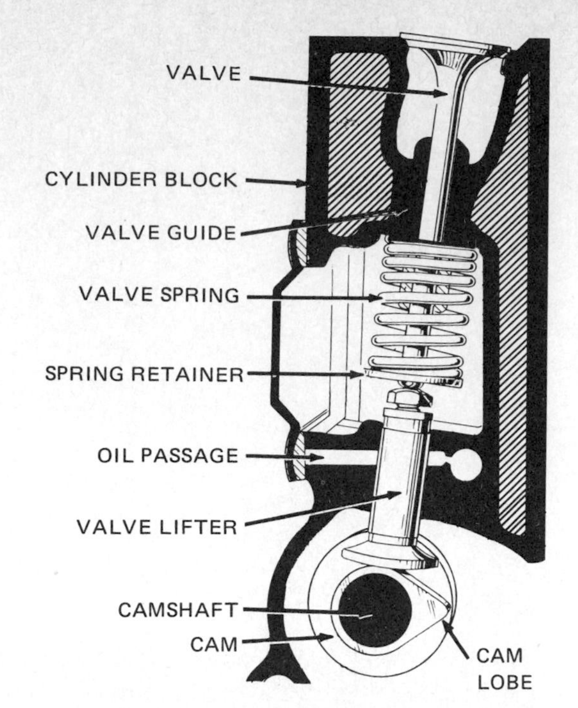

FIG. 4-6 Valve mechanism used in an L-head engine. The valve is raised up off its seat with every camshaft revolution.

valve seat in the cylinder head. The fit is so tight that no air can pass between the valve face and the seat when the valve is closed, or seated.

The intake valve is usually larger than the exhaust valve. The reason is that the only force behind the entering air-fuel mixture when the intake valve is open is atmospheric pressure. However, when the exhaust valve opens, the burned gases are under high pressure from the combustion process. Therefore, the intake valve is larger so that there will be a larger area for the air-fuel mixture to flow through to enter the cylinder.

Δ4-4 LOCATION OF THE VALVES

In most small engines, the valves are located in the cylinder block (Fig. 4-1). This arrangement forms an upside-down "L," so the engine is called a *side-valve engine,* or an *L-head engine* (Fig. 4-5). Some small engines and all larger engines, including those used in automobiles, have the valves in the cylinder head (Fig. 4-5). These engines are called *overhead-valve engines,* or *I-head engines.*

Each arrangement requires a different type of valve train. However, valve actions in opening and closing the ports are the same, regardless of the valve locations.

Δ4-5 OPERATING THE VALVES

The valves must be moved at the right times to open and close the valve ports (**Δ**4-3). Figures 4-6 and 4-7 show typical valve operating mechanisms used in small four-cycle engines. The valves move up and down in valve guides that are part of the cylinder head (Fig. 4-5) or cylinder block (Fig. 4-8).

Each valve has a *valve spring* (Fig. 4-6) that puts tension on the valve and tries to keep it closed, or seated, on the valve seat. When the valve is seated, it closes the port. The valve spring is held between the cylinder block and a *valve-spring retainer* (Fig. 4-6). The spring retainer is attached to the valve stem by a valve-spring-retainer lock or pin, which fits into the valve stem (Fig. 4-9). Below the valve stem is a *valve lifter,* or *valve tappet.* It moves up and down in a bore, or hole, in the cylinder block (Figs. 4-10 and 4-11).

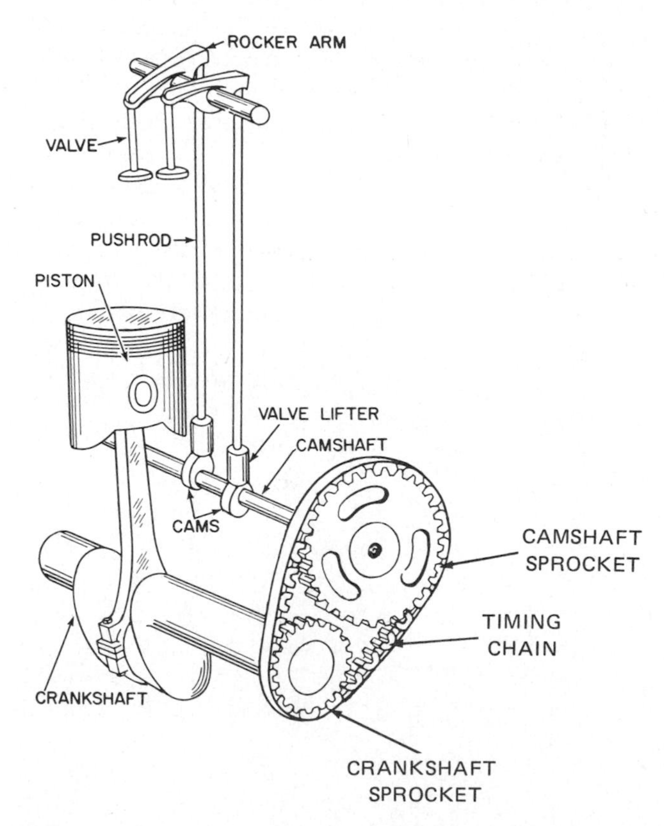

FIG. 4-7 Valve-operating mechanism for an I-head, or overhead-valve, engine.

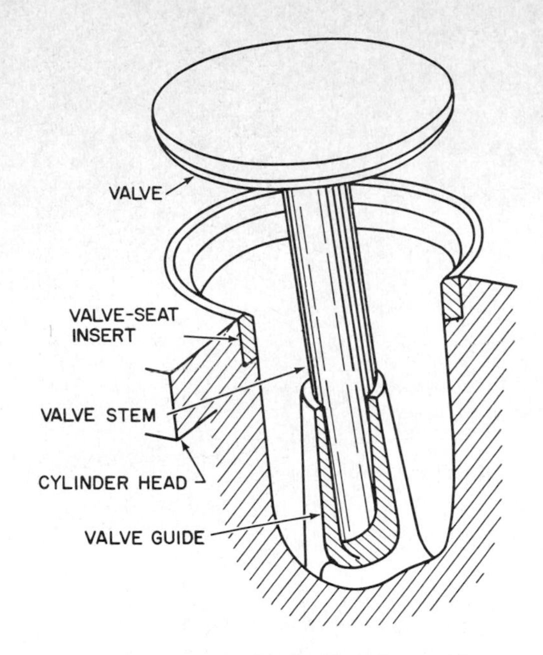

FIG. 4-8 Cylinder head and valve guide, partly cut away to show the valve, valve guide, and valve seat.

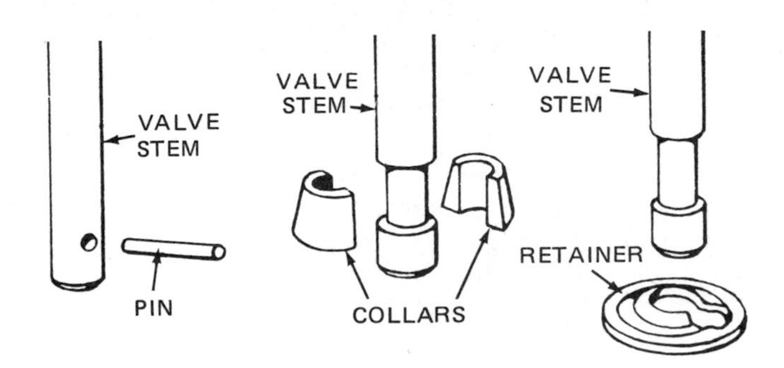

FIG. 4-9 Different types of valve-spring retainers and locks. *(Briggs & Stratton Corporation)*

The valve lifter rests on a cam which has one high spot, or *lobe*. The cam is usually mounted on a *camshaft*, which has one cam for each valve. The camshaft is driven by a gear (Fig. 4-10) or by a chain and sprockets (Fig. 4-7) from the crankshaft. This causes the camshaft to rotate with the crankshaft.

As the camshaft rotates, the cam lobe moves around under the valve lifter, causing it to be pushed upward (Fig. 4-10). This upward push overcomes the valve-spring tension so that the valve is raised off the valve seat, opening the port. Now gas can pass through. When the intake valve is open, the air-fuel mixture from the carburetor flows through the valve opening and into the cylinder. When the exhaust valve is open, the exhaust gases from the cylinder flow through the valve opening and into the exhaust port.

Further rotation of the camshaft causes the cam lobe to rotate out from under the valve lifter (Fig. 4-11). Now the valve-spring tension forces the valve back on its seat, closing the port.

Notice in Fig. 4-11 that the camshaft gear is twice as large as the crankshaft gear. In the four-stroke-cycle engine, each valve should be open for only one of the four piston strokes. During the other three strokes, the valve should be closed. Therefore, for every two complete revolutions of the crankshaft, the camshaft turns only once. This means that the camshaft rotates at one-half crankshaft speed.

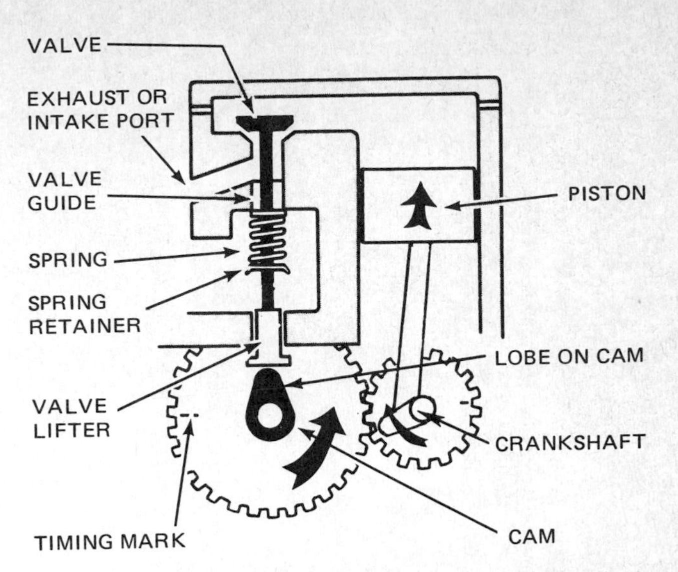

FIG. 4-10 As the camshaft is driven by the crankshaft, the cam lobe moves under the valve lifter. This forces the valve to move up off its seat.

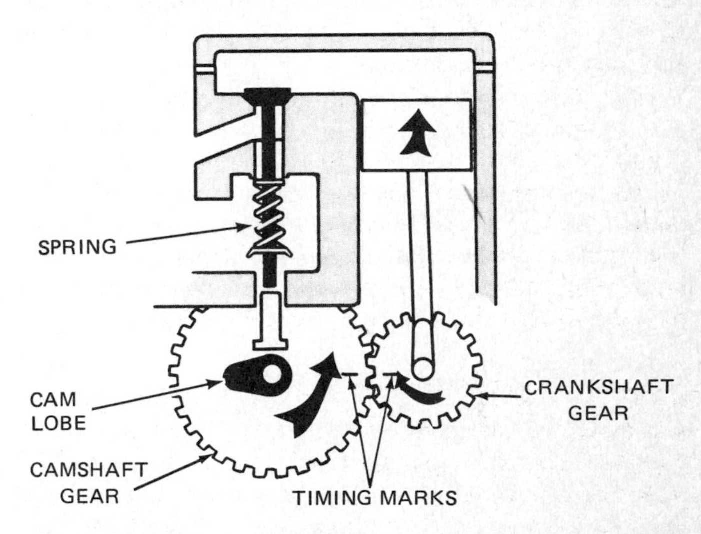

FIG. 4-11 Further rotation of the crankshaft and camshaft moves the lobe out from under the valve lifter. This allows the spring to pull the valve down and reseat it.

△4-6 LOCATION OF THE CAMSHAFT

Almost all small four-cycle engines have the camshaft located in the cylinder block (Fig. 4-3). The camshaft in most L-head engines is driven by gears from the crankshaft (Fig. 4-12). In other engines, the camshaft is driven by sprockets and chain (Fig. 4-7). Notice that the location of the camshaft is basically the same (in the cylinder block) for both L-head and I-head engines.

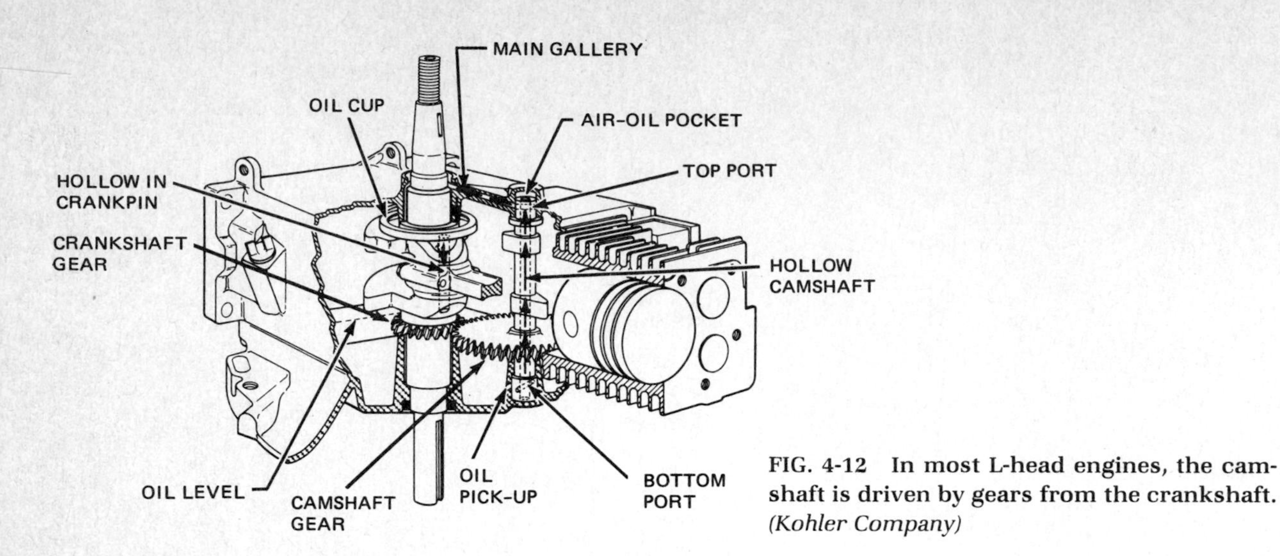

FIG. 4-12 In most L-head engines, the camshaft is driven by gears from the crankshaft. *(Kohler Company)*

Some high-performance small engines with I-heads also have the camshaft located in the cylinder head (Fig. 4-13). This type of construction is typical of small four-cycle motorcycle, outboard, and snowmobile engines. If a single camshaft is located in the head (Fig. 4-13, left), *rocker arms* are often used to operate the valves. This arrangement is called a *single-overhead camshaft* (SOHC) engine.

When intake and exhaust valves each have a separate camshaft, then there are two camshafts in the cylinder head. These engines have *double-overhead camshafts* (DOHC). Rocker arms are not required. A cam follower, or bucket tappet, transmits the cam-lobe action directly to the valve stem to open the valve (Fig. 4-13, right).

Overhead camshafts may be driven from the crankshaft by gears or by sprockets and a chain or toothed belt. With a chain or toothed belt, some type of *tensioner* is often used. The tensioner is usually a slide of some sort or a spring-loaded idler sprocket. Its job is to take up all looseness in the belt or chain. This prevents erratic valve action and reduces noise caused by stretch and wear.

Gears that drive the camshaft are called the *timing gears.* The chain or belt is called the *timing chain* or the *timing belt.* They "time" the operation of the valves.

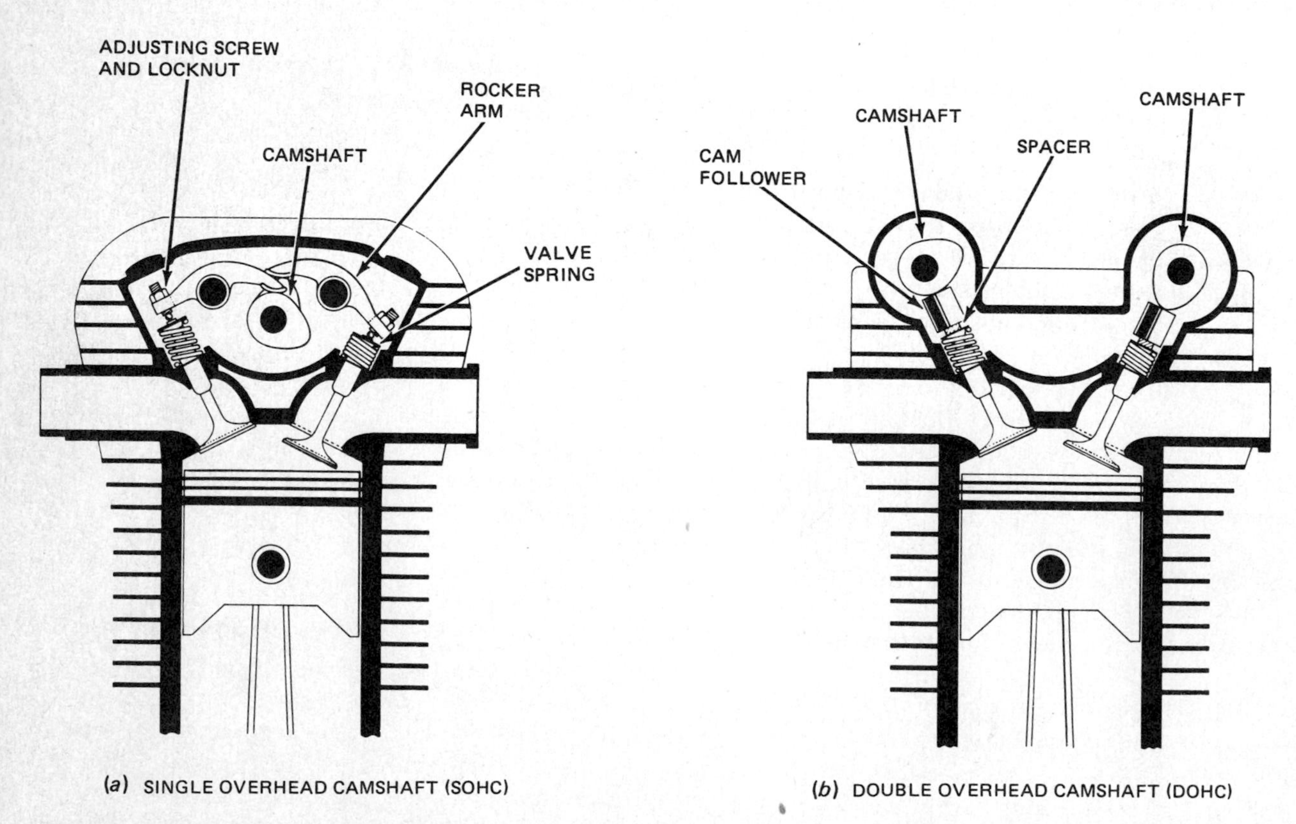

FIG. 4-13 Valve and camshaft arrangements for a single-overhead-camshaft engine and for a double-overhead-camshaft engine.

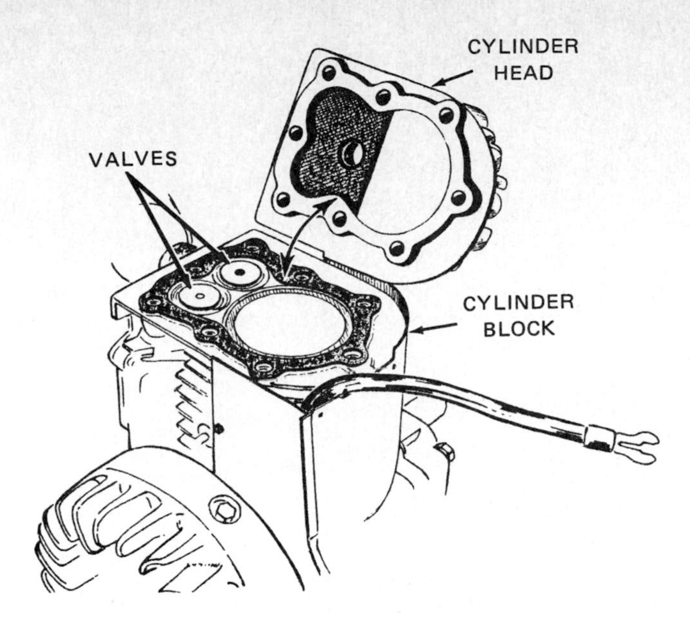

FIG. 4-14 In an L-head engine, the valves are below the cylinder head in the cylinder block.

Δ4-7 OVERHEAD-VALVE ENGINE OPERATION

An engine with overhead valves operates on the same four-stroke cycle of intake, compression, power, and exhaust as the L-head engine (Δ4-2). The basic difference in the two engines is in the placement of the intake and exhaust valves. In the L-head engine, the valves are below the cylinder head in the cylinder block (Fig. 4-14). In the overhead-valve engine, the valves are located in the cylinder head (Fig. 4-15). As the valves open and close, they control the flow of air-fuel mixture and exhaust gas through the ports in the cylinder head.

Figure 4-15 shows the four piston strokes in the overhead-valve engine. The valves are operated either by pushrods from a camshaft located in the cylinder block (Fig. 4-7) or by a camshaft located on top of the cylinder head (Fig. 4-13). The engine with the pushrods and the camshaft in the cylinder block is often called a *pushrod engine.*

Figure 4-16 shows how the valves are operated in an overhead-valve engine. When the camshaft lobe moves up under the valve lifter, the lobe raises the lifter. This causes the pushrod to move up so that it pushes up on one end of the rocker arm. This causes the rocker arm to rock, or pivot, so that the other end of the rocker arm pushes down on the valve stem. This forces the valve down off its seat, opening the valve.

After the cam lobe moves out from under the valve, the valve spring pushes the valve up so that it closes. This causes the rocker arm to rock back, forcing the pushrod and valve lifter down into the closed-valve position.

In the overhead-camshaft engine, the camshaft is mounted on top of the cylinder head and more directly operated the valves. Figure 4-13 shows how overhead camshafts can be used to operate the valves. In Fig. 4-13*a*, rocker arms are used between the cam lobe and the valve stem. As the lobe moves under one end of the rocker arm, it pivots and forces the valve stem down, opening the valve.

In Fig. 4-13*b*, each cam is directly over a small cup, or cam follower. It fits over the end of the valve stem. As the

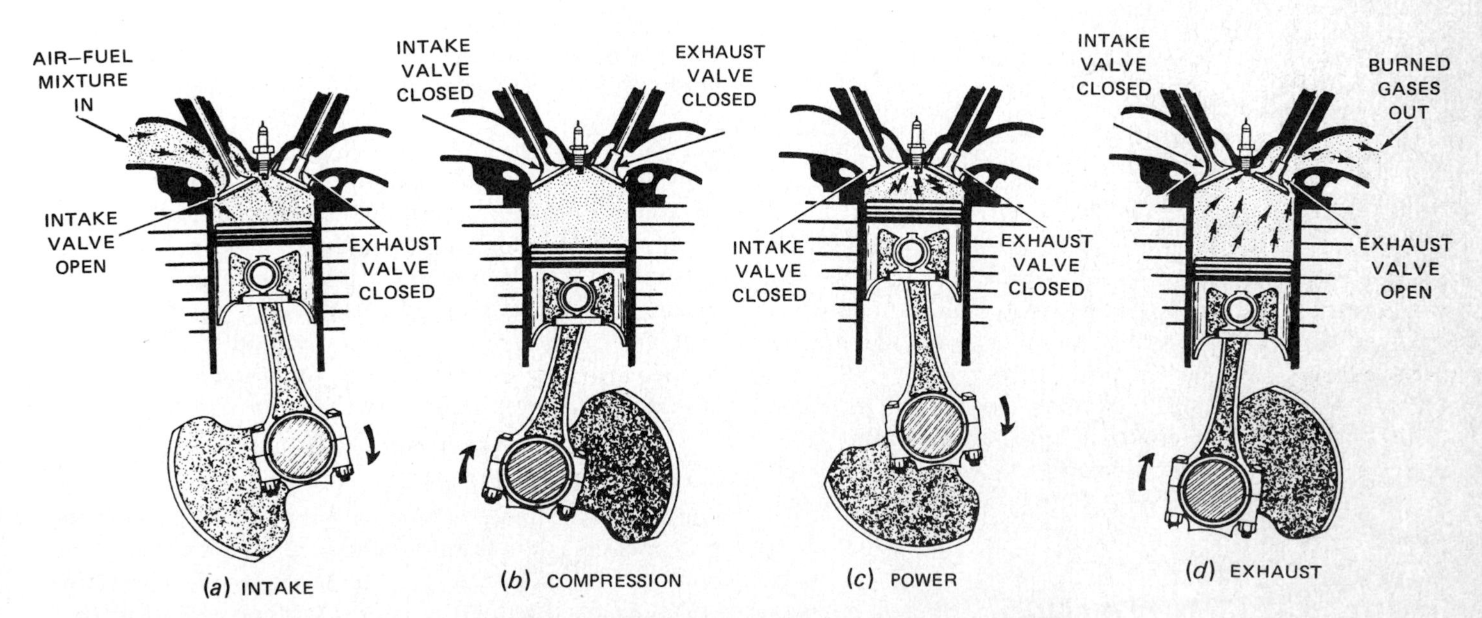

FIG. 4-15 *(a)* Intake stroke. The intake valve (at left) has opened. The piston is moving downward, allowing a mixture of air and gasoline vapor to enter the cylinder. *(b)* Compression stroke. The intake valve has closed. The piston is moving up, compressing the mixture. *(c)* Power stroke. The ignition system has delivered a spark to the spark plug that ignites the compressed mixture. As the mixture burns, high pressure is created, pushing the piston down. *(d)* Exhaust stroke. The exhaust valve (at right) has opened. The piston is moving upward, as the burned gases escape from the cylinder.

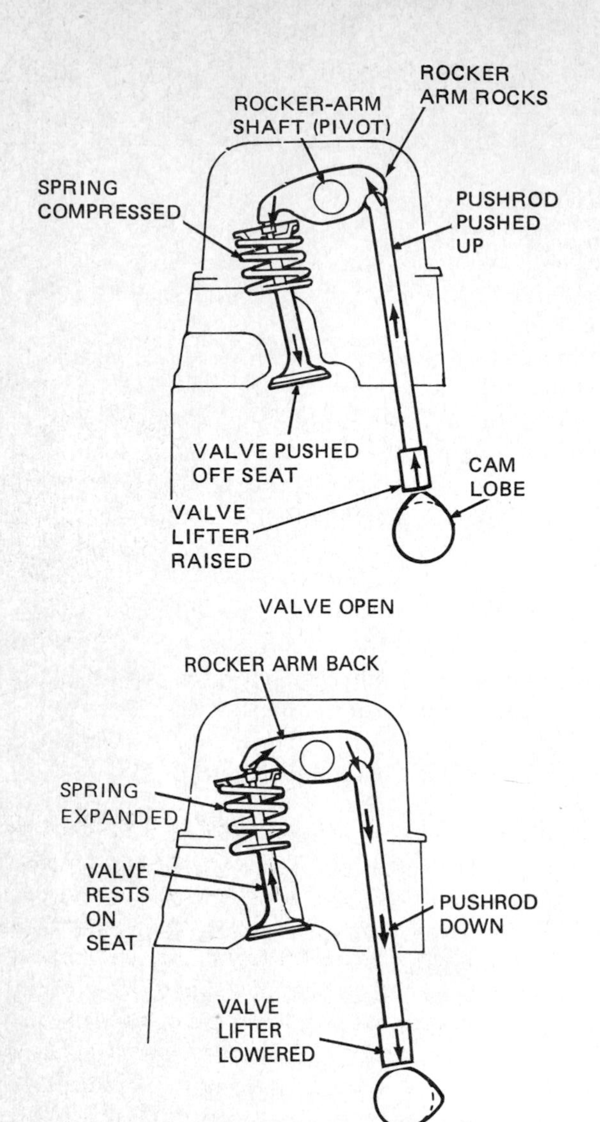

FIG. 4-16 Operation of the valve train in an overhead-valve engine.

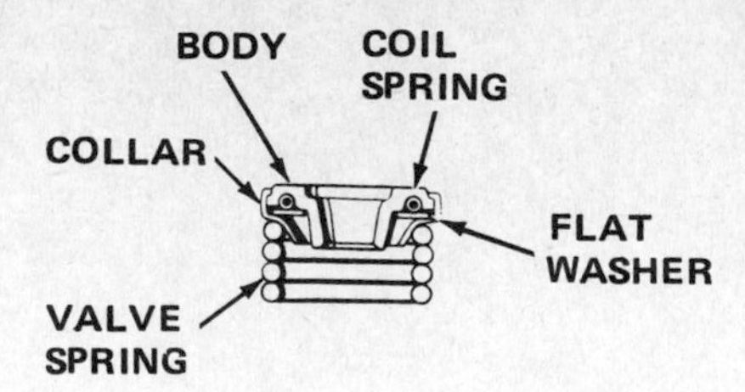

FIG. 4-17 A valve rotator is installed in place of the valve-spring retainer (especially on the exhaust valve) in some four-cycle engines.

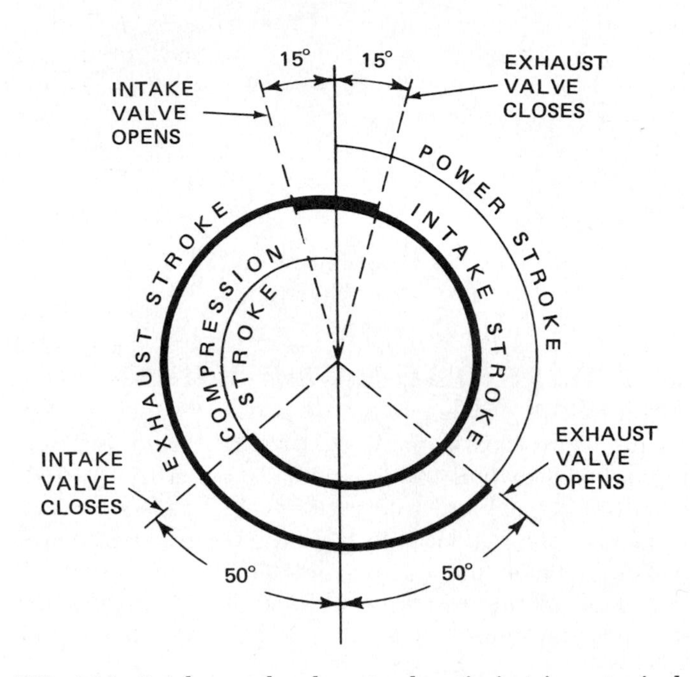

FIG. 4-18 Intake- and exhaust-valve timing in a typical small engine. The complete cycle of events is shown as a 720° spiral, which represents two complete crankshaft revolutions. The timing of valves differs for different engines.

cam lobe moves to open the valve, the lobe pushes down on the cam follower. This forces the cam follower to push down on the valve, opening it. When the lobe on the cam moves off the cam follower, the valve spring closes the valve.

Most four-cycle small engines are L-head engines. They are simpler in construction than overhead-valve engines. However, an overhead-valve engine runs more efficiently and is capable of producing more power per cylinder than an L-head engine of the same size.

Δ4-8 VALVE ROTATORS

Some engines have valve rotators, especially on the exhaust valves (Fig. 4-17). These devices are installed in place of the valve-spring retainers. Rotators turn the valve slightly as it opens. The turning motion is carried through the valve locks to the valve stem.

By rotating the valve slightly each time it opens, the valve seat and face are kept clean so that good seating continues. Without valve rotation, particles of carbon and dirt might cling to the valve face and seat. These particles could cause valve leakage and burning. However, valve rotation tends to wipe away any particles, and so the valve and seat last longer. Valve rotators are used in engines where long service life is required.

Δ4-9 VALVE TIMING

Valve timing is the relation between valve action and piston position. The timing is expressed as the number of crankshaft degrees before or after the TDC or BDC position of the piston. This number indicates the open or closed position of the valve in relation to piston travel. A valve-timing diagram, such as shown in Fig. 4-18, is used to show valve timing of a camshaft.

The valves open and close, not at TDC or at BDC, but sometime before or after the piston reaches the upper or lower limit of travel. The intake valve normally opens several degrees of crankshaft rotation before TDC on the exhaust stroke. That is before the exhaust stroke is finished. This gives the valve enough time to reach the fully open position before the intake stroke begins. Then, when the intake stroke starts, the intake valve is already wide open and air-fuel mixture can start to enter the cylinder immediately.

The intake valve remains open for several degrees of crankshaft rotation after the piston has passed BDC at the end of the intake stroke. This allows additional time for the air-fuel mixture to continue to flow into the cylinder. The fact that the piston has already passed BDC and is moving up on the compression stroke while the intake valve is still open does not affect the movement of air-fuel mixture into the cylinder. Actually, air-fuel mixture is still flowing in as the intake starts to close.

The reason for this is that the air-fuel mixture has inertia. It tends to keep on flowing after it once starts through the carburetor and into the engine cylinder. The momentum of the mixture then keeps it flowing into the cylinder even though the piston has started up on the compression stroke. This packs more of the air-fuel mixture into the cylinder and results in a stronger power stroke.

For a somewhat similar reason, the exhaust valve opens before the piston reaches BDC on the power stroke. As the piston nears BDC, most of the push on the piston has ended. No power is lost by opening the exhaust valve toward the end of the power stroke. This gives the exhaust gases additional time to start leaving the cylinder so that exhaust is under way by the time the piston passes BDC and starts up on the exhaust stroke.

The exhaust valve then stays open for several degrees of crankshaft rotation after the piston has passed TDC and the intake stroke has started. This makes good use of the momentum of the exhaust gases. They are moving rapidly toward the exhaust port. Leaving the exhaust valve open for a few degrees after the intake stroke starts gives the exhaust gases some additional time to leave the cylinder. This allows more air-fuel mixture to enter on the intake stroke so that a stronger power stroke results.

Actual timing of the valves varies with different four-cycle engines. A typical example for a small engine is shown in Fig. 4-18. The intake valve opens 15° of crankshaft rotation before TDC on the exhaust stroke, and it stays open until 50° of crankshaft rotation after BDC on the compression stroke. The exhaust valve opens 50° before BDC on the power stroke and stays open 15° after TDC on the intake stroke. This gives the two valves an overlap of 30° at the end of the exhaust stroke and beginning of the compression stroke.

Valve timing varies for different engines. It depends on the shape of the cam lobe that operates the valves. It also depends on the relationship between the gears or sprockets on the crankshaft and camshaft.

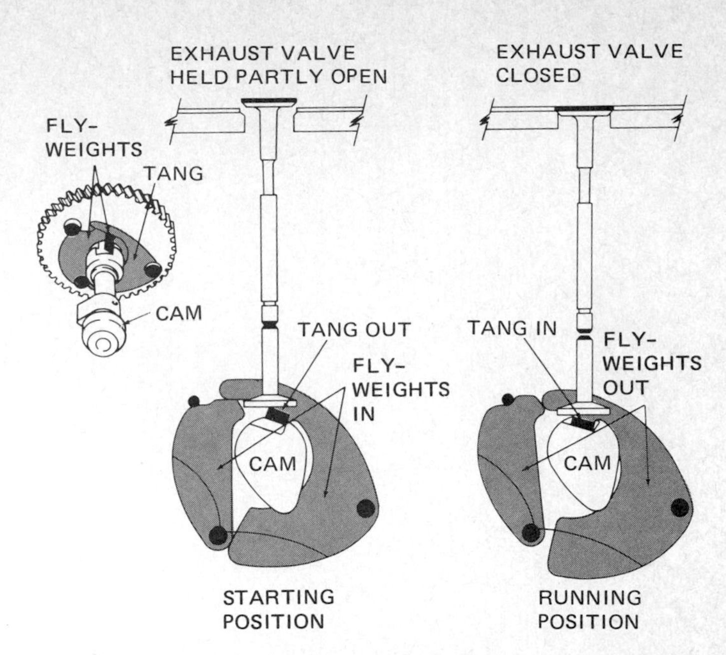

FIG. 4-19 **Operation of the automatic compression release.**

Δ4-10 AUTOMATIC COMPRESSION RELEASE

Cranking an engine is sometimes difficult. To pull the engine through the compression stroke requires some force, either muscle power or starting-motor power. One way to reduce the force required is to partly release the compression pressure during cranking. One method of doing this is shown in Fig. 4-19. The mechanism consists of a pair of flyweights on the camshaft drive gear. When the engine is not running, the flyweights are held in their inner position by springs, as shown to the left in Fig. 4-19. In this position, a tang on the end of one of the flyweights has moved out of a notch in the base circle of the exhaust cam. When the engine is cranked for starting, this tang prevents the exhaust valve from closing completely. Every time the base circle of the cam comes around under the valve tappet of the exhaust valve, the tang prevents the tappet from moving all the way down. With the exhaust valve held partly open, some of the compression pressure is relieved.

After the engine starts and engine speed increases, centrifugal force acting on the two flyweights forces them to move out into the running position (Fig. 4-19, right). This movement allows the tang on the end of one of the flyweights to move into the notch in the base circle of the cam, as shown to the lower right in Fig. 4-19. When the base circle of the cam for the exhaust valve moves under the valve tappet, the valve tappet can move all the way down. This allows the exhaust valve to close completely. Engine operation continues in a normal manner as long as the engine speed is maintained. However, if the engine is stopped, then the springs will cause the flyweights to move into the starting position (Fig. 4-19, left).

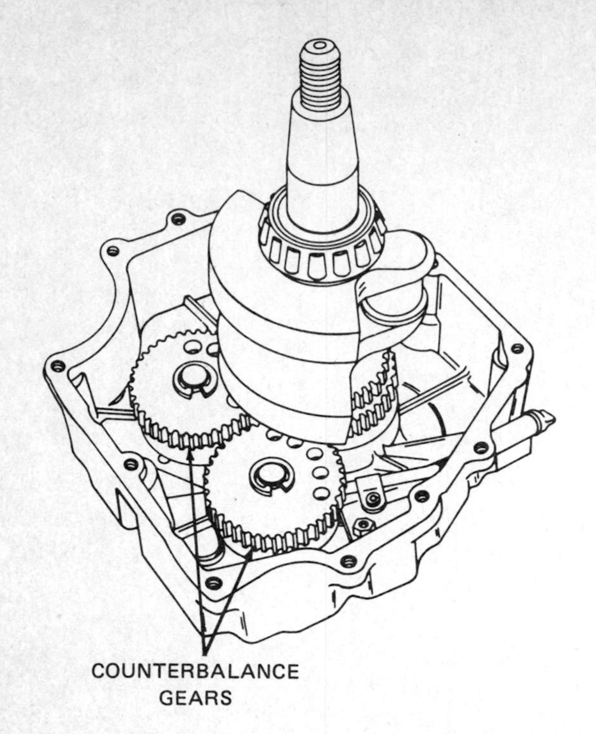

FIG. 4-20 A counterbalancing system in the crankcase of a four-cycle engine. *(Tecumseh Products Company)*

Δ4-11 COUNTERBALANCERS

Many small engines have a high level of vibration while running. This can easily tire a rider on a mower or tractor, and places additional stress on the frame, fasteners, and other parts.

To help reduce engine vibration, some manufacturers build a counterbalancing system into the crankcase. One type of counterbalancer is shown in Fig. 4-20. Basically, the counterbalancer is made up of one or more weights on shafts that are driven by the crankshaft. The weights are designed and positioned so that as they rotate, they tend to balance out the vibrations that occur as each cylinder fires. The result is a smoother-running engine.

Δ4-12 COMPARING TWO-CYCLE AND FOUR-CYCLE ENGINES

The four-cycle engine is very similar in many ways to the two-cycle engine. In both engines, a piston moves up and down in the cylinder. The piston is attached to a crank on the crankshaft by a connecting rod. When ignition of the compressed air-fuel mixture takes place, the high pressure forces the piston down. This force, carried through the crankshaft, causes the crankshaft to rotate.

To this point, the actions are similar in both engines (Figs. 4-21 and 4-22). However, in the two-cycle engine, the air-fuel mixture is admitted to the cylinder, and the burned gases leave the cylinder through openings, or ports, in the cylinder wall (Chap. 3). Also, in the two-cycle engine, the air-fuel mixture is compressed in the cylinder every time the piston reaches TDC. Then there is combustion and resulting high pressure to push the piston down . Only one crankshaft revolution and two pis-

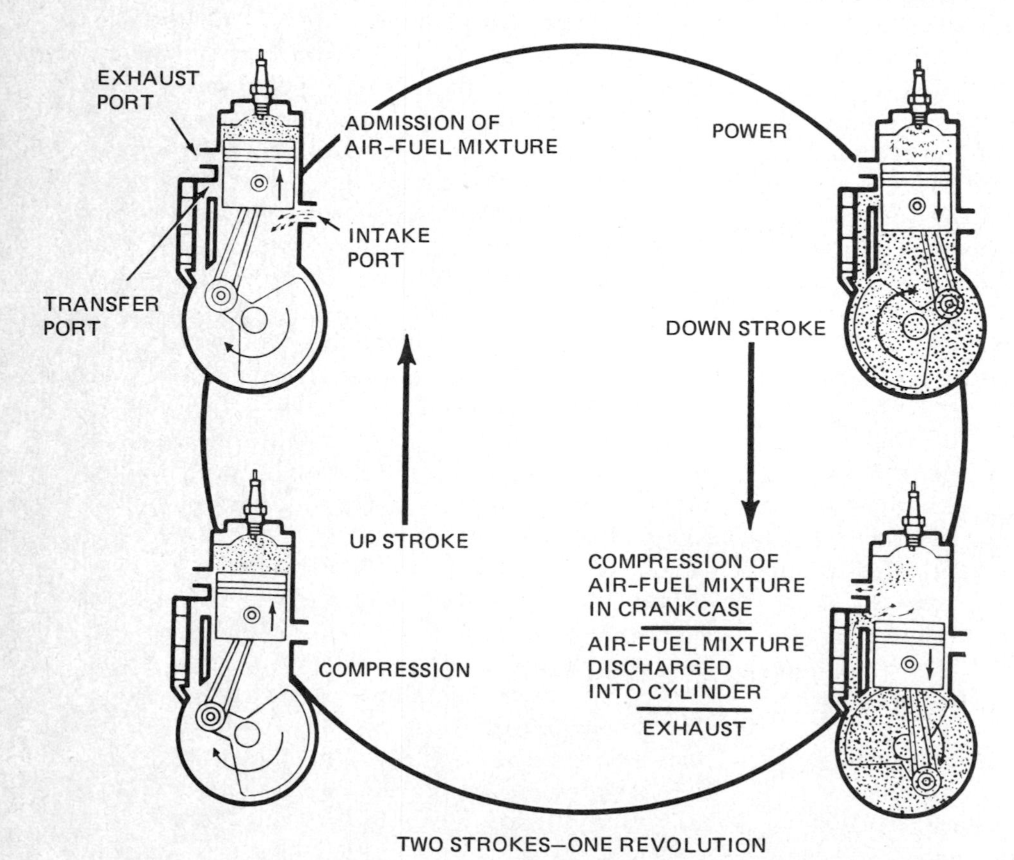

FIG. 4-21 Piston-port two-stroke-cycle engine. Compare with Fig. 4-22.

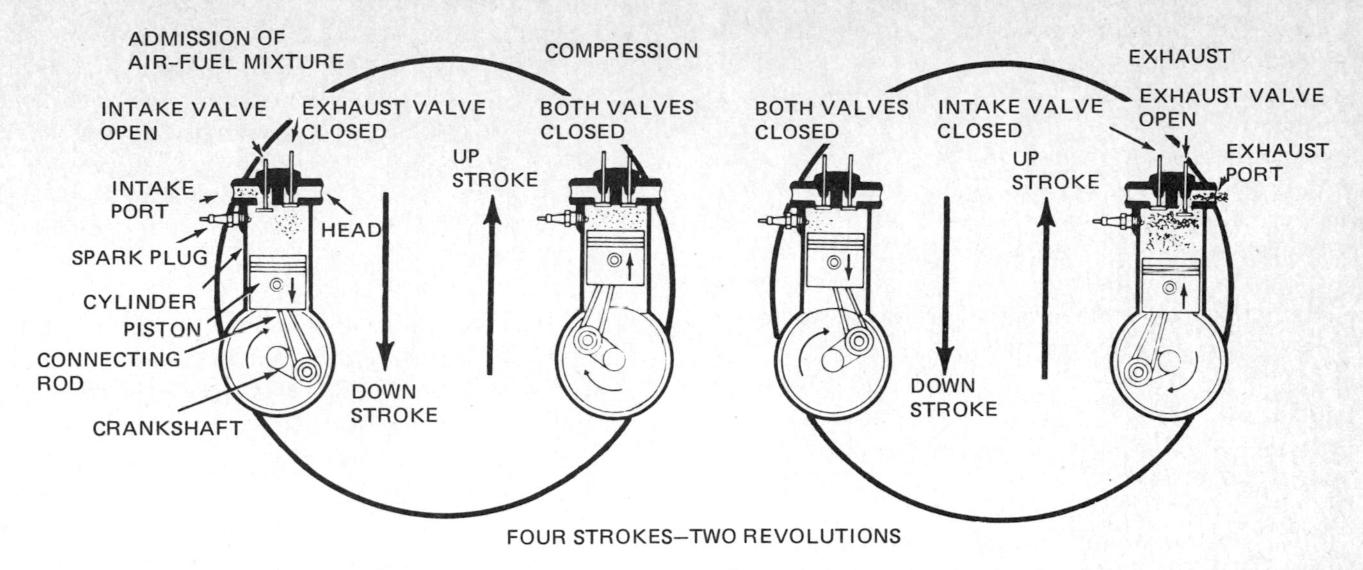

FIG. 4-22 Overhead-valve four-stroke-cycle engine. Compare with Fig. 4-21.

ton strokes are required to complete the cycle of engine operation in the two-cycle engine.

The four-cycle engine does not have valve ports in the cylinder wall. Instead, this engine has movable metal valves. These valves are opened by cams on a camshaft. The intake valve opens to allow the air-fuel mixture to enter the cylinder. The exhaust valve opens to allow the burned gases to escape from the cylinder.

It takes two revolutions of the crankshaft to complete the four strokes in a four-cycle engine (Fig. 4-22). In the two-cycle engine (Fig. 4-21), a power stroke occurs every two piston strokes—each crankshaft revolution. Every downward movement of the piston is a power stroke. In effect, the intake and compression strokes are combined. Also, the power and exhaust strokes are combined.

△4-13 IDENTIFYING TWO-CYCLE AND FOUR-CYCLE ENGINES The small-engine mechanic can usually tell at a glance whether an air-cooled engine is a two-cycle or a four-cycle engine. The four-cycle engine has an oil sump and an oil-filler plug (Fig. 4-23). The two-cycle engine does not. The four-cycle en-

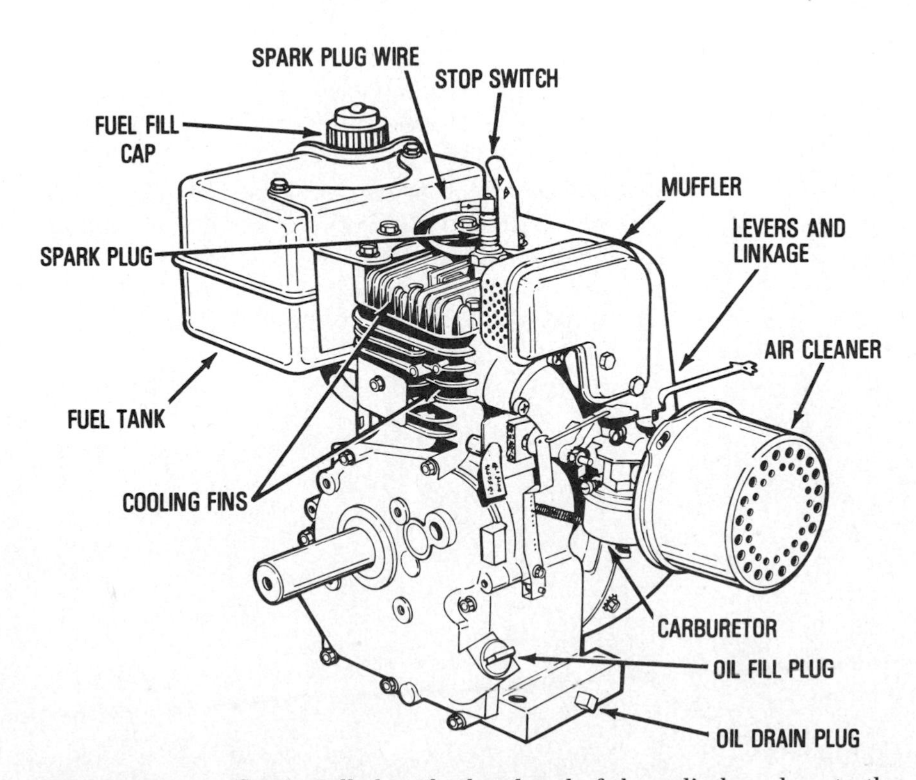

FIG. 4-23 A muffler installed at the head end of the cylinder, close to the exhaust-valve location, is one identifying feature of a four-cycle engine. *(Tecumseh Products Company)*

gine requires oil drains and refills periodically. In the two-cycle engine, the oil is usually added to the gasoline so that a mixture of gasoline and oil enters the crankcase with the air.

Another distinguishing feature of the four-cycle engine is that the muffler is installed at the head end of the cylinder, close to the exhaust-valve location (Fig. 4-23). The muffler on the two-cycle engine is installed toward the middle of the cylinder, at the exhaust-port location. Notice the location of the muffler on the power-lawn-mower engine shown in Fig. 3-17.

CHAPTER 4: REVIEW QUESTIONS

Select the *one* correct, best, or most probable answer to each question. You can find the answers in the section indicated at the right of each question.

1. In the four-cycle engine, the valves are located in the (**Δ4-4**)
   a. block or crankcase
   b. block or head
   c. cylinder wall or crankcase
   d. crankcase or crankshaft
2. The valves are operated by a (**Δ4-5**)
   a. crankshaft
   b. pull rod
   c. camshaft
   d. connecting rod
3. Two types of valve arrangements are (**Δ4-4**)
   a. I-head and J-head
   b. L-head and M-head
   c. L-head and I-head
   d. I-head and overhead valve
4. The L-head valve train includes a (**Δ4-5**)
   a. cam on the camshaft
   b. valve lifter
   c. valve spring
   d. all of the above
5. In the four-cycle engine, the four piston strokes occur in this order: (**Δ4-2**)
   a. intake, exhaust, compression, power
   b. intake, compression, power, exhaust
   c. power, intake, compression, exhaust
   d. exhaust, compression, power, intake
6. Mechanic A says that the exhaust valve opens just before the piston reaches TDC on the intake stroke. Mechanic B says that the exhaust valve opens just before the piston reaches BDC on the power stroke. Who is right? (**Δ4-2**)
   a. mechanic A
   b. mechanic B
   c. both A and B
   d. neither A nor B
7. The intake valve normally opens (**Δ4-9**)
   a. several degrees before TDC on the exhaust stroke
   b. several degrees before BDC on the power stroke
   c. at TDC on the power stroke
   d. at BDC on the power stroke
8. The intake valve normally closes (**Δ4-9**)
   a. several degrees after BDC after the intake stroke
   b. at BDC on the compression stroke
   c. at TDC on the intake stroke
   d. at BDC on the power stroke
9. The exhaust valve opens before the (**Δ4-9**)
   a. piston reaches TDC on the power stroke
   b. piston reaches BDC on the power stroke
   c. piston reaches BDC on the exhaust stroke
   d. intake stroke opens
10. Valve overlap occurs when (**Δ4-9**)
    a. both valves are closed
    b. both valves are open
    c. neither valve is open
    d. only the intake valve is open
11. The purpose of the automatic compression release is to (**Δ4-10**)
    a. make starting the engine easier
    b. reduce the force required to start the engine
    c. relieve compression pressure during starting
    d. all of the above
12. The automatic compression release is inactivated as engine speed increases and (**Δ4-10**)
    a. the intake valve closes
    b. the exhaust valve opens
    c. centrifugal force acts on the flyweights
    d. none of the above

CHAPTER

5 Small Diesel Engines

After studying this chapter, you should be able to:

1. ***Explain the operation of the four-cycle diesel engine.***
2. ***Describe the differences between diesel and spark-ignition engines.***
3. ***Describe the operation of the diesel-engine fuel-injection system.***
4. ***Discuss the purpose of the governor.***
5. ***Define* glow plug *and* precombustion chamber.**

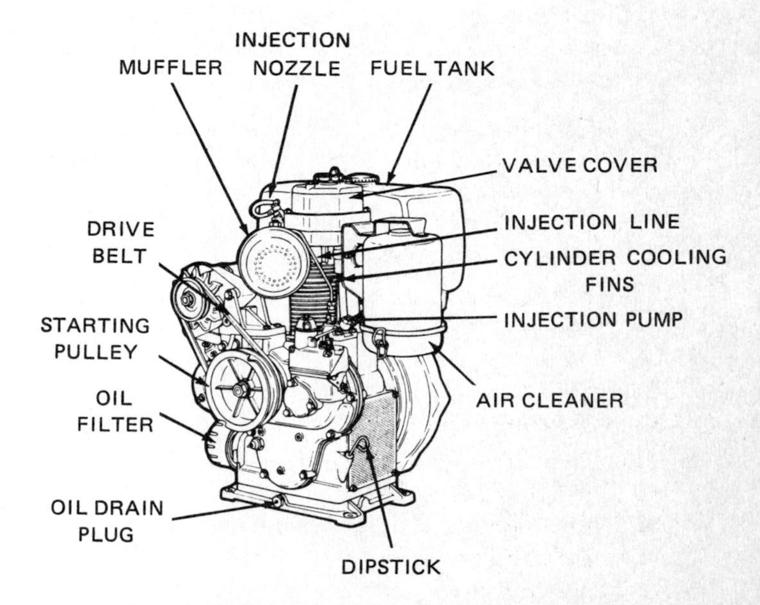

FIG. 5-1 A small diesel engine, which has a single cylinder and is air-cooled. *(Deutz Corporation)*

Δ5-1 HISTORY OF DIESEL ENGINES

The diesel engine was invented during the 1890s by Dr. Rudolf Diesel, a German engineer. Dr. Diesel had a theory that almost any fuel could be ignited by the heat generated when air was quickly compressed to a high pressure. This meant that no spark-plug or electric-ignition system was required.

Early diesel engines were big and heavy when compared with spark-ignition engines of the same power. This was one of the reasons that the diesel engine operated at a much slower speed than the spark-ignition engine. However, the diesel engine was greatly improved during the 1920s. During this time, Robert Bosch developed a new type of mechanical fuel-injection system for the diesel engine. Bosch was a German inventor and manufacturer. His diesel fuel-injection system was so successful that it is still used today.

Over the years, diesel engines have been used as a source of power for cars, locomotives, and ocean liners. Some diesel engines have even powered airplanes. In recent years, new lightweight models have been developed (Fig. 5-1). These new small diesel engines are now used in lawn and garden tractors and in other types of small power equipment.

Δ5-2 DIESEL-ENGINE CONSTRUCTION

The diesel engine is mechanically similar to the spark-ignition engine in design (Fig. 5-2). The arrangements of the pistons, connecting rods, and crankshafts are similar. However, diesel-engine parts are usually heavier in construction than spark-ignition engine parts. This is because of the greater pressures in the cylinder of the diesel engine during combustion.

Both types of engines are internal-combustion engines that use air, fuel, compression, and ignition to produce power. The power is produced by extracting energy from the fuel in an air-fuel mixture that is burned inside the engine cylinder. The piston strokes of intake, com-

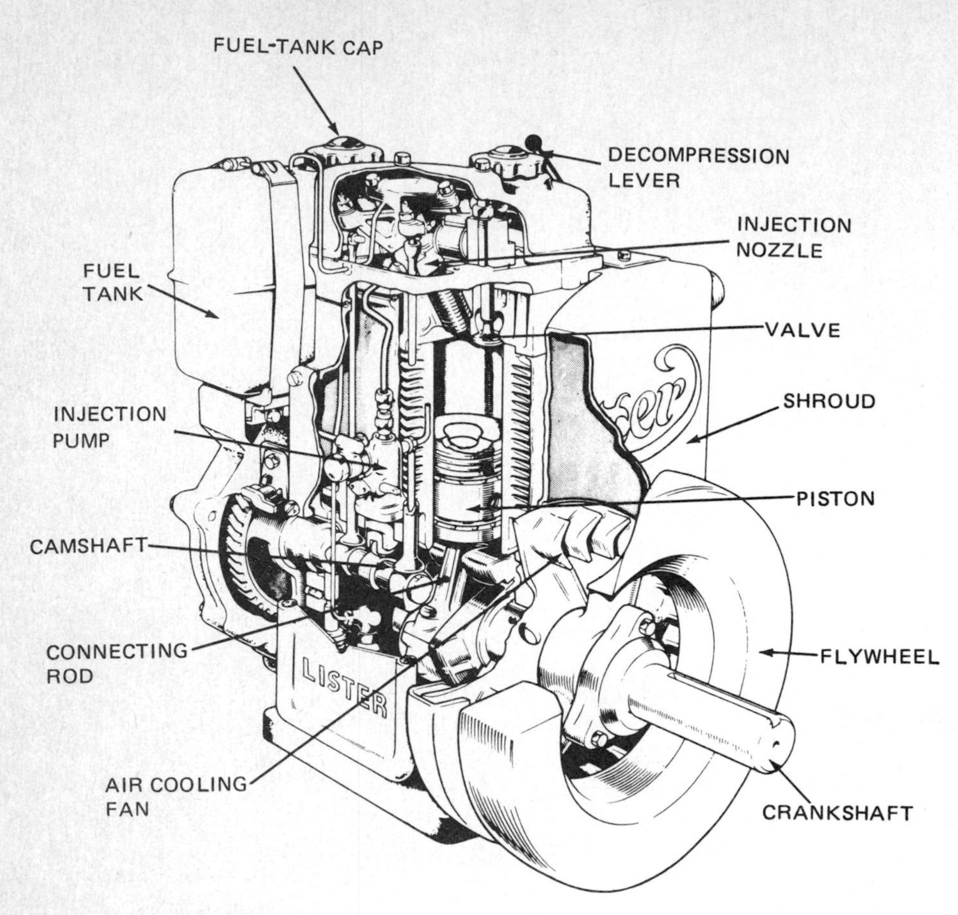

FIG. 5-2 A cutaway air-cooled one-cylinder diesel engine. *(Lister Diesels Division of Hawker Siddeley, Inc.)*

pression, power, and exhaust occur in the same sequence.

Δ5-3 DIFFERENCES BETWEEN DIESEL AND SPARK-IGNITION ENGINES Diesel engines can be built to work on either the two-stroke cycle (Chap. 3) or the four-stroke cycle (Chap. 4). However, most small diesel engines are four-cycle engines. Many outboard, motorcycle, and other small spark-ignition engines are two-cycle engines.

In principles of operation, there are two basic differences between diesel and spark-ignition engines. These are (1) how the fuel is introduced into the cylinder and (2) how the resulting air-fuel mixture is ignited.

Δ5-4 THE DIESEL CYCLE During the intake stroke, only air enters the cylinder of the diesel engine (Fig. 5-3*a*). The air is compressed as the piston moves up the cylinder toward TDC on the compression stroke (Fig. 5-3*b*). This heats the air to a high temperature. As the piston nears the top of the compression stroke, the diesel fuel is injected (sprayed) into the hot air. The heat ignites the fuel. The fuel burns almost as soon as it is injected, while the piston passes TDC and is forced down the cylinder on the power stroke (Fig. 5-3*c*).

No spark plug is used. Ignition is by the "heat of compression." This is the contact of the fuel with the heated air. As the fuel burns, the pressure in the cylinder forces the piston down. Then the exhaust valve opens, and the burned gases escape from the cylinder into the exhaust port (Fig. 5-3*d*). The four-stroke cycle is repeated as long as the engine runs.

In the diesel engine, there is continuous combustion during almost the entire length of the power stroke. Therefore, the pressure resulting from combustion (and the force pushing against the top of the piston) remains almost constant throughout the stroke. For this reason, the diesel cycle is sometimes referred to as having *constant-pressure combustion.*

Δ5-5 HEAT OF COMPRESSION In the diesel engine, a compression ratio of at least 16:1 is required. This high compression ratio (Fig. 5-4) produces a compression pressure of about 500 psi [3447 kPa].

As the air is compressed during the compression stroke, each pound-per-square-inch increase in pressure increases the temperature of the air about 2°F. Therefore, at the top of the compression stroke when the fuel is injected, the air temperature will be about 1000°F [538°C] or higher. This high temperature ignites the fuel

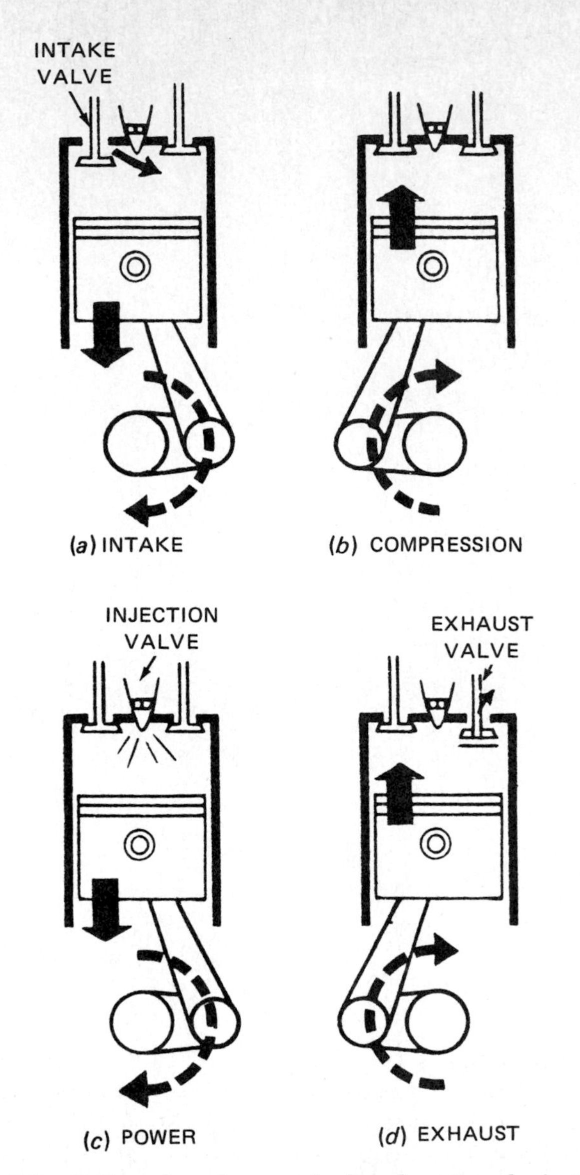

FIG. 5-3 Actions in a four-cycle diesel engine during the four piston strokes.

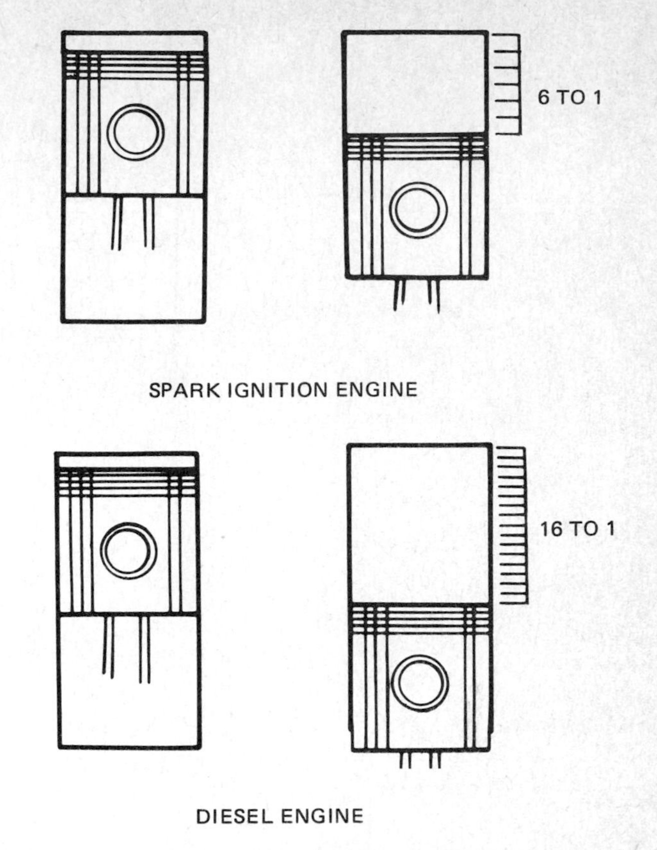

FIG. 5-4 Comparison of typical compression ratios for a spark-ignition engine and for a diesel engine. *(ATW)*

almost as soon as it is injected into the cylinder. The resulting combustion and expansion of the burning gases force the piston down the cylinder on the power stroke.

The diesel engine is called a *compression-ignition* (CI) *engine*. This is because it is the heat of compression that ignites the fuel as it is sprayed into the engine cylinders. No electric ignition system, such as used by the *spark-ignition* (SI) *engine*, is needed. All diesel engines are overhead-valve engines (Figs. 5-1 to 5-3). The high compression ratio required cannot be attained in an L-head engine, which has the valves in the cylinder block.

Δ5-6 DIESEL FUEL Diesel fuel is a light oil with certain special characteristics. It has lubricating qualities that lubricate the various moving parts in the diesel fuel system. (A different and separate system is used to lubricate the moving engine parts such as the crankshaft, pistons, and valve train.) The diesel fuel oil also must ignite easily and must burn cleanly, leaving little ash or residue.

Δ5-7 DIESEL FUEL SYSTEMS A gravity-feed fuel supply system is used on some small diesel engines (Fig. 5-5). This system is similar to the fuel system on many small spark-ignition engines. A fuel tank is mounted to the engine. Beneath the fuel tank, there is a fuel filter. The fuel leaving the tank passes through the filter and then flows on to the fuel-injection pump. The injection pump times the injection and then pressurizes and meters the fuel. Therefore the correct amount of fuel required by the engine is sprayed from the injection nozzle into the combustion chamber at the proper time. Any excess fuel flows through the fuel-return line back to the tank.

Multicylinder diesel engines frequently use a fuel system that includes the fuel tank, two (or more) fuel filters, a fuel transfer pump (or lift pump), a fuel-injection pump and governor, and fuel lines and an injection nozzle for each cylinder. Figure 5-6 shows this type of fuel system for a two-cylinder engine.

Diesel fuel must be clean and free of water. Even almost invisible dirt particles can clog the injection nozzles and cause poor engine performance. Also, water can rust interior injection-pump and injection-nozzle parts.

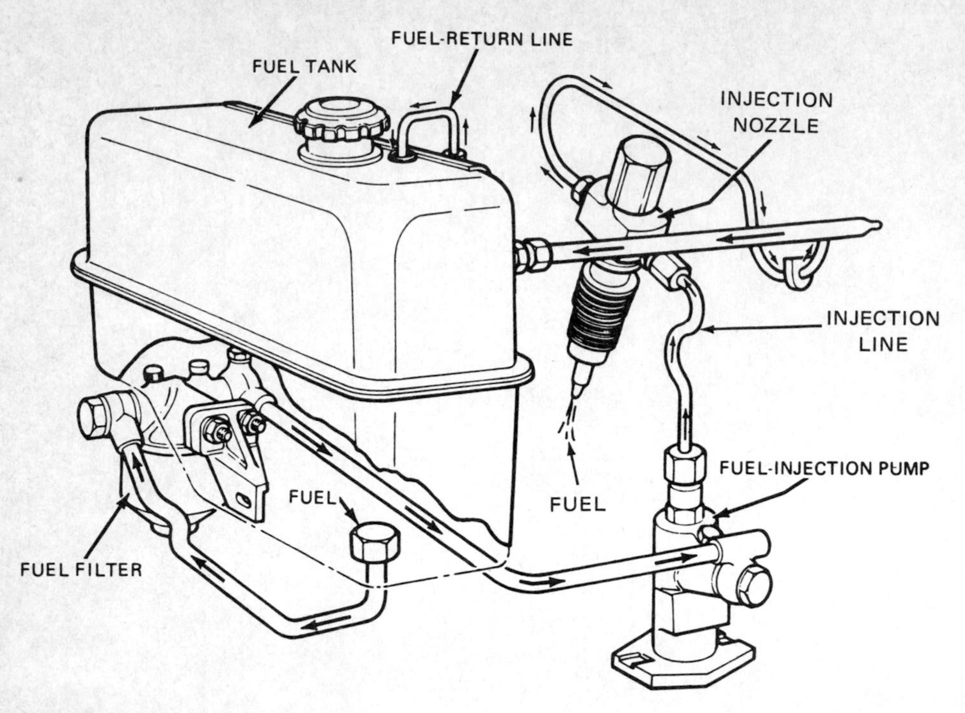

FIG. 5-5 A gravity-feed fuel supply system for a one-cylinder diesel engine. *(Lister Diesels Division of Hawker Siddeley, Inc.)*

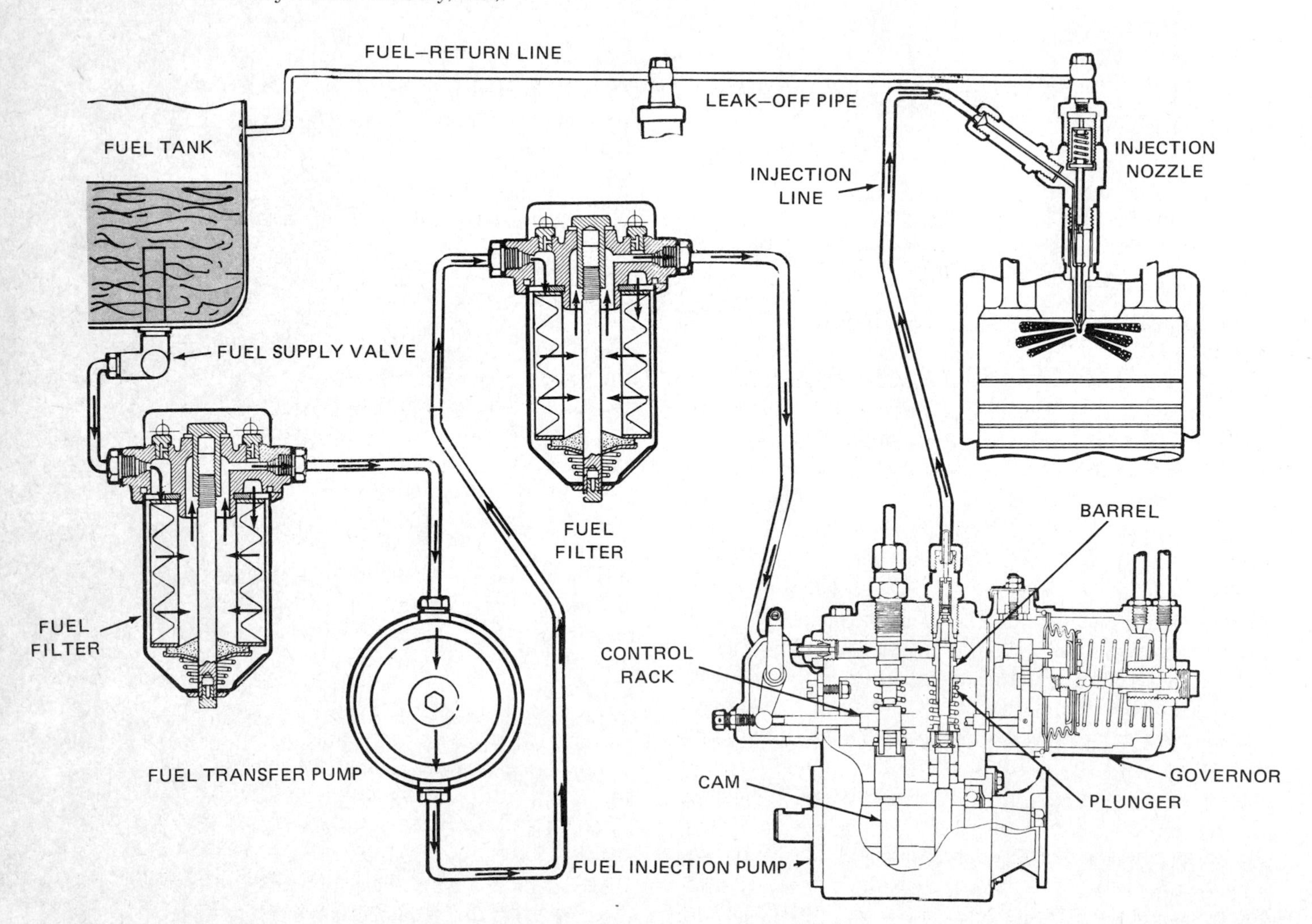

FIG. 5-6 Fuel supply system for a multicylinder diesel engine.

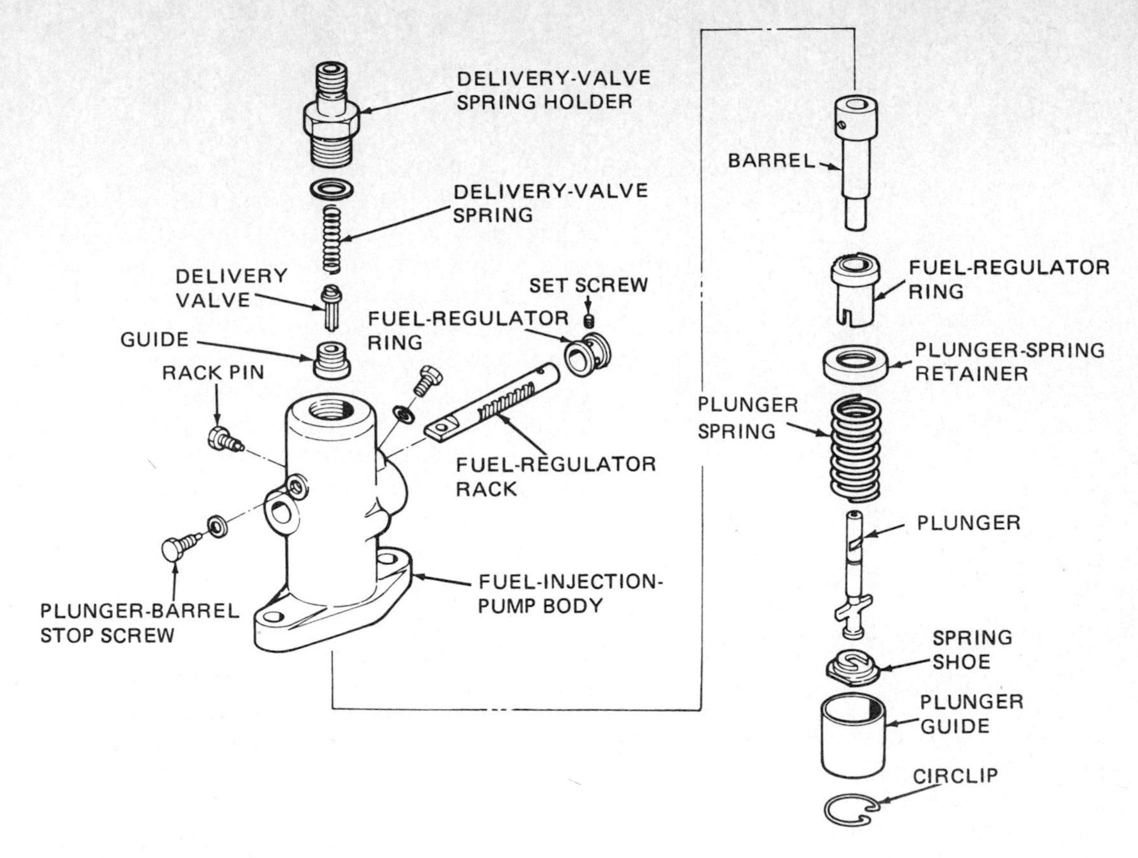

FIG. 5-7 A disassembled fuel-injection pump for a single-cylinder diesel engine. *(Yanmar Diesel Engine Company, Ltd.)*

Many diesel engines have some type of warning system to alert the operator that there is water in the fuel.

Δ5-8 DIESEL FUEL-INJECTION PUMP

The injection pump has a small barrel with a plunger for each engine cylinder. Each barrel-and-plunger assembly is connected by steel tubing to an injection nozzle in each cylinder. Figure 5-2 shows a one-cylinder diesel engine. In it, a cam on the engine camshaft operates the plunger in the injection pump.

Figure 5-7 shows a disassembled single-cylinder fuel-injection pump. It is operated by a cam on the engine camshaft, which determines injection timing. Metering of the fuel is done by movement of the rack. It moves right and left through the pump to vary the amount of fuel metered by the plunger.

Δ5-9 INJECTION-PUMP OPERATION

Figure 5-6 shows the fuel system for a two-cylinder engine. Two barrel-and-plunger assemblies are combined into a single injection-pump housing. A separate camshaft at the bottom of the pump operates the plungers. This is called an *in-line plunger pump.* The top of each barrel is connected by a separate injection tube to an injection nozzle in each engine cylinder (Fig. 5-6). Most plunger pumps are gear-driven from the engine crankshaft.

Figure 5-8 shows the construction of a plunger pump. A fuel gallery in the pump keeps the space between the end of the plunger and the delivery valve filled with fuel. When the cam lobe comes up under the plunger, the plunger is raised. This forces the fuel out of the barrel at high pressure. The fuel flows through the delivery valve and tubing to the injection nozzle in the engine cylinder (Fig. 5-6).

Δ5-10 ADVANCING INJECTION TIMING

Many diesel-engine injection pumps have a speed-advance mechanism. It advances the time of injection as engine speed increases. The speed advance supplies the fuel earlier in the compression stroke (toward the end of the stroke). This gives the fuel enough time to ignite and begin burning so that an even pressure rise pushes against the piston head.

Without this advance, the piston would be up over the top (TDC) and moving down before the fuel was well ignited. The result would be that the piston would be moving down away from the pressure rise. A weak power stroke would result, and fuel would be wasted. The advance is controlled by an engine *governor.* It is built into the in-line plunger pump (Fig. 5-6).

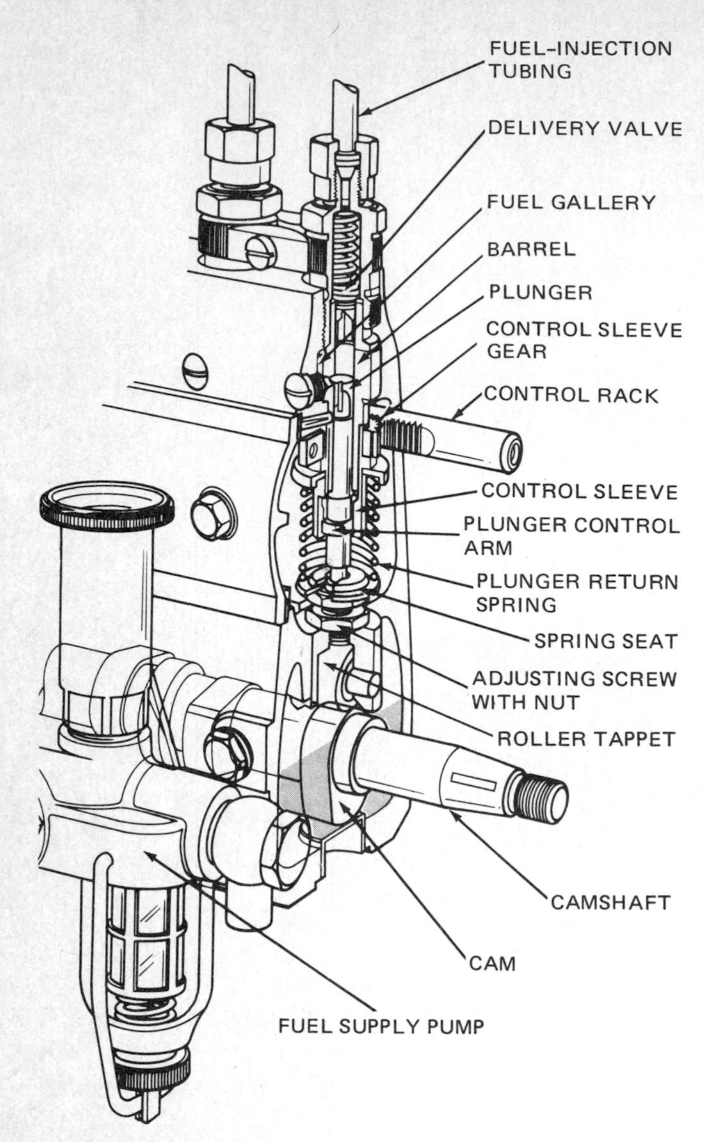

FIG. 5-8 **Construction of an in-line plunger type of fuel-injection pump.** *(Robert Bosch Corporation)*

Δ5-11 METERING THE AMOUNT OF FUEL INJECTED

To *meter* means to *measure.* In the fuel-injection pump, the amount of fuel injected is metered by varying the effective length of the plunger strokes. When less engine power is needed, the effective length of the plunger strokes is short. Relatively small amounts of fuel are delivered. When more power is needed, the effective length of the plunger strokes is increased. More fuel is delivered. The actual length of plunger travel is the same. However, the quantity of fuel injected is changed by turning the pump plungers.

The top of each plunger has an inclined *helix* machined into it (Fig. 5-9). As the plunger is turned, its effective stroke is varied. The plungers are turned by movement of the control rod, or fuel rack. It is connected to the throttle control through the governor and linkage. The movement of the throttle control is converted into a corresponding control-rod travel. This rotates the plungers to change the amount of fuel metered.

Δ5-12 DIESEL-ENGINE GOVERNORS

The speed and power output of a diesel engine are controlled by the quantity of fuel injected into the cylinder. This varying amount of fuel is mixed with a constant amount of compressed air inside the cylinder. In the diesel engine, a full charge of air enters the cylinder on each intake stroke. Because the quantity of air is constant, the amount of fuel injected determines power output and speed. However, the amount of fuel injected must always be less than the maximum amount established by the manufacturer in designing the engine. Then there will always be enough air in the cylinder to burn all the fuel injected.

To prevent an excessive amount of fuel from being injected, the diesel engine always has a governor (Fig. 5-6). It provides automatic control of fuel delivery to meet operating conditions. Actually, it is the governor that directly controls the amount of fuel injected into the engine. When you move the throttle control (**Δ5-11**), you are only changing the setting of the governor.

Without a governor, a diesel engine can stall at low speeds or run so fast it will self-destruct. Diesel engines in cars, trucks, and tractors use either a mechanical (centrifugal) governor or a pneumatic governor. Both are variable-speed governors. They can be set to control the engine at varying speeds within its operating-speed range. Constant-speed governors are used on engines that must run at a constant speed. These include diesel engines driving electric generators, and some pumps and compressors.

Many diesel engines have some type of mechanical governor. It has flyweights that spin with the camshaft in the injection pump (Fig. 5-10). The faster they spin, the further out they move. This acts on the fuel rack (or control rod) to turn the plungers and adjust the fuel delivery.

Δ5-13 DIESEL AIR-FUEL RATIOS

In operation, the governor matches the fuel delivery with the amount of air in the cylinders. Since no restriction is used in the diesel engine to limit (or *throttle*) the amount of air that enters, the cylinders are always completely filled. Air-fuel ratios in a diesel engine range from about 100:1 at idle to about 20:1 under full load. The governor keeps the air-fuel ratio within these limits. An air-fuel ratio richer than 20:1 can cause unacceptable amounts of exhaust smoke.

Δ5-14 INJECTION NOZZLES

The fuel-injection nozzle (Fig. 5-11) is a pressure-operated valve. It serves to feed and atomize the fuel injected into the cylinders of a diesel engine. Each fuel-injection nozzle has a spring-loaded check valve, or delivery valve. The valve is closed except when high pressure is applied to the fuel. Then the check valve opens, allowing fuel to pass through. As the valve is pushed down off its seat, fuel

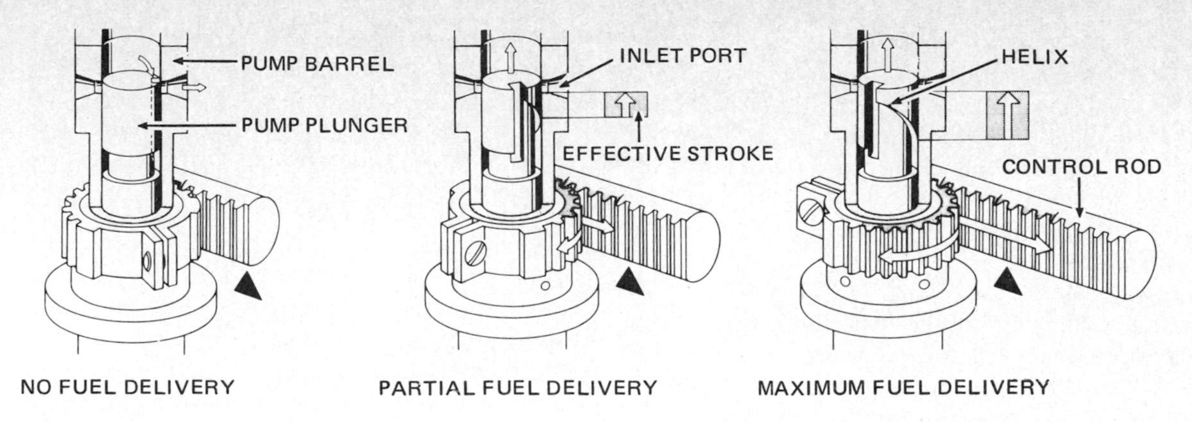

FIG. 5-9 Movement of the toothed control rod turns the pump plungers to vary the amount of fuel injected. *(Robert Bosch Corporation)*

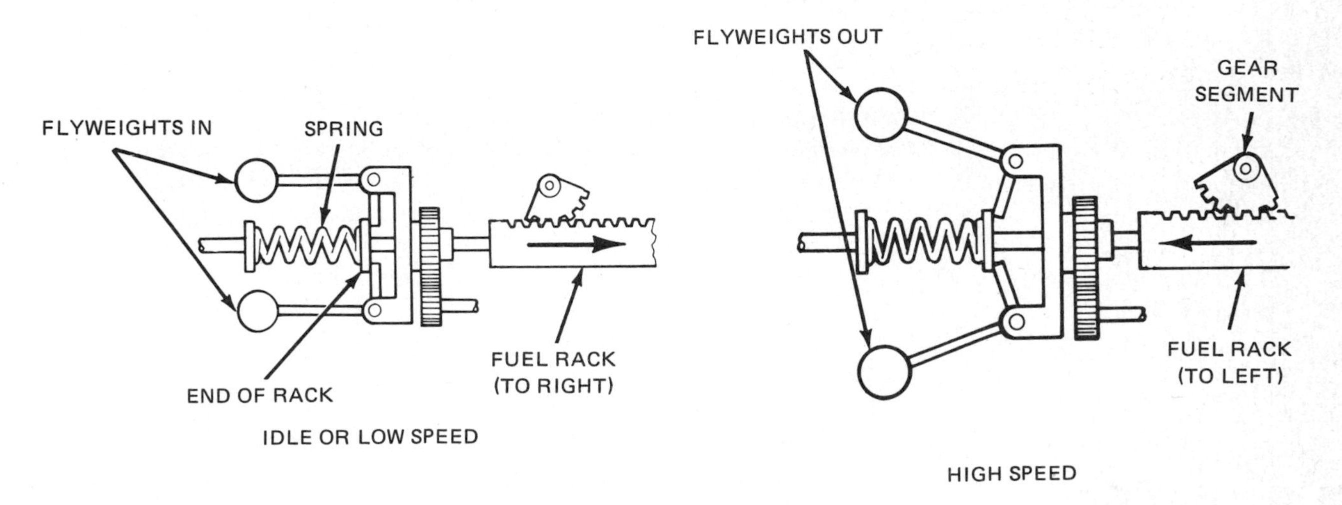

FIG. 5-10 Actions of a mechanical governor, used on many diesel engines.

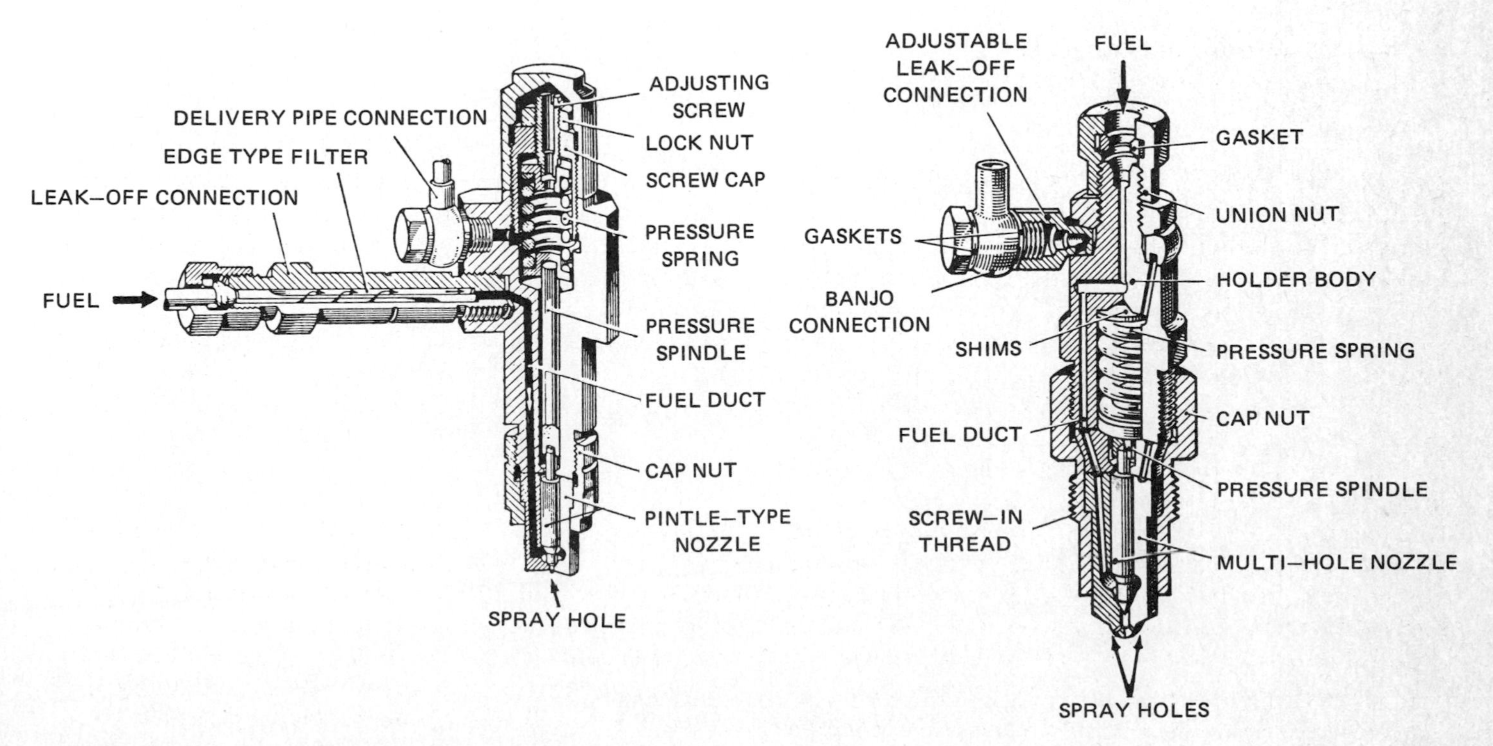

FIG. 5-11 Two types of fuel-injection nozzles. *(Robert Bosch Corporation)*

begins to spray into the engine cylinder. In the cylinder the fuel is ignited by the hot compressed air (heat of compression).

When the fuel pressure drops, the spring closes the delivery valve. This stops injection of the fuel. The nozzle must close quickly to prevent "dribble." This is unwanted leakage of fuel into the cylinder following the end of the main injection.

There are two types of injection nozzles (Fig. 5-11), the multihole type and the pintle (single-hole) type. Direct-injection diesel engines (described later) have an open combustion chamber. They use the multihole nozzle. Indirect-injection diesel engines, including those with a *precombustion chamber* (Δ5-16), use the pintle nozzle. It has only one hole from which fuel discharges. Figure 5-12 shows the difference in spray patterns for the two types of injection nozzles.

Δ5-15 GLOW PLUGS For easier starting, especially in cold weather, many diesel engines have a *glow plug* in each cylinder (Fig. 5-13). The construction of a glow plug is shown in Fig. 5-14. It has an electric heating element that quickly gets very hot when connected to the battery. When the engine is cold and the air temperature is low, the glow plug is turned on. As the heating element gets hot, it adds heat to the air in the precombustion chamber (Δ5-16). Then the fuel is sprayed into air that has already been preheated by the glow plug. This action greatly improves starting of a cold engine in cold weather.

Δ5-16 PRECOMBUSTION CHAMBERS

In some diesel engines, the fuel is sprayed directly into the combustion chamber (Fig. 5-6). This is called *direct injection*. However, many diesel engines (including automotive diesel engines) spray the fuel into a precombustion chamber. This is called *indirect injection*. Figure 5-13 shows the location of the injection nozzle and glow plug in the precombustion chamber of a diesel engine.

The precombustion chamber is a small cup into which the fuel is injected and in which the combustion starts. As the piston moves up and compresses the air trapped above it, much of the air is pushed into the precombustion chamber.

When the piston nears TDC, fuel is injected into the precombustion chamber. The hot air ignites the fuel. Then the burning mixture streams out through a connecting passage into the main combustion chamber above the piston. There, the burning mixture mixes with the combustion-chamber air for more complete combustion. The resulting high pressure pushes the piston down the cylinder on the power stroke.

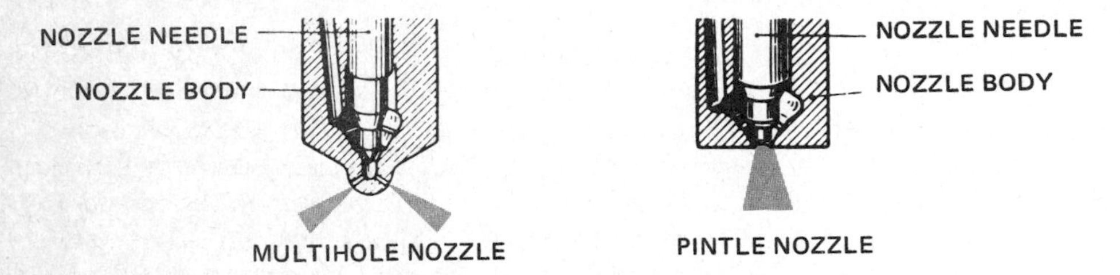

FIG. 5-12 Correct spray patterns for a multihole nozzle and for a pintle (single-hole) nozzle. *(Robert Bosch Corporation)*

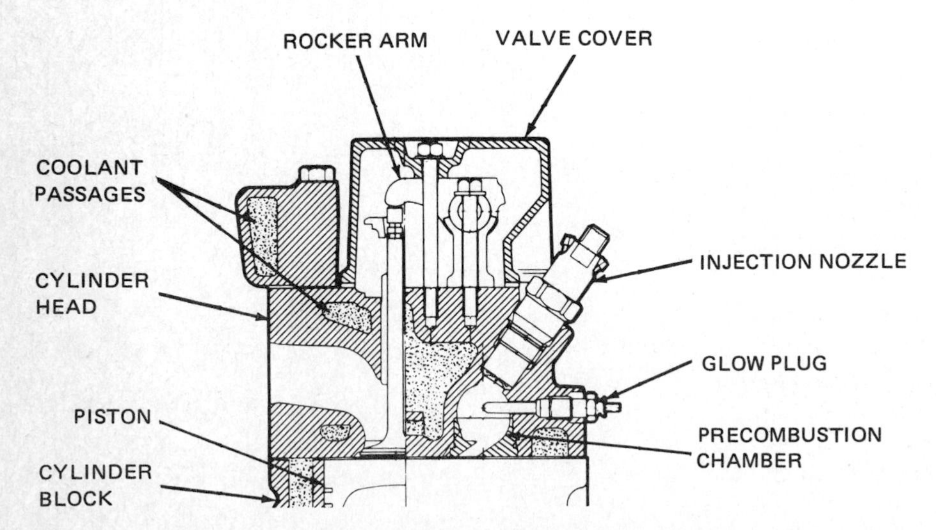

FIG. 5-13 Locations of the glow plug and the precombustion chamber in the cylinder head of a diesel engine. *(Teledyne Continental Motors)*

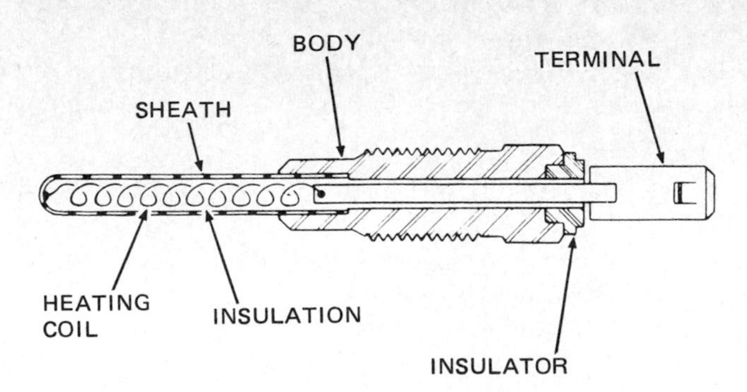

FIG. 5-14 Construction of a glow plug.

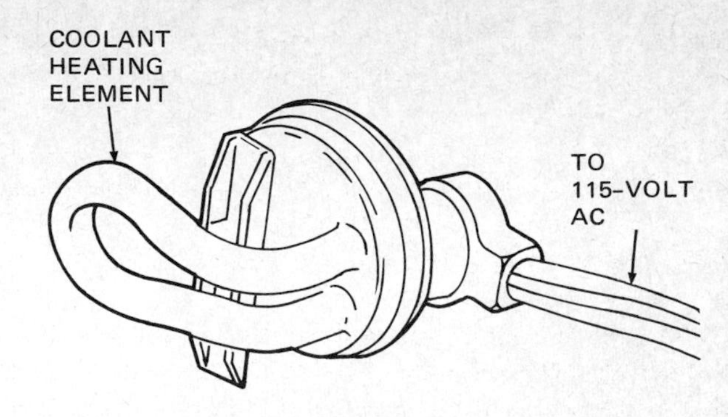

FIG. 5-15 Block heater, which is installed in the water jacket of a liquid-cooled engine. *(Phillips Temro, Inc.)*

The advantage of using a precombustion chamber is that the engine runs more smoothly and quietly. It also produces less exhaust emissions than a direct-injection engine. However, some direct-injection diesel engines produce slightly more power than an indirect-injection diesel engine of the same size.

Δ5-17 COOLANT AND FUEL HEATERS

For very cold weather operation, where temperatures get down to zero or below (−18°C or under), coolant and fuel heaters are often used to assist in starting a diesel engine. One type of coolant heater (for liquid-cooled engines) has an electric heating element that works from a 115-volt outlet (Fig. 5-15). The heating element is installed into the engine cooling-system water jacket. When the electric cord is plugged in, the heater warms the engine coolant.

At low temperatures, wax will form in diesel fuel. The wax will plug the fuel filter and prevent starting. To prevent wax formation, some type of diesel fuel heater is used. The in-line fuel heater consists of a metal heating strip wound around the fuel pipe from the tank. When the ignition switch is turned on, the heater operates if the fuel temperature is so low that wax may form. Warming the fuel flowing through the fuel line reduces the possibility that wax will plug an engine-mounted fuel filter.

> **CAREFUL** Do not use "starting aids" such as ether, gasoline, or similar materials in the air intake of a diesel engine. These so-called aids can actually delay starting. They can also damage the engine.

Δ5-18 FUEL FILTERS

The fuel for a diesel engine must be clean and free of contaminants and water. Tiny, almost invisible specks of dirt can clog the injection nozzles and damage injection-pump parts. Water can clog filters and passages and can rust parts which will cause them to stick. Therefore, the fuel filter is an essential part of the diesel-engine fuel system.

The diesel fuel filter works like the filter used in engine lubricating systems (Chap. 6). It contains a cartridge of filtering material (such as special pleated paper or fiber mat) through which the fuel must pass. The filter traps any particles and keeps them from entering the fuel system.

Figure 5-6 shows the fuel system for a multicylinder diesel engine. Notice that two fuel filters are used. One filters the fuel before it reaches the transfer pump. Then the fuel flows through the second fuel filter before reaching the injection pump.

The location of the fuel filter in the fuel system for a single-cylinder diesel engine is shown in Fig. 5-5. Because the volume of fuel that the filter must handle is small, only one fuel filter is required.

Δ5-19 INJECTION LINES AND FITTINGS

Injection lines in diesel fuel-injection systems are also called *high-pressure fuel lines, injection tubes,* and *injection pipes* (Fig. 5-6). They carry the fuel, under high pressure, from the injection pump to the injection nozzles in the cylinders. The lines must withstand pressures of several thousand pounds per square inch and must be noncorrosive. Figure 5-16 shows two types of fittings used to connect fuel lines to injection nozzles.

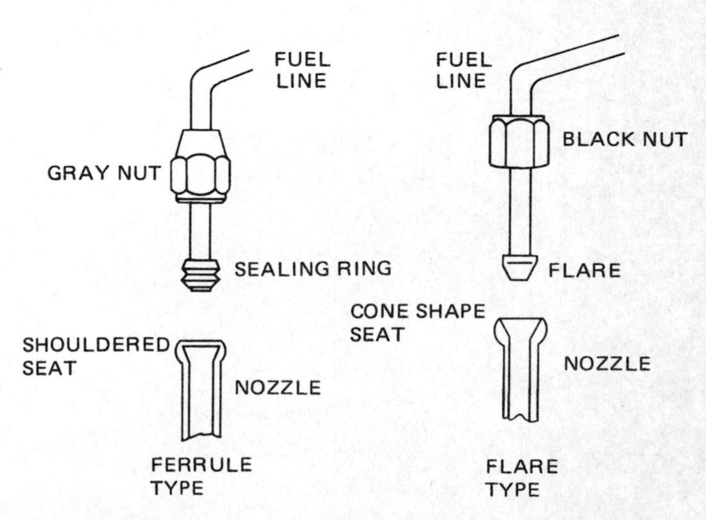

FIG. 5-16 Two types of fittings used to connect the fuel line to the injection nozzle.

CHAPTER 5: REVIEW QUESTIONS

Select the *one* correct, best, or most probable answer to each question. You can find the answers in the section indicated at the right of each question.

1. A basic difference between the diesel engine and the spark-ignition engine is that the diesel engine (**Δ*5-4***
   a. uses a spark plug
   b. uses compression ignition
   c. uses an extra stroke to compress the air
   d. compresses a mixture of air and fuel
2. Direct injection means the fuel is (**Δ*5-16***)
   a. mixed directly with the air during the intake stroke
   b. injected directly into the precombustion chamber
   c. injected directly into the combustion chamber
   d. directly mixed with the air after the power stroke
3. To obtain an adequate heat of compression, the diesel engine must (**Δ*5-5***)
   a. have a compression ratio of at least 16:1
   b. produce a compression pressure of about 500 psi [3447 kPa]
   c. both a and b
   d. neither a nor b
4. The purpose of the glow plug is to (**Δ*5-15***)
   a. ignite the fuel as it is injected
   b. warm the injection nozzle for easier starting
   c. warm the precombustion chamber for easier starting
   d. warn that the engine is overheating
5. When you move the throttle control on a diesel engine, you are changing the (**Δ*5-12***)
   a. amount of air entering the cylinders
   b. amount of fuel in the fuel tank
   c. time of injection
   d. setting of the governor

ENGINE SYSTEMS

To operate and to produce usable power, an engine requires more than just the ignition of a charge of air-fuel mixture inside the cylinder. The engine must have fuel, lubricating, and cooling systems to run. While the engine runs, it burns fuel and creates a certain amount of air pollution.

Part 3 describes the engine support systems that every engine must have.

There are four chapters in Part 3. They are:

Chapter 6: Engine Lubrication
Chapter 7: Fuel Systems for Small Spark-Ignition Engines
Chapter 8: Fuel-System Service
Chapter 9: Engine Cooling Systems

CHAPTER 6

Engine Lubrication

After studying this chapter, you should be able to:

1. ***List five jobs that the lubricating oil does in the engine.***
2. ***Explain how engine oil is rated.***
3. ***Describe the various engine-oil additives and their purpose.***
4. ***Compare the operation of the two-cycle-engine lubricating system with the four-cycle-engine lubricating system.***
5. ***List the components in the four-cycle-engine lubricating system and describe how each works.***

Δ6-1 PURPOSE OF ENGINE LUBRICATION The engine has many moving parts. If these parts rub against one another, they will wear out quickly. A "frozen," or seized, engine may result. The purpose of the engine lubricating system is to minimize friction and wear. The lubricating system performs this job by supplying oil to prevent metal from rubbing against metal. This chapter explains how oil works as a lubricant and how it gets to the moving parts in the engine.

In the two-cycle engine, the oil is mixed with the gasoline. The oil enters the engine along with the air-fuel mixture and lubricates the engine. In the four-cycle engine, there is a supply of oil in the lower part of the crankcase. When the engine is running, oil from this reserve supply is either pumped or splashed on all moving parts in the engine. In either type of engine, the result is that the engine parts get the lubrication they need to reduce friction and minimize wear.

Δ6-2 PURPOSE OF ENGINE OIL Oil flows in slippery layers onto engine parts, preventing metal-to-metal contact. Figure 6-1 shows two surfaces that are highly magnified. If you look at a part that is made of very smooth metal (such as a crankpin) under a microscope, you will see tiny rough edges sticking up from the surface of the metal. If two such dry metal surfaces rub against each other, the irregular edges catch on each other. Particles of metal are worn off. This action consumes energy and wastes power.

What is worse is that the rubbing and tearing action, or friction, produces heat. In the engine, if it ran without oil, metal parts would move against each other with great force. Without oil, these parts would quickly get so hot that the metal would melt. Then the engine would seize.

The oil prevents this because layers of oil cover the metal surfaces (Fig. 6-1). The layers of oil hold the metal surfaces away from each other. Only the oil layers must

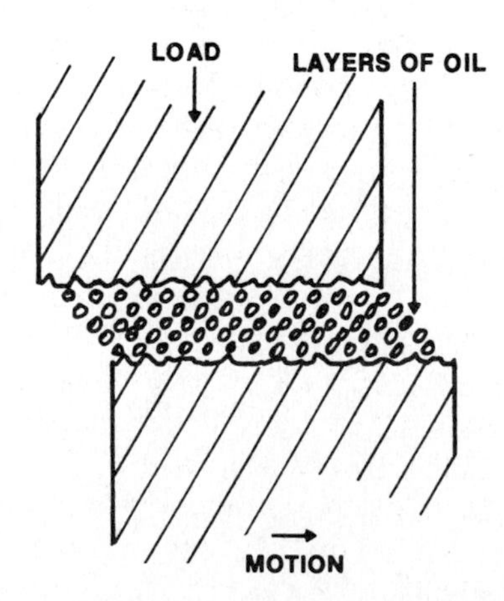

FIG. 6-1 These two highly magnified surfaces show irregularities that you could not see without a microscope. When the surfaces move in relation to each other, the layers of oil prevent metal-to-metal contact.

slip as the two metal parts move. The oil greatly reduces power loss and wear of moving metal parts. In addition, oil does other jobs which are described below.

∆6-3 OTHER JOBS FOR ENGINE OIL

The oil reduces friction and wear. It also does other jobs in the engine:

- It removes heat from the engine.
- It absorbs shocks between bearings and other engine parts.
- It forms a seal between the piston rings and the cylinder wall.
- It acts as a cleaning agent.

Let's look at each of these jobs in detail.

1. Removes heat from the engine (four-cycle engines). In four-cycle engines, the oil circulates between the hot engine parts and the cool oil pan or reservoir. Therefore, the oil continuously removes heat from the engine.
2. Absorbs shock between bearings and other engine parts. Layers of oil between metal parts, such as rotating journals and bearings (Fig. 6-2), resist squeezing out when loads are applied. The layers act as cushions to absorb shock loads (for example, when combustion starts and pressures suddenly go up).
3. Forms a seal between the piston rings and the cylinder wall. The oil fills up little irregularities in the cylinder wall. The oil also clings to the cylinder wall and piston rings to form a good seal.
4. Acts as a cleaning agent (in four-cycle engines). The oil circulates between the engine's moving parts and the oil pan or reservoir. The oil picks up particles of carbon and other dirt and carries them to the oil pan or reservoir. Therefore, the oil acts as a cleaning agent.

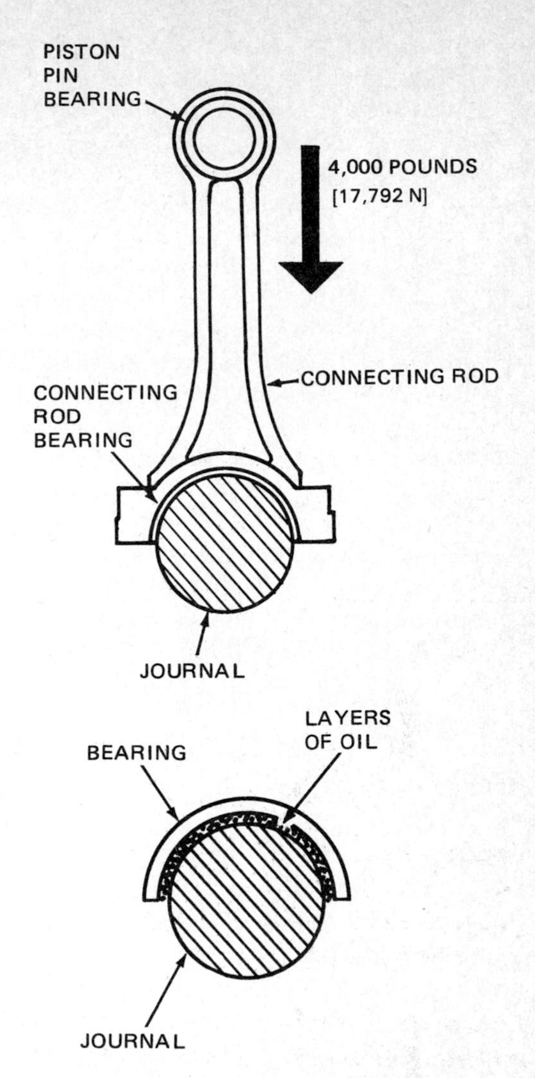

FIG. 6-2 The layers of oil between the journal and the rod bearing help to absorb the shock of the sudden load when combustion starts.

∆6-4 PROPERTIES OF ENGINE OIL

Oil is the liquid used in the engine lubricating system. There are four basic types of oil: animal, vegetable, mineral, and synthetic. The difference is in the source of the raw material from which the oil is refined. For many years, the only oil used in engines was made from natural crude oil which came from oil wells drilled deep into the earth. Much of the engine oil used today still comes from crude oil. This crude oil was formed underground millions of years ago in various parts of the world. It must be refined to make it usable. In the refining process, gasoline, kerosene, lubricating oil, and many other products are made.

In recent years, synthetic oils have been developed. These are oils made by chemical processes and do not necessarily come from petroleum. Some oil manufacturers claim that these synthetic oils have superior lubricating properties. Actually, there are several types of synthetic oils. The type most widely used now is produced from organic acids and alcohols (from plants of various types). A second type is produced from coal and crude oil. Tests have shown that the synthetics do have certain superior qualities.

Not all oil is the same. There are several grades of oil and several ratings. Oil for four-cycle engines contains a number of additives (chemical compounds that are added to the oil) that improve the performance of the oil. Oil ratings and additives are described below.

∆6-5 OIL VISCOSITY

Viscosity is a measure of an oil's resistance to flow. An oil with high viscosity is very thick and flows slowly. An oil with low viscosity flows easily. Oil gets thicker as it becomes colder. Therefore, starting an engine in cold weather is more difficult than starting it in warm weather. The cold has increased the viscosity of the oil.

Oil viscosity is rated in two ways by the Society of Automotive Engineers (SAE). It is rated for (1) winter use and (2) summer use. Winter-grade oils are packaged in three grades: SAE5W, SAE10W, and SAE20W. The "W" stands for winter grade. For other than winter use, the grades are SAE20, SAE30, SAE40, and SAE50. The higher the number, the higher the viscosity (the thicker the oil). All these grades are called *single-viscosity oils.*

Many oils are rated as *multiple-viscosity oils.* For example, one oil of this type is SAE10W-40. It has the same viscosity as an SAE10W oil when cold and as an SAE40 oil when hot.

Engine manufacturers specify the viscosity of the oil that should be used in their engines. Figure 6-3 is a chart showing how outside temperature affects the viscosity requirements of an air-cooled four-cycle engine. For example, a 5W-30 oil is recommended if the temperature will be 20°F [−8°C] or less. This oil is good for starting and operation in these low outside temperatures. If the weather is going to warm up, a different oil should be used, such as a 10W-30 or 10W-40. These oils will hold their viscosities in the higher temperatures. They will not thin out too much. However, they are thin enough for easy initial starting and cold-engine operation.

Δ6-6 SERVICE RATINGS FOR SPARK-IGNITION-ENGINE OILS

The service rating indicates the type of service for which the oil is best suited. For spark-ignition (SI) engines, the service ratings are SA, SB, SC, SD, SE, and SF. Here is a brief description of each of these ratings.

SA Lubricating oil without additives (but may contain pour and/or foam depressants); formerly used for older utility gasoline and diesel engines operating under very mild conditions. *Not* to be used in any engine unless specifically recommended by the manuacturer.

SB Lubricating oil with only antiscuff and antioxidant additives; formerly used for older minimum-duty gasoline engines operating under mild conditions. *Not* to be used in any engine unless specifically recommended by the manufacturer.

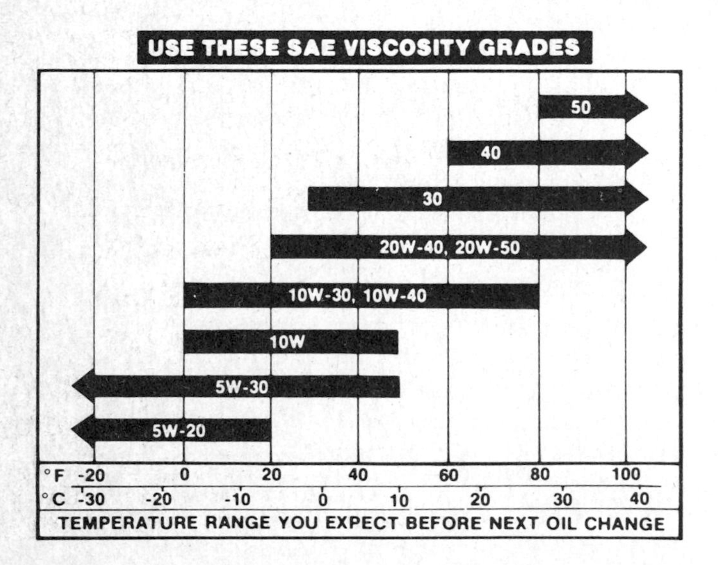

FIG. 6-3 Oil viscosity recommendations for an Onan four-cycle air-cooled engine. *(Onan Corporation)*

SC Lubricating oil that meets requirements for use in gasoline engines in 1964-to-1967 passenger cars and trucks; provides control of high- and low-temperature deposits (varnish and sludge), wear, rust, and corrosion in gasoline engines.

SD Lubricating oil that meets requirements for use in gasoline engines in 1968-to-1971 (and some later) passenger cars and some trucks; provides *more* protection against high- and low-temperature deposits, wear, rust, and corrosion than SC oils. May be used in any engine for which SC oil is recommented.

SE Designation for lubricating oil that meets requirements for use in gasoline engines in 1972-to-1979 (and some 1971) passenger cars and some trucks; provides *more* protection against oil oxidation (high-temperature thickening), high- and low-temperature deposits (varnish and sludge), rust, and corrosion than either SC or SD oils. May be used in any engine for which SC or SD oil is recommended.

SF Designation for lubricating oil that meets requirements for use in gasoline engines in 1980 and later passenger cars and some trucks; provides protection against engine deposits (varnish and sludge), rust, and corrosion, along with increased oxidation stability (reduced high-temperature thickening) and improved antiwear performance than with SE oil. May be used in any engine for which SC, SD, or SE oil is recommended.

Notice that this is an open-end series. When the engine manufacturers and oil producers see the need for other types of oil, they can bring out SG and SH service-rated oils. SA and SB oils are not recommended for use in most engines. These are nondetergent oils. Detergent oils (Δ6-8) are now required in most engines.

Δ6-7 SERVICE RATINGS FOR DIESEL-ENGINE OILS

Diesel-engine oils must have different properties than oils for spark-ignition engines. Diesel-oil service ratings are CA, CB, CC, and CD. A brief description of each is given below.

CA Designation for lubricating oil that meets requirements for use in naturally aspirated (neither supercharged nor turbocharged) light-duty diesel engines (and some gasoline engines) operating with high-quality (low-sulfur) fuel; provides protection against bearing corrosion and piston ring-belt deposits. *Not* to be used in any engine unless specifically recommended by the manufacturer.

CB Designation for lubricating oil that meets requirements for use in naturally aspirated moderate-duty diesel engines (and some gasoline engines in mild service) operating with lower-quality (high-sulfur) fuel, which necessitates more protection against wear and deposits.

CC Designation for lubricating oil that meets requirements for use in lightly supercharged moderate-duty diesel engines (and some heavy-duty gasoline engines); provides protection from high-temperature deposits in lightly supercharged diesel engines, and from rust, corrosion, and low-temperature deposits in gasoline engines.

CD Designation for lubricating oil that meets requirements for use in supercharged severe-duty (high-speed, high-output) diesel engines, requiring highly effective control of wear and deposits; provides protection against bearing corrosion and high-temperature deposits while operating on fuels of a wide quality range.

Δ6-8 OIL ADDITIVES

Engine oil is much more than a natural product refined from crude oil pumped from the ground. Certain chemical compounds, called *additives,* are blended into the oil. The additives give the oil qualities the refining process cannot provide. The refining process determines the viscosity and other basic properties of the oil. But additives give the oil the other qualities it needs to perform a specific job. Commonly used additives in today's engine oils include:

1. *Antiwear Agents* When the oil film is penetrated (or "breaks") between parts, antiwear agents react chemically with the metal surfaces. The agents immediately form a protective coating to reduce wear.
2. *Detergent Dispersants* In a running engine, dirt and particles from combustion get into the oil. Detergent dispersants surround the contaminants with a "protective shell" of molecules. This keeps the contaminants mixed throughout the oil (homogenized) to prevent sludge formation and varnish buildup.
3. *Oxidation Inhibitors* When oil gets hot, it reacts with oxygen and gets thicker. The right amount of oxidation inhibitors prevents this and keeps the hot oil flowing.
4. *Rust and Corrosion Inhibitors* When a gallon of gasoline is burned, it produces slightly more than a gallon of water vapor. Some of this vapor (and blowby) seeps past the piston and rings and into the crankcase. Byproducts of combustion also contain certain contaminants. Together they form an acid condition in the oil. These inhibitors protect the metal parts from rust and corrosion.
5. *VI Improvers* VI (viscosity index) improvers enable an oil to remain thicker at higher temperature ranges than it would without VI improvers.
6. *Foam Inhibitors* These additives weaken the surface tension of the oil so that any air bubbles "break" more rapidly. If this does not happen, moving parts of the engine will whip the oil into a foam. The foam could cause loss of oil flow, hydraulic lifter noise, and inadequate cooling of engine parts.
7. *Pour-Point Depressors* Just as salt lowers the freezing temperature of water, these additives allow oil to work in cold temperatures. They coat the wax crystals in the oil to stop their growth. If the wax crystals are allowed to grow, the oil will become too thick for use.
8. *Friction Modifiers* These additives reduce the friction between moving engine parts. This has the potential to improve engine fuel economy.

NOTE Additional additives poured into the engine crankcase can cause engine troubles if the additives are not compatible with the crankcase oil. No engine-oil additive can recondition used oil that is contaminated with dirt, combustion deposits, gasoline, and acids. Old oil should be drained and replaced with the proper quantity of fresh oil specified by the engine manufacturer.

Δ6-9 TWO-CYCLE ENGINE OIL

Two-cycle engines use a "total-loss" lubrication system. The oil is not recovered as in four-cycle engines but is burned in the combustion chamber. For this reason, the lubricating oil used in two-cycle engines does not require all the additives listed in Δ6-8. However, the oil must be clean-burning and leave a minimum of ash and carbon.

Special oils have been developed for two-cycle and outboard engines. Outboard engine manufacturers, for example, recommend a BIA TC-W oil. *BIA* means *B*oating *I*ndustry *A*ssociaton, and *TC-W* means *T*wo-*C*ycle *W*ater-cooled. Outboard-engine manufacturers caution against using automotive oils, which can produce ash and cause engine trouble. There also are special oils for small engines. Always use the oil recommended by the engine manufacturer, and follow the directions on the oil container.

Mineral oil and synthetic oil are the most frequently used base stocks for two-cycle-engine oil. Some two-cycle engines have an *oil-injection* type of lubricating system (described later). These engines usually require the use of synthetic oil. One claim made for synthetic oil is that it does not leave carbon deposits in the combustion chamber or on the spark plug.

However, care must be taken to avoid trouble when switching an engine from a petroleum-base (mineral) oil to a synthetic oil. The two may not mix satisfactorily. Before putting synthetic oil in an engine using any type of mineral oil, always drain the old oil. Then flush out the engine with a suitable flushing fluid.

Δ6-10 TYPES OF TWO-CYCLE-ENGINE LUBRICATING SYSTEMS

Two different types of lubricating systems are used in two-cycle engines. In one type, the oil is mixed with the gasoline (Δ6-11). In the other, the oil is injected into the air-fuel mixture as it flows into the engine (Δ6-12).

Each type of two-cycle-engine lubricating system requires a different type of oil. Both oils are basically similar because they have the same job to do. However, when the oil is mixed with the gasoline in the fuel tank, the oil is diluted by the gasoline. Therefore the oil is not affected by the outside temperature. But in the injector system, the oil is stored in a separate tank. In cold weather, the oil can get so cold that it thickens and then will not flow properly. To avoid this problem, solvent is added to the injector oil to improve the flow characteristics. This type of two-cycle oil is called *prediluted oil.*

Not all engine manufacturers recommend the use of prediluted oils in their two-cycle engines with oil injection. Follow the recommendations in the operator's manual for the engine.

Δ6-11 PREMIX LUBRICATING SYSTEM

Many two-cycle engines are lubricated by the premix system. In this system, a certain amount of oil is premixed with the gasoline before it is added to the fuel tank (Fig. 6-4). Then the oil enters the crankcase as an oil mist carried along with the air-fuel mixture from the carburetor (Fig. 6-5).

After compression in the crankcase, the air-fuel mixture travels to the combustion chamber, where the mixture is burned. But part of the oil mist stays in the crankcase so the oil droplets can lubricate the bearings, piston, rings, and cylinder walls. Some of the oil gets into the combustion chamber, where the oil is burned along with the air-fuel mixture.

The grade and amount of oil to be mixed with the gasoline are very critical. The manufacturer's recommendations should be carefully followed. Adding too much oil will cause the exhaust ports to become clogged very quickly. Also, carbon deposits will form on the piston and rings. This causes poor engine performance. Adding too little oil will deprive the engine of adequate lubrication. Then the engine will wear out much sooner and may seize during hard running.

Typical mixture ratios for two-cycle engines are shown in Fig. 6-6. Usually the manufacturer will specify the type and octane rating of the gasoline that should be used and the type and viscosity of the oil. Mix the gasoline and the oil by pouring the oil into an empty gasoline can. Add the amount of gasoline necessary to make a premix of the correct properties. Shake the can several times.

This procedure will result in the gasoline and oil being thoroughly mixed. Oil that is not fully mixed with the gasoline can block the carburetor jet, and the engine will not run.

One disadvantage to premix is that the lubricating qualities of the fuel-oil mixture do not last long. Do not leave the premix in the fuel tank for a long time. If the engine is to be stored, or not used, completely drain the premix from the fuel system.

FIG. 6-4 To lubricate many two-cycle engines, the lubricating oil is mixed with the gasoline.

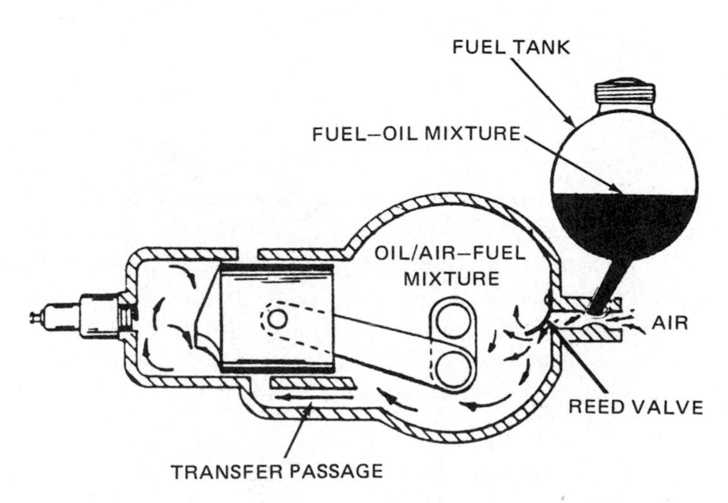

FIG. 6-5 Premix lubricating system for a two-cycle engine. *(ATW)*

Engine manufacturers also caution against mixing a mineral-base or petroleum oil with a vegetable oil. The two oils will not always mix. This could cause a premix that is too thick. Then the engine will lose power and possibly be damaged from lack of lubrication.

A two-cycle engine that is lubricated by premix usually will have needle bearings for the piston pin and the crankpin, and ball or roller bearings for the crankshaft. Figure 6-7 shows the parts that require lubrication in an engine lubricated with premix.

In an engine that is lubricated with premix, the big end of the connecting rod usually has slots around the center of the bearing area. Also, there are notches spaced around the sides of the rod. This allows the oil droplets contained in the air-fuel mixture to pass through the bearing. The drops of oil coat the bearing and provide it with the needed lubrication.

Δ6-12 OIL-INJECTION SYSTEM

A second method of lubricating a two-cycle engine uses a separate oil tank. Figure 6-8 shows this system, called the *oil-in-*

16:1 mixture		
Gasoline	Oil	
1 gallon	½ pint	8 oz
3 gallons	1½ pints	24 oz
5 gallons	2½ pints	40 oz
6 gallons	3 pints	48 oz
1 liter	0.06 liter	63 cc
5 liters	0.31 liter	313 cc
10 liters	0.63 liter	625 cc
20 liters	1.25 liters	1250 cc
24:1 mixture		
1 gallon	⅓ pint	5.3 oz
3 gallons	1 pint	16 oz
5 gallons	1⅔ pints	26.5 oz
6 gallons	2 pints	32 oz
1 liter	0.04 liter	42 cc
5 liters	0.21 liter	208 cc
10 liters	0.42 liter	417 cc
20 liters	0.83 liter	833 cc
32:1 mixture		
1 gallon	¼ pint	4 oz
3 gallons	¾ pint	12 oz
5 gallons	1¼ pints	20 oz
6 gallons	1½ pints	24 oz
1 liter	0.03 liter	31 cc
5 liters	0.16 liter	156 cc
10 liters	0.31 liter	312 cc
20 liters	0.63 liter	625 cc
50:1 mixture		
1 gallon	⅙ pint	3 oz
3 gallons	½ pint	9 oz
5 gallons	13/16 pint	15 oz
6 gallons	1 pint	16 oz
1 liter	0.02 liter	20 cc
5 liters	0.10 liter	100 cc
10 liters	0.20 liter	200 cc
20 liters	0.40 liter	400 cc

FIG. 6-6 Typical gasoline:oil mixture ratios for two-cycle engines. *(Tecumseh Products Company)*

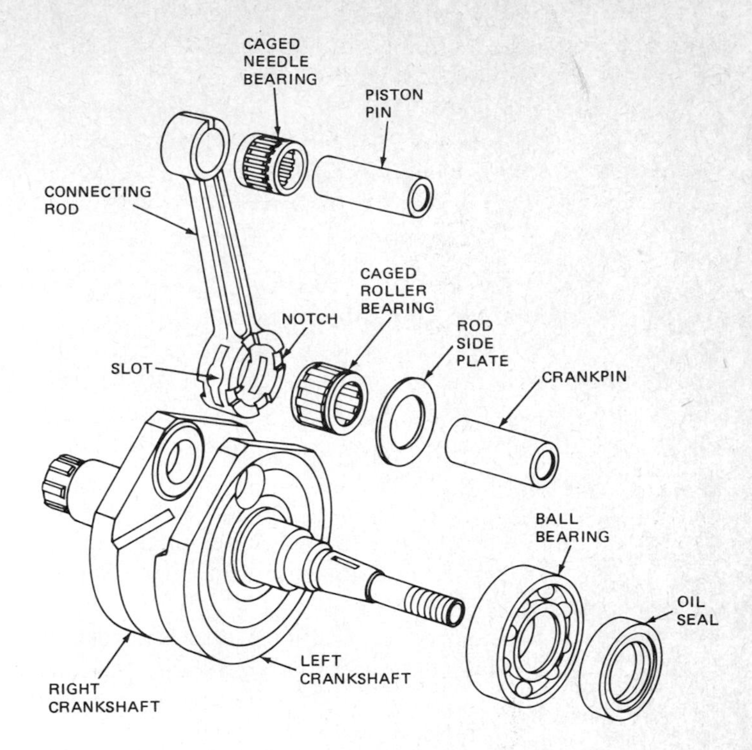

FIG. 6-7 Crankshaft, connecting rod, and bearings for an engine lubricated with premix. *(American Honda Motor Company, Inc.)*

jection system, on a one-cylinder two-cycle engine. Oil from the tank is sent by the pump to the nozzle in the intake port. There the amount of oil needed by the engine sprays out into the passing air-fuel mixture. On some engines, this is all the lubrication that the engine receives. In operation, the system is little different from the mixing achieved by the use of a premix, or a mixture of gasoline and oil, in the fuel tank. However, in some engines, a second oil line from the pump also provides direct full-pressure lubrication to the crankshaft and bearings.

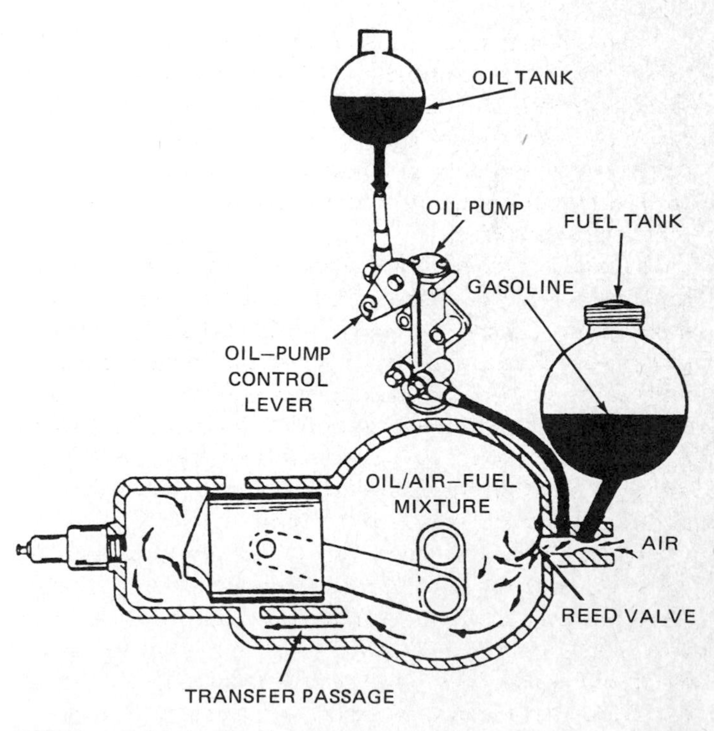

FIG. 6-8 Oil-injection lubricating system for a two-cycle engine. *(ATW)*

Δ6-13 LUBRICATING SYSTEMS FOR FOUR-CYCLE ENGINES

The four-cycle engine has a lubricating-oil reservoir from which some oil continually circulates through the engine. In some engines, the oil is splashed onto the moving parts by the crankshaft and connecting rod passing through the oil in the reservoir, or sump. In other engines, an oil pump sends oil from the reservoir to the moving parts.

There are two ways to classify four-cycle-engine lubricating systems. These are according to where the oil is stored (*wet sump* or *dry sump*) and how the oil reaches the moving parts (by *splash* or by *pressure-feed*).

Δ6-14 WET-SUMP LUBRICATING SYSTEM

Automobile engines and most small four-cycle engines use the wet-sump system. The *sump* is the oil pan or crankcase at the bottom of the engine (Fig. 6-9). It is a wet sump because the engine oil is stored under the crankshaft. In many engines, an oil pump sends oil from the oil pan up through the engine oil lines to the moving engine parts. After lubricating, cleaning, and cooling these parts, the oil drops back down into the oil pan. The oil pump keeps the oil circulating continuously while the engine is operating.

However, the wet-sump lubricating system is not satisfactory for engines that are moving over rough terrain and operating at various angles from the vertical. Under these conditions, the oil would splash up all over the lower part of the engine. Too much oil would get past the piston rings into the combustion chamber. When the engine was tilted at an angle, the oil would be over on one side of the sump. Then the oil-pump intake would be above the oil level in the sump. With the oil pump not taking in any oil, the engine would soon fail. To prevent this problem, a dry-sump system (Δ6-15) is used on engines that run over rough ground and are often tilted at sharp angles to the vertical.

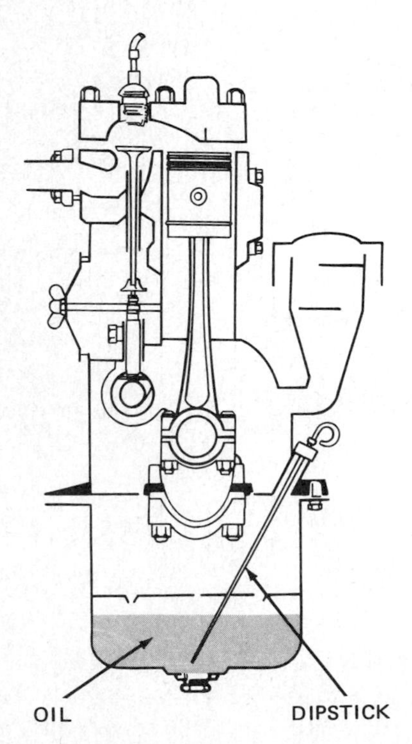

FIG. 6-9 A four-cycle engine using a wet-sump lubricating system.

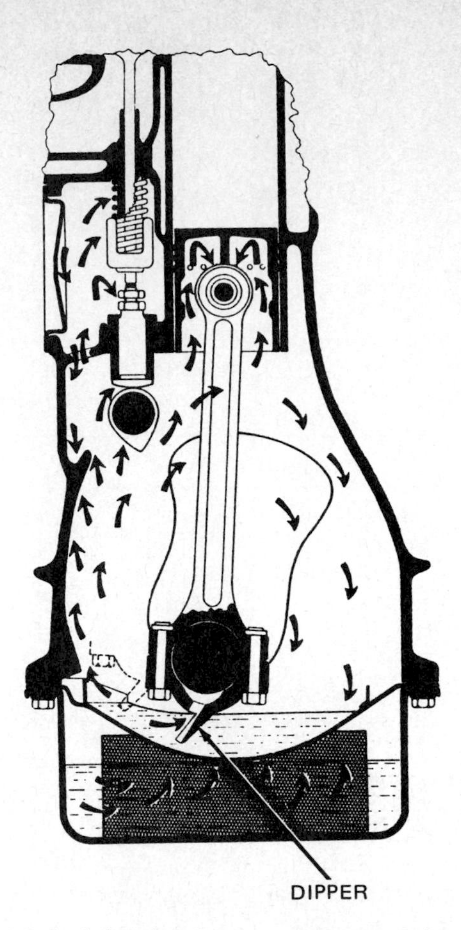

FIG. 6-10 Splash lubricating system on an L-head engine. The dipper on the connecting rod splashes oil (as shown by the arrows) every time the piston passes through BDC.

Δ6-15 DRY-SUMP LUBRICATING SYSTEM

The dry-sump lubricating system uses a "double" pump that works like two separate oil pumps. One is a scavenge pump, and the other is the pressure (or delivery) pump. As oil drops down from the engine into the crankcase, one pump picks up the oil and sends it through an oil line to the oil tank. From there, the second pump pressurizes the oil and delivers it to the moving engine parts.

With this system, the second oil pump always has a tank full of oil to draw from. There is no danger that the oil level will fall below the oil-pump intake. Therefore, all moving engine parts are properly lubricated at all times, regardless of the engine movement or tilt. For this reason, some four-cycle motorcycle engines are dry sump.

Δ6-16 TYPES OF FOUR-CYCLE-ENGINE LUBRICATING SYSTEMS

Small four-cycle engines use several different methods of getting oil to the moving parts in the engine. On engines that are either stationary or moving over fairly level terrain, the wet-sump system (Δ6-14) is used. The simplest means of lubrication is to splash the oil about so that all parts are drenched (Fig. 6-10). The splashing is produced by a dip-

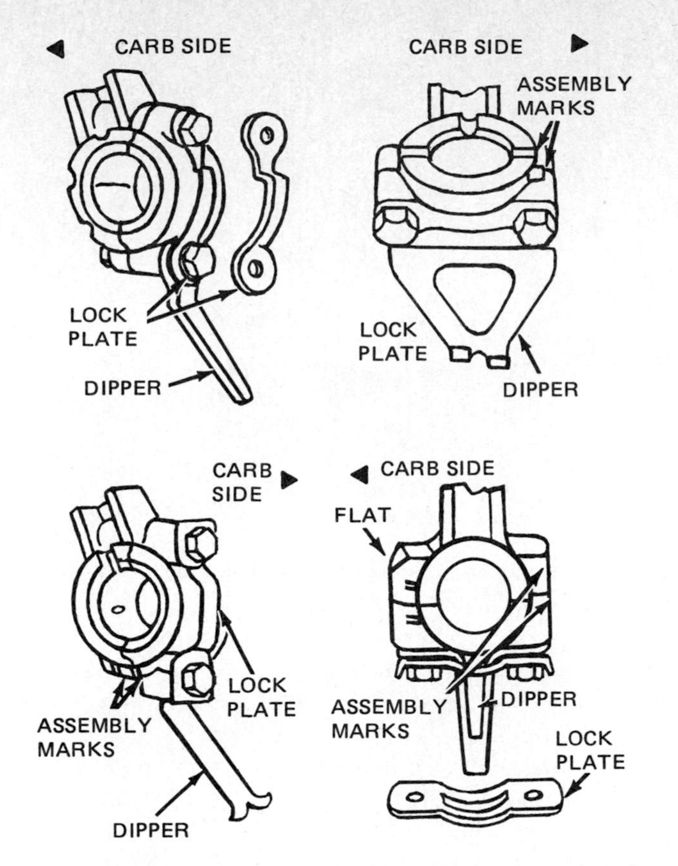

FIG. 6-11 Different ways in which the dipper is mounted on the connecting-rod cap.

per of the lower end of the connecting rod (Fig. 6-11). The oil in the crankcase is splashed every time the dipper reaches into it on the down stroke of the piston. Splash may also be accomplished by the use of an oil slinger which is rotated by the camshaft gear (Fig. 6-12).

A cam-operated barrel type of plunger pump is used in some small engines to provide a pressure lubricating system. Three typical examples are shown in Figs. 6-13 to 6-15. A small gear-type pump operated by the camshaft gear is sometimes used on small vertical-shaft engines (Fig. 6-16).

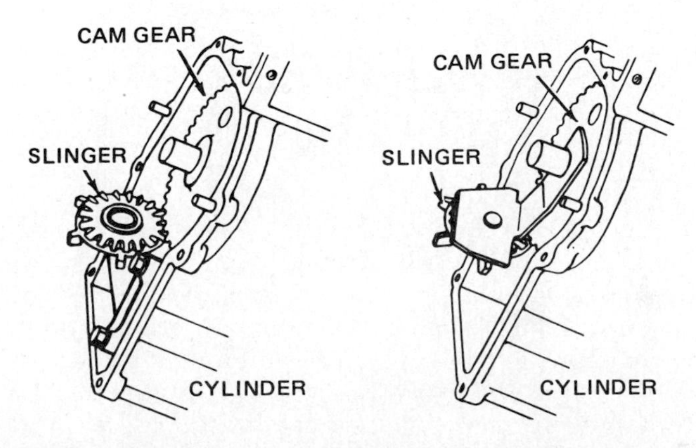

FIG. 6-12 Oil slinger is rotated by the camshaft gear.

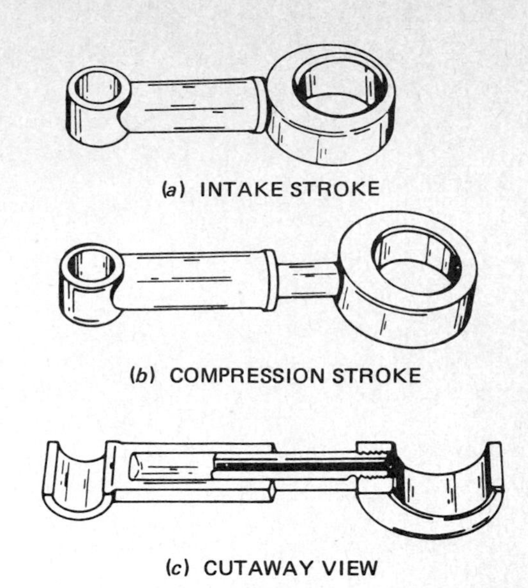

FIG. 6-13 Barrel-type lubricating pump used on the Lauson vertical-shaft engine.

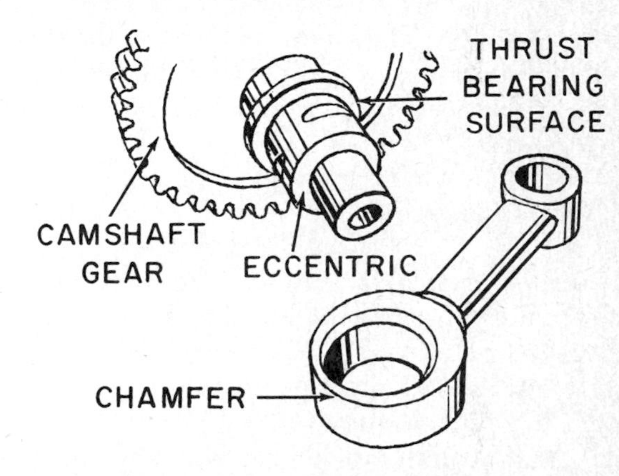

FIG. 6-14 The plunger-type oil pump is assembled on an eccentric on the camshaft. *(Tecumseh Products Company)*

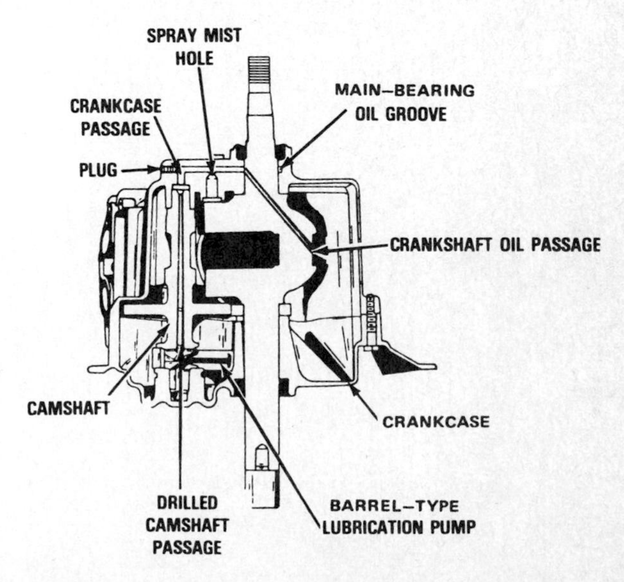

FIG. 6-15 Location of the barrel-type lubricating pump and the oil passages in the engine. *(Tecumseh Products Company)*

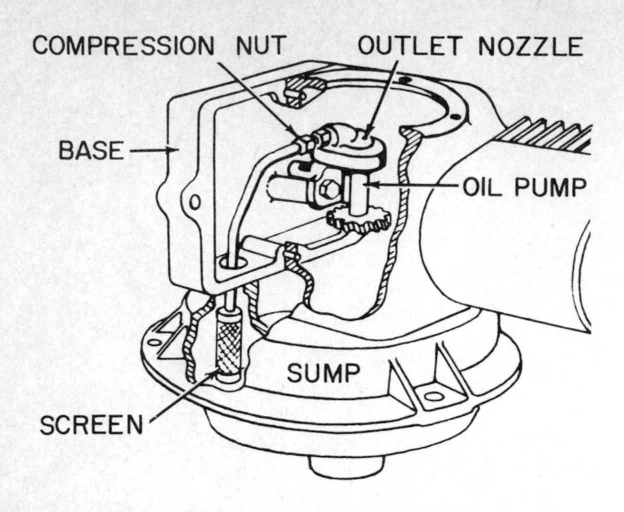

FIG. 6-16 Gear-driven oil pump. *(Briggs & Stratton Corporation)*

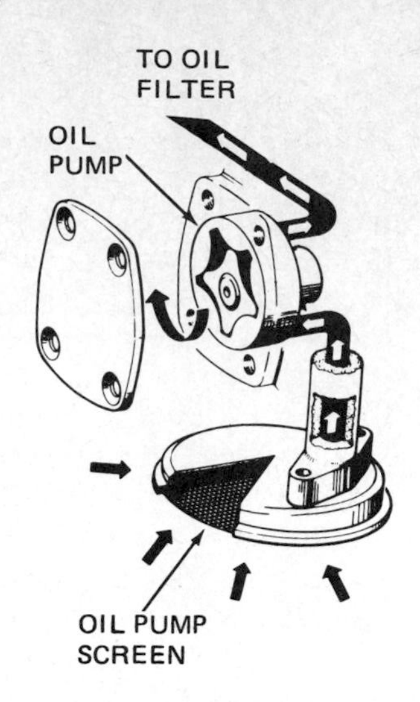

FIG. 6-18 In a pressure lubricating system, an oil pump sends oil to lubricate the moving parts.

Δ6-17 PRESSURIZED LUBRICATING SYSTEMS Small multicylinder four-cycle engines often have some type of pressurized lubricating system. In a pressurized system, an oil pump sends oil from the crankcase to holes or nozzles from which oil sprays onto the moving parts.

Figure 6-17 shows an engine with a pressurized-spray lubricating system. A rotor-type oil pump (Fig. 6-18) picks up oil from the crankcase and forces the oil through passages called *galleries* machined into the crankcase. Openings in the galleries provide oil to lubricate the engine parts. The oil pickup is located in the center of the crankcase. This allows proper lubrication with the engine operating at angles up to 30°.

In Fig. 6-17, the oil pump delivers oil to the crankshaft main bearings and camshaft bearings at a pressure of about 5 psi [35 kPa]. The connecting-rod journals are lubricated by oil that sprays from two small holes drilled in the camshaft.

Some engines have a full-pressure lubricating system (Fig. 6-19). It sends oil to the crankshaft bearings, camshaft bearings, and connecting-rod bearings at a pressure ranging from about 25 to 50 psi [172 to 345 kPa]. The oil pump is driven by the crankshaft. To prevent high engine speed causing an excessively high oil pressure, the system includes a *pressure-relief valve.* When the pressure gets too high, the valve opens to allow some of the oil from the pump to return to the crankcase.

The engine is not factory-equipped with an oil-pressure gauge. However, after a pipe plug in the crankcase is

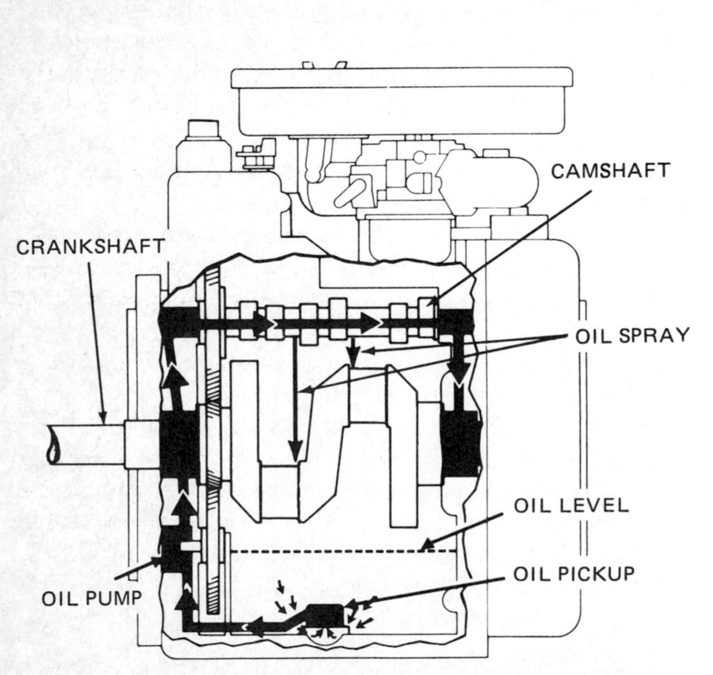

FIG. 6-17 A two-cylinder four-cycle engine with a pressurized-spray lubricating system. *(Kohler Company)*

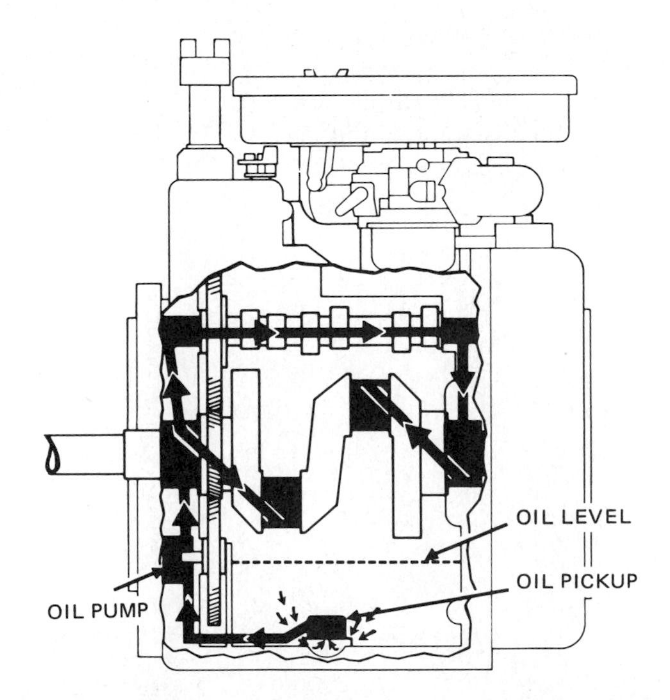

FIG. 6-19 A full-pressure lubricating system. *(Kohler Company)*

removed, an oil-pressure gauge can be installed to check the oil pressure in the engine.

Some vertical-crankshaft four-cycle engines are lubricated by a pressure system that uses pulsating crankcase pressures as an oil pump. Figure 4-12 shows in cutaway view an engine with this type of lubricating system. The crankcase is sealed, similar to the way that the crankcase for a two-cycle engine is sealed.

When the piston is at BDC, pressure in the crankcase is highest. In this position, the bottom port in the camshaft aligns with the oil-pickup passage in the crankcase. The pressure above the oil forces it to fill the hollow camshaft. As the camshaft continues turning, the bottom port is closed off, trapping the oil under pressure in the hollow camshaft. Then the top port in the camshaft aligns with the main gallery in the crankcase. As this occurs, the piston has reached TDC, and pressure in the crankcase is lowest.

The combined effect of the oil trapped under pressure in the hollow camshaft and of the lower pressure in the crankcase causes some oil to flow out of the camshaft. The oil flows through the main gallery to lubricate the top main bearing and then flows down to fill the oil cup. From the oil cup, oil flows into the hollow crankpin and from there through another hole to the connecting-rod bearing.

One advantage to this type of lubricating system is that it permits the engine to be operated at an angle. The maximum angle possible changes with the oil level and position of the carburetor-side of the engine (up or down). Various types of these engines can operate at angles of 30 to 45°.

Δ6-18 OIL FILTERS

Many four-cycle engines with a pressurized lubricating system use an oil filter (Fig. 6-20). Part or all of the oil from the oil pump flows through the oil filter. Inside the filter, there is a *filter element* made of pleated paper or fibrous material (Fig. 6-21). As the oil passes through the filter, the paper or fibers trap any carbon, dirt, and metal particles. As the engine runs hour after hour, the filter element tends to become plugged with contaminants. Then a new filter or filter element must be installed.

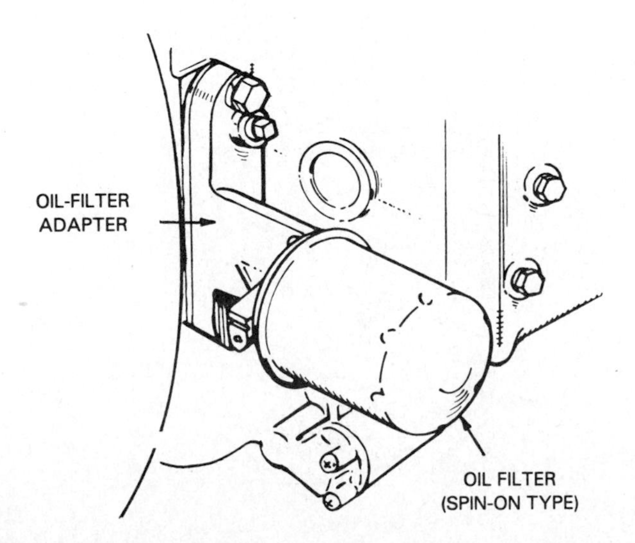

FIG. 6-20 A spin-on type of oil filter. *(Onan Corporation)*

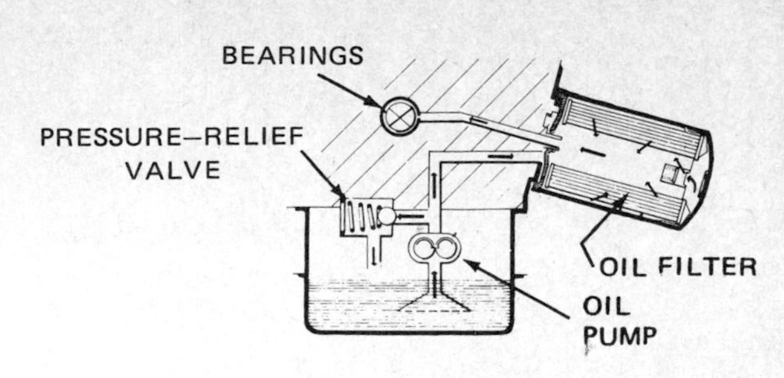

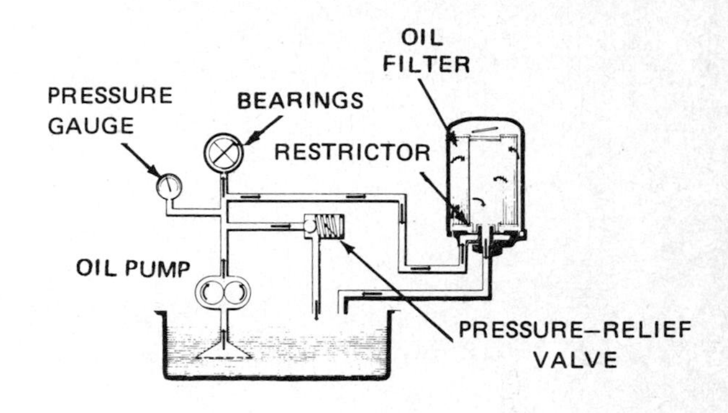

FIG. 6-21 Two types of engine lubricating-oil filter installations. *(Fram Corporation)*

There are two types of engine lubricating-oil filters (Fig. 6-21). Those which filter part of the oil from the oil pump are called *bypass filters.* Those which filter all the oil in circulation through the system are called *full-flow filters.* The full-flow filter includes a spring-loaded bypass valve. If the oil cannot pass through the filter, the pressure buildup in the filter causes the bypass valve to open. This allows oil from the pump to bypass the filter and go directly to the bearings. Therefore, engine damage from lack of lubrication is avoided.

Engine oil filters also differ in construction. Figure 6-20 shows a spin-on, or replaceable can, type of oil filter. The filter element is housed in a disposable can. It is removed and thrown away at periodic intervals specified by the engine manufacturer. This type of filter construction is used for most full-flow oil filters.

Figure 6-22 shows an oil filter using a replaceable, or cartridge-type, filter element. To remove the filter element, the top is removed from the housing. Then the filter element is pulled out. After cleaning the housing, the new filter element is installed. This type of filter is often used with bypass-filter installations.

Δ6-19 OIL-LEVEL INDICATORS

A dipstick is used to check the level of the oil in the crankcase (Fig. 6-23). To use the dipstick, pull it out, wipe it off, and put it back in place. Then pull it out again and check the level of the oil shown on the dipstick markings.

Δ6-20 CRANKCASE VENTILATION

In all piston engines, some blowby gases get past the piston and rings and enter the crankcase. In addition, some water and liquid fuel appear in the crankcase of four-cycle engines during cold-engine operation. These must be cleared from the crankcase before they cause acids, rust, and corrosion. If pressure builds up in the crankcase, oil may be forced out past the crankshaft seals.

To prevent these problems, four-cycle engines have some method of crankcase ventilaton. Figure 6-24 shows one type of *crankcase breather*. It has a small reed valve placed over an opening to the crankcase. Above the valve is a filter pack and a tube which connects to the engine air cleaner. As the engine runs, the air flowing through the air cleaner creates a slight vacuum in the tube. This helps pull the blowby gases from the crankcase and into

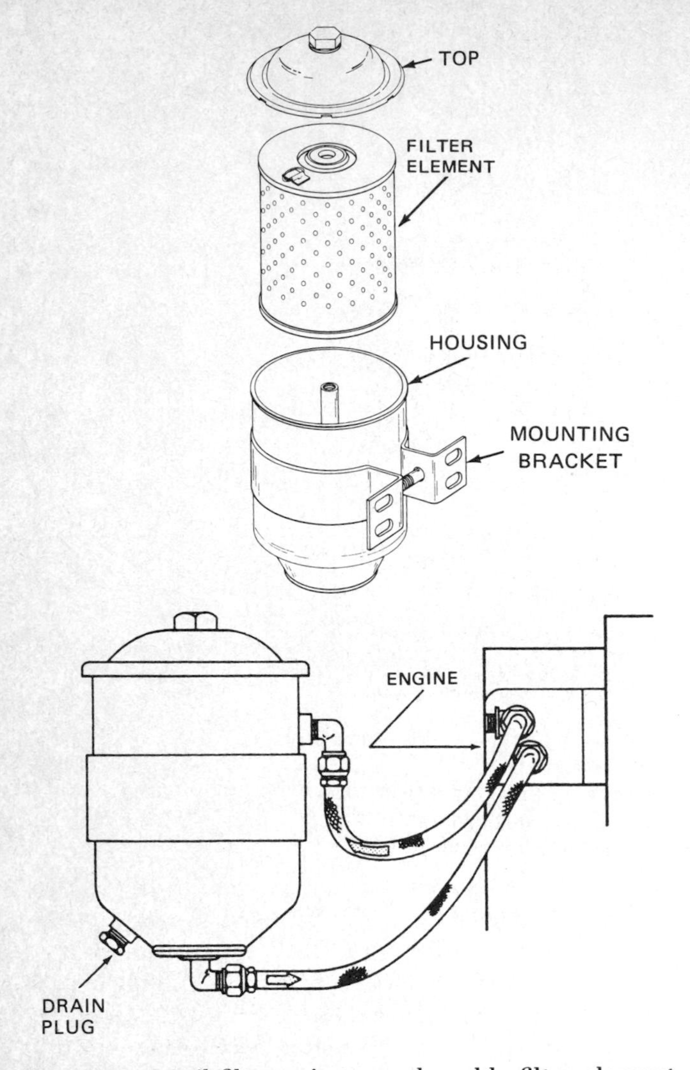

FIG. 6-22 An oil filter using a replaceable filter element. *(Kohler Corporation)*

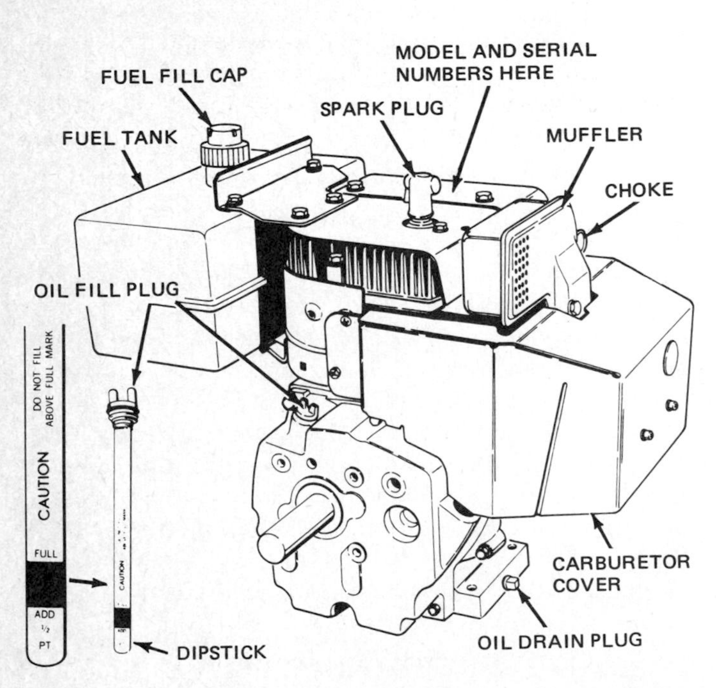

FIG. 6-23 An oil-fill plug and dipstick for a four-cycle engine. *(Tecumseh Products Company)*

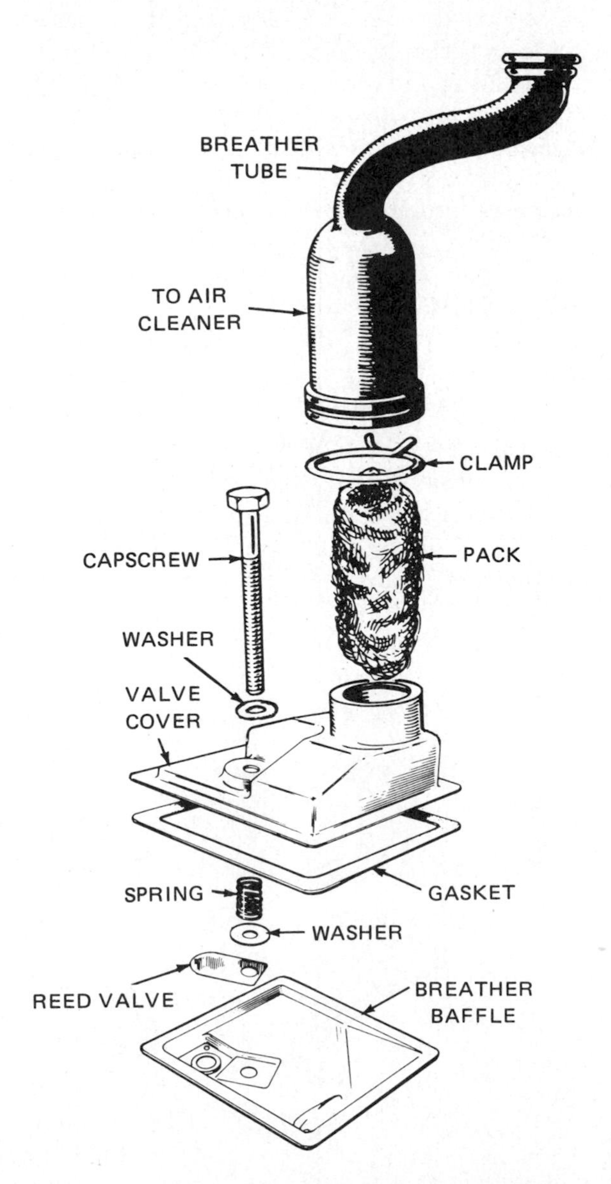

FIG. 6-24 A crankcase breather. *(Onan Corporation)*

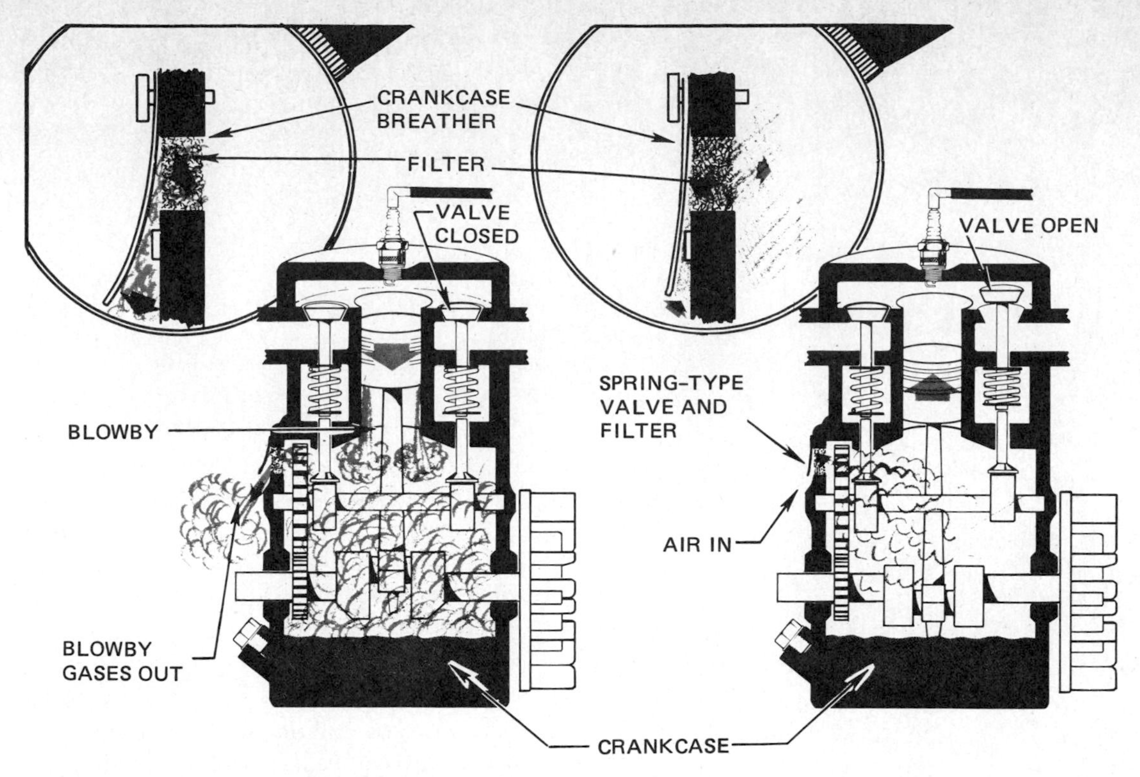

FIG. 6-25 A crankcase breather for a small four-cycle engine. Left, when the piston is moving down, pressure in the crankcase forces the reed valve to open so blowby gases are forced out of the crankcase. Right, when the piston moves up, the partial vacuum created in the crankcase causes the reed valve to close.

the engine to be burned. As a result, a properly operating engine will have a slight vacuum in the crankcase.

Figure 6-25 shows another type of crankcase breather. It consists of a mesh-type filter element and a reed valve. The reed valve is a flexible metal plate which rests against one or more openings in the crankcase. Figure 6-26 shows the crankcase breather disassembled.

When the piston moves down on either the intake or power stroke, a slight pressure is created in the crankcase. This pressure pushes the reed valve open so that the blowby gases are forced out of the crankcase (Fig. 6-25, left). Then, when the piston moves up on either the compression or exhaust stroke, a slight vacuum is produced in the crankcase. This permits atmospheric pressure to push fresh air into the crankcase.

The reed valve in Fig. 6-25 is designed so that it will cause a slight vacuum to be retained in the crankcase. The vacuum helps prevent oil leakage through the oil seals and gaskets. To achieve this, the reed valve either has a small hole in it or does not close completely. Either

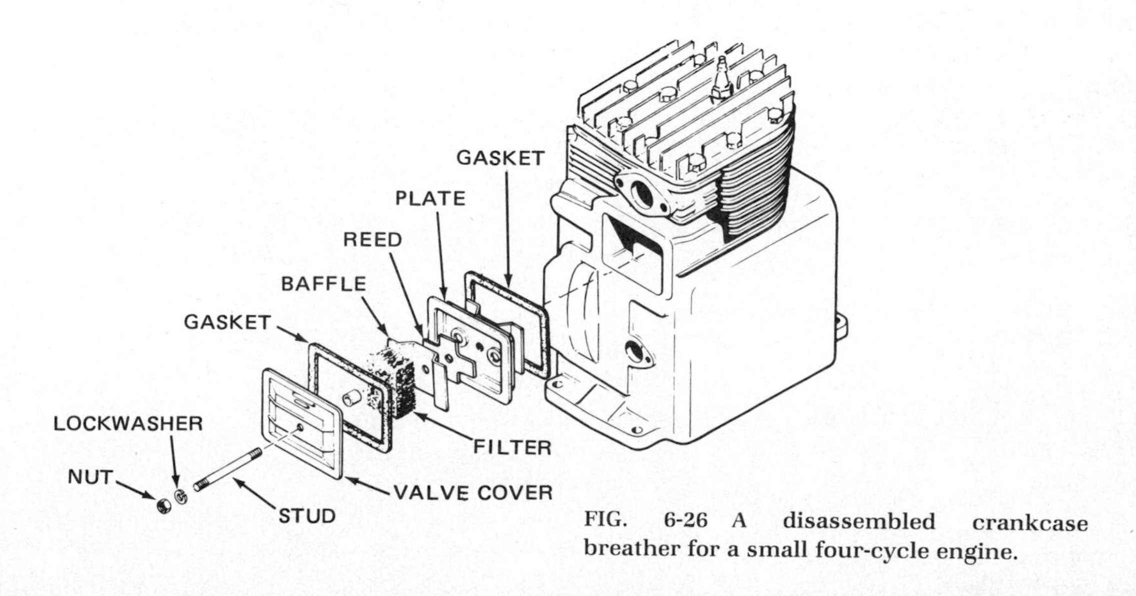

FIG. 6-26 A disassembled crankcase breather for a small four-cycle engine.

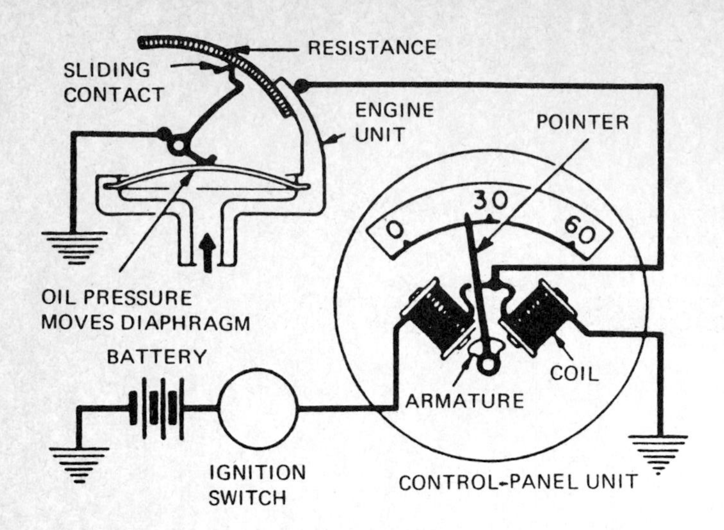

FIG. 6-27 Electric circuit of an electric-resistance oil-pressure indicator.

arrangement restricts the air entering the crankcase. This causes a slight vacuum to remain at the end of the compression or exhaust stroke.

Other types of crankcase breathers include a ball-check type and a floating-disk type. These are opened by pressure in the crankcase and partly closed by gravity and by atmospheric pressure. Some crankcase breathers include oil-drain holes. They allow any trapped oil to drain back into the crankcase.

Δ6-21 INDICATORS OF OIL PRESSURE

Some engines are equipped with an indicator to notify the operator if the engine oil pressure is too low for safe engine operation. Three types of indicators are used:

1. A light that comes on to warn of low oil pressure
2. A gauge with a needle that swings around to indicate the pressure (Fig. 6-27)
3. A buzzer that sounds when oil pressure is low

Any type of low-oil-pressure indicator may be used with an automatic engine-shutdown device. After the engine is running, low oil pressure (or overheating) will cause the shutdown device to stop the engine.

CHAPTER 6: REVIEW QUESTIONS

Select the *one* correct, best, or most probable answer to each question. You can find the answer in the section indicated at the right of each question.

1. The two-cycle engine is lubricated by (**Δ*6-1***)
   a. oil poured into the crankcase
   b. oil mixed with the gasoline
   c. fuel oil
   d. the fuel
2. Besides lubricating moving parts in the four-cycle engine, the oil also (**Δ*6-3***)
   a. removes heat from engine parts
   b. absorbs shocks between bearings and other engine parts
   c. forms a seal between the piston rings and cylinder wall
   d. all of the above
3. In the oil-injection system used in some two-cycle engines (**Δ*6-10***)
   a. the oil is premixed with the gasoline
   b. oil is sprayed into the air-fuel mixture
   c. oil is injected into the engine cylinder
   d. none of the above
4. Additives in lubricating oil for spark-ignition engines include (**Δ*6-8***)
   a. viscosity improvers
   b. pour-point depressants
   c. detergent dispersants
   d. all of the above
5. Viscosity (**Δ*6-5***)
   a. measures an oil's resistance to flow
   b. means the temperature at which a liquid flows
   c. refers to how slippery an oil is as a lubricant
   d. determines how much oil should be mixed with gasoline
6. A "total-loss" lubricating system (**Δ*6-9***)
   a. is used in two-cycle engines
   b. means the lubricating oil is burned in the combustion chamber
   c. means the lubricating oil is not recovered and stored
   d. all of the above
7. Basic lubricating systems for four-cycle engines are
   a. wet sump and crankcase (**Δ*6-13***)
   b. wet sump and dry sump
   c. pump type and compressor
   d. full flow and no flow
8. The dry-sump lubricating system (**Δ*6-15***)
   a. assures lubrication of all engine parts regardless of engine tilt
   b. uses a double oil pump
   c. has a pump that removes any oil that accumulates in the sump
   d. all of the above
9. Service ratings of lubricating oil for spark-ignition engines (**Δ*6-6***)
   a. indicate the oil viscosity
   b. run from SA to SF
   c. run from CA to CD
   d. range from SAE5W to SAE50
10. The purpose of the dipstick is to (**Δ*6-19***)
    a. check the quality of the oil
    b. measure the oil pressure
    c. check the oil level in the crankcase
    d. control the bypass-flow oil filter

CHAPTER

7

Fuel Systems for Small Spark-Ignition Engines

After studying this chapter, you should be able to:

1. ***List the types of fuel systems used on small spark-ignition engines and explain how they work.***
2. ***List the parts of fuel systems and the purpose and operation of each.***
3. ***Describe the construction of a basic carburetor and explain how the venturi works.***
4. ***Explain the purpose of the governor and describe the operation of each type.***

Δ7-1 INTRODUCTION TO FUEL SYSTEMS The air-fuel mixture must have the proper proportions of air and gasoline for good engine operation. If the mixture does not have enough gasoline vapor (mixture too lean) or if the mixture has too much gasoline vapor (mixture too rich), the engine will not run properly. Also, to start a cold engine, the mixture must be enriched. It must have a higher proportion of gasoline vapor in it.

A variety of fuel systems and carburetors are used in small engines. This is because small engines are used in so many different ways. Some small engines, such as those used in lawn mowers, are designed to run at one speed in an upright position. Other small engines, such as those used in chain saws, are designed to be used in almost any position.

Δ7-2 TYPICAL SMALL-ENGINE FUEL SYSTEMS The fuel system (Fig. 7-1) includes a fuel tank, fuel filter, carburetor, air cleaner, and (on some engines) a fuel pump. Three types of fuel systems are used on small engines:

1. Gravity-feed fuel system
2. Suction-feed fuel system
3. Pressure-feed fuel system

The two most widely used fuel systems for small engines are the gravity-feed and the suction-feed systems (Fig. 7-2). Some engines use the pressure-feed system (Fig. 7-3), which has a fuel pump.

A gravity-feed system for a two-cycle engine in a power lawn mower is shown in Fig. 7-1. Four-cycle engines using a gravity-feed fuel system are shown in Figs. 7-2 and 7-4.

In the gravity-feed system, gasoline flows down by gravity from the fuel tank, through a fuel filter, to the carburetor. When the engine is running, air passes through the carburetor where it picks up a charge of gasoline. Then the mixture enters the engine.

Essential parts of the carburetor (Fig. 7-1) include the air cleaner, float bowl, choke valve, throttle valve, and (on some carburetors) fuel nozzle with an adjustment needle. Carburetors are described later.

Δ7-3 FUEL TANK The fuel tank (Fig. 7-4) is made of sheet metal or plastic. The tank has a filler cap which is removed to add gasoline. The cap has a small hole through which air can enter the tank as gasoline is withdrawn (Fig. 7-4). This prevents a vacuum from forming in the tank and stopping fuel flow.

Δ7-4 FUEL STRAINERS AND FILTERS A fuel strainer or filter removes any dirt that might have entered the tank from the fuel. This prevents dirt from entering the carburetor. There, the dirt could clog fuel passages and stall the engine.

All small engines have some type of fuel filter or strainer. It is installed on the end of the fuel pickup in the tank (Fig. 7-1), in the fuel line between the tank and the carburetor (Fig. 7-2, left), or in the carburetor fuel inlet.

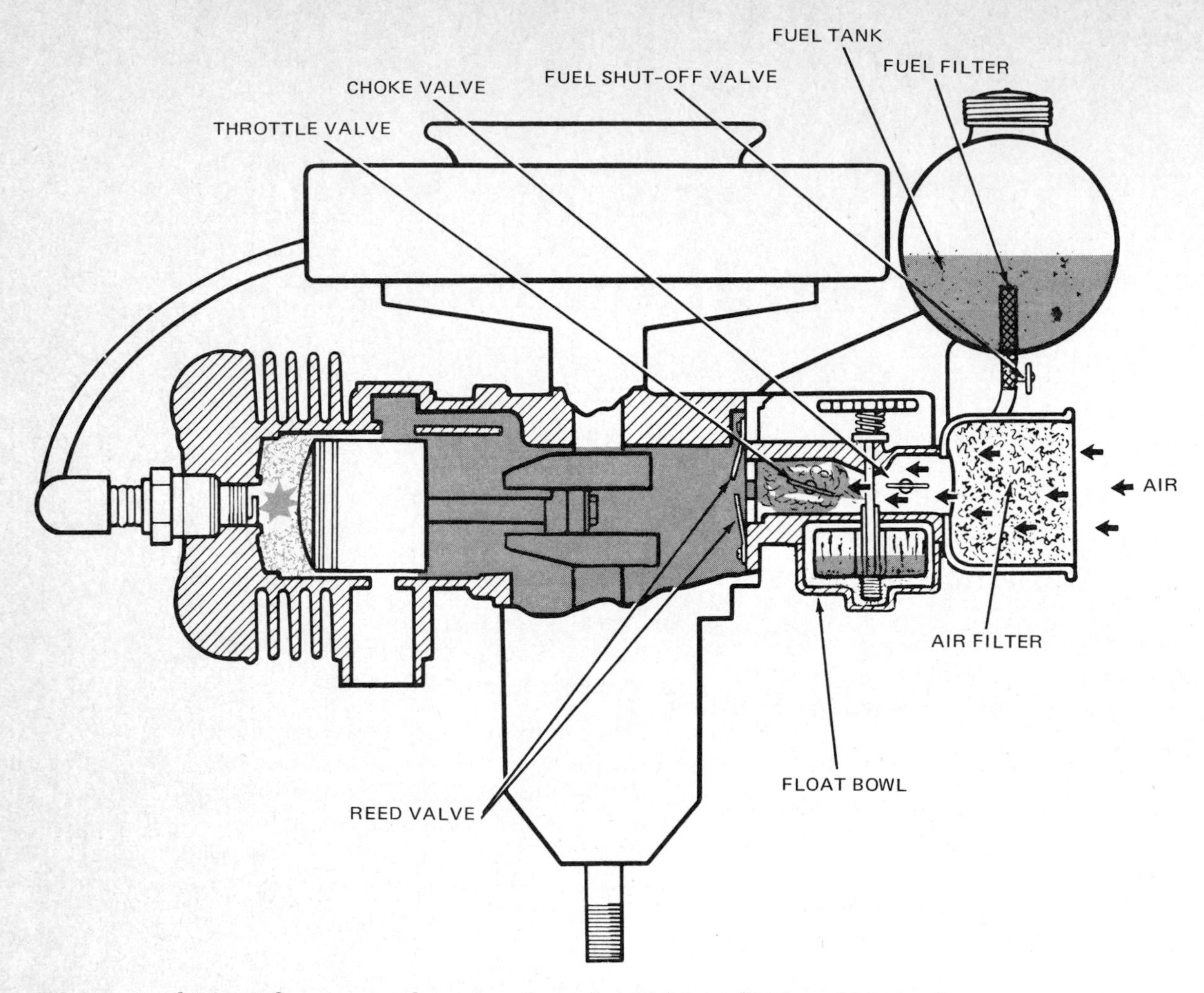

FIG. 7-1 Fuel system for a two-cycle engine. *(Lawn Boy Division of Outboard Marine Corporation)*

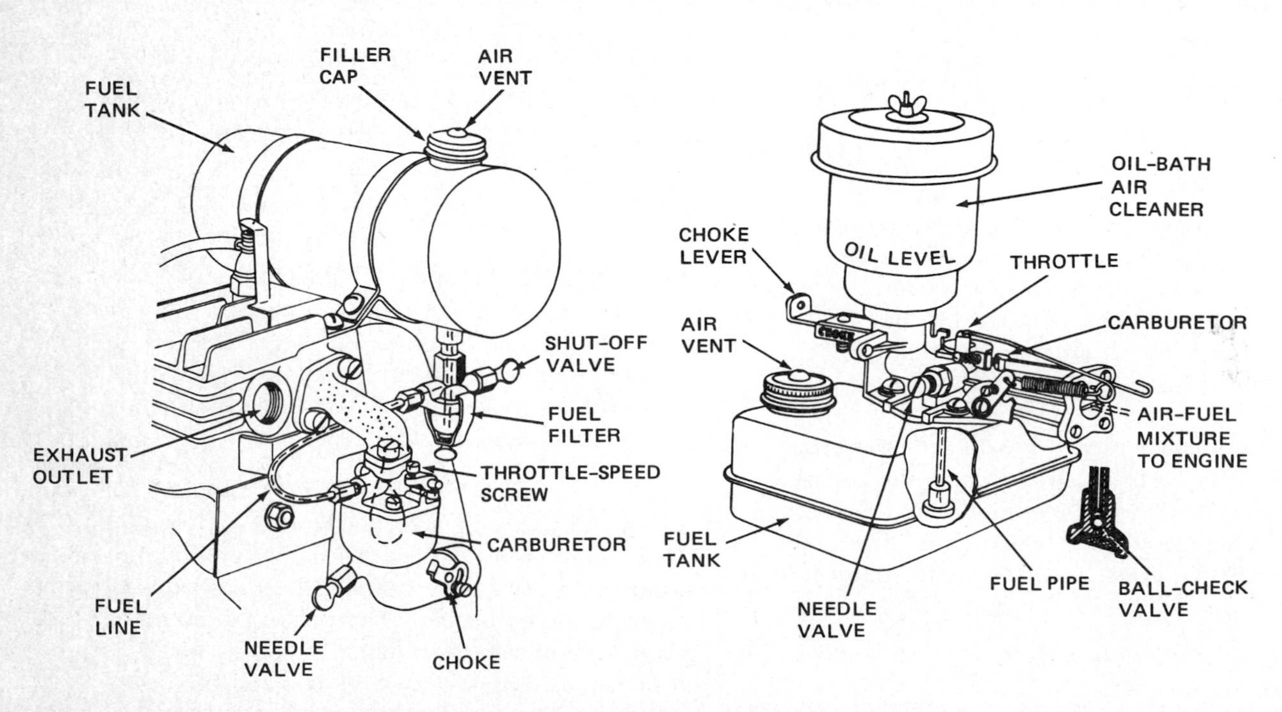

FIG. 7-2 Two types of small-engine fuel systems. Left, a gravity-feed system. Right, a suction-feed system. These fuel systems are used on single-cylinder four-cycle engines. *(Briggs & Stratton Corporation)*

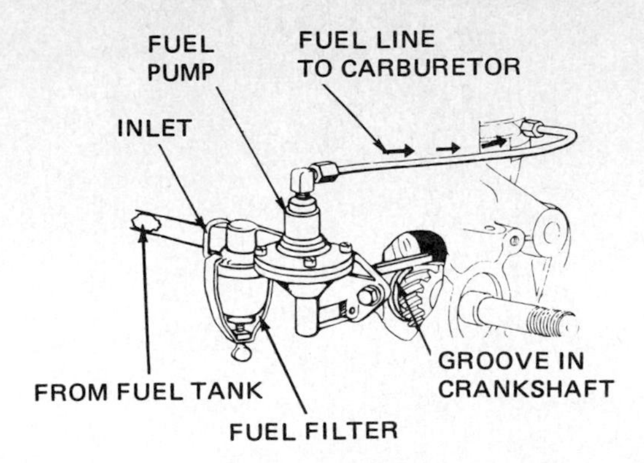

FIG. 7-3 Pressure-feed fuel system for a small engine. The engine has been partly cut away so the position of the pump lever on the groove in the crankshaft eccentric can be seen.

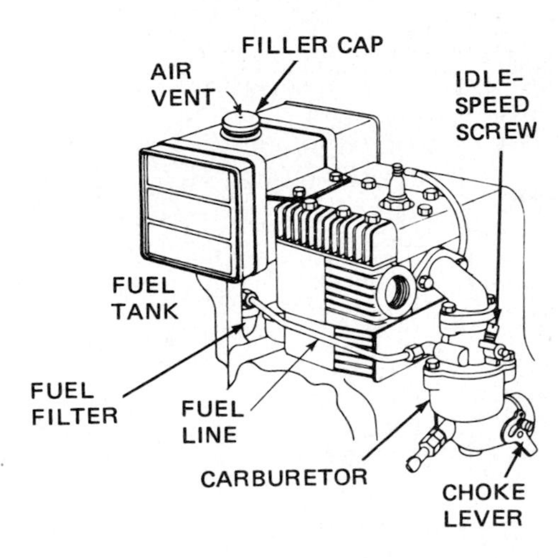

FIG. 7-4 Engine with a gravity-feed fuel system. *(Briggs & Stratton Corporation)*

Some engines have both a strainer in the tank and a filter between the tank and the carburetor.

Regardless of type and location, the fuel filter must be cleaned or replaced at periodic intervals. If this is not done, the filter will restrict fuel flow until the engine loses power and stalls. Fuel-filter service is covered in Chap. 8.

Δ7-5 AIR CLEANER A relatively large volume of air passes through an engine while it is running. The air usually contains dust and grit particles which float around in it. This dirt, if it gets into the engine, can seriously damage and rapidly wear moving engine parts. Dust on bearings can gouge them and scratch the shaft journals rotating in the bearings. This leads to early bearing and shaft failure. Dust on the piston and piston rings can scratch the rings, the piston, and the cylinder wall. This will result in rapid wear of the rings and wall and in loss of engine performance.

Dirt in the air entering the carburetor can clog the very small fuel and air-bleed passages. In addition, the dirt can cause wear and sticking of carburetor parts and linkage. To keep the dust and dirt from entering the engine, an air cleaner or air filter is installed on the carburetor (Fig. 7-1). Then all air entering the engine through the carburetor must first pass through the air cleaner.

The air cleaner also muffles the noise that the intake air makes as it passes through the carburetor and ports. In addition, the air cleaner acts as a flame arrester if the engine backfires through the carburetor. Backfiring may occur if the air-fuel mixture is ignited before the intake valve or port closes. When this happens, there is a momentary flashback of flame through the carburetor. The air cleaner prevents the flame from leaving the carburetor and igniting any gasoline fumes or other flammable material nearby.

The engine shown in Fig. 7-1 uses a metal-mesh air cleaner. The mesh fills the case and is moistened with oil. This traps particles of dirt that enter with the air. When the filter is loaded with dirt, the filter is removed and washed in clean solvent. Then the filter mesh is reoiled and reinstalled.

In addition to metal mesh, other types of small-engine air cleaners are oil-bath, oiled-foam, and dry-element. These are shown in Fig. 7-5. The filter element in the oiled-foam air cleaner (Fig. 7-5*a*) is a polyurethane-foam pad, soaked in oil. The intake air must pass through the foam, which traps the dirt particles on the oily surfaces.

The oil-bath air cleaner (Fig. 7-5*b*) has a reservoir of oil past which the incoming air must flow. The air picks up particles of oil and carries them up into the filter mesh. The oil washes off the dirt particles and drains back into the oil reservoir. On these, when the filter element is cleaned, the oil is changed in the reservoir.

The dry-element air cleaner (Fig. 7-5*c*) has a pleated-paper or fiber filter element through which the air must pass. The element has extremely small pores, or openings, through which the intake air must pass. These tiny holes will not permit dust particles to pass through. Therefore, any dirt and dust is trapped on the surface of the filter element.

Figure 7-5*d* shows a dry-type air cleaner with a *precleaner*. The precleaner is a foam sleeve that fits over the dry element to trap most of the dirt in the intake air. With cleaner air passing through the dry-filter element, its useful life is extended. In many dry-filter air cleaners, the precleaner fits over the dry element with no modification to the air cleaner.

Δ7-6 FLOAT BOWL The purpose of the float bowl is to prevent the delivery of too much gasoline to the carburetor. Without the float system, all the fuel in the fuel tank would run down into the carburetor. The float system is made up of a small bowl; a float of metal, plastic, or cork; and a needle valve that is operated by the float. Figure 7-6 is a simplified drawing of a float system. When gasoline from the fuel tank enters the float bowl, the float is raised. As the float moves upward, it lifts the needle valve into the inlet hole (called the *valve seat*).

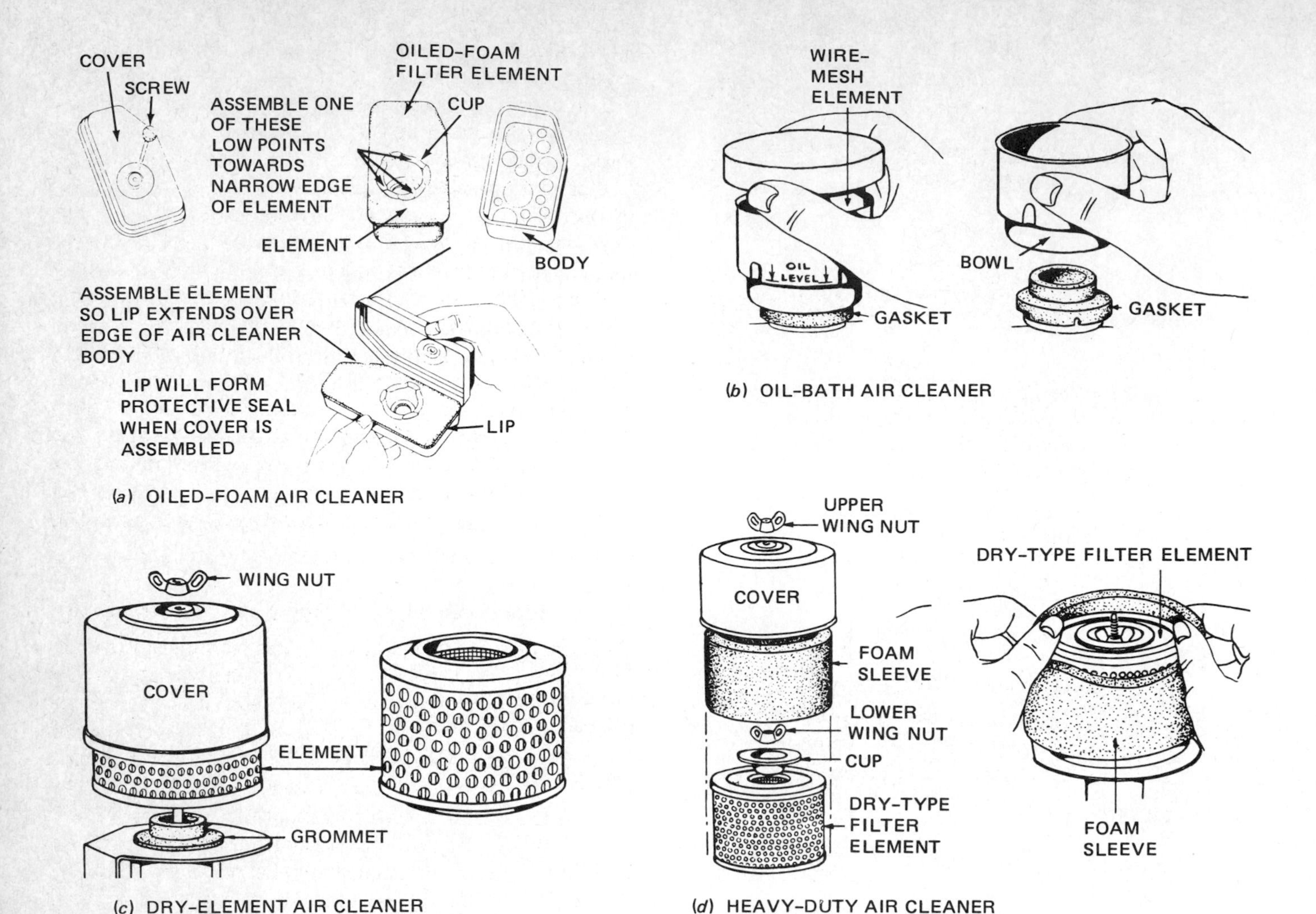

FIG. 7-5 Various types of small-engine air cleaners. *(Briggs & Stratton Corporation)*

When the gasoline is at the proper height in the bowl, the needle valve is pressing tightly against its seat so that no more gasoline can enter. As the carburetor withdraws gasoline to operate the engine, the gasoline level in the float bowl falls. Then the float and needle drop down, and more gasoline can enter. In operation, the needle valve holds a position that allows gasoline to enter at the same rate that the carburetor withdraws it. This keeps the level of gasoline in the float bowl at the same height.

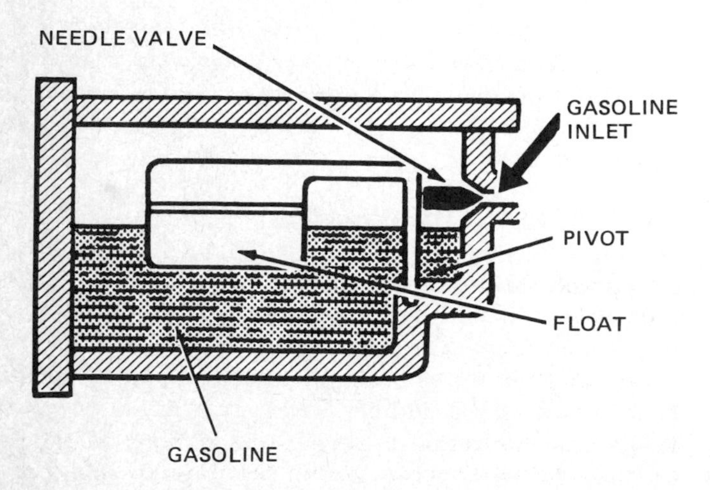

FIG. 7-6 A basic carburetor float system.

Not all carburetors for small engines have a float bowl. For example, a chain saw must be operated in many different positions. A float bowl does not permit this. Therefore, chain saws and certain other small-engine-powered equipment use a type of carburetor that does not have a float bowl.

Δ7-7 THE BASIC CARBURETOR The carburetor has three basic parts besides the float system. These are the air horn, the fuel nozzle, and a throttle valve (Fig. 7-7). The throttle valve is a round plate fastened to a shaft. When the shaft is turned, the throttle valve is tilted more or less in the air horn. The throttle valve can be tilted to allow air to flow through freely or be turned to block the passage of air. This is the basic control of engine speed and power. Turning the throttle valve allows more or less air-fuel mixture to enter the engine so it can produce more or less power.

The air horn has a restriction, or venturi, at the point where the fuel nozzle enters the air stream (Fig. 7-7). The

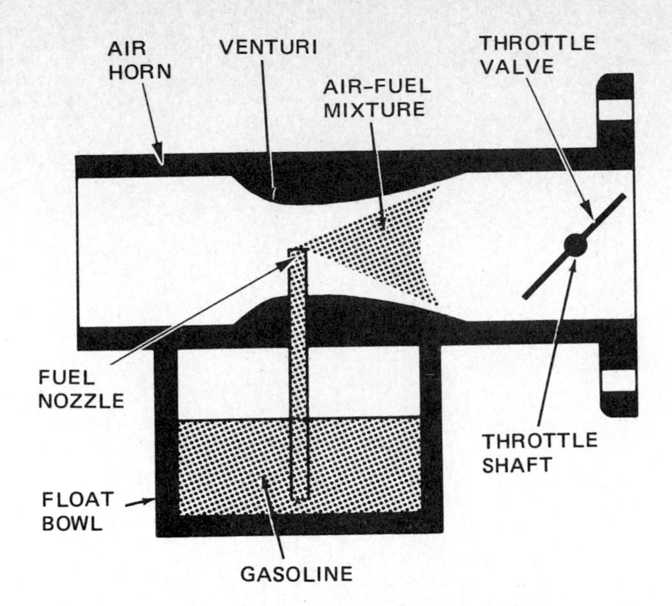

FIG. 7-7 If the air horn is turned to a horizontal position, the carburetor will work just as well.

venturi creates a partial vacuum (low-pressure area) when air is passing through the air horn.

In the carburetor venturi, this vacuum is located around the end of the fuel nozzle. Then air pressure on the fuel in the float bowl pushes fuel up the tube and out the fuel nozzle (Fig. 7-8). The fuel sprays into the passing

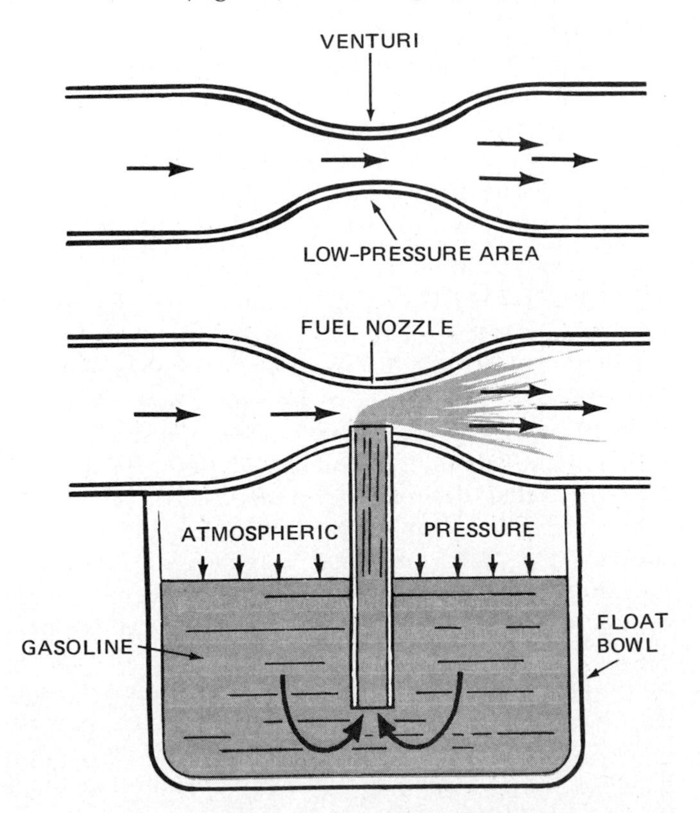

FIG. 7-8 Top, the venturi produces a vacuum, or low-pressure area, when air flows through it. Bottom, a nozzle leading up from the float bowl will pass gasoline upward into the airstream as atmospheric pressure pushes it up toward the low-pressure area. *(Lawn Boy Division of Outboard Marine Corporation)*

air, mixing with it to form the air-fuel mixture that the engine needs to operate.

The more air that passes through the air horn, the higher the vacuum at the venturi and the greater the amount of fuel that feeds from the fuel nozzle. Therefore, the proper proportions of air and fuel are maintained throughout the full range of throttle positions.

When the throttle is partly opened, only a small amount of air flows through and only a small amount of fuel feeds from the fuel nozzle. But when the throttle is wide open, a large amount of air flows through the venturi and a large amount of fuel feeds from the fuel nozzle.

The carburetor shown in Fig. 7-7 is a horizontal carburetor. The air horn lies in the horizontal position. This is the type of carburetor used on many small engines and motocycles. However, the air horn can be turned to a vertical position and the carburetor will work just as well. This type of carburetor is used on many automobile engines.

Δ7-8 ADJUSTING THE AIR-FUEL MIXTURE

On some carburetors, the richness of the air-fuel mixture can be varied by turning an adjustment knob or screw to raise or lower the adjustment needle (Fig. 7-9). If the adjustment needle is lifted away from its seat, the fuel passage around the needle tip is enlarged and more fuel can flow. This means that there will be more fuel and that the air-fuel mixture will be richer. But if the adjustment knob or screw is turned to move the needle tip toward its seat, then the passage is smaller. Less fuel can flow (Fig. 7-10). Now the air-fuel mixture will be leaner.

Δ7-9 CHOKE VALVE

The choke valve is located between the air filter and the carburetor venturi (Fig. 7-11). The purpose of the choke is to help start the engine. During starting, especially when the engine is cold, only part of the gasoline will evaporate to form a combustible mixture. This means that the carburetor nozzle must deliver more gasoline to the air passing through. The choke valve has this job.

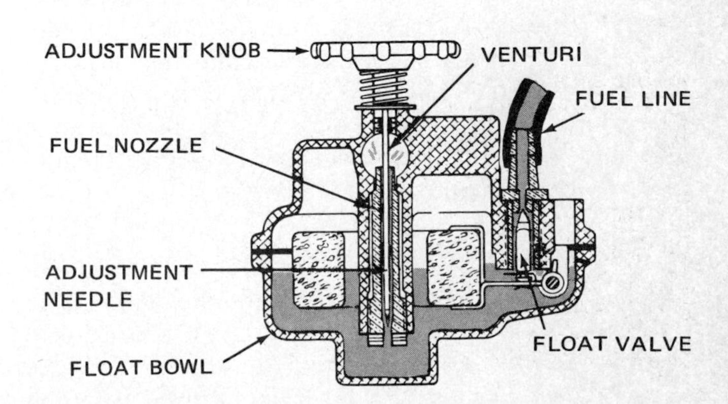

FIG. 7-9 Sectional view of a carburetor for a two-cycle engine used on a lawn mower. *(Lawn Boy Division of Outboard Marine Corporation)*

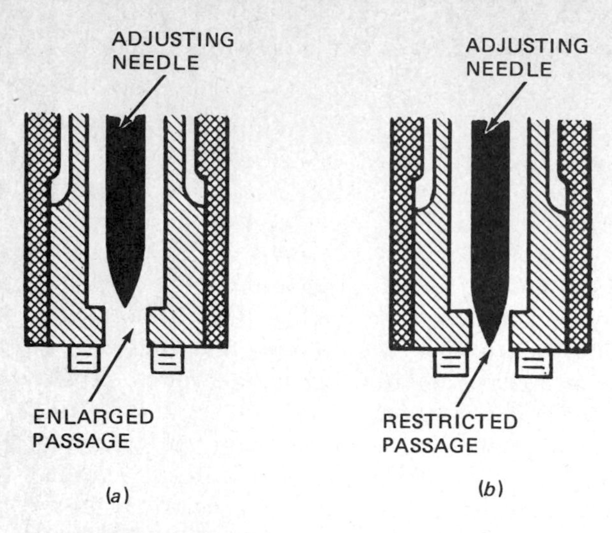

FIG. 7-10 *(a)* Carburetor adjustment needle and seat showing enlarged passage, allowing more fuel to flow. *(b)* Restricted passage, allowing less fuel to flow.

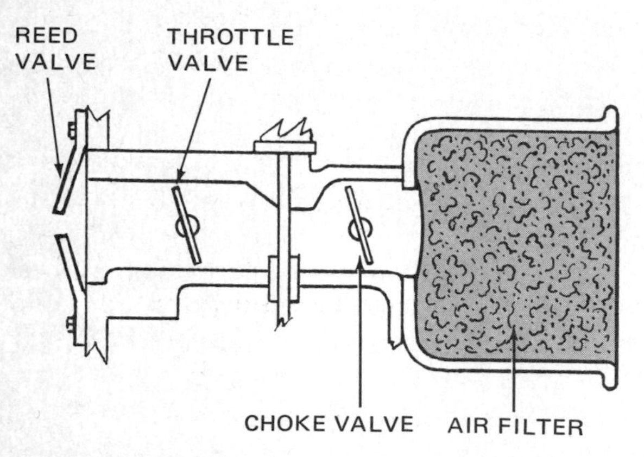

FIG. 7-11 The choke valve is located between the air filter and the venturi.

The choke valve is a round plate, like the throttle valve. When the choke valve is closed, it partially blocks the air horn so that less air can get through. This produces a partial vacuum in the air horn when the engine is cranked and the piston pulls air from the air horn. This partial vacuum, added to the vacuum caused by the venturi, results in a greater vacuum at the fuel nozzle. More fuel feeds from the fuel nozzle, and the resulting mixture is enriched. After the engine starts, the choke valve must be opened to prevent delivery of an excessively rich mixture to the running engine.

Δ7-10 AUTOMATIC CHOKE

Some small-engine carburetors have an automatic choke. Figure 7-12 shows one type. The choke valve is connected by a link to a flexible diaphragm. A spring holds the choke valve closed when the engine is not running. However, when the engine is started, vacuum produced in the engine cylinder during the intake strokes pulls down on the flexible diaphragm. This causes the choke valve to open.

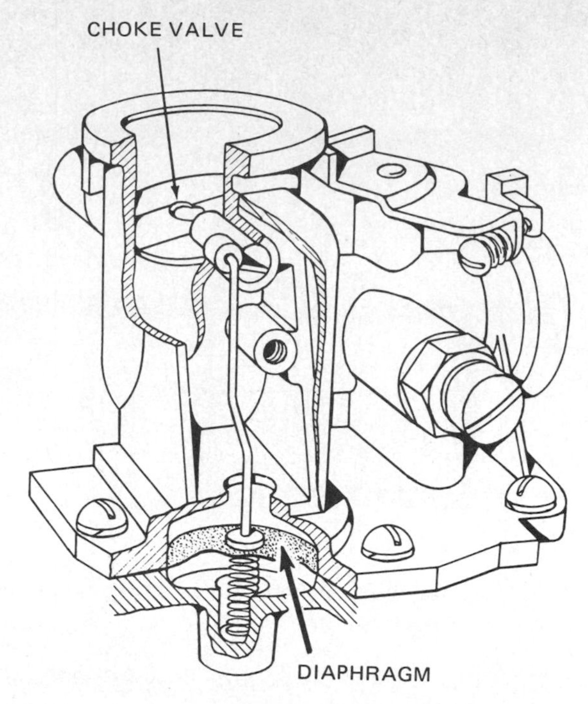

FIG. 7-12 Linkage between the diaphragm and the choke valve in an automatic-choke system. *(Briggs & Stratton Corporation)*

During cranking, with the choke valve closed, the engine is getting an extra-rich air-fuel mixture. This aids in starting (**Δ**7-9).

This automatic choke also causes an enriched mixture whenever the load on the engine causes it to slow down. Then the vacuum decreases. This allows the choke spring to partly close the choke. The result is a richer mixture to improve low-speed, full-load performance.

Δ7-11 PRIMER

Many small-engine carburetors have a *primer* instead of a choke valve. One type of primer (or *tickler*) is shown in Fig. 7-13. When operated, it supplies extra fuel to the carburetor discharge holes. When the primer bulb is pressed down, it shuts off the vent hole and forces air into the float bowl above the fuel. This forces fuel up through the fuel discharge hole.

Another primer which is actually a small pump is shown in Fig. 7-14. The primer has a cup-shaped disk on the bottom of a spring-loaded rod. When the plunger is pushed down, fuel passes around the edge of the disk (Fig. 7-14*b*). When the plunger is released, the spring pulls it back up. This lifts fuel upward into the carburetor (Fig. 7-14*c*). The fuel sprays out into the carburetor air horn. Then, when the engine is cranked, the air passing through the air horn is enriched by the additional fuel for easier starting.

Δ7-12 GRAVITY-FEED SYSTEM

In the gravity-feed fuel system (Figs. 7-1, 7-2, and 7-4), the fuel tank is located above the carburetor. The fuel feeds down to the carburetor float bowl by gravity. Figure 7-15 is a

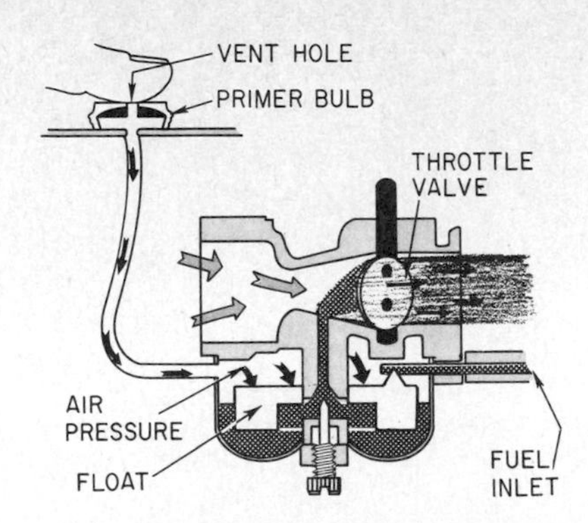

FIG. 7-13 Bulb-type primer. When the primer bulb is pressed down, it pushes air into the carburetor float bowl, causing the float bowl to discharge fuel into the airstream through the carburetor.

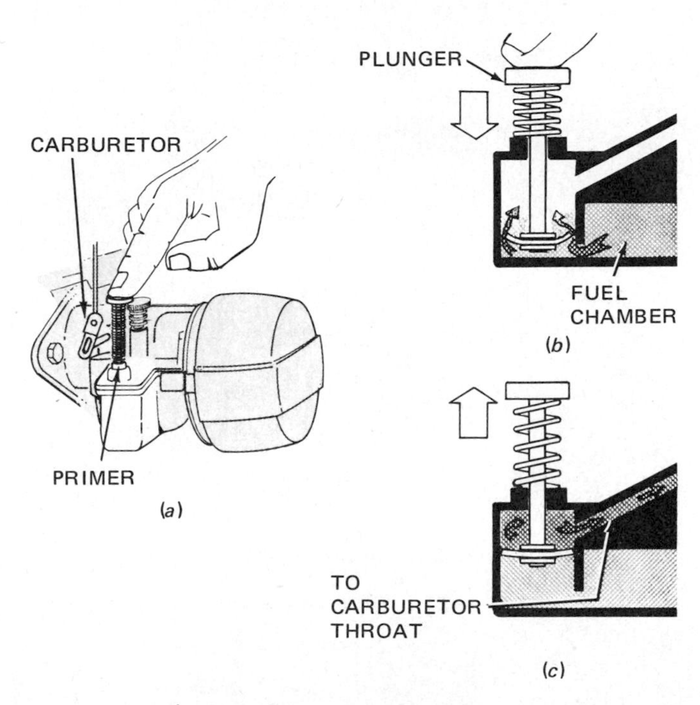

FIG. 7-14 *(a)* Primer for a small engine. *(b)* When the plunger is pushed down, fuel passes by the cup-shaped disk. *(c)* When the plunger is released, the spring raises the disk and lifts fuel up into the carburetor.

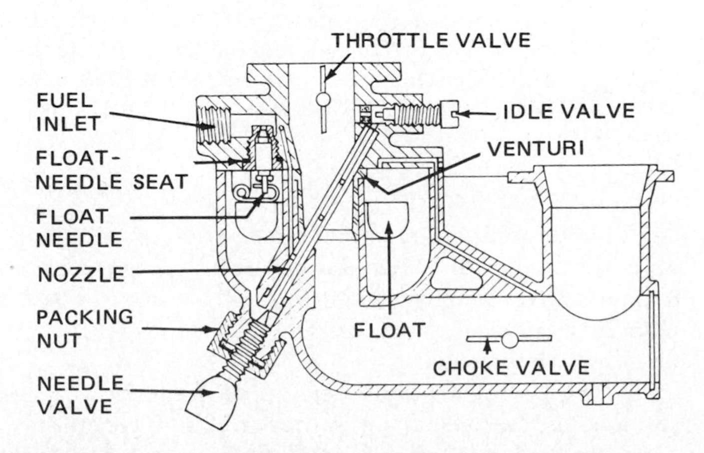

FIG. 7-15 Sectional view of a carburetor for a small engine. *(Briggs & Stratton Corporation)*

sectional view of a carburetor used with this system. The carburetor uses a choke valve (Δ7-9).

Δ7-13 SUCTION-FEED FUEL SYSTEM

In the suction-feed system (also called the suction-lift system), the fuel tank is below the carburetor. Fuel feeds upward directly from the fuel tank to the carburetor discharge holes. No separate float bowl is needed. Figure 7-16 shows how this carburetor works. Figure 7-17 is a cutaway suction-feed carburetor. The fuel pipe from the fuel tank is connected to the two discharge holes below

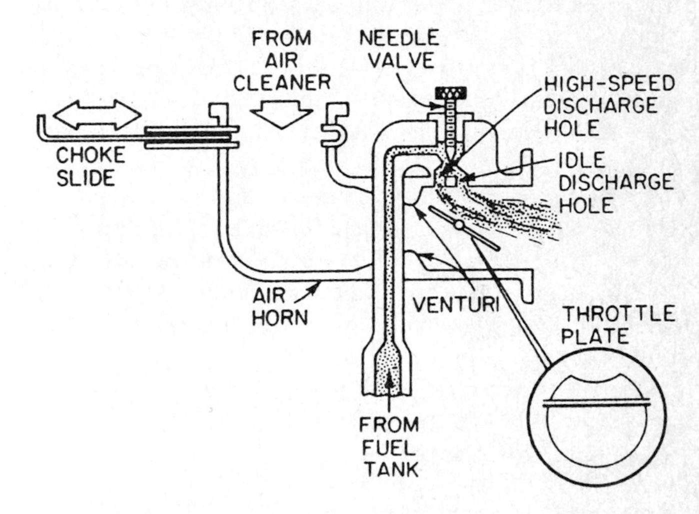

FIG. 7-16 Schematic view of a suction-type carburetor.

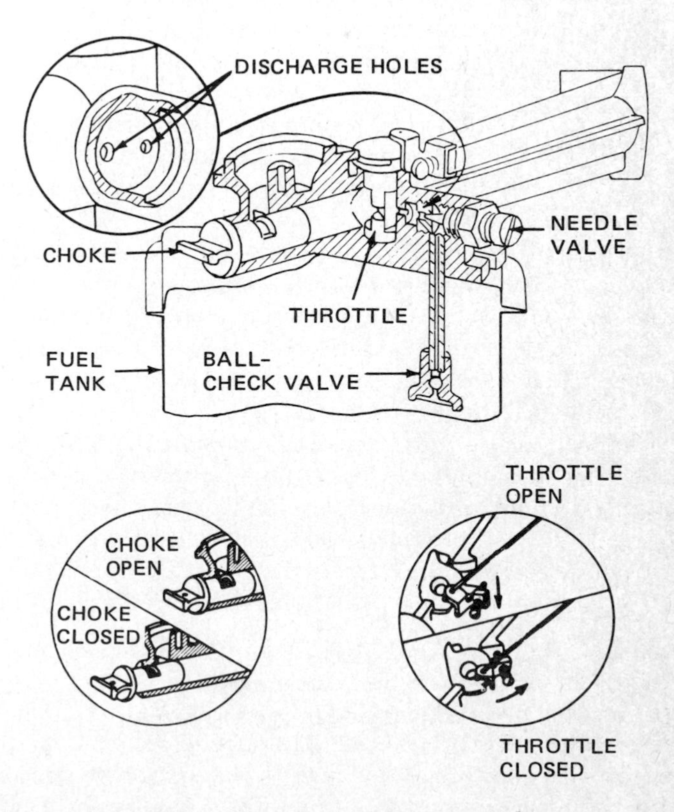

FIG. 7-17 Cutaway view of a suction-type carburetor. Note the round slide-type carburetor choke to the lower left. *(Briggs & Stratton Corporation)*

the point where the needle valve is located. In operation, the partial vacuum produced by the passage of air through the air horn causes atmospheric pressure to push fuel upward from the fuel tank and out through the discharge holes. The throttle plate (in Fig. 7-16) has a notch cut out of it at the top. This is for idle operation.

When the throttle plate is closed, the only passage for the ingoing air is through this notch. The passing air produces a sufficient vacuum close to the idle discharge hole to cause fuel to flow. This ensures that the idle mixture contains sufficient fuel.

Some suction-feed carburetors have a ball check in the fuel pipe (Fig. 7-17). The purpose of the ball-check valve is to prevent fuel in the pipe from flowing back down into the fuel tank when the engine stops. This improves subsequent starting, because the fuel pipe is already filled. Therefore, it begins to feed fuel just as soon as the engine is cranked.

The choke used in the carburetor shown in Figs. 7-16 and 7-17 is the slide type. It can be slid into the air horn to restrict air flow and thereby increase the vacuum at the discharge holes. This produces additional fuel feed during cranking so that an adequately rich mixture is delivered to the engine.

Figure 7-16 shows a suction-type carburetor that has a flat slide-type choke. The carburetor in Fig. 7-17 has a round slide-type choke.

Δ7-14 SUCTION-FEED CARBURETOR WITH SINGLE CONTROL

One model of suction-feed carburetor uses a single control valve. It provides for choking, running, and stopping the engine. A carburetor of this type is shown in Fig. 7-18 with the three positions of the valve indicated. The valve is in the shape of a half cylinder set in a hollow cylinder (Fig. 7-19). The valve can be rotated into the three positions to provide for engine control.

When the control valve is positioned as shown in Fig. 7-19*a*, the air flow through the carburetor is choked off. Therefore, a high vacuum will develop on the intake stroke as the engine is cranked. This high vacuum will cause a heavy flow of fuel from the tank so the engine receives a rich mixture for starting.

After the engine has started, the control valve is turned to the position shown in Fig. 7-19*b*. Now it is up out of the way, and normal engine operation results. Then, when the engine is shut off, the control valve is turned to the position shown in Fig. 7-19*c*. In this position, the fuel pipe from the fuel tank is blocked off. No fuel can reach the carburetor.

Δ7-15 SUCTION-FEED CARBURETOR WITH DIAPHRAGM

The suction-lift, or suction-feed, carburetor described in the previous section works satisfactorily on smaller engines of ½ to 3 horsepower. It does not work well on larger engines. The reason is that the vacuum will not provide sufficient fuel when the tank is nearly empty. It is easier for the vacuum to lift fuel up to the carburetor when the tank is full. But when the fuel tank is nearly empty, the fuel must be raised considerably farther. Therefore, less fuel will be fed into the air passing through the carburetor.

To provide for a more nearly even fuel feed, many suction-feed carburetors for larger engines have an auxiliary fuel tank or reservoir. It is very similar to the float bowl in carburetors previously described.

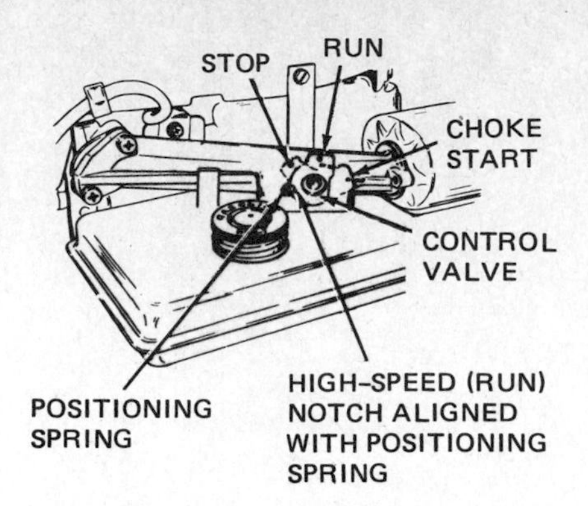

FIG. 7-18 Exterior view of a suction-feed carburetor with a single control for starting, running, and stopping.

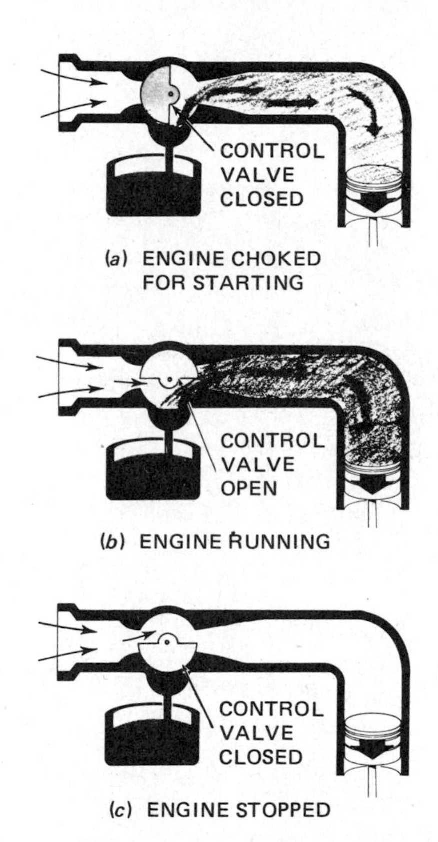

FIG. 7-19 For starting, the control valve is closed, and so the engine is choked. When the engine starts, the control valve is turned to the running position. This leaves the control valve out of the way so that air and fuel can feed into the engine. To stop the engine, the control valve is turned to the closed position, so that the fuel passage from the fuel tank is closed off.

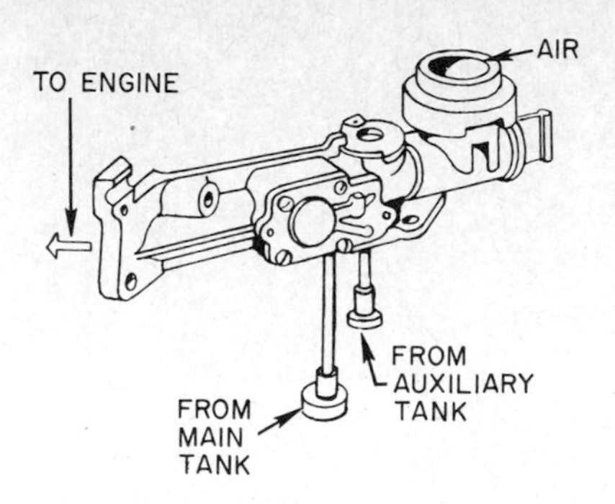

FIG. 7-20 A suction-feed carburetor with a diaphragm. *(Briggs & Stratton Corporation)*

Figure 7-20 shows a carburetor of this type with the fuel tanks removed. There are two separate fuel pipes: one from the main fuel tank and one from the auxiliary tank (Fig. 7-21).

Just above the main fuel tank in Fig. 7-21 is a fuel pump operated by engine vacuum. When the piston is moving down on the intake stroke, the vacuum pulls air from the smaller chamber in which the pump spring is located. The pump diaphragm is pulled upward. This produces a vacuum under the diaphragm.

Atmospheric pressure on the fuel in the main fuel tank then forces fuel up through the fuel pipe, as shown by the arrow in Fig. 7-21. The inlet valve is opened by the vacuum, and at the same time the outlet valve is closed by the vacuum. Fuel flows into the pump chamber.

When the intake stroke is completed, vacuum is lost in the carburetor. The spring can then push the pump diaphragm down. The pressure from the spring closes the inlet valve and opens the outlet valve (Fig. 7-22). As a result, the pressure can push fuel from the pump chamber upward and into the auxiliary tank. This action keeps the auxiliary tank filled. Therefore, fuel flow is unaffected by the level of fuel in the main fuel tank.

The intake stroke of the piston in the four-cycle engine produces the vacuum that causes the pump to work. In two-cycle engines, the vacuum is developed in the crankcase as the piston moves up the cylinder on the compression stroke. With both types of engines, the vacuum operates the pump diaphragm.

Δ7-16 DIAPHRAGM CARBURETOR

The carburetors discussed previously will not work with a chain saw. It and other types of power equipment must run reliably in almost any position. Therefore, a carburetor which depends on either a float bowl or a fuel tank under the carburetor cannot be used.

The diaphragm carburetor (Fig. 7-23) is used on chain saws and other engines that must operate in various positions. It provides a uniform fuel feed to the engine regardless of the position in which it must run.

Figure 7-24 shows the construction of a diaphragm carburetor. When the engine piston moves down during the intake stroke in a four-cycle engine (or when the

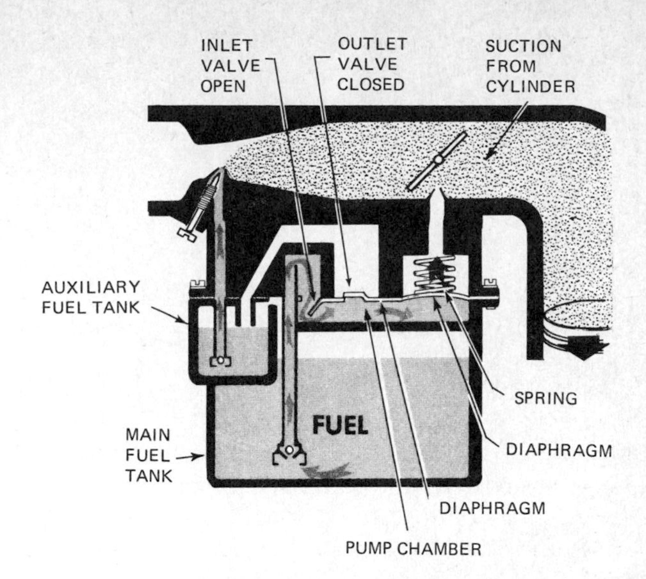

FIG. 7-21 The actions taking place in the diaphragm-type suction-feed carburetor when the piston is moving down.

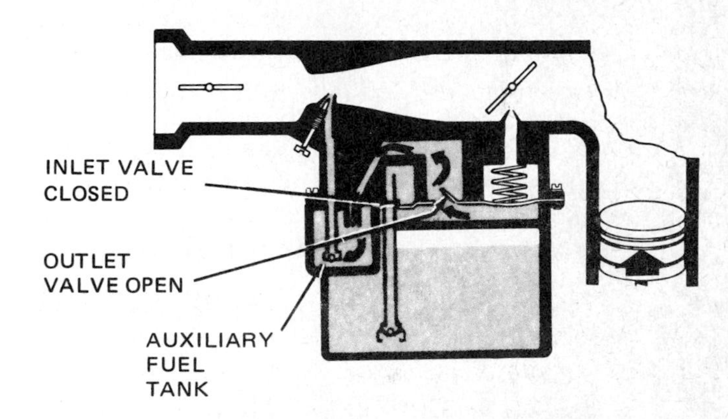

FIG. 7-22 A diaphragm-type suction-feed carburetor, showing the actions when the piston is moving up.

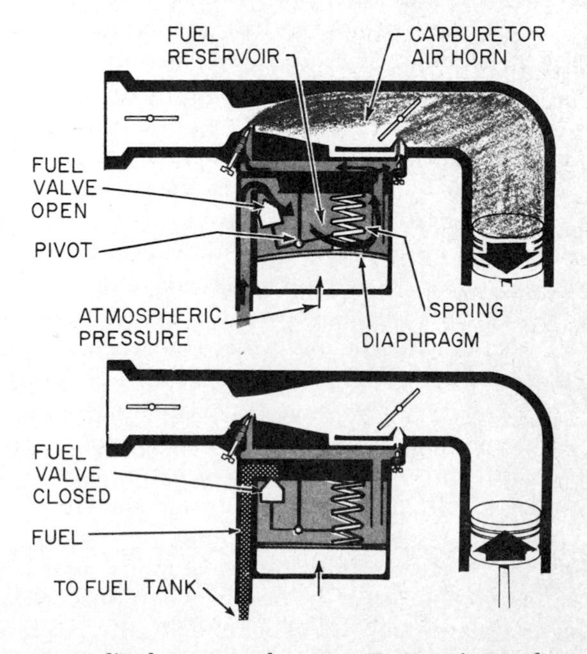

FIG. 7-23 A diaphragm carburetor. Top, actions when piston is moving down on the intake stroke (four-cycle engine). Bottom, actions when piston is moving up.

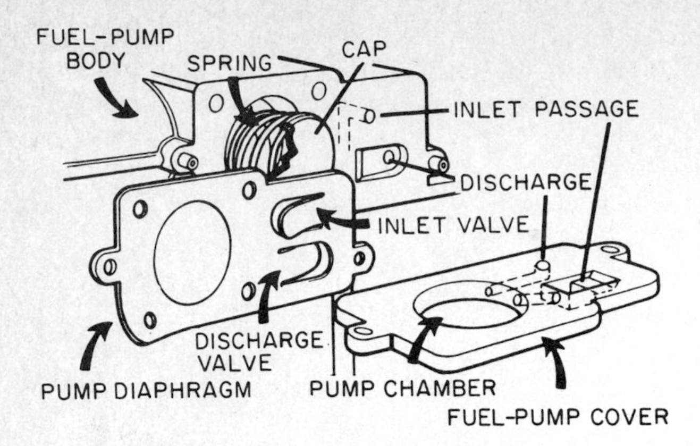

FIG. 7-24 Partially disassembled diaphragm carburetor showing the details of the pump diaphragm. *(Briggs & Stratton Corporation)*

piston moves up in a two-cycle engine), a partial vacuum is created in the carburetor air horn. This causes fuel to discharge from the fuel reservoir into the air horn (Fig. 7-23). The partial vacuum also causes the diaphragm to move up against spring tension.

When the vacuum is lost (when the piston stroke ends), the spring pushes the diaphragm down. This creates a partial vacuum in the fuel reservoir. Atmospheric pressure then pushes fuel from the fuel tank into the reservoir. This fresh fuel replaces the fuel withdrawn during the piston intake stroke.

The actions described above continue as long as the engine runs.

∆7-17 PRESSURE-FEED FUEL SYSTEMS

On engines where the fuel tank must be mounted level with or below the carburetor, a gravity-feed fuel system will not work. A suction-feed fuel system is often not satisfactory. It works only on "small" engines. To obtain proper fuel delivery to the carburetor, a *fuel pump* is used (Fig. 7-3). The system using a fuel pump is called a *pressure-feed system.*

A fuel pump will deliver fuel to the carburetor float bowl, regardless of the relative positions of the fuel tank and float bowl. The fuel pump withdraws fuel from the tank and pumps the fuel into the carburetor float bowl. This always keeps the float bowl filled to the proper level.

Most small-engine fuel pumps are mechanically operated (Fig. 7-3). A cam, or eccentric, on the engine crankshaft forces a pump lever to move up and down. This produces the pumping action in the fuel pump. In larger engines, the pump lever is actuated by an eccentric on the engine camshaft, instead of the crankshaft. However, the principle of operation is the same.

Figure 7-25 shows how the mechanical fuel pump works. When one end of the pump lever is pushed down by the lobe on the cam, the other end lifts the diaphragm against the force of the diaphragm spring. This produces a vacuum in the pump chamber which lifts both the inlet and outlet valves. The upward movement of the outlet

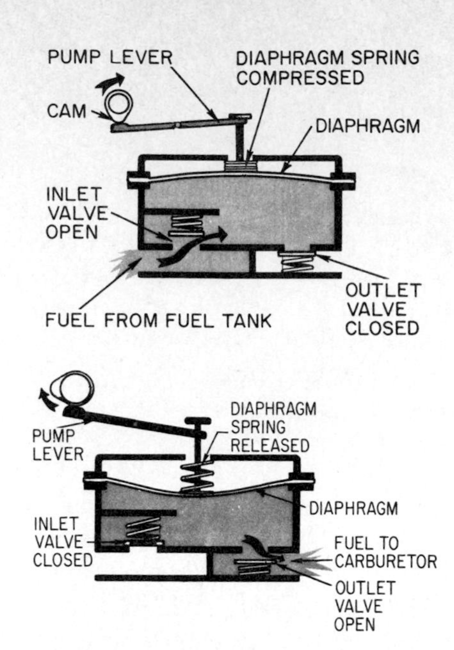

FIG. 7-25 Top, action in the fuel pump when the pump lever is pushed down by the cam lobe on the eccentric. Bottom, action in the fuel pump when the lobe has moved out from under the pump lever.

valve closes it. The upward movement in the inlet valve opens it. Now the vacuum in the pump chamber draws fuel from the fuel tank. This fills the pump chamber with gasoline (Fig. 7-25, top).

When the lobe of the cam moves out from under the pump lever, the diaphragm spring pushes down on the diaphragm. This creates pressure on the fuel in the pump chamber (Fig. 7-25, bottom). The pressure pushes down on both valves. This closes the inlet valve and opens the outlet valve. Now the pressure pushes fuel from the pump chamber into the carburetor float bowl.

As the float bowl fills with gasoline, the float rises. This pushes the needle valve into the seat, shutting off any further delivery of fuel (Fig. 7-6). Since no more fuel can enter the float bowl, the fuel is trapped under pressure in the pump chamber and fuel line. The pressurized fuel holds the diaphragm in its upper position (Fig. 7-25, top). The diaphragm is held up even though the pump lever releases it and spring force is trying to push it down.

When some fuel passes from the float bowl into the engine, the float drops. This opens the float needle valve. Now as fuel flows into the float bowl again, the pressure is released. The diaphragm moves down (Fig. 7-25, bottom), and pumping action resumes.

∆7-18 GOVERNORS FOR SPARK-IGNITION ENGINES

When the load on the engine varies but a steady speed is required, a *governor* prevents engine stalling or overspeeding. The governor controls the opening of the carburetor throttle valve.

When an engine is running at a constant speed with a normal load, any decrease in load allows engine speed to increase. As this happens, the governor causes the throttle valve to close slightly. Now the engine is receiving less air-fuel mixture. The effect is to counter the tendency of the engine to speed up. Therefore engine speed remains constant.

If the load increases, the governor causes the throttle valve to open wider. This allows more air-fuel mixture to enter the engine. As a result, the engine develops more power which handles the load while maintaining the same engine speed.

Two general types of governors are used with small spark-ignition engines. These are the *air-vane governor* (Δ7-19) and the *centrifugal governor* (Δ7-20). The air-vane governor works on the flow of air from vanes on the engine flywheel. The centrifugal governor is operated by flyweights which move in and out according to engine speed.

Δ7-19 AIR-VANE GOVERNOR

The air vane is located under the flywheel shroud close to the flywheel (Fig. 7-26). This places the air vane in the path of the air coming off the flywheel blades. The air vane is connected by linkage to the throttle valve. When the air vane is moved, the throttle valve will open or close. A spring in the linkage tries to pull the throttle into the opened position. Therefore, when the engine is stopped, the throttle valve is open.

When the engine is running, a flow of air from the flywheel blows against the air vane, pushing it toward the right (in Fig. 7-26). As the air vane moves, the linkage tends to close the throttle. The faster the engine rotates, the stronger the air blast from the flywheel grows and the farther the air vane moves. Therefore, the engine cannot overspeed. The air-vane governor will close the throttle sufficiently to prevent it.

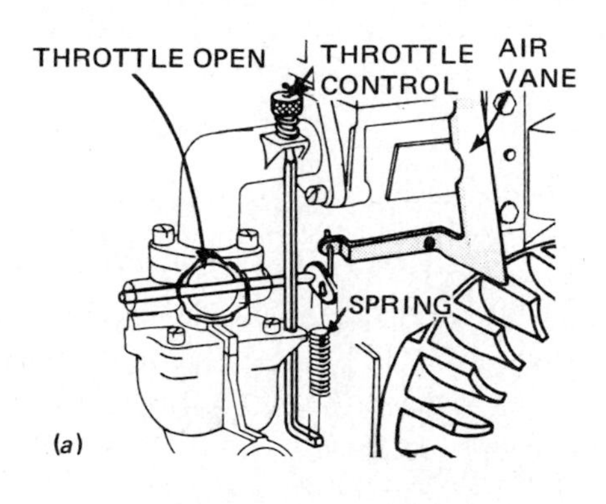

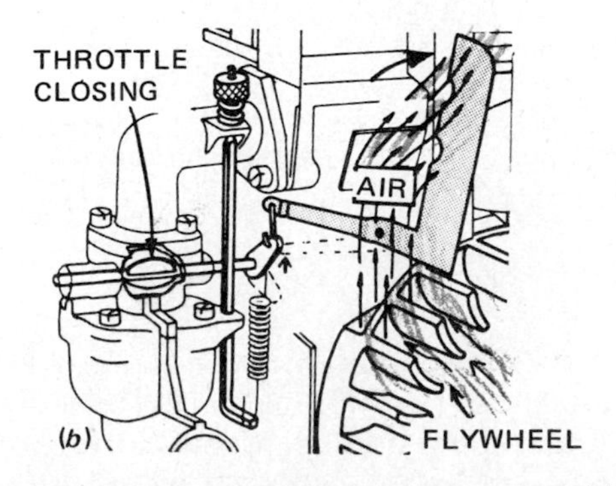

FIG. 7-26 Details of an air-vane-type governor. *(a)* When the engine is not running, the spring holds the throttle open. *(b)* When the engine is running, air from the blades on the rotating flywheel causes the vane to move, thereby partly closing the throttle.

Δ7-20 CENTRIFUGAL GOVERNOR

The centrifugal governor has a lever that is linked to the throttle through an arm and a rod (Fig. 7-27). Figure 7-28 shows how the governor works. As engine speed increases, the two flyweights move out and push against

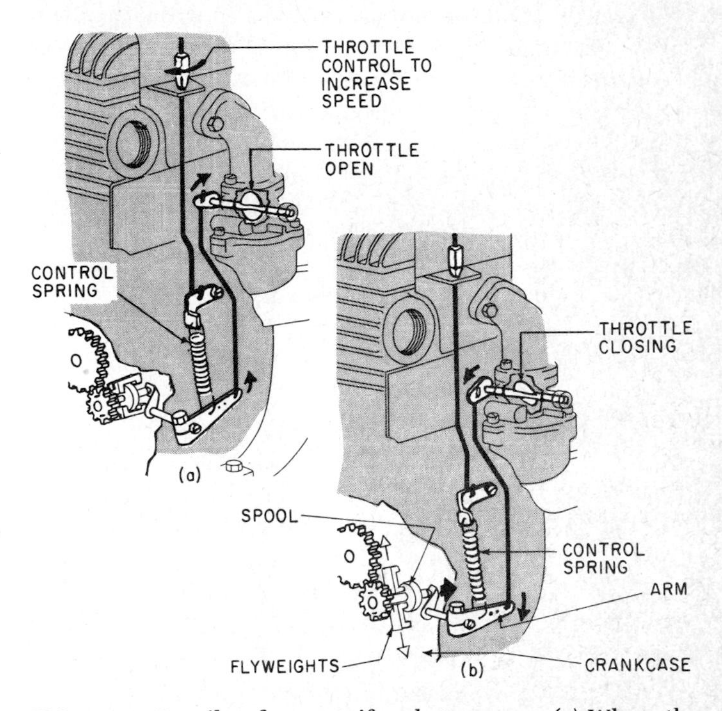

FIG. 7-27 Details of a centrifugal governor. *(a)* When the engine is not running, the spring holds the throttle open and also holds the spool at the "in" position so that the flyweights are retracted. *(b)* When the engine runs, the flyweights move out. This causes the throttle to partly close.

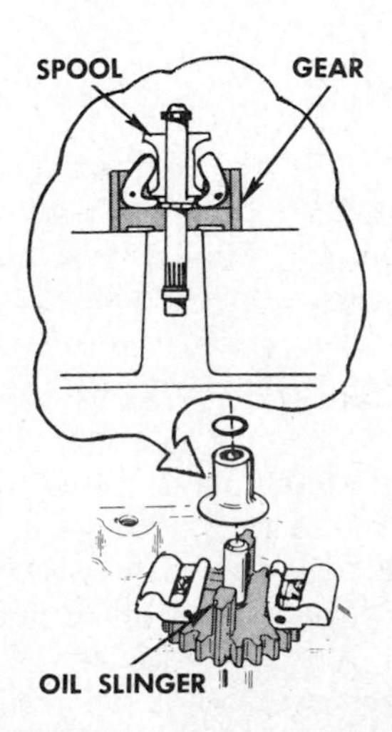

FIG. 7-28 Details of the centrifugal governor using flyweights.

the spool. This motion is carried to the arm so that the rod pulls down on the throttle lever and tends to close the throttle.

When the engine is started, the operator opens the throttle to get the engine speed desired. This puts a certain tension on the control spring. Then the throttle is opened through the pull of the spring on the governor arm. This gives the engine the preset speed that the operator wants. Now, if the engine tends to speed up, the governor arm is turned, as shown to the right in Fig. 7-27. This tends to close the throttle valve to prevent overspeeding.

If the engine tends to slow down because of a heavy load, the flyweights move inward to allow the throttle to open wider. This allows the engine to produce more power. Therefore, the engine maintains the speed at which the operator has set the throttle.

Instead of flyweights in the governor, some governors have flyballs (Fig. 7-29). The flyballs are located under a curved plate. As engine speed increases, the flyballs tend to move outward. This causes them to press against the angled part of the plate, raising the plate. As the plate is raised, it also raises the spool. The spool, which is linked to the throttle, then causes the throttle to partly close. This prevents overspeeding of the engine.

Most small engines should be operated in the high-speed range. At high speed, the engine has the capacity to adjust to a wide range of power demands. If the throttle setting is high enough, the engine is ready to start pulling hard the instant the governor calls for more power. If the throttle setting is too low, there is not enough tension on the control spring to allow the engine to deliver full power quickly.

NOTE The governor should never be adjusted to allow the engine to run above rated speed. Even though the engine might temporarily operate at the excessive speed and temporarily handle excessively heavy loads, it would quickly wear out. Never operate an engine having an air-vane governor with the engine shroud removed. With the shroud off, the air flow from the flywheel is not directed against the air vane. As a result, there is no governor control. The engine could greatly overspeed and self-destruct.

A typical centrifugal governor used on a lawn-mower engine is shown in Fig. 7-30. In this governor, the lower collar is fastened to the crankshaft. The upper collar is attached to the lower by a pair of pivoted links. A spring holds the two collars apart. When the engine runs, the pivoted links move out owing to centrifugal force. This action moves the upper collar down toward the lower, partly compressing the spring. The faster the engine runs, the greater the centrifugal force on the pivoted links and the farther down the upper collar moves by further compressing the spring.

The upper collar is connected by linkage to the carburetor throttle valve. As the upper collar moves up or down, it opens or closes the throttle valve. If the engine slows down, for example, because of heavy loads, then the collar starts to move up. This causes the throttle valve to open and supply additional air-fuel mixture to the engine so it can produce added power. If the engine starts to speed up, the collar moves down to cause the throttle valve to partly close and reduce the amount of air-fuel mixture to the engine. This governor action is shown in Fig. 7-31.

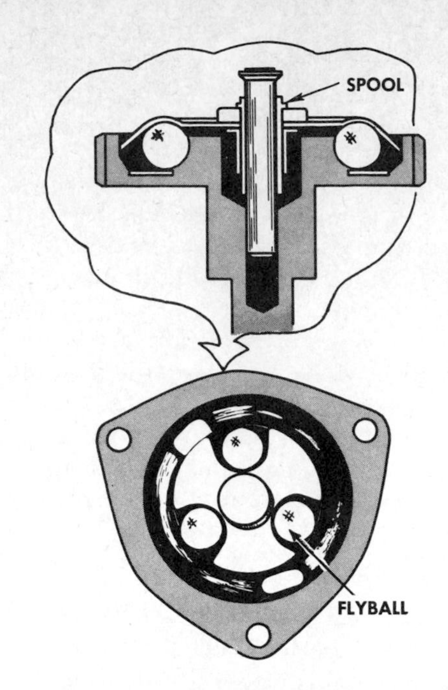

FIG. 7-29 Details of the centrifugal governor using flyballs.

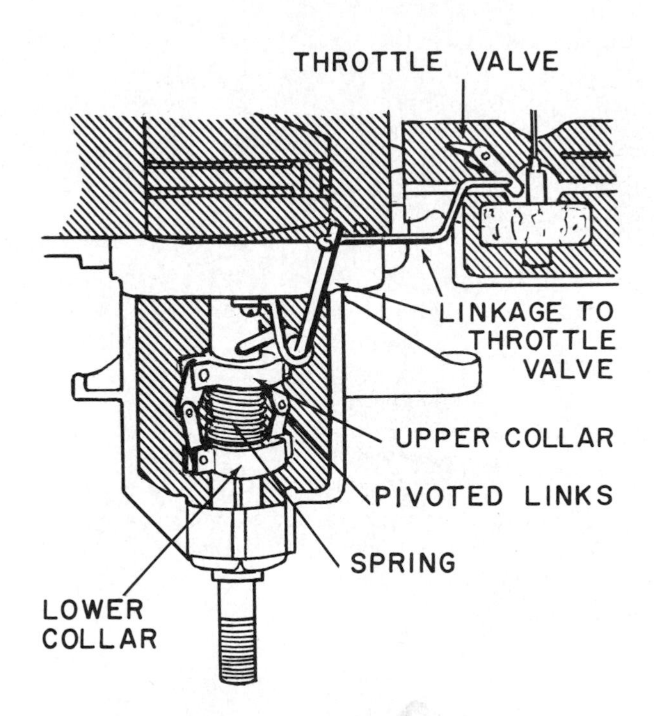

FIG. 7-30 Lower part of the engine, showing the governor and the linkage to the throttle valve. *(Lawn Boy Division of Outboard Marine Corporation)*

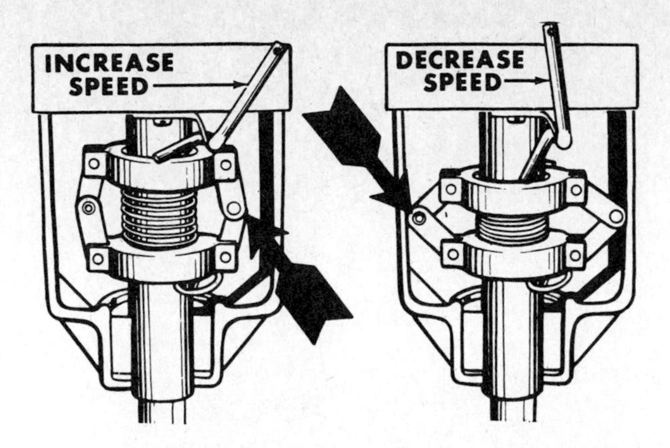

FIG. 7-31 **As the engine speed increases, the pivoted links fly out, causing the control arm to move the throttle valve toward the closed position.** *(Lawn Boy Division of Outboard Marine Corporation)*

There are other types of governors. Some are mounted at the upper end of the crankshaft, above the magneto. However, they all work in a similar manner.

Δ7-21 EXHAUST SYSTEM

After the air-fuel mixture has been burned in the engine cylinder, the burned exhaust gases are forced out of the cylinder and into the exhaust port. From there they pass through the muffler and into the open air. Figure 4-22 shows the muffler in place on a small engine.

The muffler provides a series of passages and chambers through which the exhaust gases must pass before being discharged into the air (Fig. 7-32). These passages and chamber muffle the exhaust noise, quieting the engine.

FIG. 7-32 **A cutaway small-engine muffler, which threads into the exhaust-port opening.**

Δ7-22 LPG FUEL SYSTEM

Liquefied petroleum gas requires a special fuel system. Figure 7-33 shows an LPG fuel system that uses propane. The system consists of a vacuum fuel lock and filter, a vaporizer, a fuel regulator, and a gas carburetor connected by flexible hose. A pulse-balance line is needed between the carburetor and regulator for one- and two-cylinder engines. A vacuum line is needed between the intake manifold and the fuel lock and filter. Some systems have a vacuum or solenoid-operated fuel-cutoff valve, or both. The solenoid-operated valve may be connected to the ignition system.

The fuel regulator is a two-stage regulator. The primary regulator reduces the pressure from the tank so that the liquid fuel begins to turn to gas. LPG is a liquid only under pressure. When the pressure is released, the liquid turns to gas.

After leaving the primary, or high-pressure, regulator, the LPG enters the secondary, or low-pressure, regulator. Here it is reduced to slightly *below* atmospheric pressure. Now when the engine is operating, the partial vacuum in the carburetor causes the gas to flow into the air passing through the carburetor venturi. A combustible air-gas mixture is formed.

The reason for reducing the gas pressure to *below* atmospheric pressure is that when the engine is stopped and there is no venturi vacuum, no gas will flow into the carburetor.

Δ7-23 KEROSENE FUEL SYSTEM

Some small engines operate on kerosene instead of gasoline. These engines are low-compression engines. Standard engines can be changed to use kerosene as follows:

1. A low-compression head must be installed, or two or more cylinder-head gaskets must be installed to lower the compression.

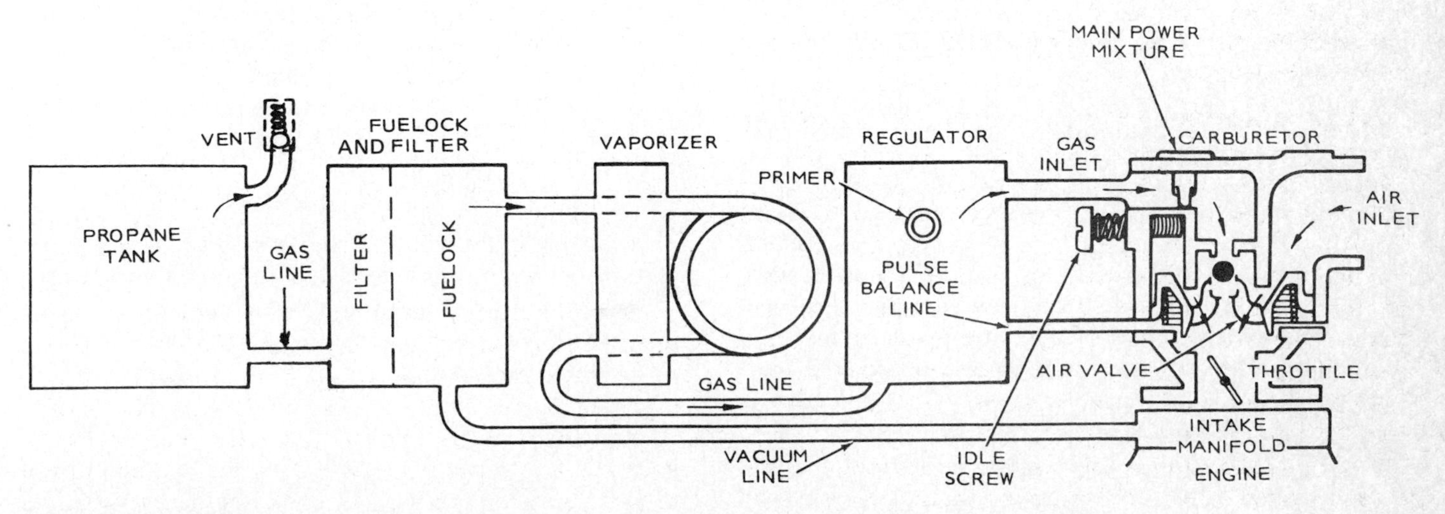

FIG. 7-33 **An LPG fuel system that uses propane for a small air-cooled engine.** *(Onan Corporation)*

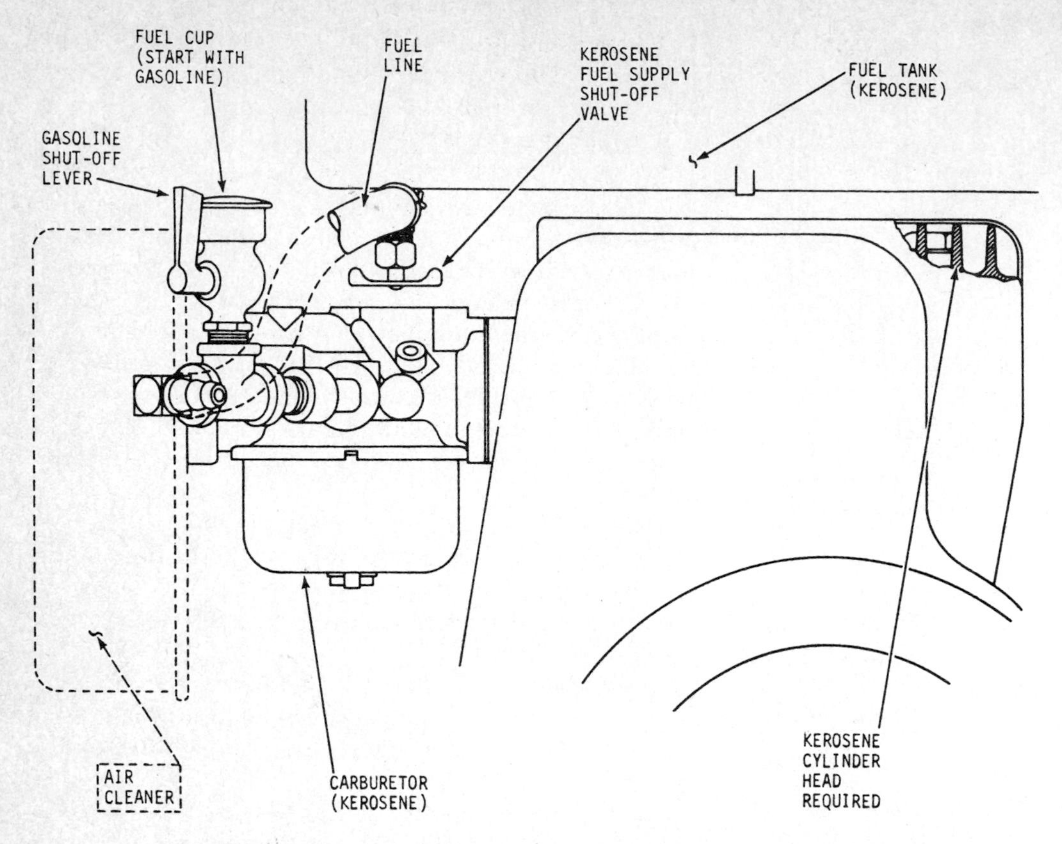

FIG. 7-34 A kerosene fuel system for a small air-cooled engine. To start the engine, the kerosene must be drained from the carburetor and then it must be refilled with gasoline. *(Kohler Company)*

2. A special spark plug must be used.
3. The contact-point gap must be reduced, and the ignition must be retimed.
4. Due to the low volatility of kerosene, a dual fuel system must be used to start cold engines (Fig. 7-34). With this system, gasoline is used to start the engine and warm it up. Then the engine is switched to kerosene. When the engine is shut down, the carburetor must be emptied of kerosene so that the engine can be started on gasoline.

CHAPTER 7: REVIEW QUESTIONS

Select the *one* correct, best, or most probable answer to each question. You can find the answer in the section indicated at the right of each question.

1. Three general types of fuel systems used on small engines are (Δ*7-2*)
   a. injection feed, gravity feed, and pressure feed
   b. viscosity feed, gravity feed, and temperature feed
   c. gravity feed, suction feed, and pressure feed
   d. suction feed, injection feed, and pressure feed
2. The gravity-feed fuel system (Δ*7-2*)
   a. uses a fuel pump
   b. does not require a fuel pump
   c. has the carburetor above the fuel tank
   d. does not require a venturi
3. The purpose of the choke valve is to (Δ*7-9*)
   a. choke off the engine to stop it
   b. enrich the air-fuel mixture for starting
   c. raise the level of the fuel in the float bowl
   d. allow the engine speed to increase
4. The purpose of the primer is to (Δ*7-11*)
   a. supply extra fuel for starting the engine
   b. speed up the engine when more power is needed
   c. operate the fuel pump
   d. help the engine carry a full load
5. In the suction-feed fuel system (Δ*7-13*)
   a. the fuel tank is below the carburetor
   b. no separate float bowl is needed
   c. a partial vacuum causes atmospheric pressure to push fuel up to the discharge holes
   d. all of the above
6. The venturi works because (Δ*7-7*)
   a. air passing through produces a partial vacuum
   b. the vacuum produced causes atmospheric pressure to push fuel up to the fuel nozzle
   c. as air flow increases, so does fuel discharge
   d. all of the above

7. The diaphragm carburetor (**Δ**7-16)
   a. is designed for engines that are operated in various positions
   b. does not require a float bowl
   d. operates from the partial vacuum in the air horn produced by the piston movement
   d. all of the above

8. Two types of governor are (**Δ**7-17)
   a. air vane and pressure
   b. suction feed and centrifugal
   c. air vane and centrifugal
   d. pressure feed and suction feed

9. The pressure-feed fuel system (**Δ**7-17)
   a. uses an eccentric to operate the fuel-pump lever
   b. has an air-vane venturi
   c. does not require a venturi
   d. uses the suction-lift principle

10. The centrifugal governor uses (**Δ**7-20)
   a. either air vanes or fly vanes
   b. either flyweights or flyballs
   c. both air vanes and flyballs
   d. neither air vanes nor flyweights

CHAPTER

8

Fuel-System Service

After studying this chapter, you should be able to:

1. ***Service each type of air cleaner.***
2. ***Service each type of fuel filter.***
3. ***Describe the conditions caused by a clogged crankcase breather.***
4. ***Service and adjust a carburetor.***
5. ***Adjust various types of spark-ignition-engine governors.***

Δ8-1 AIR-CLEANER SERVICE

Recommendations vary on how frequently the air filter should be cleaned or replaced. Under normal operating conditions, follow the engine manufacturer's recommendations. When the engine is operating in dirtier than normal or extremely dusty conditions, service may be required twice daily, or even hourly.

The condition of the air cleaner should be checked every time an engine is brought in for service. In addition, a bent mounting stud should be straightened or replaced. Also, a worn or damaged air-cleaner gasket or mounting gasket should be replaced. This prevents dirt and dust from entering the engine through improper sealing.

Under normal conditions, Briggs & Stratton recommends cleaning the oiled-foam air-cleaner element after every 25 hours of engine operation, or at 3-month intervals. For average home use, this type of air cleaner should last through the entire mowing season without requiring cleaning. However, in extremely dusty conditions the air cleaner should be cleaned every few hours.

Kohler recommends cleaning dry elements after about 50 hours of engine operation. Under good operating conditions, the dry element should be replaced after every 100 to 200 hours of engine operation. However, under extremely dusty or dirty operating conditions, these services should be performed more frequently.

Service procedures for various types of small-engine air cleaners are given below.

Oil-Bath Air Cleaner To remove an oil-bath air cleaner, disconnect the spark-plug wire to make sure the engine will not start. The air cleaner will be secured either by a bail wire, by screw threads on the filter element itself, or by a wing nut. The air cleaner can be removed by taking off the bail wire or wing nut or by unscrewing the filter element (Fig. 8-1). If there is any chance that dirt or dust might fall into the carburetor, cover the opening with a clean cloth or plastic film.

After the air-cleaner parts have been removed, separate them and pour out the old oil. Clean the oil cup,

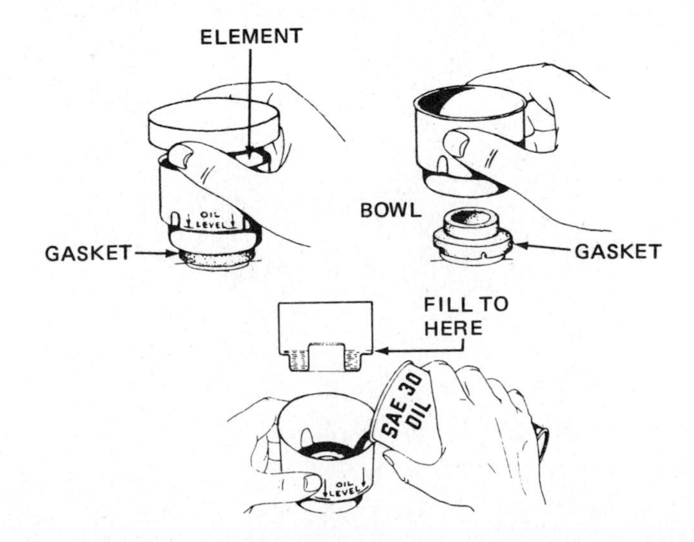

FIG. 8-1 Removing an oil-bath air cleaner for cleaning and replacement of the oil. *(Briggs & Stratton Corporation)*

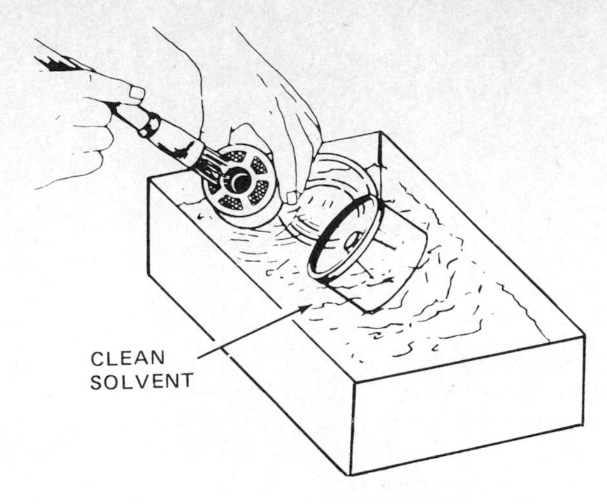

FIG. 8-2 Cleaning an oil-bath air cleaner in solvent.

filter, and cap with solvent and a brush (Fig. 8-2). Be sure to remove any caked dirt in the bottom of the oil cup. Refill the cup to the oil-level mark with the oil specified, which is usually the same oil used in the engine crankcase.

Examine the condition of the air-cleaner gasket. If the gasket is damaged, it should be replaced. Then reinstall the oil cup and filter element.

Oiled-Foam Air Cleaner These are held in place by a snap-on cover, a wing nut, or a screw. One type of oiled-foam air cleaner is shown in Fig. 8-3. The steps in servicing an oiled-foam filter are shown in Fig. 8-4. To service this type of air cleaner, remove it from the engine. Then wash the filter element in detergent and water. Squeeze the foam repeatedly with your hands to get out all the old oil and dirt (Fig. 8-4*a*).

Wrap the foam in a clean cloth, and squeeze the foam until it is dry (Fig. 8-4*b*). Coat the foam with fresh clean engine oil. Finally, squeeze the foam between your hands and let the excess oil drip off. Failure to do this may cause the excess oil to choke the engine, and it may fail to start.

The metal-mesh element is cleaned by washing in solvent, as shown in Fig. 8-2. The element is dried by blow-

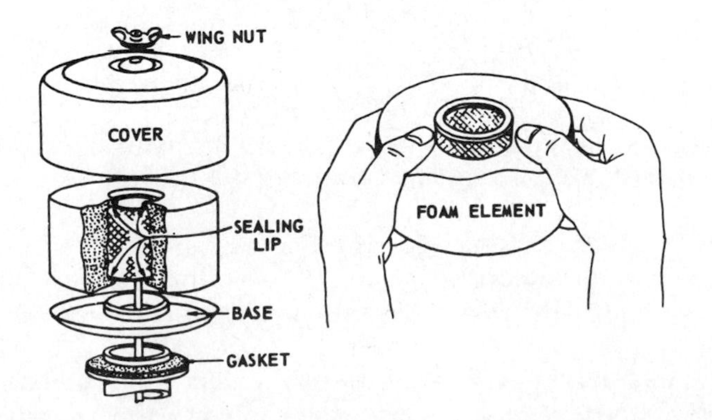

FIG. 8-3 An oiled-foam air cleaner. *(Briggs & Stratton Corporation)*

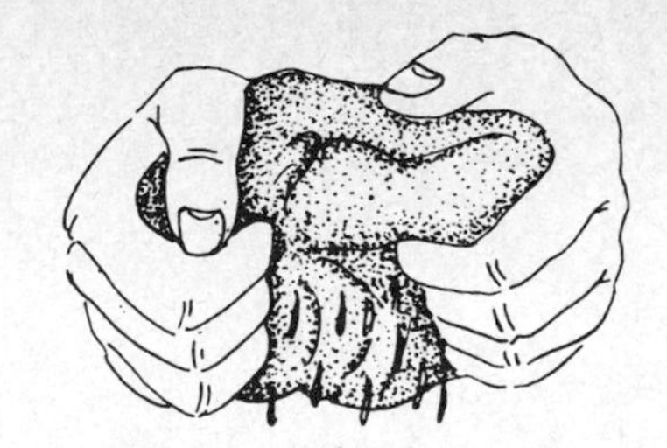

(*a*) SQUEEZE IN DETERGENT AND WATER

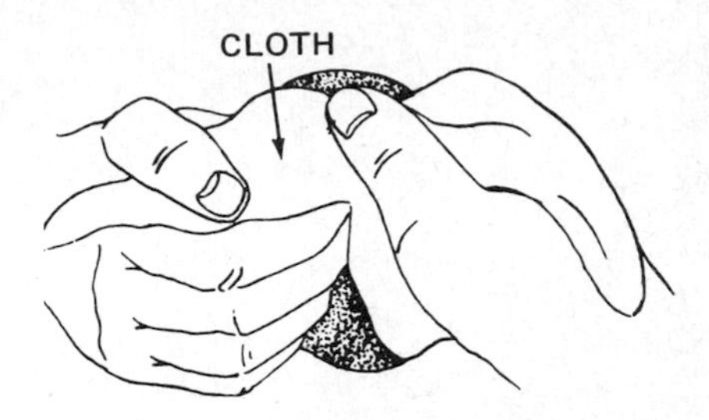

(*b*) DRY IN A CLEAN CLOTH

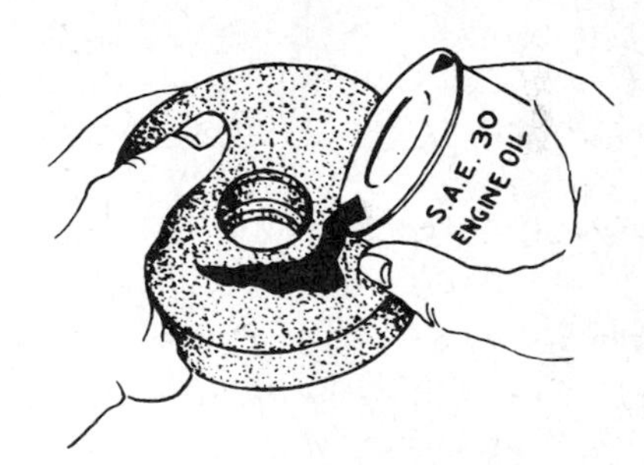

(*c*) COAT WITH CLEAN OIL

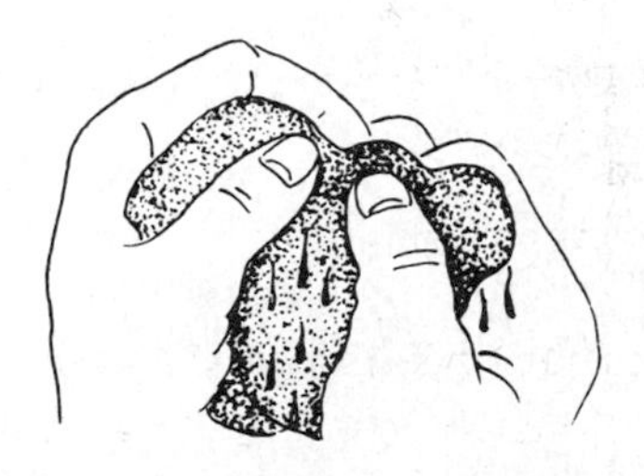

(*d*) LET EXCESS OIL DRIP OFF

FIG. 8-4 Steps in servicing an oiled-foam filter element. *(Briggs & Stratton Corporation)*

ing compressed air through it or swishing it in the air several times (Fig. 8-5). It should be dipped in engine oil to reoil it.

Reinstall the filter-element and air-cleaner assembly. Some polyurethane elements have a coarse filter on the outside and a fine filter on the inside. When installing this type, make sure that the coarse side faces out.

Dry-Element Air Cleaner Remove the filter-element cover if it is a separate part. Some air cleaners of this type are made in one piece without a separate cover. Cover the carburetor intake with a cloth or plastic film. The paper filter element can be cleaned by tapping it lightly

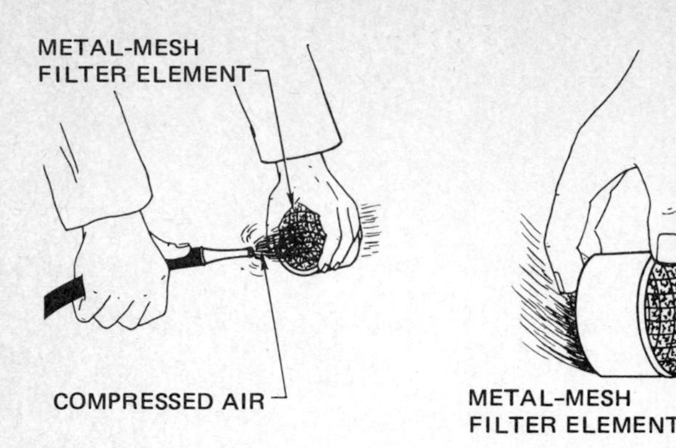

FIG. 8-5 Cleaning and drying the metal-mesh filter element.

on a flat surface (Fig. 8-6). Do not wash a dry filter unless the manufacturer's instructions specify doing so. Wetting the paper will clog the paper pores and ruin the element. If the dust does not drop off easily or if the element is damaged, throw it out and install a new element. Even one pin hole in the paper element can let in enough dust to wear out the engine prematurely.

If the element is fiber or moss, clean it by blowing compressed air through from the inside (Fig. 8-7). Wash the element in soap and water. Do not use an oily solvent. It could clog the element and prevent air from passing through.

Δ8-2 FUEL-FILTER SERVICE

Typical service procedures for various types of fuel filters and fuel strainers (Δ7-4) are given below.

Sediment-Bowl Fuel Filter A fuel filter that has a strainer screen and a detachable sediment bowl, usually made of glass, is shown in Fig. 8-8. To clean the strainer and sediment bowl, close the shutoff valve, if so equipped. This will prevent fuel from flowing out of the fuel tank while the bowl is off. Loosen the thumb nut on the wire bail, or bowl retainer. Then remove the bail, or swing it to one side out of the way. Remove the bowl with a twisting motion. This reduces the chances of breaking the gasket.

Remove the gasket and the strainer screen. The screen is usually held in place by the clamping action between the bowl and gasket. On some, the screen is held in place by a retainer clip. Wash the screen and dry it. Wash out the sediment bowl and make sure it is clean. Open the shutoff valve and allow about a cupful of gasoline to drain out into a container. This will remove any dirt in the line between the tank and filter. If the fuel flows out very slowly, the air vent in the fuel-tank cap may be clogged. Remove it to see if the fuel flows more freely. If it does, then clean the cap vents by soaking the cap in solvent.

Install the filter screen, gasket, and sediment bowl. Use a new gasket if needed. If a new gasket is not available, turn the old gasket over upon reinstallation to get a better seal. Before tightening the thumb nut, open the shutoff valve to fill the sediment bowl. This eliminates any air that might otherwise cause an air lock in the line.

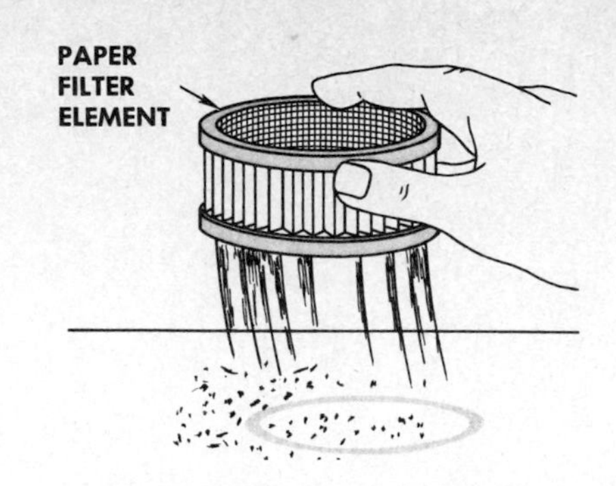

FIG. 8-6 Tapping the paper filter to knock dirt loose.

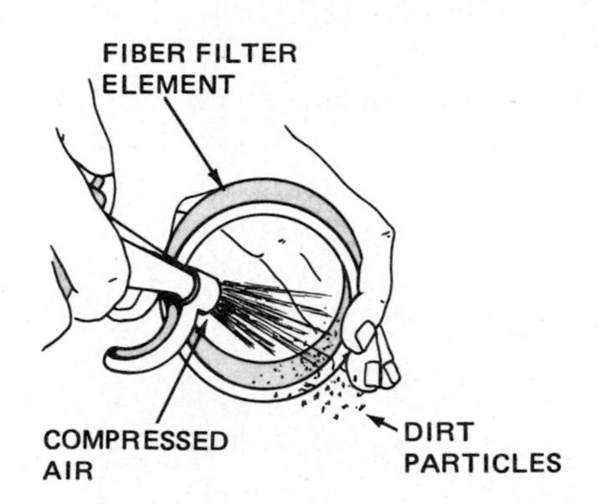

FIG. 8-7 Using compressed air to clean the fiber-type filter element. The compressed air is blown from inside out, or in the direction opposite normal air flow during engine operation.

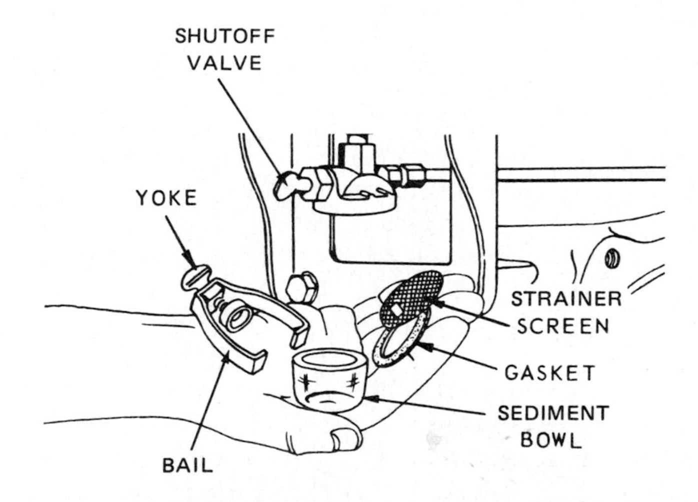

FIG. 8-8 Sediment-bowl type of fuel filter with the bowl removed. *(Briggs & Stratton Corporation)*

In-Line Fuel Filter Figure 8-9 shows an engine that has an in-line fuel filter in the fuel line between the fuel tank and the carburetor. This filter cannot be cleaned. Service is performed by installing a new filter at periodic intervals.

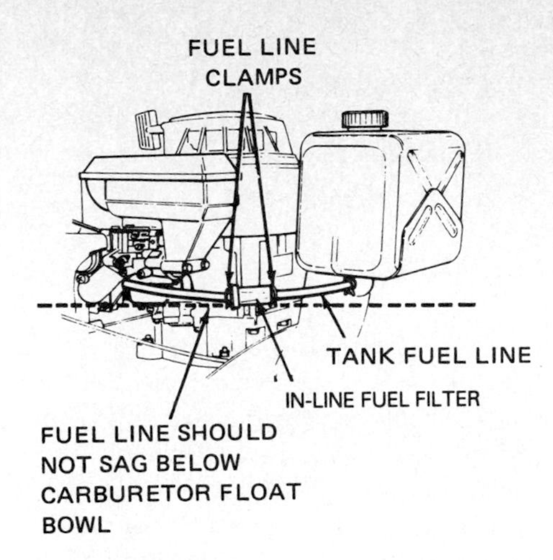

FIG. 8-9 An in-line fuel filter in the fuel line between the fuel tank and the carburetor. *(Tecumseh Products Company)*

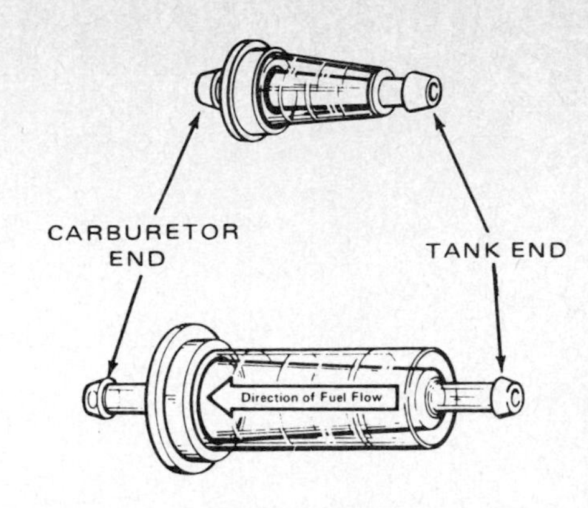

FIG. 8-10 When installing an in-line fuel filter, position the filter so the large end is toward the carburetor. *(Tecumseh Products Company)*

To change the filter, drain the fuel system. Then remove the fuel-line clamps on each side of the filter. Discard the old filter. Position the new filter so that the large end is toward the carburetor (Fig. 8-10). Install the tank fuel line first. The filter must be installed so that the direction of fuel flow is the same as indicated on the filter. Then install the carburetor fuel line.

When the installation is completed, check that no point in the fuel line is lower than the carburetor float bowl (Fig. 8-9). If necessary, remove the clamps and trim the fuel line. After the filter is installed, the fuel line should have only a very slight sag.

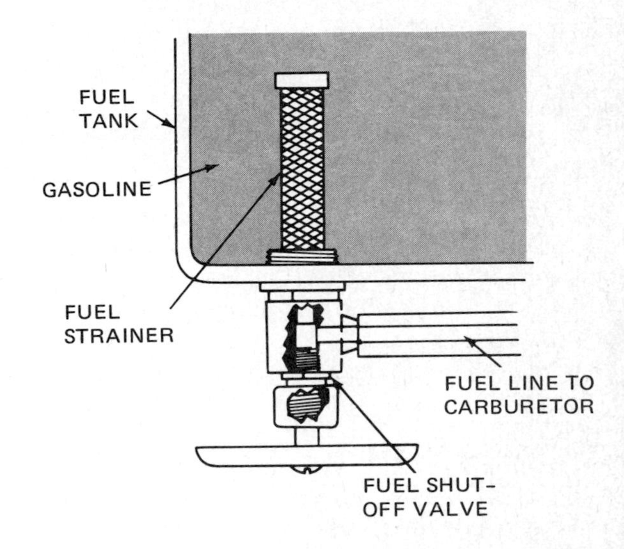

FIG. 8-11 An in-tank fuel strainer, mounted over the end of the pickup tube in the tank.

In-Tank Fuel Strainer The in-tank fuel strainer may or may not be removable from the tank (Fig. 8-11). The removable type may be removed by unscrewing the fuel shutoff valve or tank fitting on which the strainer is mounted. Clean the strainer in solvent. Then dry the strainer with compressed air.

To clean a fuel strainer that is permanently mounted in the tank, remove the tank from the engine. Wash out the tank with solvent several times. This will clean the tank and strainer.

Figure 8-12 shows a weighted strainer that is attached to the end of a flexible hose. The hose is inserted inside the fuel tank. This type of fuel strainer is used on engines that must operate in any position, such as chain saws. Regardless of the position of the fuel tank, the weighted strainer will fall to the low part of the tank which contains the fuel.

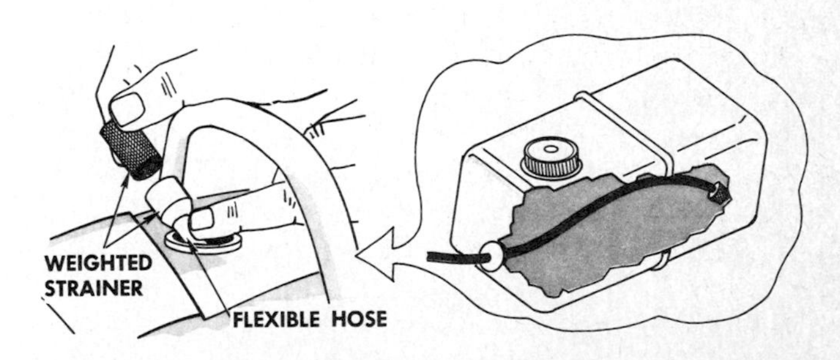

FIG. 8-12 A weighted strainer attached to the flexible hose inside the tank. Left, the flexible hose partly pulled from the fuel tank so the strainer can be removed for cleaning. Right, the tank partly cutaway to show the position of the hose and strainer inside the tank.

To clean a weighted strainer (Fig. 8-12), fish the strainer out of the tank with a bent wire. Remove the strainer from the end of the hose. Clean the strainer in solvent. Then dry it with compressed air.

Δ8-3 CRANKCASE-BREATHER SERVICE

Oil leakage through an oil seal is a good indication of a clogged crankcase breather (Δ6-20). If the crankcase breather becomes clogged, excessive pressure will build up in the crankcase. This will cause oil leaks and may cause the oil seals to rupture.

Figure 8-13 shows a floating-disk type of crankcase breather for a four-cycle engine. The filter element is either metal mesh, fiber, or polyurethane. To clean the

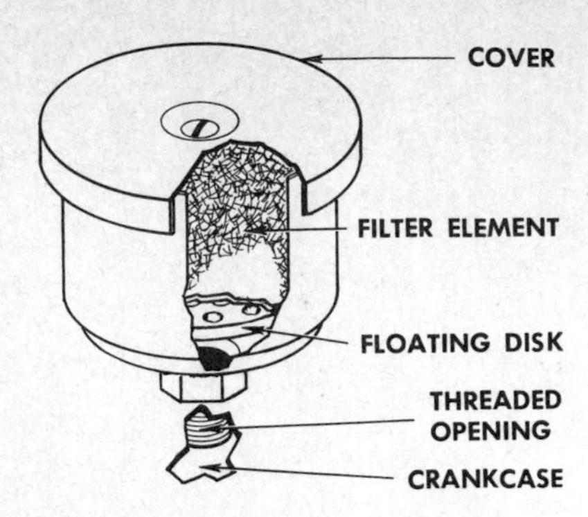

FIG. 8-13 Floating-disk type of crankcase breather for a small four-cycle engine.

filter, remove the nut, screw, or other fastener which holds it in position. Then clean the element in solvent and dry it. Reassemble the parts in their original positions.

Another type of crankcase breather is shown disassembled in Fig. 6-26. It is serviced in the same way.

Δ8-4 FUEL-TANK SERVICE Fuel tanks (Fig. 8-14) are made in a variety of sizes and shapes. They require very little service. The tanks all have a vent to admit air when fuel is taken out. Generally, the vent is in the fuel cap. Sometimes the vent will get plugged. This prevents air from entering, and so fuel cannot flow out. The result is that the engine starves for fuel and stops running.

Some fuel tanks have a cap-and-gauge combination. The fuel gauge is made of a float on a twisted blade fastened to the indicating needle. The float moves down as the fuel tank loses fuel. This twists the blade so that the indicating needle moves to indicate the lowered fuel level in the tank.

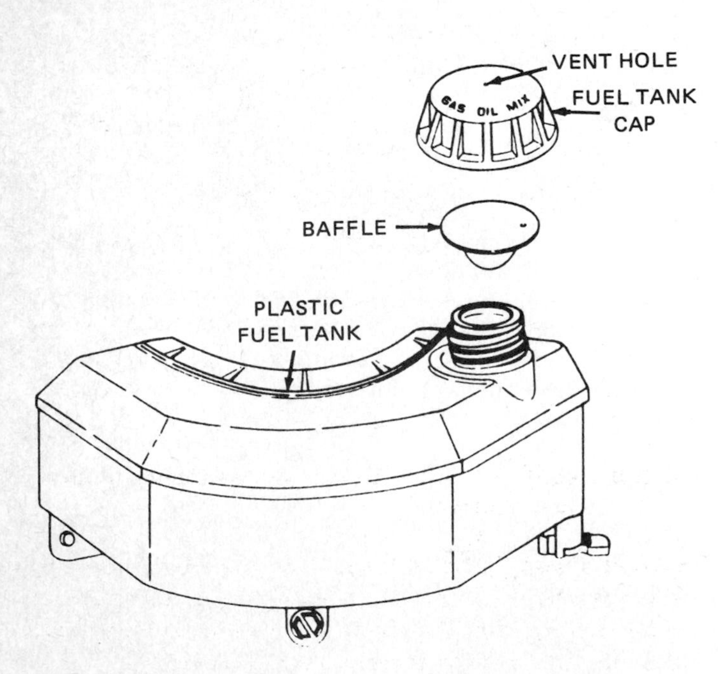

FIG. 8-14 Plastic fuel tank and cap with a vent hole in it for a two-cycle engine. *(ATW)*

If the fuel tank is damaged in any way, it should be replaced with a new tank. The tank often has a bracket which attaches to the engine by screws. Disconnecting the fuel line and removing the screws permits removal of the tank.

Δ8-5 FUEL-PUMP SERVICE Most fuel pumps (Δ7-17) are serviced by complete replacement. They are relatively cheap. It may cost more in labor to repair an old pump than to buy a new one. You can check a fuel pump to see if it works by disconnecting the spark-plug wire and the fuel line at the carburetor. Then crank the engine while holding a small container under the fuel line to catch any fuel. If fuel flows out strongly and in regular squirts, the fuel pump is working properly. If fuel flow is weak or erratic, something is wrong with the fuel pump and it should be replaced.

To remove the old pump, disconnect the fuel lines and take out the screws holding the fuel pump on the engine. Lift the fuel pump off, observing the position of the rocker arm (above or below the eccentric). Install the new pump, making sure the rocker arm goes on the correct side of the eccentric. Then attach the fuel lines and tighten the attaching screws. Figure 7-3 shows the installation of a fuel pump on an engine. The pump lever fits into an eccentric groove on the crankshaft. The part of the lever that rides in the groove should be greased when the pump is installed.

Some manufacturers supply fuel-pump repair information and repair kits. As an example, Fig. 8-15 shows a

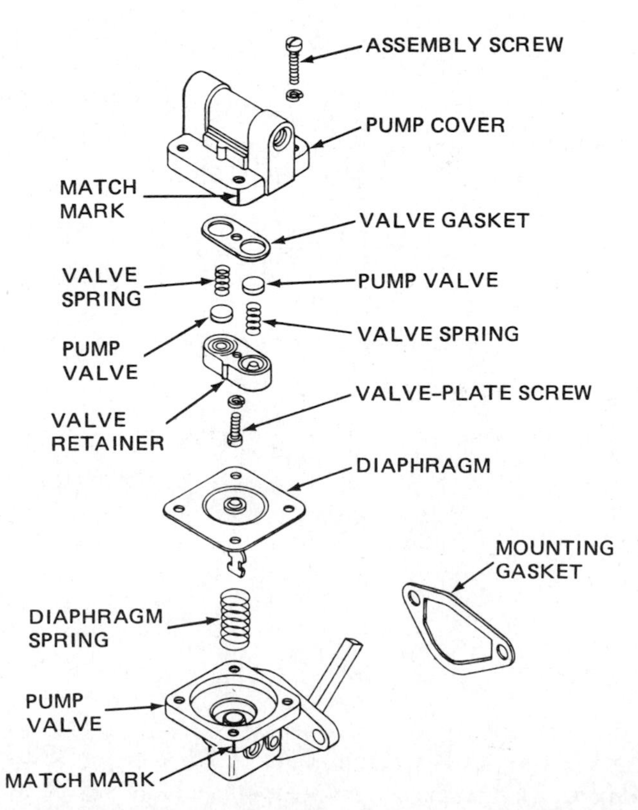

FIG. 8-15 A disassembled fuel pump for a small engine. *(Kohler Company)*

disassembled fuel pump. Before disassembling this fuel pump, mark the pump cover and pump body with a file as shown in Fig. 8-15. Then on reassembly, you can match the marks and not reverse the cover as it goes on the body.

Δ8-6 CARBURETOR SERVICE

The carburetor has the job of mixing air and gasoline vapor in the proper ratio to provide good engine operation. If the carburetor is properly adjusted to give this correct air-fuel ratio, it is not likely to go too far out of adjustment in normal operation. However, screws can loosen and throw the adjustment off. In addition, fuel lines and jets in the carburetor can clog. This can mean a partial disassembly of the carburetor for cleaning, which then means a carburetor adjustment. Carburetor servicing is divided into two parts: (1) adjustments and (2) removal and rebuilding.

To adjust the carburetor correctly, there are certain preliminary steps you should take.

1. Fill the fuel tank full except on engines with suction-feed carburetors, which you should fill only half full. Then, when you adjust the suction-feed carburetor, you will be working with an average air-fuel ratio. If you started with a full fuel tank on the suction-feed carburetor, you would adjust correctly for a full tank. Then, as the tank emptied, the mixture would tend to lean out and might become too lean when the tank is nearly empty. With a nearly empty tank, the fuel must be lifted farther. Less fuel would flow into the passing airstream in the carburetor.
2. Be sure the throttle and governor linkages are free and move easily.
3. On four-cycle engines, check the oil level in the crankcase. Add oil, if necessary.
4. Clean the air cleaner (Δ8-1) and the fuel strainer or filter (Δ8-2).
5. Make sure the fuel-tank cap vent is open. If the vent is clogged, it will prevent normal flow from the tank to the carburetor.
6. Check the ignition and spark plugs.

Δ8-7 CARBURETOR ADJUSTMENTS

A great variety of carburetors have been used on small engines. The carburetor may have three adjustments (Fig. 8-16): idle speed, idle mixture, and high-speed load mixture. Sometimes the idle-mixture screw and the high-speed load screw are difficult to identify. Two examples are shown in Fig. 8-17. The idle-mixture screw is usually closest to the engine. However, this is not always true.

To identify the screws, start the engine and operate it at idle speed. Then turn the screw you think might be the idle-mixture screw clockwise, or in toward the closed position. If the engine slows down or stops, you know you have found the idle-mixture screw. If the engine speed changes little or not at all, increase engine speed to about three-fourths throttle. Now if you get a difference in speed as you turn the screw one way or the other, you have located the high-speed screw.

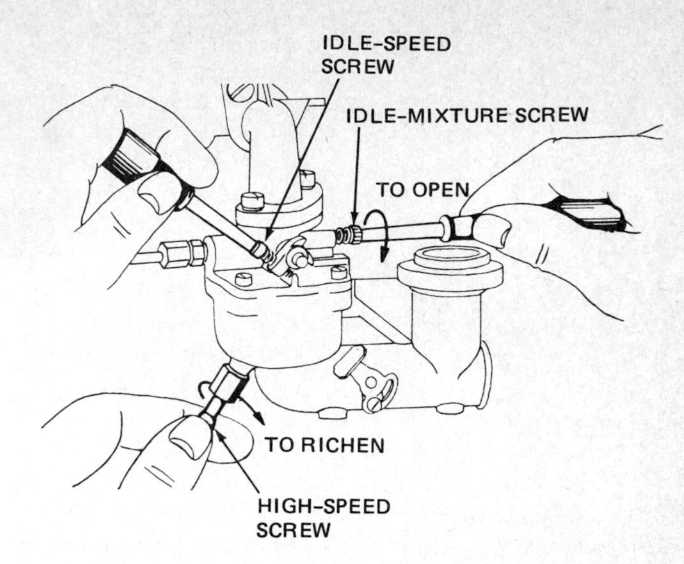

FIG. 8-16 The three adjustment screws on a typical small-engine carburetor. *(Briggs & Stratton Corporation)*

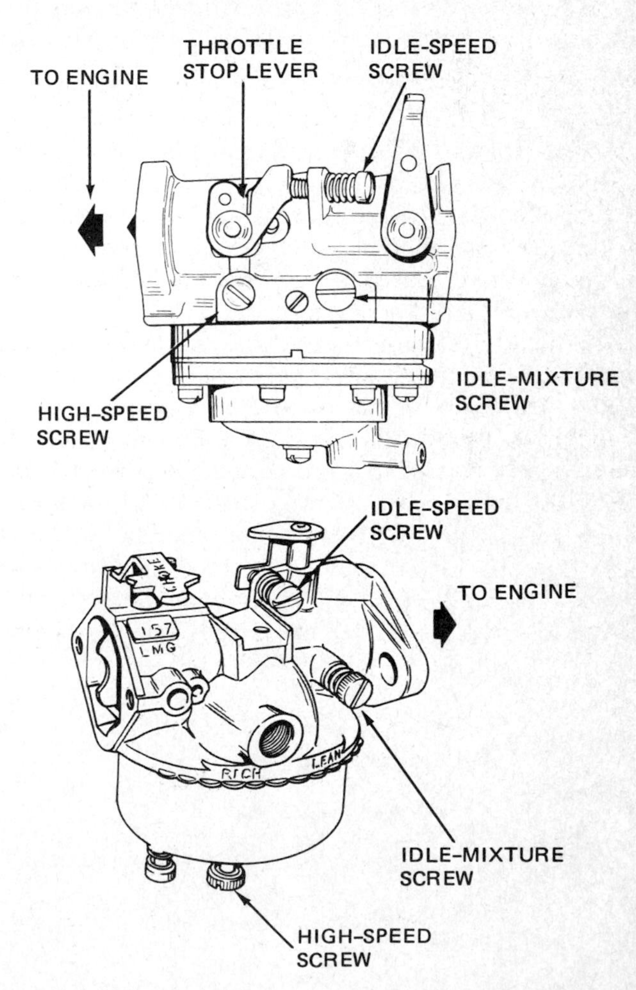

FIG. 8-17 Locations of adjustment screws on two types of small-engine carburetors.

Many carburetors do not have an idle-mixture adjustment. For these, the idle mixture is preset during manufacture and is determined by the size of the discharge port in the carburetor.

To make the initial adjustments of a carburetor, turn the adjustment screws in until the needles bottom. Next, back them off about one turn. This gives an approximate adjustment that should enable you to start the engine and run it until it warms up. Then the final adjustments can be made. These are described later.

Never tighten the adjusting screws more than finger tight. Excessive tightening can cause the needle to jam down into the seat so tightly that both the needle and seat are damaged.

If adjustment cannot be made to give good engine operation, then the carburetor should be either replaced or rebuilt. Carburetor repair kits are available which contain all the necessary new parts. These kits also include the instructions and specifications for rebuilding the carburetor.

Always install new gaskets when repairing a carburetor. The old gaskets may be hardened and may not provide a good seal. Leakage of fuel or air can occur when old gaskets are reused.

Δ8-8 CHOKE ADJUSTMENTS Most chokes in carburetors for small engines are directly controlled by manually operating them. Other chokes are controlled by linkage that connects a remote-control lever with the choke lever (Fig. 8-18). This type of choke occasionally may get out of adjustment.

To adjust the choke, remove the air cleaner so you can see the action of the choke valve. Move the control lever to the choke position (Fig. 8-18). Now look at the choke valve in the carburetor. The choke valve should be closed. If it is not, adjust the control linkage or cable. Loosen the adjusting screw and place the control lever in the choke position. Next, check that the choke valve is fully closed. Tighten the adjusting screw. Now move the control lever to the fully open position. Make sure that the choke valve also moves to the fully open position.

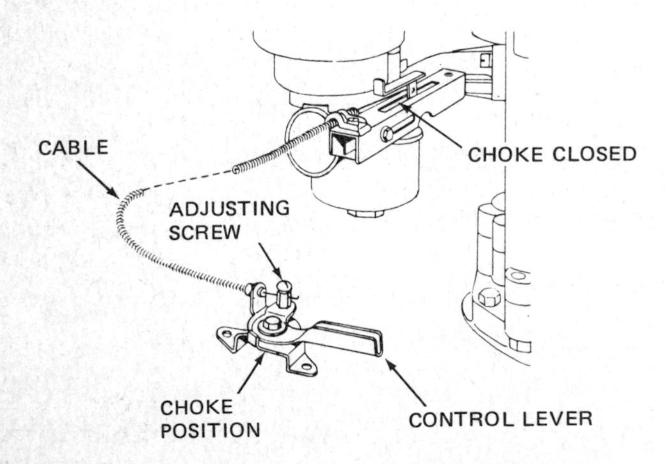

FIG. 8-18 Adjusting screw on the linkage for a manually operated choke. *(Briggs & Stratton Corporation)*

To adjust an automatic choke that uses an electric solenoid to control the choke valve, refer to the engine manufacturer's service manual for the adjustment procedure. The thermostatic-spring type of automatic choke can be checked by noting the choke position with the engine cold (it should be partly to fully closed) and then with the engine hot (it should be wide open). If the choke does not work this way, then the thermostat is faulty or adjustment is required. If the engine uses an electric starter and solenoid, operate the starter and note whether or not the solenoid is actuated. If it is not, then the solenoid is faulty or the starter switch or wiring circuit is defective.

Δ8-9 HIGH-SPEED LOAD ADJUSTMENT The high-speed screw in two different carburetors is shown in Fig. 8-17. The high-speed load adjustment affects the air-fuel mixture ratio when the engine is operating at rated speed and under full load. The engine must be warmed up to normal operating temperature and under full load when this adjustment is made. If you cannot load the engine to make the adjustment, then make the adjustment without load. But after the adjustment, be sure to check the engine operation under normal full load. Use a tachometer to measure engine speed during the adjustment. The operation and use of the tachometer are discussed in later chapters.

With the engine operating at full speed, turn the high-speed screw (Fig. 8-16) in slowly until the engine begins to slow down. When this happens, the air-fuel mixture has leaned out so much that engine power is reduced. Now slowly turn the adjustment screw out until the engine slows down or the exhaust begins to turn black. At this position, the needle is passing too much gasoline, and so the mixture is too rich. Not all the gasoline burns, and this turns the exhaust black. Next, slowly turn the adjustment screw in until the engine runs smoothly and at full speed.

Make adjustments of about one-eighth turn at a time. Then wait a few seconds between turns for the engine to adjust to the changed air-fuel mixture.

Δ8-10 IDLE-MIXTURE ADJUSTMENT This adjustments affects the mixture richness when the engine is idling. Many carburetors do not have this adjustment, because the idle-mixture port is usually fixed. If the carburetor does have the adjustment (Fig. 8-16), it is made with the engine running and warmed up.

First, turn the idle-speed screw, with the engine idling, to get the lowest engine speed possible without stalling. Next, turn the idle-mixture screw in until the engine begins to slow down or roll. This means that the mixture is too lean to support normal engine operation. Now turn the idle mixture screw back out slowly until the engine idles smoothly.

Recheck the high-speed load adjustment to make sure it is still correct. Then operate the throttle several times from idle to full speed to make sure the engine will go from idle to full speed and back again without hesitation. Finally, adjust the idle speed (Δ8-11).

Δ8-11 IDLE-SPEED ADJUSTMENT

This adjustment is controlled by a stop screw (Fig. 8-16) which can be turned in or out to change the idle speed. Its basic purpose is to prevent the throttle valve from closing completely and causing the engine to stall.

Small engines usually idle at fairly high speeds, from 1200 to 3000 rpm. Always check the manufacturer's service manual to determine the specified speed before attempting to set the idle speed. Most specifications call for setting the idle speed at about one-half full speed. If the engine is idled too slowly, spark plugs, piston, and exhaust ports on two-cycle engines will soon foul from carbon, due to only partly burned gasoline.

Idle speed should be set with the engine warmed up. Therefore, the other settings above should be made first. Then a tachometer should be used to measure the speed while the idle-speed screw is turned to obtain the specified speed.

Δ8-12 FLOAT ADJUSTMENT

The float in the carburetor should be adjusted so that the proper level of gasoline will be maintained in the float bowl. Normally, this adjustment will not change. However, if the carburetor requires repair, then this adjustment should be checked. Checking the float level on one model of carburetor is shown in Fig. 8-19. The float should be parallel to the body mounting surface with the body gasket in place and the needle valve and float installed. Bend the tang on the float, if necessary, to bring the float to parallel.

Δ8-13 GOVERNOR SERVICE

Two types of governors are used on small spark-ignition engines: air vane and centrifugal. The air-vane governor works on a blast of air from the blades on the engine flywheel. The centrifugal type is operated by a centrifugal device which is actuated by engine speed.

If the engine is not governed at the correct speed, the governor should be adjusted. On some engines, this can be done by bending the link between the throttle and the governor or making some similar linkage adjustment. On others, the spring can be changed to change the engine speed (Fig. 8-20). Do not attempt to adjust the governor speed by stretching a spring. This will not work, because you may weaken the spring and it will probably go back to its original set after a while.

Figure 8-21 shows various methods of adjusting different types of governors. When adjusting a governor, never increase engine speed above its specified maximum rpm. Excessive engine speed greatly shortens engine life. If you cannot find the type of governor linkage you are working on in Fig. 8-21, refer to the service manual covering the engine on which the governor is mounted.

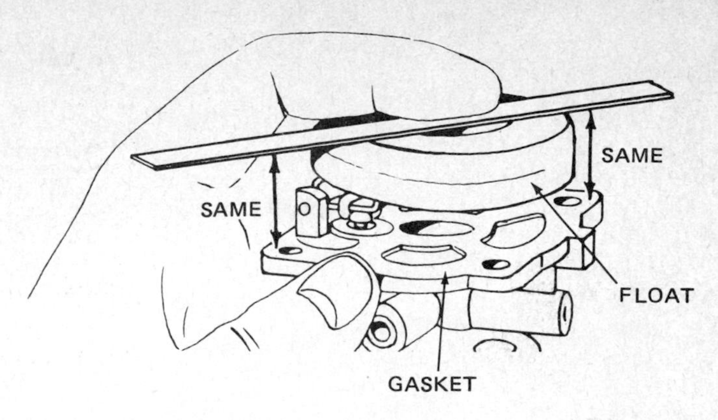

FIG. 8-19 Checking the float level on one model of small-engine carburetor. *(Briggs & Stratton Corporation)*

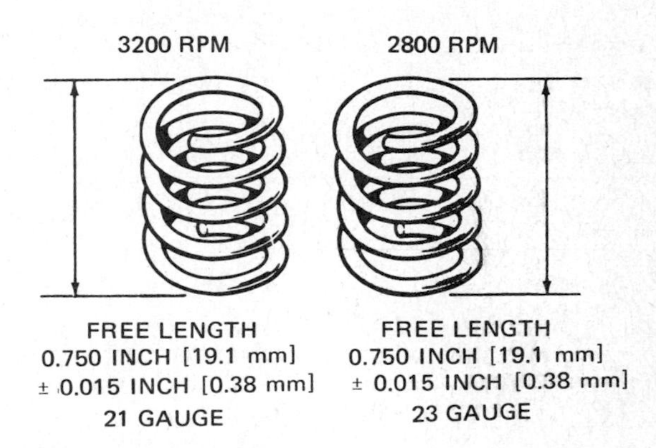

FIG. 8-20 On some engines, the governor spring must be changed to change governed engine speed. *(Lawn Boy Division of Outboard Marine Corporation)*

From the standpoint of governor service, there are three basic types: the air-vane type, the externally mounted centrifugal type, and the internally mounted centrifugal type.

Air-Vane Governor Figure 7-26 shows an air-vane governor. There is little in the way of service this governor requires. To gain access to the governor, remove the engine shroud. Make sure the air vane is not bent. It must be free to move when air blows on it.

Externally Mounted Centrifugal Governor Servicing this type of governor, which is mounted under the flywheel (upper left in Fig. 8-21), requires removal of the flywheel. If you must take the governor apart, be sure to notice how the parts fit together for reassembly.

Internally Mounted Centrifugal Governor If the governor is mounted inside the engine crankcase (Fig. 8-22), disassemble the engine to gain access to the governor. These governors sometimes are more complicated in design. Be sure to note the relationship of all parts before disassembling the governor.

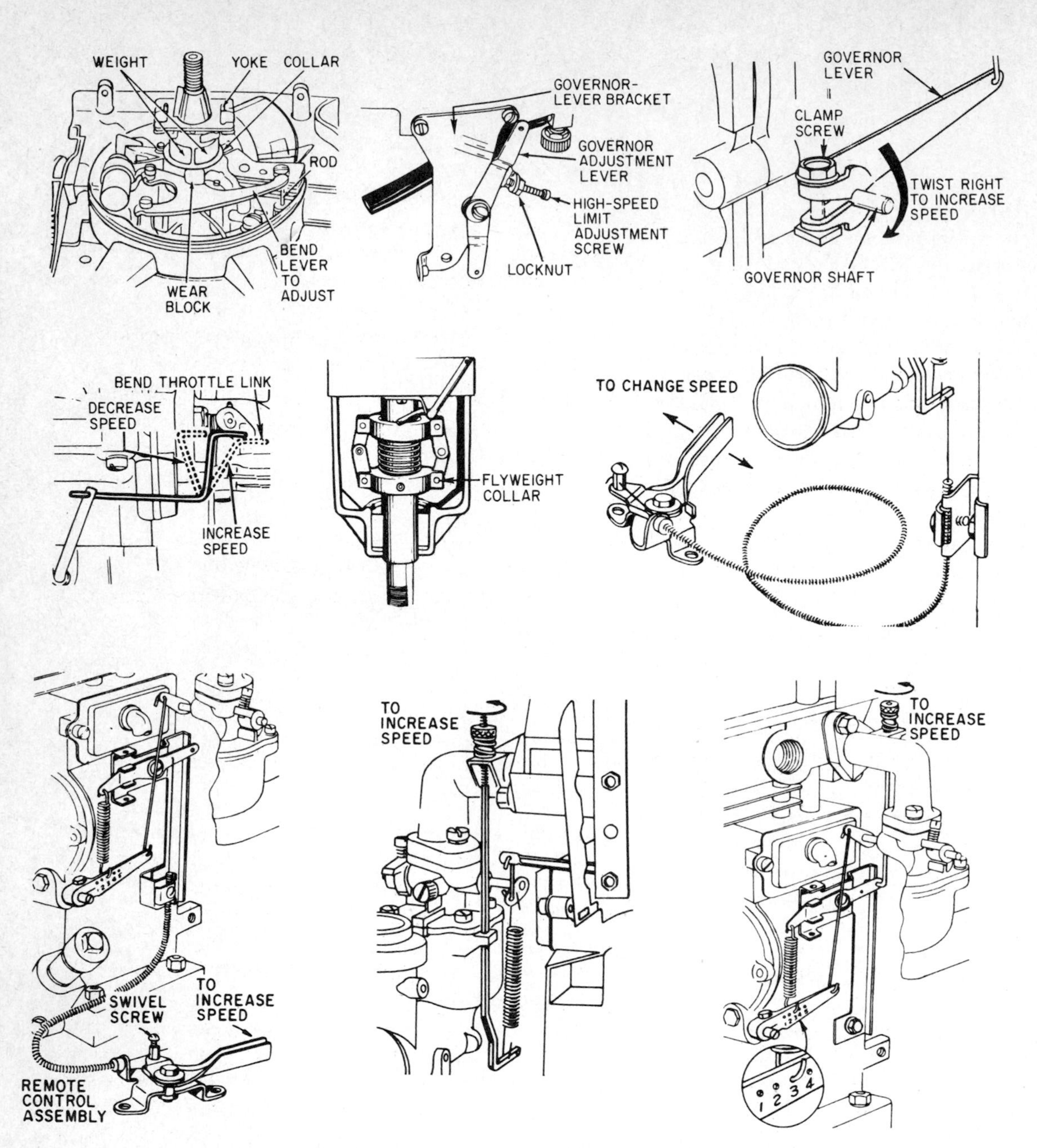

FIG. 8-21 Nine methods of adjusting the governor on different engines.

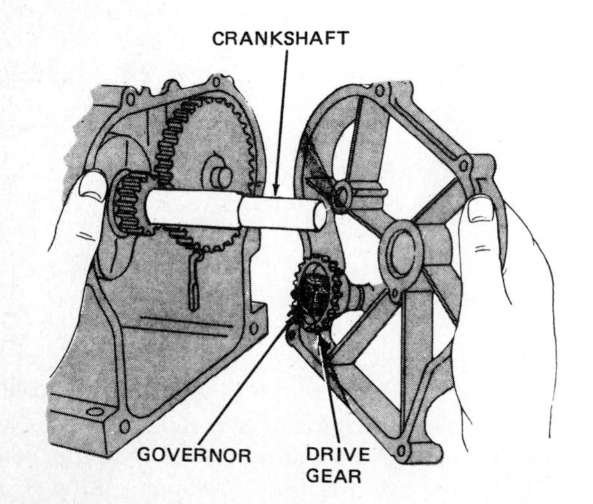

FIG. 8-22 Location of an internally mounted governor on one engine.

CHAPTER 8: REVIEW QUESTIONS

Select the *one* correct, best, or most probable answer to each question. You can find the answer in the section indicated at the right of each question.

1. The three types of carburetor air cleaners for small engines are (Δ*8-1*)
   a. oil-bath, air-bath, and foamed
   b. dry-element, oiled-element, and wet
   c. oil-bath, oiled-foam, and dry-element
   d. air-bath, oil-bath, and dry-bath
2. Crankcase breathers (Δ*8-3*)
   a. allow blowby gases to escape
   b. prevent pressure buildup in the crankcase
   c. are used in four-cycle engines
   d. all of the above

3. The oil-bath air cleaner (**Δ*8-1***)
   a. has an oil cup
   b. uses oiled foam
   c. has a dry element
   d. none of the above
4. The fuel filter may use (**Δ*8-2***)
   a. a sediment bowl
   b. a strainer mounted inside the fuel tank
   c. a strainer installed in the fuel line
   d. any of the above
5. Chain saws use (**Δ*8-2***)
   a. a bowl-type fuel filter
   b. a filter on the end of a flexible hose
   c. a filter mounted rigidly in the fuel tank
   d. the dry-element fuel filter
6. Most fuel pumps are serviced by (**Δ*8-5***)
   a. replacing the diaphragm
   b. replacing the valves
   c. replacing the pump assembly
   d. cleaning the pump
7. The three adjustment screws found in most carburetors for small engines are (**Δ*8-7***)
   a. idle-speed, idle-mixture, and choke
   b. idle-mixture, high-speed load mixture, and choke
   c. idle-speed, idle-mixture, and high-speed load mixture
   d. low-speed, idling, and high-speed
8. All small-engine carburetors have (**Δ*8-7***)
   a. an idle-mixture adjustment
   b. an idle-speed adjustment
   c. a low-speed adjustment
   d. no means of adjusting the choke
9. On some small-engine carburetors, it is difficult to see any difference between the (**Δ*8-7***)
   a. high-speed screw and the idle-mixture screw
   b. choke-adjustment screw and the high-speed screw
   c. throttle-stop screw and idle screw
   d. idle-mixture screw and idle-speed screw
10. Three general types of governors used in small engines include the internally mounted centrifugal governor and (**Δ*8-13***)
    a. the air-vane governor and the rotary governor
    b. the centrifugal governor and the rotary governor
    c. the air-vane governor and the externally mounted centrifugal governor
    d. the idling governor and the full-speed governor

CHAPTER

9

Engine Cooling Systems

After studying this chapter, you should be able to:

1. ***Describe the purpose of the engine cooling system.***
2. ***List the types of cooling systems and describe the operation of each.***
3. ***Explain how to service the cooling system on an air-cooled engine.***
4. ***Explain the operation of the thermostat in the liquid-cooling system.***
5. ***Explain the purpose of pressurizing the liquid-cooling system and tell how it is done.***

Δ9-1 ENGINE COOLING The combustion process in the engine cylinder produces a large amount of heat. Part of the heat is useful. It causes the push on the piston which makes the piston move and the crankshaft rotate. Part of the heat is lost in the hot exhaust gases. Some of the heat passes into the cylinder wall, cylinder head, and piston. This heat must be disposed of to prevent the cylinder walls, head, and piston from getting too hot. Excessive temperature will cause the oil film on the cylinder wall to fail. Without adequate means of disposing of the excess heat, the oil would char or burn and its lubricating properties would be lost. The result would be engine failure.

But even before seizure occurred, there could be serious difficulty in the engine. As the heat accumulated and the cylinder-head and spark-plug temperature went up, there would soon be a point at which preignition would occur. The spark plug would be so hot that it would ignite the air-fuel mixture prematurely. This condition could also cause engine failure, because preignition can melt holes in pistons and cause other internal damage.

To prevent such troubles, there has to be some means of getting rid of the excess heat from the cylinder walls, head, and piston. *Engine cooling systems* have been developed to do just that. The two basic types of small-engine cooling systems are air cooling (Fig. 9-1) and liquid cooling (Fig. 9-2). One method uses air to carry away the heat (air cooling). The other uses a liquid, such as a mixture of water and antifreeze (liquid cooling).

Most automotive engines and some small engines are liquid-cooled. However, most small engines are air-cooled. These include the engines for power lawn mowers and other small power equipment such as string trimmers, chain saws, blowers, vacuums, pumps, and generators. Both types of cooling systems are described later.

Δ9-2 PURPOSE OF ENGINE COOLING SYSTEM The purpose of the cooling system is to keep the engine at its most efficient operating temperature at all engine speeds and under all operating conditions. During the combustion of the air-fuel mixture in the engine cylinders, temperatures as high as 6000°F [3316°C] may be reached by the burning gases. Some of

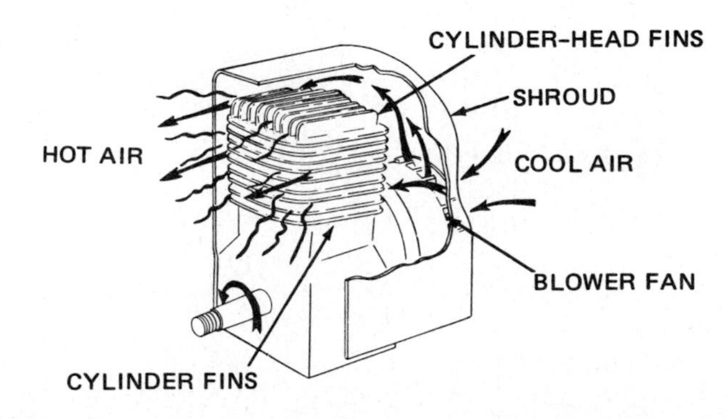

FIG. 9-1 **Forced-draft type of air-cooling system used on most small engines. This is an example of heat transfer by convection.** *(Briggs & Stratton Corporation)*

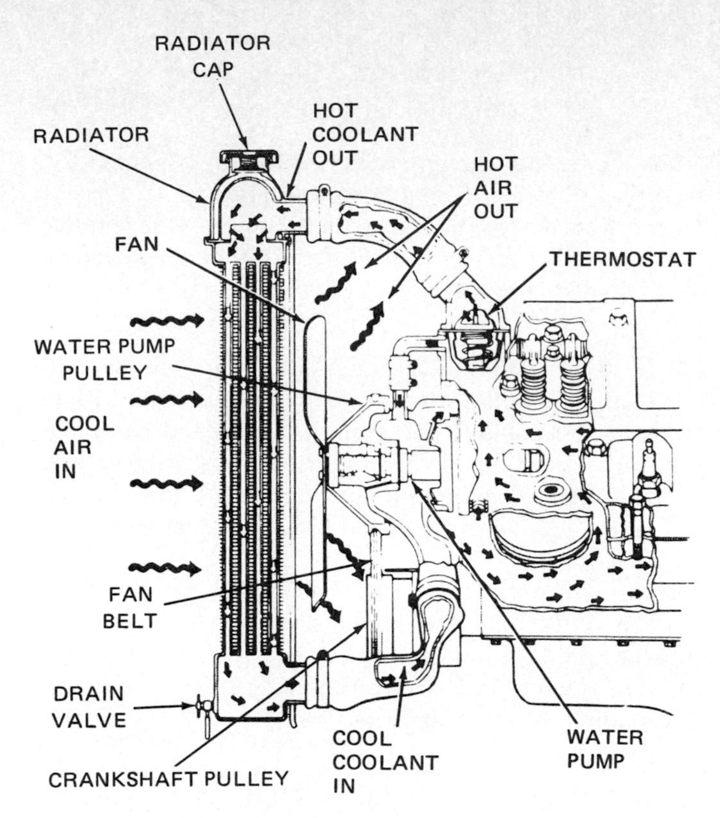

FIG. 9-2 Engine with a liquid-cooling system. Wavy arrows show the path of air. Solid arrows show the path of liquid coolant.

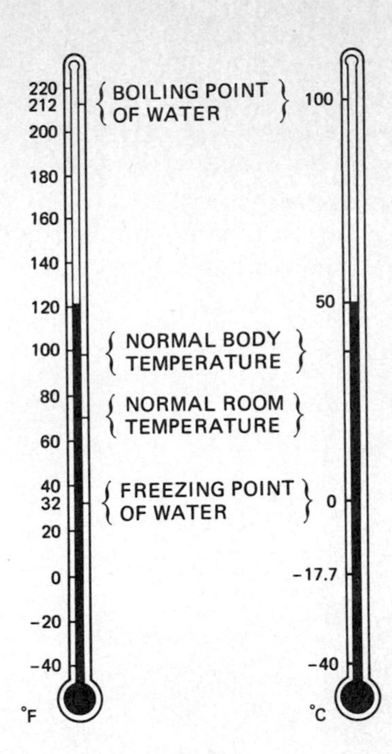

FIG. 9-3 Thermometers comparing Fahrenheit and Celsius (also called *centigrade*) readings.

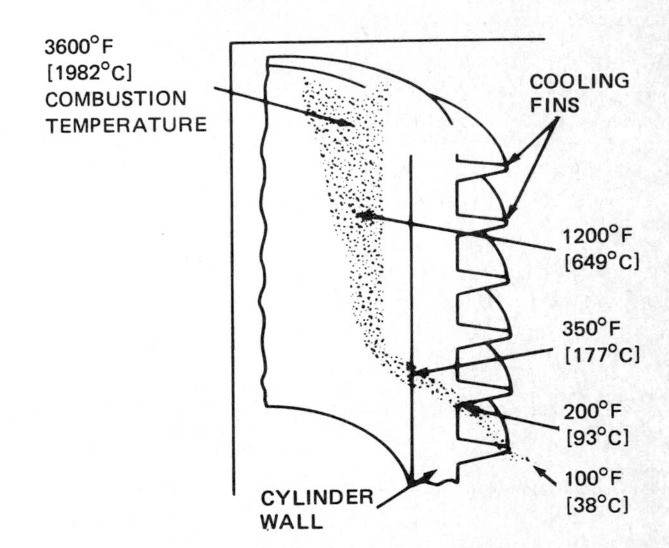

FIG. 9-4 Heat-travel path from combustion gases to the lower cooling fins. This is an example of heat transfer by conduction.

this heat is absorbed by the cylinder walls, cylinder head, and pistons. They, in turn, must be provided with some means of cooling so that their temperatures will not reach excessive values that melt them.

Cylinder-wall temperature must not increase beyond about 400 to 500°F [204 to 260°C]. Temperatures higher than this will cause the lubricating-oil film to break down and lose its lubricating properties. However, it is desirable to operate the engine at temperatures as close to the limits imposed by oil properties as possible. Removing too much heat through the cylinder walls and head would lower engine thermal efficiency. Cooling systems are designed to remove 30 to 35 percent of the heat produced in the combustion chambers by the burning of the air-fuel mixture.

Δ9-3 HEAT Heat keeps a person warm, raises the temperature, and makes water boil or iron melt. Heat is a form of energy, and as such, has the ability to do work. It is released by the burning of any fuel. In an engine, the heat energy is converted into mechanical energy. But the combustion process releases more heat than can be converted. This excess heat must be removed by the engine cooling system.

Heat is measured by *temperature.* A hot object has a high temperature. The temperature of an object is given in *degrees* and is read on a scale on a thermometer such as shown in Fig. 9-3. There are two types: Fahrenheit and Celsius (also called *centrigrade*). On the Fahrenheit scale, water freezes at 32°F and boils at 212°F. On the celsius scale, water freezes at 0°C and boils at 100°C.

Δ9-4 HEAT TRANSFER The cooling system removes excess heat from the engine. To do this, the heat must travel from the burning combustion gases through the cylinder to the fins. The movement of heat from one place to another is called *heat transfer.*

There are three modes of heat transfer, *conduction, convection,* and *radiation.* Conduction is the transfer of heat through a solid object. Heat travels, by conduction, from the cylinder walls through the cylinder to the cooling fins (Fig. 9-4).

Liquids and gases do not conduct heat very well. But they are used to produce convection cooling. In convection cooling, the gas or liquid moves from one place to another, carrying heat with it. In air-cooled engines, a gas (the air) moves past the engine fins, picking up heat. The air gets hot and rises. When the air rises from the fins simply because it gets hot (Fig. 9-5), the system is called an *open-draft* or *natural-convection system.* When a fan is used to force cool air over the hot fins (Figs. 9-1 and 9-6), the system is called a *forced-draft* or *forced-convection system.*

Radiation is the movement of heat through space — for example, the heat you feel from sunlight or from a fire in the fireplace across the room. Radiation provides no significant amount of heat transfer from either air-cooled or liquid-cooled engines.

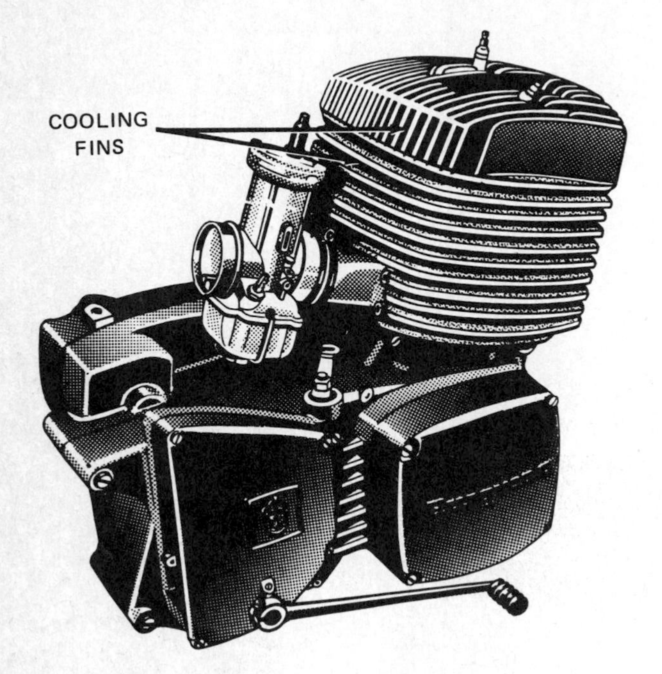

FIG. 9-5 Open-draft or natural-convection system of air cooling. *(Husqvarna)*

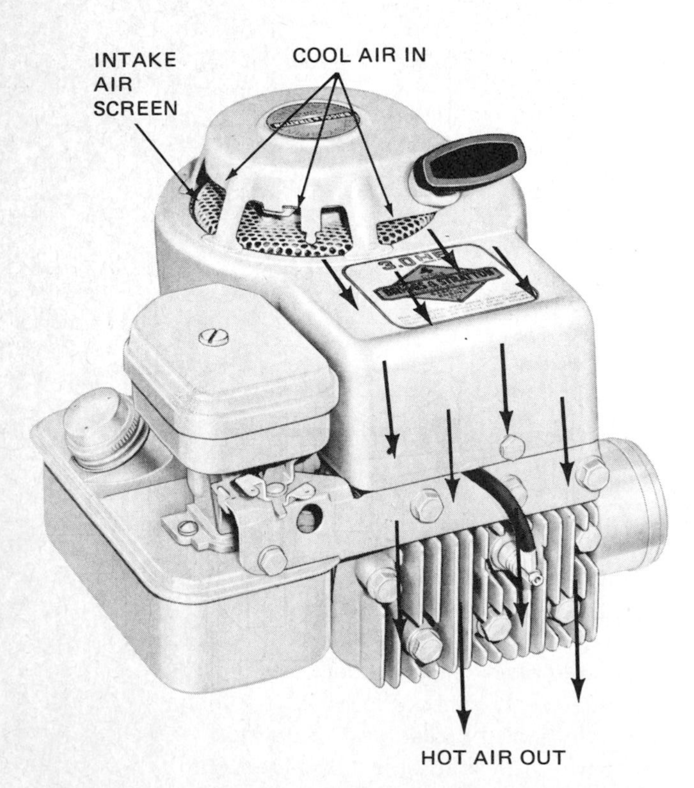

FIG. 9-6 Forced-draft or forced-convection system of air cooling, which is used on most small engines. *(Briggs & Stratton Corporation)*

Δ9-5 AIR-COOLED ENGINES Almost all small single-cylinder engines, both two-cycle and four-cycle, are air-cooled. The cylinder and head have fins (Fig. 9-1). These fins are actually part of the head and cylinder. They greatly increase the outer metal surface area of the engine from which heat can escape into the surrounding air.

There are two types of air-cooling systems. The simpler is the open, or open-draft, system used on mopeds and most motorcycle engines (Fig. 9-5). The second type is the forced-draft system (Fig. 9-6). It is used on almost all other small engines.

Δ9-6 FORCED-DRAFT COOLING SYSTEM In the forced-draft cooling system, the cylinder and head are finned to help dissipate heat. To assist in this heat escape, the engine has *shrouds* and a cooling fan (or *blower*) to force the air over and around the fins (Fig. 9-7). The fan is usually part of the flywheel. Shrouds, along with baffles and deflectors, are parts made from plastic or metal sheets. They are shaped in such a way that, when assembled on the engine, they force the air from the fan to flow between the fins. Shrouds are held in place by screws, which can be taken out so that the shroud parts can be removed.

Most single-cylinder air-cooled engines use some type of forced-draft cooling system. In addition, multiple-cylinder air-cooled small engines also have a forced-draft cooling system.

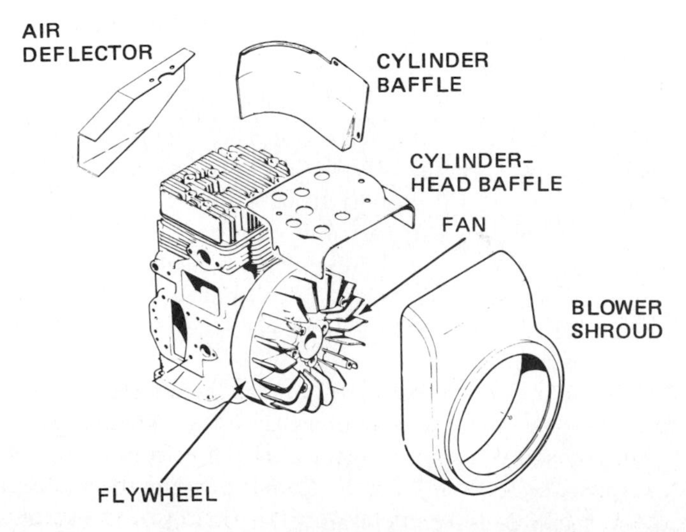

FIG. 9-7 Small engine with the shrouds removed.

Δ9-7 ARRANGEMENTS OF FINS

Fins are arranged in various ways to make the best use of the air passing by. However, the fins are always placed so that the flow of cooling air passes between them. With open-draft engines, such as used in motorcycles, the fins are usually rather long (Fig. 9-5). They are positioned so that the forward motion of the motorcycle will allow air to flow between them. In engines using a forced-draft cooling system, the shrouds are shaped so that the cooling air is forced to flow between the fins (Fig. 9-6).

A two-cycle engine usually has a larger fin area than a four-cycle engine. The reason is that the two-cycle engine has twice as many power strokes (to create heat) as the four-cycle engine when both are running at the same speed.

Δ9-8 LIQUID-COOLING SYSTEMS

A second method of engine cooling is the liquid-cooling system. It uses a liquid coolant such as water, which circulates around the cylinder and through the head to carry away excess heat. This cools the engine. Most automobile and outboard engines, and some motorcycle and small engines, are liquid-cooled.

The operation of the liquid-cooling system is shown in Fig. 9-2. A water pump circulates the coolant through *water jackets*. These are spaces surrounding the cylinder and head through which coolant flows (Fig. 9-8). As the coolant passes through the water jackets, it picks up heat and carries it to the *radiator*.

The radiator is a heat exchanger that has a series of coolant passages and a series of air passages (Fig. 9-9). Hot coolant from the engine flows through the coolant tubes in the radiator. This heats the fins, which form the air passages through the radiator.

While the engine is running, cool air is pulled through the radiator by the engine *fan* (Fig. 9-10). The passing airstream picks up heat from the fins, which cools the coolant. Coolant flows continuously between the engine water jackets and the radiator. It is this continuous circulation of coolant that prevents the engine from overheating.

Some engines have used *thermosiphon* (natural circulation) *cooling*. This is basically a liquid-cooling system without a water pump. As the coolant around the cylinder is heated, the coolant expands, becomes lighter, and rises. This causes cool, and heavier, coolant from the radiator to flow into the water jackets. At the same time, the hot coolant enters the radiator. There the coolant loses heat, becomes heavier, and sinks. This produces the circulation of coolant that continues to remove heat from the engine.

Δ9-9 ANTIFREEZE AND COOLANT

The coolant is a mixture of water and an *antifreeze* compound, usually ethylene glycol. The antifreeze lowers the freeze temperature of the coolant, raises the boiling temperature, and helps prevent corrosion of the metal surfaces in the cooling system.

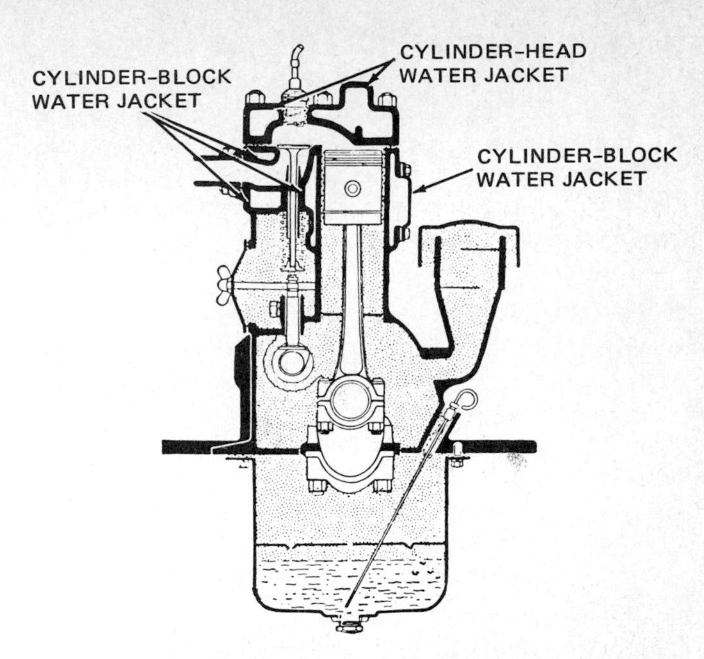

FIG. 9-8 Water jackets in the head and cylinder block of a four-cycle engine.

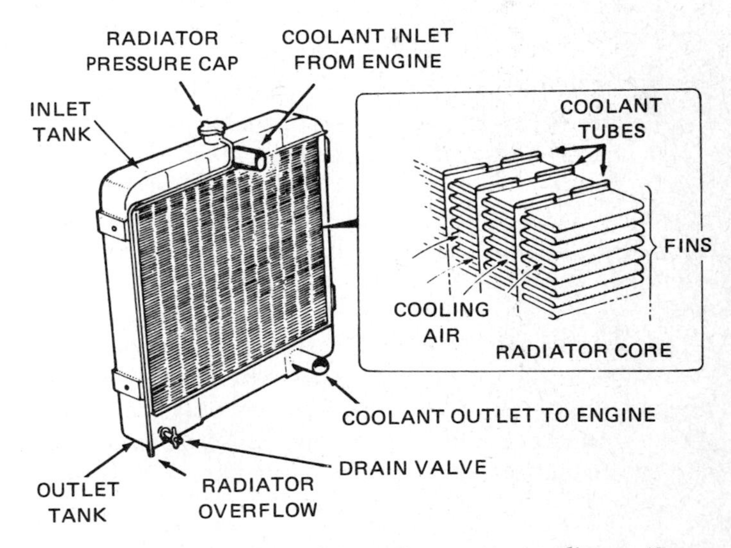

FIG. 9-9 Construction of a cooling-system radiator. *(Onan Corporation)*

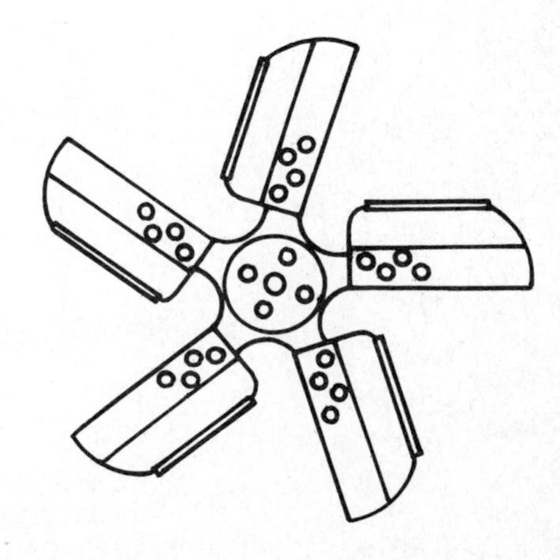

FIG. 9-10 An engine fan, used to pull cooling air through the radiator.

Δ9-10 WATER PUMP The water pump is a small centrifugal pump that forces coolant to flow through the cooling system. The pump is located at the front of the engine, between the cylinder block and radiator (Fig. 9-2). The pump is usually driven by a belt from the front of the crankshaft. The pulley that drives the pump may also carry the engine fan (**Δ**9-11).

Δ9-11 ENGINE FAN The engine fan usually mounts on the water-pump shaft and is driven by the belt that drives the water pump (Fig. 9-2). The purpose of the fan is to pull cool air through the radiator.

The fan (Fig. 9-10) usually has from four to six blades which, in turning, pull air through the radiator. Some engines have a fan shroud to direct the airflow. Use of the shroud assures that all air pulled back by the fan passes through the radiator, and not around it.

CAUTION **Fan blades can break and fly off. Whenever the engine is running, never stand directly in line with the rotating fan. Also, avoid allowing your clothing, hands, or tools to get near the fan blades or fan belt.**

Δ9-12 DRIVE BELTS The cooling-system fan is usually driven from the engine crankshaft by a *drive belt* (Fig. 9-11). An engine may have more than one drive belt, so each belt is usually identified by the name of the unit that it drives. For example, Fig. 9-2 shows how the engine fan is driven by the crankshaft through an arrangement of pulleys and a belt. Since the unit being driven is the fan, the belt is called the *fan belt.*

There are various types of belt drives. The most common type is the V-belt (Fig. 9-11). When the V-belt is properly tightened, it wedges into the pulley groove. There is friction between the sides of the belt and the sides of the groove. This friction transmits power from the crankshaft, through the belt, to the fan pulley. An engine may require several V-belts to drive the fan and other engine-driven accessories. Then a multiple-groove pulley is used on the crankshaft.

Other types of belts are the *serpentine belt* and the *flat belt.* In addition, a toothed belt is used on some engines to turn the diesel-engine fuel-injection pump. On other engines, the toothed belt may be used to turn the overhead camshaft. The advantage to the toothed belt is that it does not slip in normal use. Therefore, it can be used to drive devices which must rotate in time with the crankshaft.

Belt tension is usually adjusted by moving the unit being driven. This changes the distance between the centers of the pulleys (Fig. 9-11). However, sometimes the position of the unit being driven (also called the *load*) cannot be changed. Then an *idler pulley* is repositioned to adjust belt tension.

In a power transmission system using pulleys and a belt, speed increase or decrease is obtained by changing the diameters of the pulleys. Speed change is *not* related to the length of the belt.

Δ9-13 RADIATOR The radiator is a *heat exchanger* (Fig. 9-9). It removes heat from the coolant passing through it. The radiator does this by holding some of the hot coolant close to a large volume of cooler air. The hot coolant enters the radiator, loses heat, and comes out cooler. The heat is transferred from the coolant to the air passing through the radiator.

A filler cap, called a *radiator cap,* is removed to add coolant to the cooling system. Coolant may be lost because of evaporation or leakage.

Δ9-14 RADIATOR PRESSURE CAP Most engine liquid-cooling systems are sealed and pressurized by a *radiator pressure cap* (Fig. 9-12). There are two advantages to sealing and pressurizing the liquid-cooling

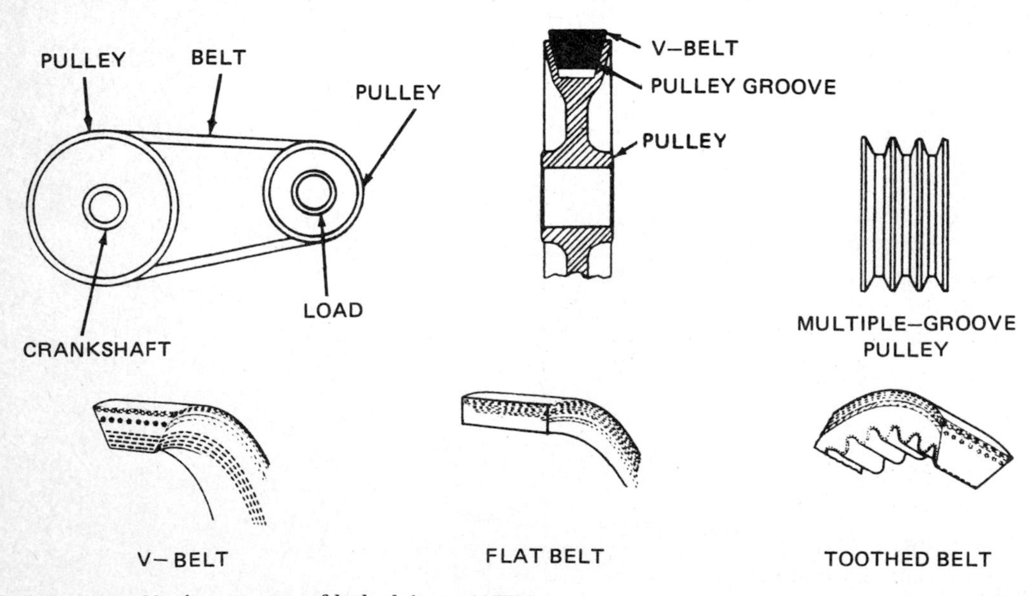

FIG. 9-11 Various types of belt drives. *(ATW)*

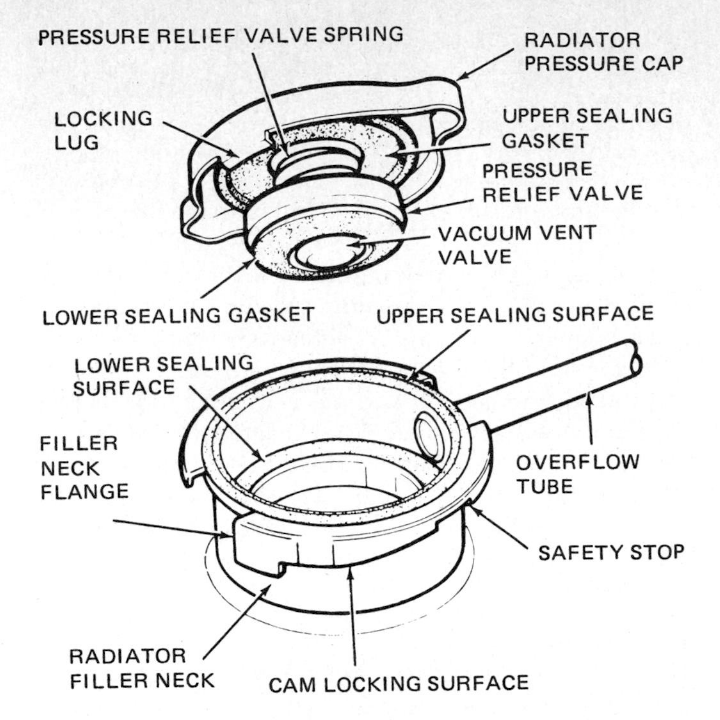

FIG. 9-12 A radiator pressure cap removed from the radiator filler neck.

system. First, the increased pressure raises the boiling temperature of the coolant. This increases the efficiency of the cooling system. Second, sealing the cooling system reduces coolant loss from evaporation and permits the use of an expansion tank.

The radiator pressure cap has two valves built into it, a pressure valve and a vacuum valve. When the cap is installed in the radiator filler neck, the top gasket seals the filler neck to prevent any leakage (Fig. 9-12). This causes pressure to build up in the cooling system as the temperature rises and the coolant expands. When the pressure rises above the rated pressure of the cap, the pressure forces the pressure relief valve up off the lower seal seat. This allows steam and coolant to escape through the overflow tube.

When the engine cools, the coolant contracts. Since the pressure cap seals the radiator, a vacuum will develop in the cooling system. This vacuum may be strong enough to collapse the hoses and possibly damage the radiator. To prevent this, the pressure cap includes a vacuum valve (Fig. 9-12). It opens as soon as a slight vacuum develops in the radiator. This allows air to enter the cooling system, preventing any greater vacuum from forming.

Δ9-15 THERMOSTAT

A *thermostat* is placed in the coolant passage between the cylinder head and the top of the radiator (Fig. 9-2). The thermostat is a valve that automatically opens and closes in response to changes in coolant temperature. This is necessary because varying engine loads cause more or less heat to be rejected to the cooling system. When the engine is cold, the thermostat valve is closed (Fig. 9-13). This prevents coolant from flowing through the upper hose to the radiator. As the engine and coolant warm up, the valve starts to open. Now coolant flows to the radiator so that the cooling system can remove heat.

Thermostats are designed to open at a specific temperature. This temperature is known as the *thermostat rating*, and it may be stamped on the thermostat. Most thermostats begin to open at their rated temperature. They are fully open about 20°F [11°C] higher than their rated temperature. For example, a thermostat with a rating of 195°F [91°C] starts to open at that temperature. It is fully open at about 215°F [102°C].

Δ9-16 TEMPERATURE INDICATORS

In a running engine, the temperature of the engine coolant should be easily determined. For this reason, a temperature indicator is installed in the engine control panel. An abnormal heat rise is a warning of abnormal conditions

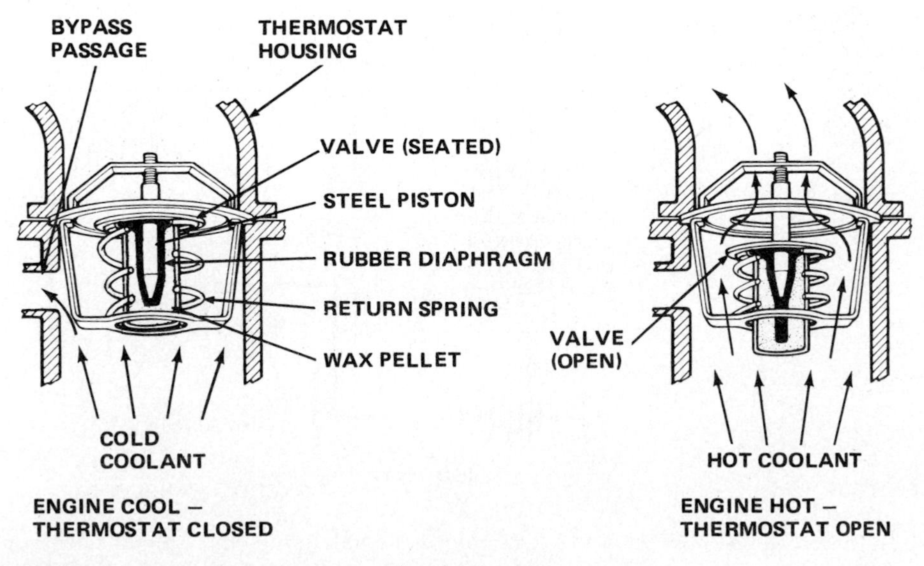

FIG. 9-13 Operation of a wax-pellet thermostat.

in the engine. The indicator warns the operator to stop the engine before serious damage is done.

There are two kinds of coolant temperature indicators: a light and a gauge. The light (Fig. 9-14) comes on when the engine temperature goes too high. The other system has a gauge on the engine or on the engine control panel (Fig. 9-15). A needle or pointer moves across the face of the gauge. This shows the temperature of the coolant in the engine.

NOTE Some engines have an automatic-shutdown device as part of the temperature-indicating system. When the engine begins to overheat and the coolant temperature rises excessively, a sensor signals for automatic shutdown. Then the ignition system stops furnishing sparks at the spark plugs, or fuel delivery to the injection nozzles is stopped (in a diesel engine).

△9-17 CLEANING COOLING-SYSTEM SHROUDS Most small engines have a shroud that must be removed before the engine can be cleaned (Fig. 9-7). The shroud is usually held in place with screws. These screws can be taken out to remove the shroud and its related parts. On a few engines, you must remove some other parts before the shroud can be removed. These parts might include the air cleaner, muffler, spark-plug wire, governor spring, or some other minor part.

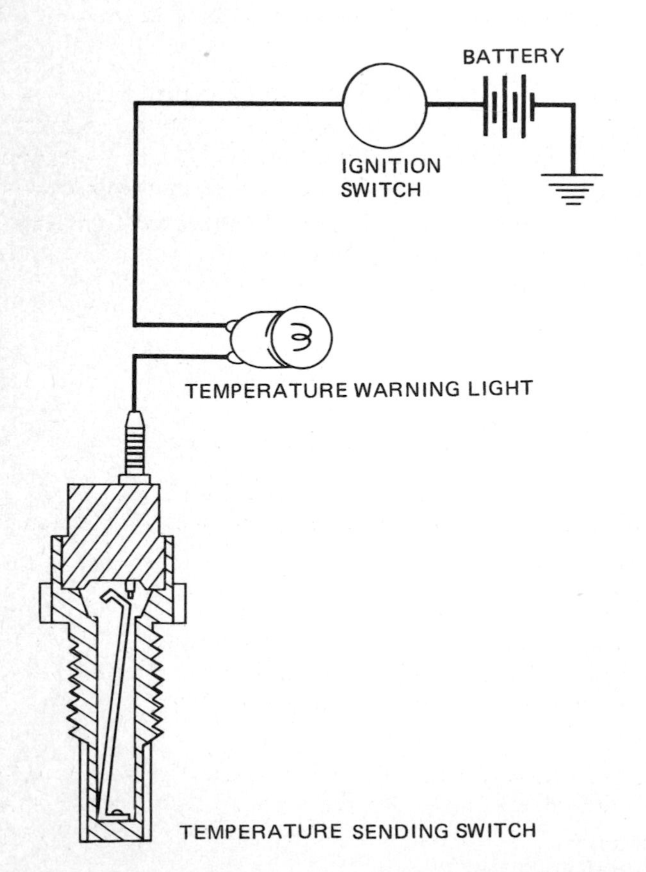

FIG. 9-14 Temperature warning light which indicates engine overheating. *(ATW)*

NOTE Never operate the engine with the shroud and baffles removed! The shroud is there to force cooling air over the engine. When the shroud is off, the engine will overheat if operated. In addition, engines which have air-vane governors that operate on air flow will not function properly with the shroud off. The possible result is that the engine can overspeed and damage itself.

If the shroud is bent or damaged, it should be straightened, repaired, or replaced. A defective shroud can cause engine overheating. A defective shroud also might interfere with the fan or other moving parts.

If the shroud is dirty and has accumulations of grass clippings or other trash, scrape it clean with a putty knife or similar tool. Use a stiff-bristled brush and solvent if needed. Clean the air-intake screen with a brush and solvent, if necessary, to get rid of all accumulations of trash that could prevent normal air flow through it.

△9-18 SERVICING OF AIR-COOLED-ENGINE COOLING SYSTEMS Fins on the head and cylinder provide large surface areas from which heat can be dissipated. The heat flows from the inside of the cylinder, through the cylinder metal, to the fins, which transfer it to the outside air (Fig. 9-4).

If the fins become dirty and covered with oil or mud, the heat cannot get through. The accumulations act as a blanket to hold heat in the engine. As a result, the engine becomes overheated. The oil film on the engine parts becomes less effective, or actually fails. The result is that engine parts wear rapidly and engine life is shortened. Therefore, it is essential for long engine life to clean the engine when needed.

Another purpose of periodically cleaning the engine is to check for loose nuts or bolts and loose, broken, cracked, or otherwise damaged parts. One way to clean the engine is to use a stiff brush and water. This will get into all the crevices where dirt can accumulate and will clean away most of the mud and oil that can cause trou-

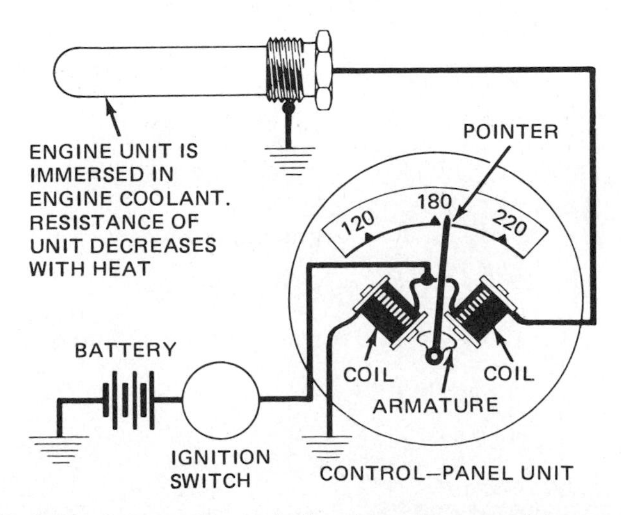

FIG. 9-15 A magnetic, or balancing-coil, type of temperature-gauge system.

ble. For a complete cleaning job, use a degreasing compound.

CAUTION **Do not clean a hot engine. Wait until it is cool. Water thrown on the hot engine can cause the head or cylinder to crack. Some cleaning solutions are flammable and could burst into flames if sprayed on a hot engine. Also, make sure that there is adequate ventilation. Some fumes from cleaning solutions are unhealthy to breathe.**

Three substances for cleaning the cylinder and head can be used—a degreaser, a solvent, or live steam. As a first step, use a wooden stick to scrape away all the accumulated trash, dirt, and grease. Do not use a metal tool because this will scratch the cylinder and head and encourage accumulations of dirt.

Then use the material you have available to finish the cleaning job. Degreasing compound comes in pressure-spray cans or in larger containers. To use live steam, you need a steam cleaner. This will be available in many shops.

While cleaning the cylinder and head, check for oil leaks which usually show up as a heavy accumulation of dirt. Check also for cracks or other damage. Then apply the solvent on the areas to be cleaned. The degreaser in the pressure can is very easy to use. Other types of solvent can be applied with a bristle brush. After about five minutes, flush off the solution with a stream of water from a hose. If you have used solvent, use a solution of soapy water brushed on and then flushed off.

∆9-19 WORKING SAFELY ON LIQUID-COOLING SYSTEMS

There are several safety hazards you must watch for when working on engines with liquid-cooling systems:

1. Keep your hand away from the moving fan! When the engine is running, the fan is turning so fast it is a blur. But it can mangle your hand and cut off fingers if your hand should get into the fan.
2. Never stand in a direct line with the fan. A fan blade could break off and fly out from the engine compartment. Anyone standing in line with the fan could be injured or killed. Before starting the engine, examine the fan for cracked or loose blades. If you find any damage, the fan must be replaced.
3. Electrically operated and thermostatically controlled fans may not be shut off when the ignition switch is turned to OFF. Because the coolant-temperature switch turns the fan motor on and off solely on the basis of coolant temperature, the fan may start and run even with the engine stopped. When working near an electric fan, disconnect the lead to the fan motor to avoid injury should the fan start.
4. Keep your fingers away from the moving belt and pulleys! Your fingers could be pinched and cut off if they are caught between the belt and pulley.
5. Never attempt to remove the radiator cap from the cooling system of an engine that is near or above its normal operating temperature. Releasing the pressure may cause instant boiling of the coolant. Boiling coolant and steam spurting from the filler neck can cause scalding and burns. Allow an engine to cool before attempting to remove the cap.
6. Coolant is poisonous! It can cause serious illness or even death if it is swallowed! Always wash your hands thoroughly if you get coolant on them.

∆9-20 LIQUID-COOLING-SYSTEM TROUBLE DIAGNOSIS

Three complaints about liquid-cooling systems are engine overheating, slow warm-up, and cooling-system leaks. Possible causes of these complaints are listed in the Liquid-Cooling-System Trouble-Diagnosis Chart.

LIQUID-COOLING-SYSTEM TROUBLE-DIAGNOSIS CHART

Condition	Possible cause	Check or correction
1. Loss of coolant	a. Pressure cap and gasket defective	Inspect. Wash gasket and test. Replace only if cap will not hold pressure specified.
	b. Leakage	Pressure test system.
	c. External leakage	Inspect hose, hose connection, radiator, edges of cooling system, gaskets, core plugs, drain plugs, oil-cooler lines, water pump, expansion tank and hoses, heater-system components. Repair or replace as required.
	d. Internal leakage	Check torque of head bolts; retorque if necessary. Disassemble engine as necessary. Check for cracked intake manifold, blown head gasket, warped head or block gasket surfaces, cracked cylinder head or engine block.

(Chart Continued on Next Page)

Condition	Possible cause	Check or correction
2. Engine overheating	a. Low coolant level	Fill as required. Check for coolant loss.
	b. Loose fan belts	Adjust.
	c. Pressure cap defective	Test. Replace if necessary.
	d. Radiator or air-conditioner condenser obstructed	Remove bugs, leaves, and debris.
	e. Thermostat stuck closed	Test. Replace if necessary.
	f. Fan-drive clutch defective	Test. Replace if necessary.
	g. Ignition faulty	Check timing and advance. Adjust as required.
	h. Temperature gauge or HOT light defective	Check electric curcuits. Repair as required.
	i. Inadequate coolant flow	Check water pump and block for blockage.
	j. Exhaust system restricted	Check for restrictions.
3. Engine failing to reach normal operating temperature; slow warm-up	a. Open or missing thermostat	Test. Replace or install as necessary.
	b. Defective temperature gauge or COLD light	Check electric circuits. Repair as required.

Δ9-21 SERVICING LIQUID-COOLING SYSTEMS Engines with a liquid-cooling system require more cooling-system service than air-cooled engines. When the liquid-cooling system is not serviced periodically, overheating and engine damage may result. Liquid-cooling-system problems are a major cause of engine failure.

Several tests can be performed on the liquid-cooling system and its components. These are listed below. Then testing and servicing procedures are described.

Cooling-System Tests Many tests can be made on the cooling system and its components. They include:

1. Checking the coolant level
2. Testing antifreeze strength
3. Testing the thermostat operation
4. Checking the condition of the hoses and connections
5. Making a Bloc-Chek for exhaust-gas leakage into the cooling system
6. Pressure-testing the cooling system
7. Pressure-testing the radiator cap
8. Testing the condition of the fan and fan belts

Checking the Coolant Level Coolant level in the cooling system can be checked on a system with an expansion tank by looking at the coolant level in the expansion tank (Fig. 9-16). No coolant in the expansion tank indicates that the radiator cap should be removed and the coolant level checked.

When the radiator cap must be removed to check the coolant level in the radiator, allow the engine to cool.

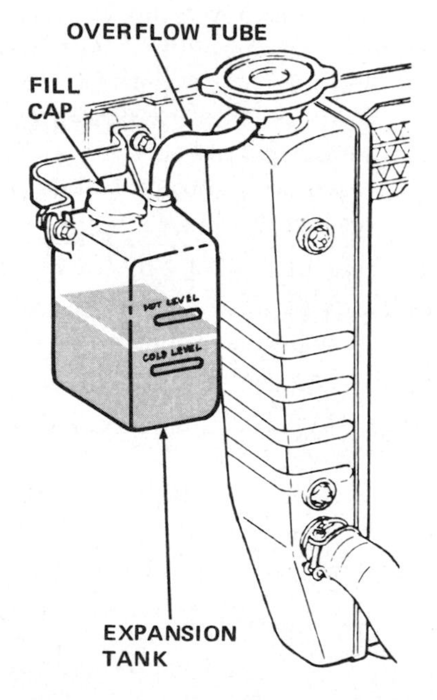

FIG. 9-16 Cooling system using an expansion tank.

Then remove the cap in two steps. Place a cloth over the radiator cap to protect your hand. Then rotate the cap against the safety stop on the radiator filler neck. This will safely release any remaining pressure. Then press down and rotate the cap until it can be removed.

Testing Antifreeze Strength The amount of antifreeze in the coolant can be checked with either a ball- or float-type hydrometer (Fig. 9-17). In the shop, practice

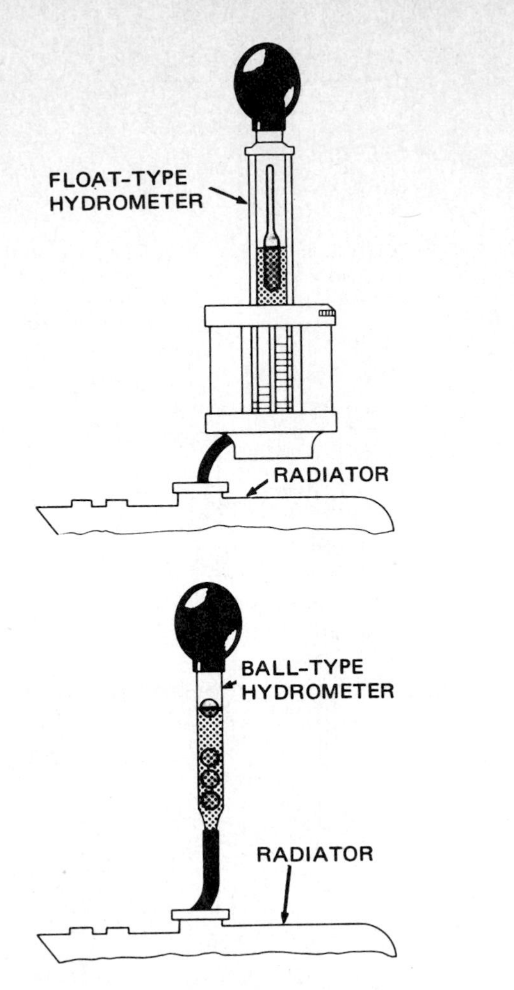

FIG. 9-17 Two types of hydrometers used to check the strength of the antifreeze in the coolant.

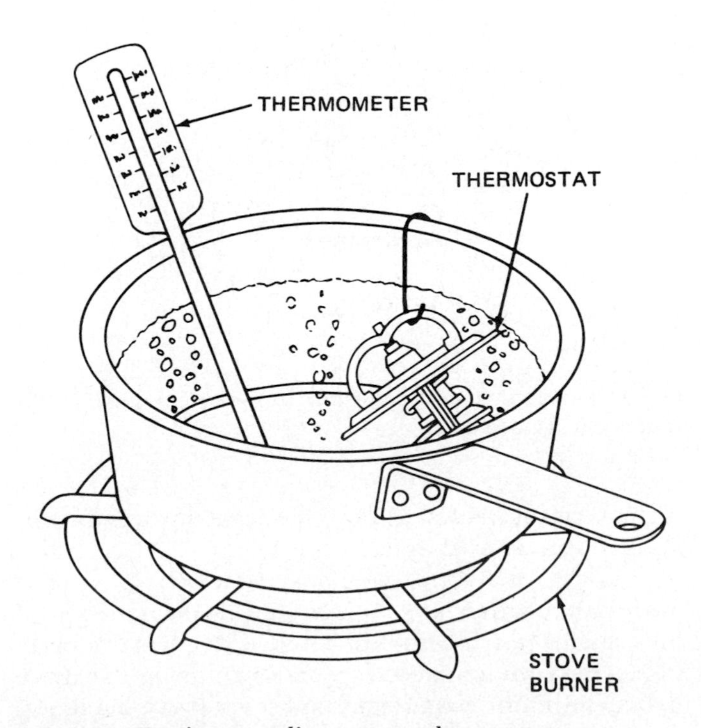

FIG. 9-18 Testing a cooling-system thermostat.

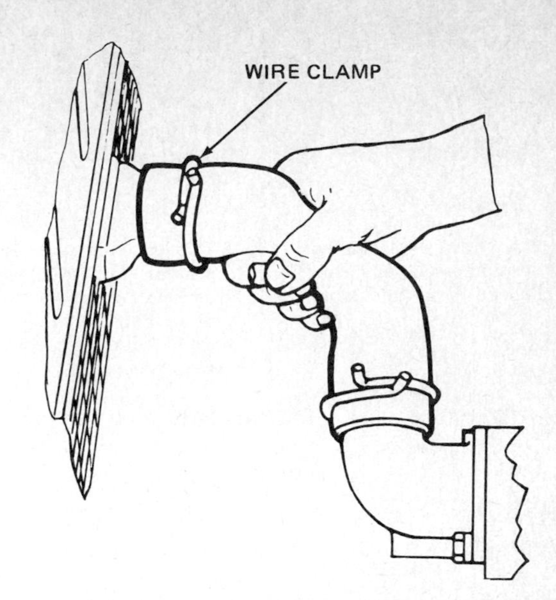

FIG. 9-19 To check the condition of a radiator hose, squeeze the hose.

using the hydrometer until you can quickly take an accurate reading.

Testing the Thermostat Different manufacturers recommend different testing methods. One method is to suspend the thermostat in hot coolant and then in cold coolant to see if the thermostat will open and close at the proper temperatures (Fig. 9-18). Another method is to check the thermostat in the engine. A third method is to immerse the thermostat in boiling water to see if it will open.

Checking the Hose and Hose Connections Hoses must be in good condition with tight connections. To check the condition of the hose, squeeze it (Fig. 9-19). The appearance and feel indicate the condition (Fig. 9-20). If the hose is soft, it may be deteriorating inside. A soft hose should be replaced.

Checking for Exhaust-Gas Leakage into the Cooling System A defective head gasket or cracked head or block can allow damaging exhaust gas to leak into the cooling system. A Bloc-Chek can be used to check for exhaust gas in the cooling system (Fig. 9-21).

Pressure-Testing the Cooling System A special tester is used to apply pressure to the radiator filler neck (Fig. 9-22). Then the pump is operated to apply 15 psi [103 kPa]. If the pressure does not remain constant, there is a leak. Check all possible leak points.

Pressure-Testing the Radiator Pressure Cap The tester can also be used to check the radiator cap (Fig. 9-23).

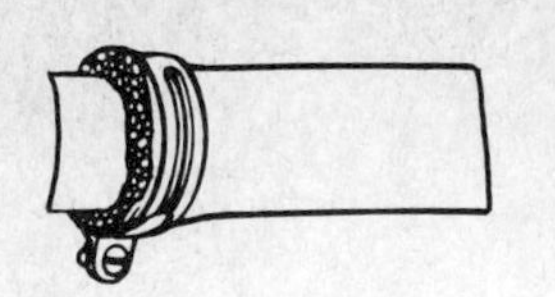
HARD hose can fail. Tightening hose clamps will not seal the connection or stop leaks.

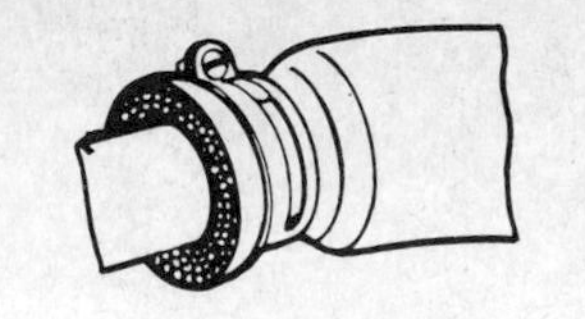
SWOLLEN hose or oil–soaked ends indicate possible failure from oil or grease contamination. Squeeze the hose to locate cracks and breaks that cause leaks.

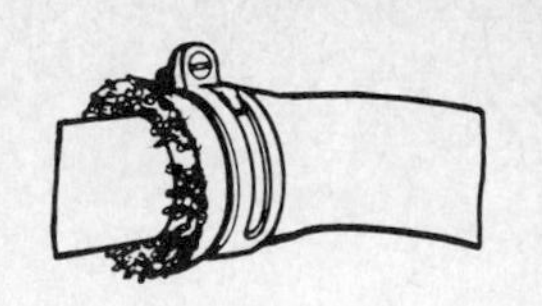
ALWAYS CHECK hose for chafed or burned areas that may fail.

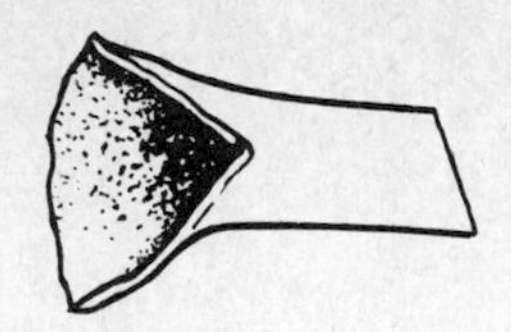
SOFT hose indicates inside deterioration. This can contaminate the cooling system and clog the radiator.

FIG. 9-20 Conditions that indicate failure of a radiator hose.

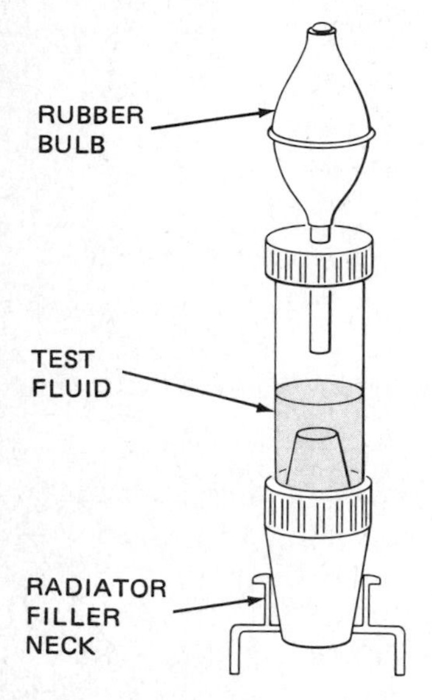

FIG. 9-21 Checking for exhaust-gas leakage into the cooling system with a Bloc-Chek tester.

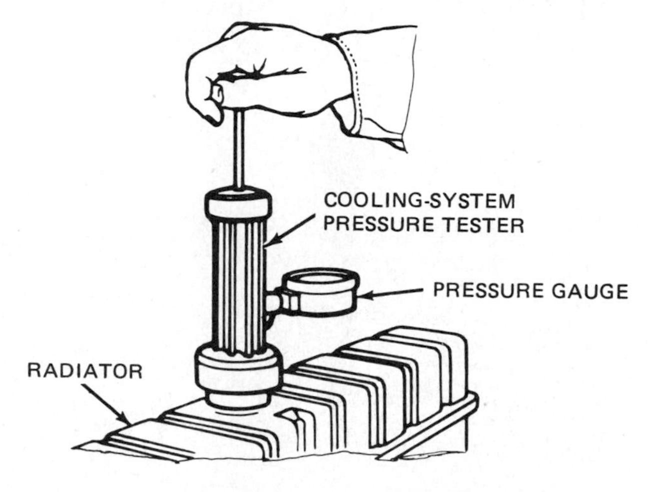

FIG. 9-22 Using a cooling-system pressure tester to check the cooling system for leaks.

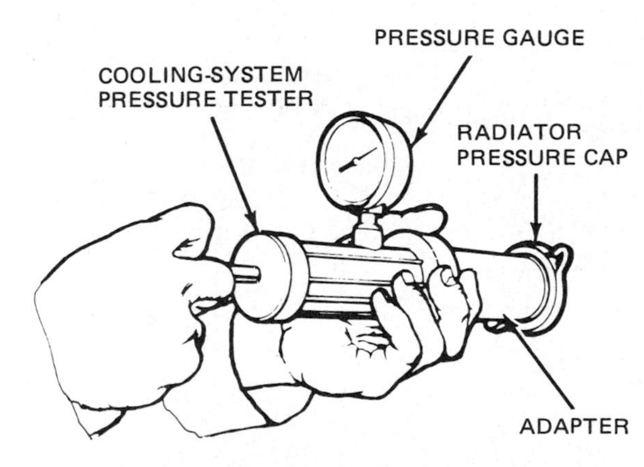

FIG. 9-23 Using a cooling-system pressure tester to check a radiator pressure cap.

Testing the Fan Belt Check for wear and tension (Fig. 9-24). On engines with dual fan belts, replace both at the same time, even if only one is worn. A faulty or slack belt (Fig. 9-25) can cause engine overheating and a run-down battery.

Testing the System for Accumulations of Rust and Scale There is no quick and accurate way to determine the amount of rust and scale in a cooling system. Rust and scale in the cooling system can be minimized by periodic cleaning and replacement of the antifreeze.

Cleaning the Cooling System Engine manufacturers recommend periodic cleaning of the cooling system. Some recommend draining, flushing, and adding new coolant with fresh antifreeze every two years. Refer to the manufacturer's service manual for the recommended cleaning procedure.

Locating and Repairing Radiator Leaks Radiator leaks usually leave telltale scale or dye marks. Leaks may often be repaired by soldering. Radiator repair is usually done in a radiator shop.

Water-Pump Service Water pumps do not require lubrication. Most pump problems are due to worn bearings. This is often caused by overtightening the drive belt. Refer to the manufacturer's service manual if you need to repair a water pump.

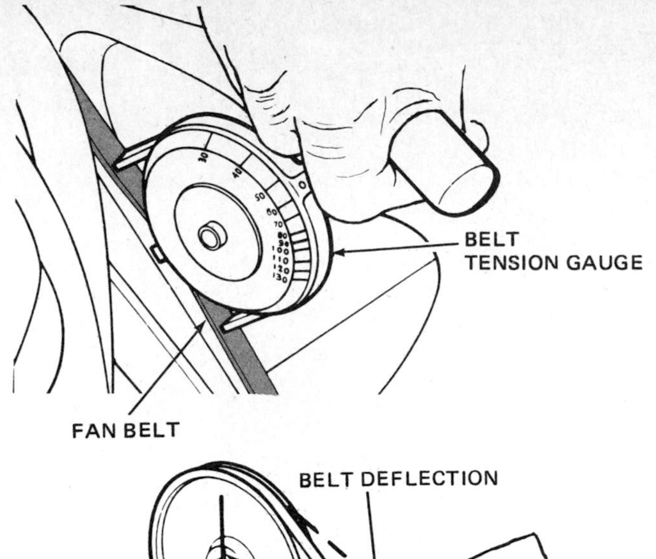

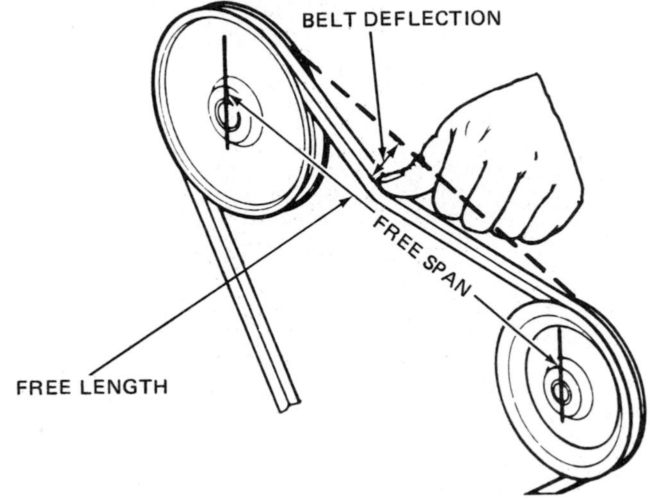

FIG. 9-24 Top, using a belt-tension gauge to check the tension of a drive belt. Bottom, checking belt deflection.

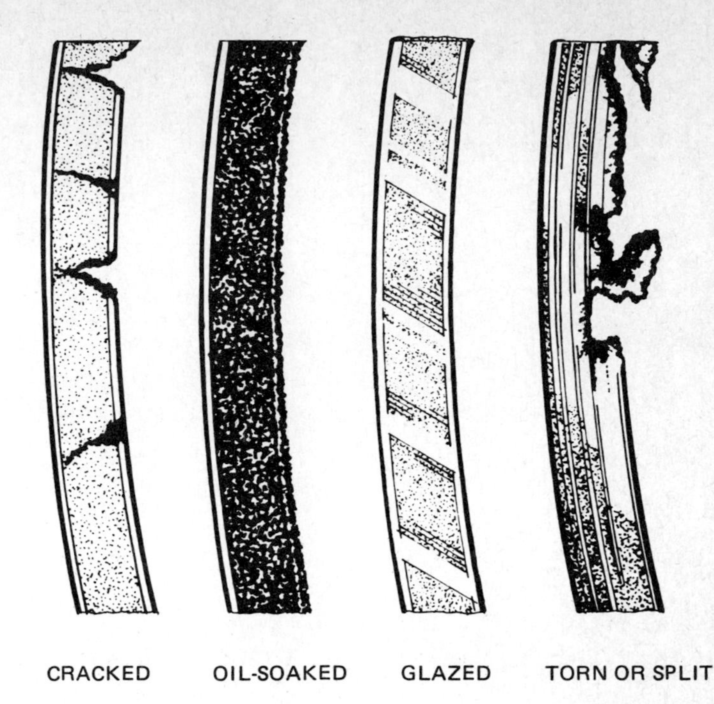

FIG. 9-25 Conditions to look for when inspecting a drive belt.

CHAPTER 9: REVIEW QUESTIONS

Select the *one* correct, best, or most probable answer to each question. You can find the answer in the section indicated at the right of each question.

1. Almost all small single-cylinder engines are (**Δ*9-1***)
   a. liquid-cooled
   b. air-cooled
   c. not cooled
   d. thermosiphon-cooled
2. Of the heat produced by the burning fuel, the cooling system is designed to remove about (**Δ*9-2***)
   a. 5 percent
   b. 50 percent
   c. 100 percent
   d. 30 percent
3. The two types of air-cooling systems are (**Δ*9-4***)
   a. open-draft and closed-draft
   b. liquid and air
   c. open-draft and forced-draft
   d. convection and conduction
4. The movement of heat from one place to another is called (**Δ*9-4***)
   a. heat transfer
   b. conduction
   c. convection
   d. radiation
5. The liquid-cooling system cools by circulating coolant between the (**Δ*9-8***)
   a. radiator and engine fins
   b. radiator fins and engine
   c. engine water jackets and radiator
   d. engine water jackets and fins
6. The coolant used in liquid-cooled engines is (**Δ*9-9***)
   a. water
   b. ethylene glycol
   c. a mixture of water and ethylene glycol
   d. alcohol and water
7. One of the most important services for the air-cooled-engine cooling system is to (**Δ*9-18***)
   a. replace the fan
   b. install a new shroud
   c. clean the fins
   d. keep enough antifreeze in the system
8. Air-cooled engines are cooled by (**Δ*9-4***)
   a. conduction and convection
   b. conduction and radiation
   c. convection and radiation
   d. none of the above

9. The purpose of the thermostat is to (*Δ9-15*)
   a. control engine temperature
   b. measure engine temperature
   c. prevent engine cooling
   d. prevent coolant circulation when the engine is hot

10. One purpose of the radiator cap on the liquid-cooled engine is to (*Δ9-14*)
   a. pressurize the system
   b. prevent the system from being pressurized
   c. allow vacuum to develop in the system
   d. aid in circulation of the coolant

ELECTRICITY AND ELECTRIC SYSTEMS

Part 4 covers the various electric and electronic systems used with small engines. It describes electricity, magnetism, and electronics, including the actions of diodes and transistors. Then, later chapters explain how the components of the small-engine electric system work and how they are serviced. These include batteries, starting systems, ignition systems, and charging systems.

There are eight chapters in Part 4. They are:

Electricity and Electronics

After studying this chapter, you should be able to:

1. ***Explain current flow and tell how electrons are made to move in a conductor.***
2. ***Describe electromagnets and explain how they produce magnetism.***
3. ***Define amperes, voltage, and resistance and explain how they are related in an electric circuit.***
4. ***Explain how diodes and transistors work.***
5. ***Explain the difference between a fuse and a circuit breaker.***

Δ10-1 ELECTRICITY AND SMALL ENGINES Electricity produces the electric sparks that ignite the air-fuel mixtures in the engine cylinder. Electric starters are used on many engines. Most of these electric starters are powered by storage batteries. A few operate from house current (115 volts). Some small-engine applications are equipped with lights and horn. These are also operated by elecricity.

Δ10-2 WHAT IS ELECTRICITY? Nobody has even seen what composes electricity. So we have to rely on the description of scientists. Electricity is composed of tiny particles. The particles are so tiny that it would take billions upon billions of them, all piled together, to make a spot big enough to be seen through the average microscope. These particles are called *electrons*. Electrons are all around us in fantastic numbers. In 1 ounce [28 g] of iron, for example, there are about 22 million billion billion electrons. Electrons are normally locked into the elements that form everything in our world.

Δ10-3 ELECTRIC CURRENT If electrons are forced to move together in the same direction, there is a flow, or current, of electrons. This flow is called an *electric current*. The job of the battery and the alternator is to get the electrons to flow in the same direction. When many electrons are moving, the current is high. When relatively few electrons are moving, the current is low.

Δ10-4 MEASURING ELECTRIC CURRENT The movement of electrons, or electric current, is measured in *amperes*, or *amps*. One ampere of electric current is a very small amount of current. A battery can put out 200 to 300 amperes as it operates the starting motor. Headlights draw 10 amperes or more. One ampere is the flow of 6.28 billion billion electrons per second.

We cannot count electrons to find out how many amperes are flowing in a wire. We must use an *ammeter* (Fig. 10-1) to find this out. The ammeter uses an effect of electron flow. This effect is that any flow of electrons, or electricity, produces magnetism.

Δ10-5 MAGNETISM There are two forms of magnetism: natural and electrical. Natural magnets are made of iron or other metals (Fig. 10-2). Electrically produced magnets are called *electromagnets*. Natural and electromagnets act in the same way. They attract iron objects. Here are two important facts about magnets:

- Magnets can produce electricity.
- Electricity can produce magnets.

Magnets and electromagnets, and how they act, are described in later sections.

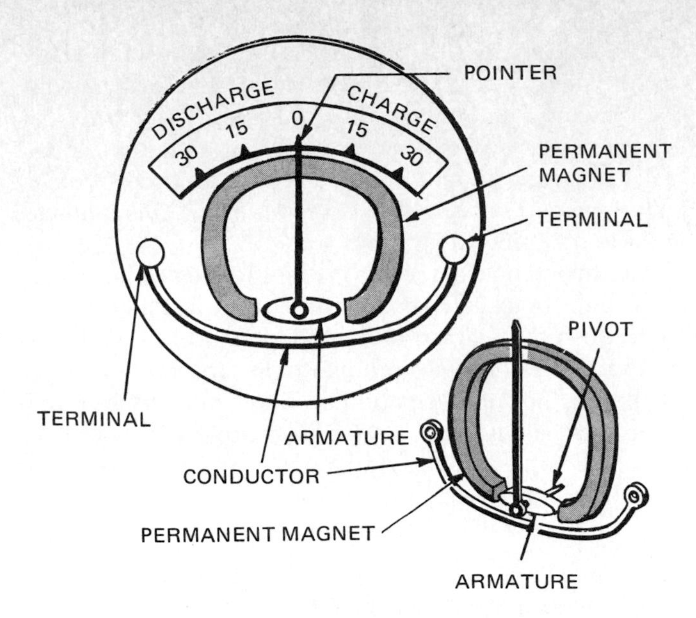

FIG. 10-1 Simplified drawing of an ammeter.

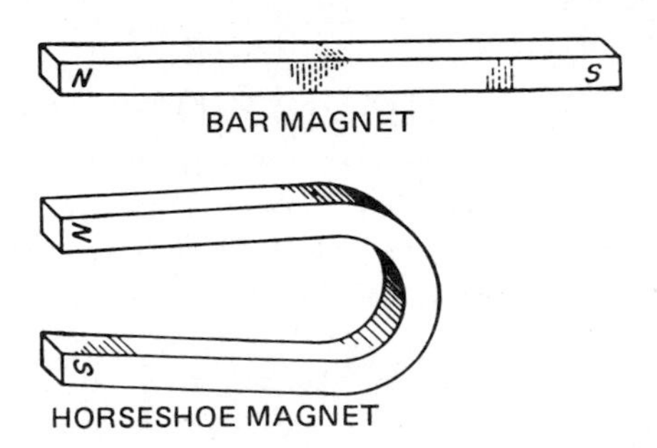

FIG. 10-2 Bar and horseshoe magnets.

Δ10-6 THE AMMETER The ammeter measures electric current. A simple ammeter is shown in Fig. 10-1. It is used in the control panel of some engines. Its purpose is to tell the operator if the charging system is operating properly.

The ammeter works by magnetism. The conductor (in Fig. 10-1) is connected in series in a circuit. One end is connected to the battery. The other end is connected to a *load.*

The pointer is mounted on a pivot. A small piece of oval-shaped iron is mounted on the same pivot. This oval-shaped piece of iron is called the *armature.* A permanent magnet, almost circular in shape, is positioned so its two ends are close to the armature. The permanent magnet attracts the armature and tends to hold it in a horizontal position. In this position, the pointer or needle points to zero. When the alternator starts sending current to the battery, the current passing through the conductor produces magnetism. The magnetism attracts the armature and causes it to swing clockwise. This moves the pointer to the "charge" side. The more current that flows, the stronger the magnetism and the farther the pointer moves. The meter face is marked to show the number of amperes flowing.

Suppose the charging system is not working and the tractor lights are turned on. Current will flow from the battery to the lights. Now current will flow in the reverse direction through the conductor in the ammeter. Therefore, the armature is attracted in the opposite direction. It swings in a counterclockwise direction. This moves the pointer to the "discharge" side of the ammeter. The more current that flows out of the battery, the farther the pointer moves across the discharge side of the ammeter.

Δ10-7 MAKING ELECTRONS MOVE Electrons on the move make up electric current. When too many electrons are gathered in one place, they try to move away. The battery and the alternator are devices that collect electrons. They collect electrons at one terminal by taking them away from the other. If the two terminals are connected by a conductor, electrons flow from the "too many" terminal to the "too few" terminal.

Δ10-8 VOLTAGE Suppose there are many electrons at one terminal. And suppose there is a shortage of electrons at the other terminal. When there is a great excess and a great shortage, we say that the electrical pressure is high. This means that there is a high pressure on the electrons to move from the "too many" terminal to the "too few" terminal.

Electrical pressure is measured in *volts.* High pressure is high voltage. Low pressure is low voltage. Most small-engine batteries are 12-volt units. Twelve volts is considered low pressure. The spark at the spark-plug gap is a flow of electrons at high pressure or voltage. The voltage at the spark-plug gap can be 20,000 volts or more.

Δ10-9 INSULATION If electrons escape from the wire in which they are flowing, electric power is lost. This is the reason that electric wires are insulated. In addition to power loss, electrons can be dangerous. For example, if the insulation on the wire to a household appliance or a lamp is damaged, a fire could result. A person who touches the wire or appliance could get an electric shock.

On the engine, the wires and cables between the battery, the alternator, and other electric devices are all covered with insulation (Fig. 10-3). The insulation is a *nonconductor.* This means that it is extremely difficult for electrons, or electric current, to flow through it. But if the insulation is cut, scraped off, or cracks, electric current will go where it is not supposed to. It could take a shortcut through the metal of the frame and the engine. Such a shortcut is called a *short circuit.* It can cause trouble.

Insulation has the job of keeping the electrons, or the electric current, moving in the proper path, or circuit. Circuits include the wires and the electrical devices on the engine.

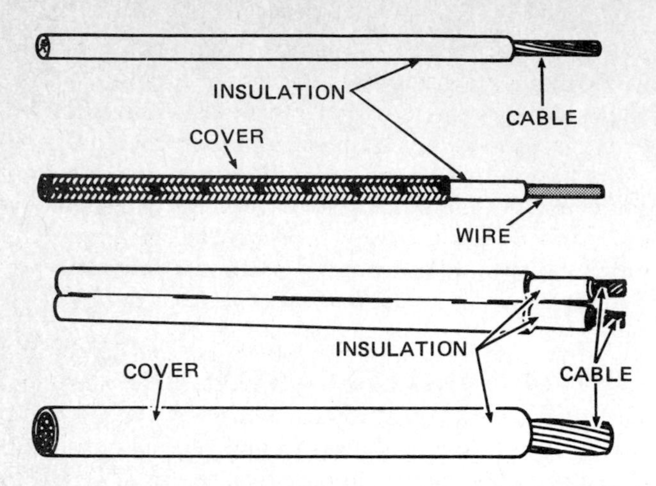

FIG. 10-3 Various types of engine wiring, showing the insulation and outer cover. *(ATW)*

Δ10-10 MAGNETS

Magnets (Δ10-5) act through *lines of force.* These lines of force stretch between the ends of the magnet. The two ends of the magnet are called the *magnetic poles,* or the *poles.* One pole is called the *north pole.* The other pole is called the *south pole.* The area surrounding the poles is called a *magnetic field.*

Δ10-11 LINES OF FORCE

The lines of force have two characteristics. First, the lines of force try to shorten themselves. If you hold the north pole of one magnet close to the south pole of another magnet, the two magents will pull together (Fig. 10-4). If we drew the lines of force between the two poles, the picture would look something like Fig. 10-5. The lines of force, stretching between the two poles, try to shorten. This tends to pull the two poles together.

The second characteristic is that the lines of force run more or less parallel to each other. At the same time they try to push away from one another. Suppose we bring like poles together—two north poles, for example (Fig. 10-6). The lines of force run parallel to one another and try to push away (Fig. 10-7). The magnet that is free to move will move away as the same pole of the other magnet is brought closer.

We can draw these conclusions:

- Like magnetic poles repel each other. North repels north. South repels south.
- Unlike magnetic poles attract each other. North attracts south. South attracts north.

Δ10-12 ELECTROMAGNETS

Electromagnets act just like natural magnets. An electromagnet can be made by wrapping a wire around a tube (Fig. 10-8). We saw what happened in the ammeter when current flowed one way or another through the conductor. The

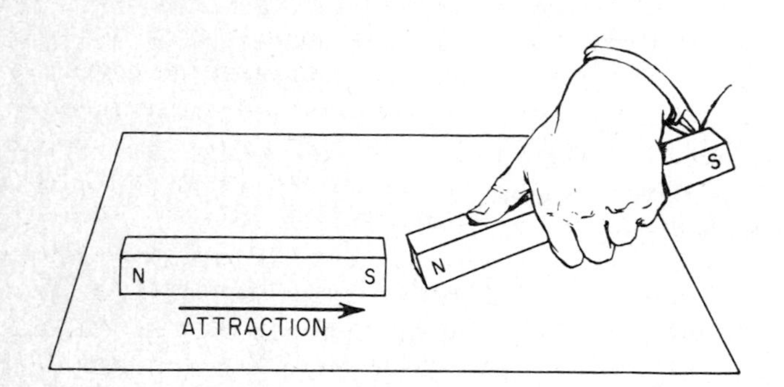

FIG. 10-4 Unlike magnetic poles attract each other.

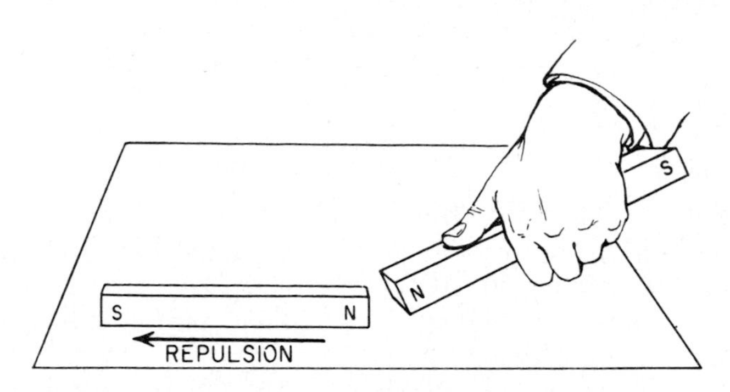

FIG. 10-6 Like magnetic poles repel each other.

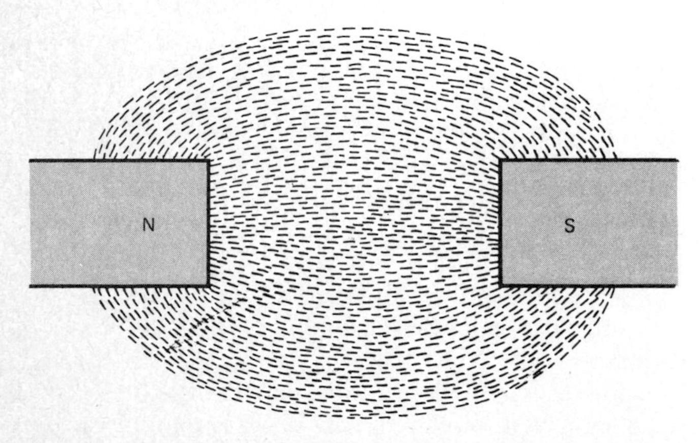

FIG. 10-5 Magnetic lines of force between two unlike magnetic poles. The magnetic lines of force tend to shorten, producing the attractive force that pulls the poles together.

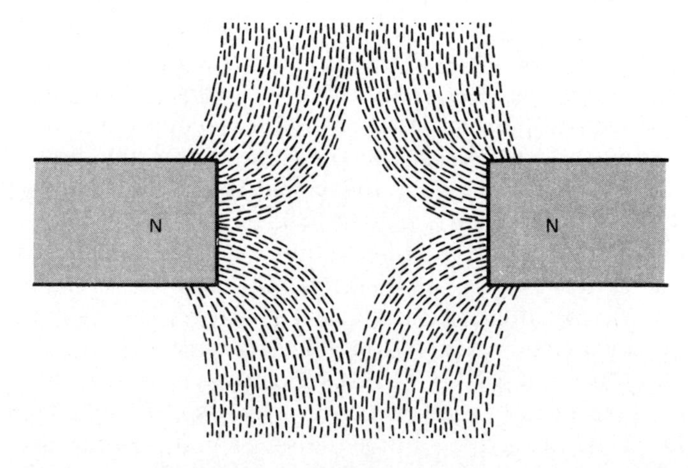

FIG. 10-7 Magnetic lines of force between two like poles. Magnetic lines of force tend to parallel each other, forcing the two like poles away from each other.

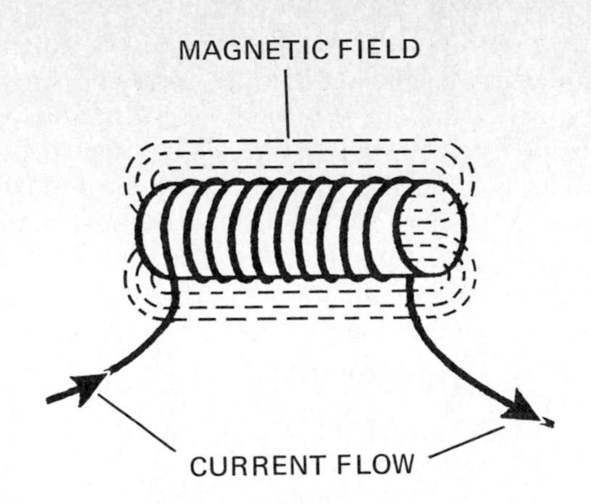

FIG. 10-8 **Current flow through a wire coiled around a tube produces a magnetic field.**

current produced magnetism, or magnetic lines of force.

Current flowing through a single wire, or conductor, will not produce much magnetism. But suppose you wind a conductor, or wire, around a rod. Then connect the ends of the wire to a source of electric current (or electrons). A strong magnetic field will develop around the coil of wire.

With current flowing through the winding, the winding acts just like a bar magnet. You can point one end of the winding toward a pole of a bar magnet. The winding will either attract or repel the bar-magnet pole. One end of the winding is a north pole. The other end is a south pole. You can change the poles by reversing the leads to the source of current. When the electrons flow through in one direction, one of the poles becomes north. But when the electrons flow through in the opposite direction, the poles reverse. The north pole becomes the south pole, and the south pole becomes the north pole.

An electromagnet such as the one made by winding wire around a tube is also called a *solenoid*. It is used in several places in the small-engine electric system. Solenoids are explained in more detail in later chapters.

Δ10-13 RESISTANCE

An insulator has a high resistance to the movement of electrons through it. A conductor, such as a copper wire, has a very low resistance. Resistance is found in all electric circuits. In some circuits a high resistance is needed to keep down the amount of current flow. In other circuits we want as little resistance as possible so that a high current can flow.

Resistance is measured in *ohms*. For example, a 1000-foot [305-m] length of wire that is about 0.1 inch [2.5 mm] in diameter has a resistance of 1 ohm. A 2000-foot [610-m] length of the same wire has a resistance of 2 ohms. The longer the path, the greater the resistance.

A 1000-foot [305-m] length of wire that is about 0.2 inch [5.1 mm] in diameter has a resistance of only ¼ ohm. The heavier the wire, the lower the resistance. The longer the path, or circuit, the farther the electrons have to travel. Therefore, the higher the resistance to the electric current. With the heavier wire, the path is larger. This means that the resistance is lower.

With most conductors, such as copper, resistance goes up with temperature. A hot copper wire will carry less current than a cold copper wire. The reverse is true for a few materials. For example, the material in the engine sending unit of the cooling-system temperature indicator has less resistance as the temperature increases.

Δ10-14 OHM'S LAW

There is a definite relationship between amperes (electron flow), voltage (electrical pressure), and resistance. As the electrical pressure goes up, more electrons flow. Increasing the voltage increases the amperes of current. However, increasing the resistance decreases the amperes of current. These relationships are summed up in a formula known as *Ohm's law:*

Voltage is equal to amperage times ohms.

$$V = I \times R$$

where V = voltage
I = current in amperes
R = resistance in ohms

The important point about Ohm's law is that increasing the amount of resistance decreases the current flowing.

Δ10-15 ONE-WIRE SYSTEMS

For electricity to flow, there must be a complete path, or circuit. The electrons must flow from one terminal of the battery or the alternator, through the circuit, and back to the other terminal. On the riding mower, for example, the engine, metal frame, and body are used to carry to electrons back to the other terminal. Therefore, no separate wires are required for the return circuit from the electric device to the battery or the alternator. The return circuit is called the *ground*. It is indicated in wiring diagrams by the symbol ⏚ or ⏚ . The ground—the engine, body, and frame—is the other half of the circuit. It is the return circuit between the source of electricity (battery or alternator) and the electric device.

Δ10-16 ALTERNATING CURRENT AND DIRECT CURRENT

Most of the electricity generated and used is alternating current (ac). The current flows first in one direction and then in the opposite direction (Fig. 10-9). It alternates. The current you use in your

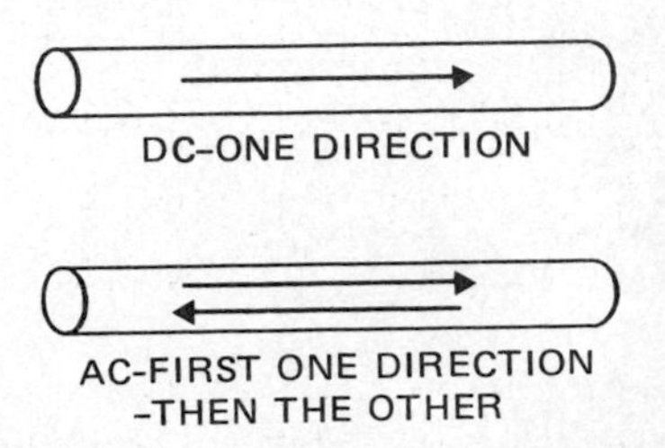

FIG. 10-9 **Directions of ac and dc current flow in a wire.**

home is ac. It alternates 60 times per second and therefore is called 60-cycle [Hz] ac. In the metric system, cycles per second are called *hertz,* abbreviated Hz.

The charging system cannot use ac. The battery is a direct-current (dc) unit. When you discharge the battery by connecting electric devices to it, you take current out in one direction only. The current does not alternate, or change directions. Most electric devices on the small-engine operate on dc only.

Δ10-17 FUNDAMENTALS OF ELECTRONICS

The words *electronics* and *solid state* are key words today in the study of electricity and electronics. *Electronics* refers to any electric assembly, circuit, or system that uses solid-state electronic devices such as transistors and diodes. A *solid-state* device is one that has no moving parts except electrons. Diodes and transistors are such devices.

Before electronics, all electric circuits were turned on and off by some type of mechanical switch. Now, electric circuits can be turned on and off electronically by devices that have no moving parts. An example is the primary circuit in the typical ignition system. The old-style system used a set of contact points, which acted as a mechanical switch. The new electronic systems use a transistor to do this. The transistor acts faster and more accurately than the old-style system and has no moving parts to wear. Ignition systems are described in Chap. 14.

Our interest is in what electronics do in the engine's electric equipment and how it is done. First, we look at semiconductors, diodes, and transistors—the electronic devices that are the heart of electronic equipment. Basically, diodes are one-way valves that permit electricity to flow through in only one direction. Transistors can be electric switches that start and stop current flow. Both diodes and transistors use materials called *semiconductors.*

Semiconductors are materials that are halfway between a conductor and a nonconductor. Sometimes they conduct; sometimes they act like insulators. This action is described when we discuss diodes and transistors in Δ10-18 and 10-19.

Δ10-18 DIODES

The *diode* is a device that permits electricity to flow through in one direction but not in the other. Figure 10-10 shows this. The alternator is the device in one type of charging system that produces current to charge the battery and operate electric devices. However, the current from the alternator is alternating. So diodes are used to change it to direct current. You cannot use alternating current to charge the battery or operate most small-engine electric equipment. Direct current is required. Chapter 16 describes alternators and the diodes used in them.

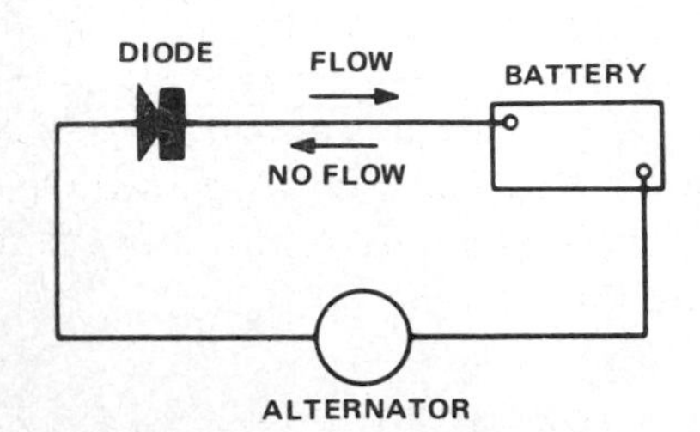

FIG. 10-10 Alternating current from an alternator can be rectified, or changed to direct current, by a diode so that the current can charge the battery.

Δ10-19 TRANSISTORS

The *transistor* is a diode with some additional semiconductor material. This added material makes it possible for the transistor to amplify current. In operation, a small signal current allows a large main current to flow. Figures 10-11 and 10-12 show this. In Fig. 10-11 a small current of 0.35 ampere is flowing to the base marked *n*. The *n* means that this slice of semiconductor material has extra electrons which have negative charges. When the extra electrons flow in (after the switch is closed), they form a path through which current can flow from one *p* slice to the other. This is shown by the arrows in Fig. 10-11. The *p* means that the material lacks electrons and therefore, in effect, has a positive charge.

When the switch is opened (Fig. 10-12), no current (electrons) flows into the *n* slice. Therefore, there is no path through the slice for electrons to flow from one *p* to the other.

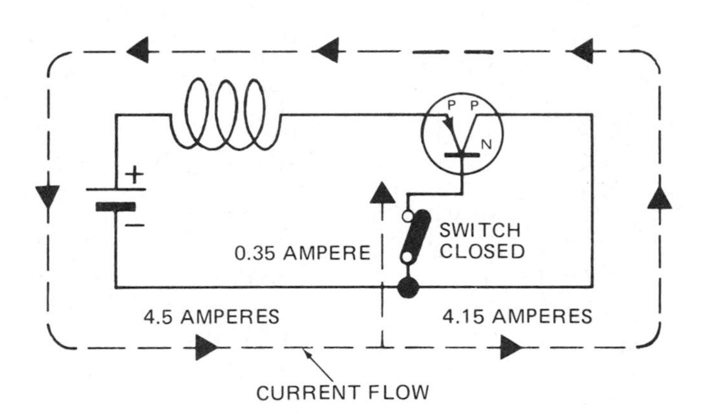

FIG. 10-11 When the switch is closed, current flows through the transistor.

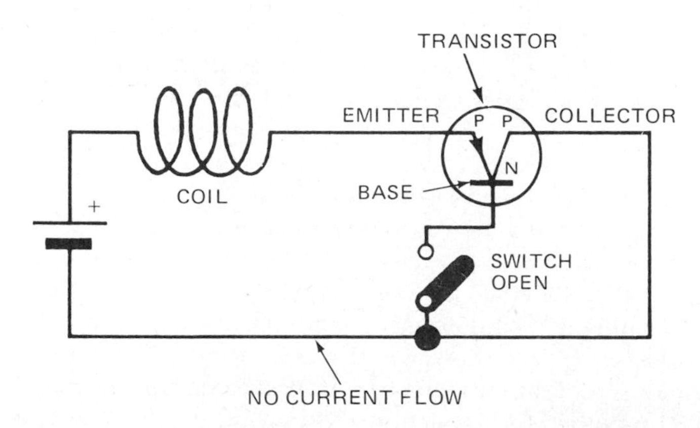

FIG. 10-12 When the switch is open, no current flows.

In the transistor, a small current to the controlling slice of semiconductor material allows a large current to flow through the transistor. When this small current flows (which is called the *signal,* or *trigger current*), the large current flows. When the signal current stops, the large current stops.

Δ10-20 INTEGRATED CIRCUITS Scientists and engineers have found ways to make diodes and transistors extremely small. This makes is possible to group large numbers of these semiconductor devices together in a small space. Such groups are called *integrated circuits* because many components are put together, or integrated, into a very small package called a *chip.* Integrated-circuit chips are used in computers and complex controlling devices. Examples in engines are the electronic ignition system, electronic fuel injection, and electronic engine-system controls.

Δ10-21 WIRING CIRCUITS With the increasing number of electrically operated devices that can be used with a small engine, the wiring circuits may be complicated. Figure 10-13 shows the wiring diagram for one model of a small engine. On some engines, the wires between the components are bound together into *wiring harnesses.* Each wire is marked by special colors in the insulation—for example, light green, dark green, blue, red, black with a white tracer, and so on. These markings make identification of the various wires easier.

The wires connecting electric components and instruments usually are fastened together by terminals attached to the ends of the wire. Figure 10-14 shows various types of terminals. Use of terminals and connectors makes disassembly and assembly operations of wires and electric devices faster and easier. Many types of connectors are available for use in the shop. One advantage of many connectors is that they may be used to splice and tap into a wire without your having to strip insulation, solder, or tape the bare connection afterward.

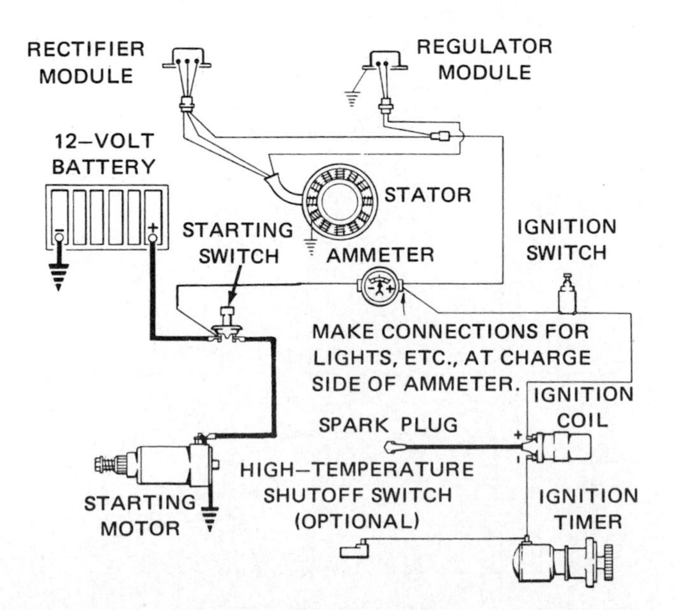

FIG. 10-13 Wiring diagram of a small engine that has a battery ignition and an alternator charging system. *(Teledyne Wisconsin Motor)*

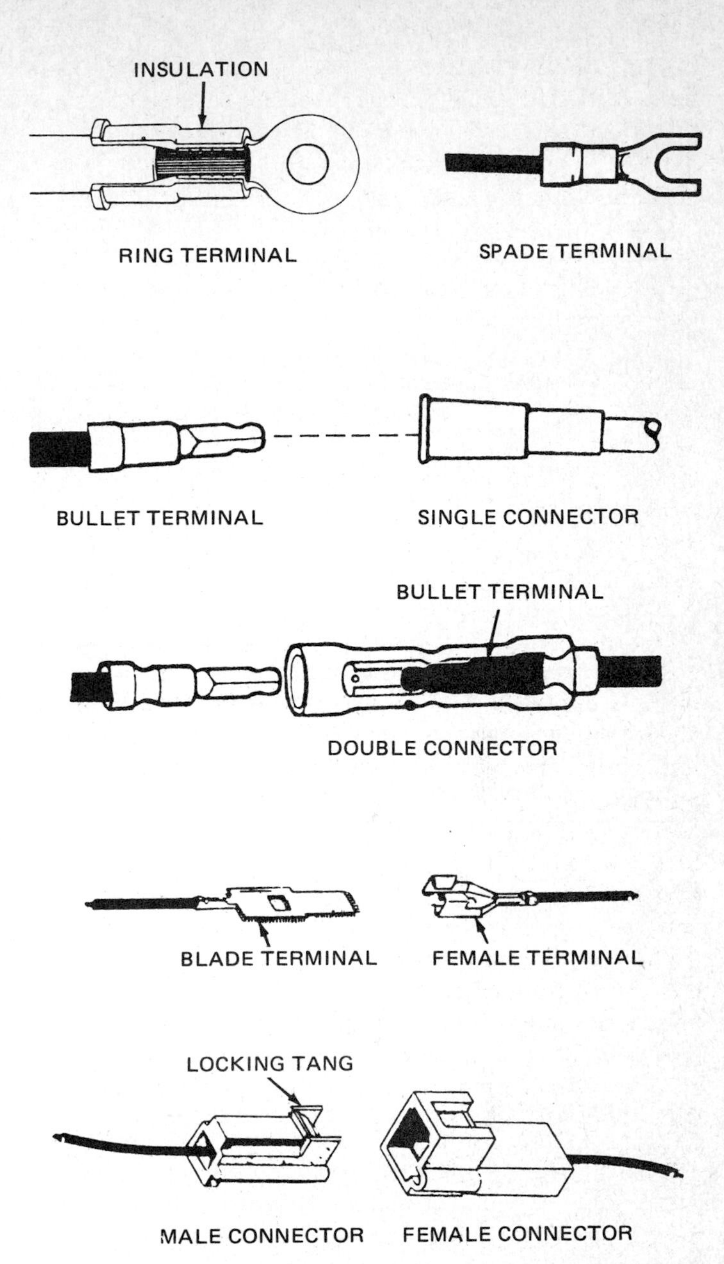

FIG. 10-14 Various types of wire terminals. *(ATW)*

In electrical work, abbreviations and symbols are used constantly. Symbols are a kind of shorthand. Their use permits large and complicated electric systems to be accurately drawn in schematic form. Figure 10-15 shows many of the widely used electrical symbols that you may find in manufacturers' manuals. If you see a symbol that is not on this list, look carefully through the manual to find the legend for symbols used.

Δ10-22 FUSES AND CIRCUIT BREAKERS A fuse is the device that is used most often to protect the components of an electric circuit from damage due to excessive current flow. A typical cartridge

ELECTRICAL SYMBOLS			
SYMBOL	REPRESENTS	SYMBOL	REPRESENTS
A	AMMETER		GROUND—CHASSIS FRAME (Preferred)
	BATTERY—ONE CELL		GROUND-CHASSIS FRAME (Acceptable)
	BATTERY—MULTICELL		LAMP or BULB (Preferred)
12 V + −	(Where required, battery voltage or polarity or both may be indicated as shown in example. The long line is always positive polarity.)		LAMP or BULB (Acceptable)
		MOT	MOTOR—ELECTRICAL
	CABLE—CONNECTED	—	NEGATIVE
or	CABLE—NOT CONNECTED	+	POSITIVE
	CAPACITOR		RESISTOR
	CONNECTOR—FEMALE CONTACT		SWITCH—SINGLE THROW
	CONNECTOR—MALE CONTACT		SWITCH—DOUBLE THROW
	CONNECTORS—SEPARABLE—ENGAGED		TERMINATION
	FUSE	V	VOLTMETER
GEN	GENERATOR	or	WINDING—INDUCTOR

FIG. 10-15 Commonly used electrical symbols.

fuse is shown in Fig. 10-16. It contains a soft metal strip, connected at the ends to the fuse caps. When too much current flows through the fuse, the current overheats the metal strip, and it melts, or blows. This opens the circuit and stops the current flow. Then the circuit should be checked to find out what caused the fuse to blow. After the trouble is fixed, a new fuse should be installed.

Circuit breakers are used in headlight and other high-current-carrying circuits. These circuits may have temporary overloads. But safe operation requires that normal circuit action be rapidly restored. In most circuit breakers, the current passes through a bimetallic strip (Fig. 10-17). There is a set of normally closed contacts at one end. When too much current flows, the heat causes

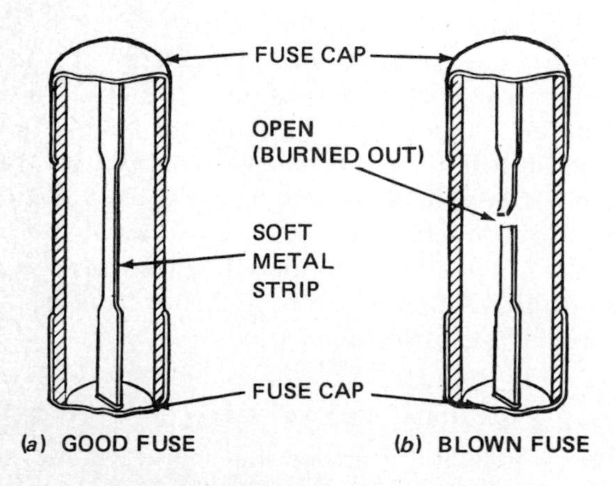

FIG. 10-16 A good and a blown cartridge fuse.

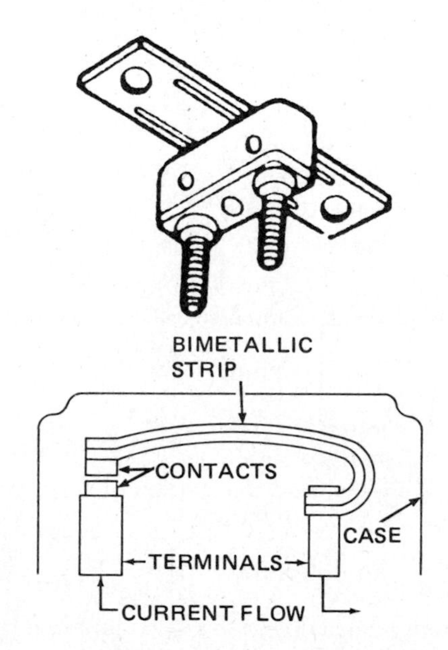

FIG. 10-17 A self-resetting circuit breaker.

the strip to bend, or warp. As the strip moves, the contacts separate. This opens the circuit and stops the current flow. However, with the contacts open, the bimetallic strip cools and closes the contacts again. If the overload is still present, the circuit breaker will cycle by opening the contacts again. When the overload is no longer present, the circuit will resume normal operation.

The type of circuit breaker that automatically reconnects the circuit is called *self-resetting* (Fig. 10-17). Some circuit breakers must be *manually reset*. On this type, the contacts remain open after the circuit breaker is tripped. This forces a reset button to pop out on the case of the circuit breaker. To manually reset the circuit breaker, push in on the button.

CHAPTER 10: REVIEW QUESTIONS

Select the *one* correct, best, or most probable answer to each questions. You can find the answer in the section indicated at the right of each question.

1. Movement of electrons in one direction in a wire is called (**Δ*10-3***)
   a. voltage
   b. electric current
   c. insulation
   d. watts
2. Electric current is measured in (**Δ*10-4***)
   a. volts
   b. current flow
   c. amperes
   d. watts
3. Materials that strongly oppose the movement of electrons through them are called (**Δ*10-9***)
   a. metals
   b. insulators
   c. conductors
   d. volts
4. Magnetic lines of force (**Δ*10-11***)
   a. are a flow of electrons
   b. never exist in a dc circuit
   c. can be seen
   d. none of the above
5. When a great many electrons crowd around one alternator terminal and there is a shortage of electrons at the other terminal, the result is (**Δ*10-8***)
   a. higher voltage
   b. lower voltage
   c. greater resistance
   d. higher alternator speed
6. Two basic facts about magnets are (**Δ*10-11***)
   a. like poles attract; unlike poles repel
   b. unlike poles attract; like poles repel
   c. lines of force cross and shorten
   d. lines of force merge and lengthen
7. A solenoid (**Δ*10-12***)
   a. is an electromagnet
   b. becomes magnetic when current flows through it
   c. includes a coil, or winding
   d. all of the above
8. Most electric devices on the engine (**Δ*10-16***)
   a. operate on dc
   b. operate on dc or ac
   c. use neither ac nor dc
   d. are electromechanical
9. The purpose of a fuse is to (**Δ*10-22***)
   a. open the circuit if a short or ground occurs
   b. protect the system from excessive current flow
   c. blow out and open the circuit if high current flows
   d. all of the above
10. When the insulation fails on wiring in the small-engine electric system, the result could be (**Δ*10-9***)
    a. a short circuit
    b. a long circuit
    c. no current
    d. high voltage
11. If you wrap wire around a tube and send electric current through the wire, the assembly becomes (**Δ*10-12***)
    a. an electromagnet
    b. a solenoid
    c. both a and b
    d. neither a nor b
12. Ohm's law is (**Δ*10-14***)
    a. $I = V \times R$
    b. $R = A \times I$
    c. $V = I \times R$
    d. $A = R \times C$
13. The reason that the small-engine electric system is a one-wire system is that the (**Δ*10-15***)
    a. battery has only one terminal
    b. engine and frame furnish the return circuit
    c. alternator can charge in only one direction
    d. system uses direct current
14. The device that acts as a one-way check valve for electric current is called a (**Δ*10-18***)
    a. triode
    b. diode
    c. monode
    d. transistor
15. The device that can act like a high-speed electronic switch is called a (**Δ*10-19***)
    a. triode
    b. diode
    c. monode
    d. transistor

CHAPTER

11

Batteries and Battery Service

After studying this chapter, you should be able to:

1. ***Explain the purpose and operation of the battery.***
2. ***Describe the construction and operation of the battery.***
3. ***Define* ni-cad battery.**
4. ***List the steps in battery maintenance.***
5. ***Check electrolyte level and specific gravity and interpret the specific-gravity readings.***
6. ***List battery troubles and their possible causes.***
7. ***Connect the battery charger and charge a battery.***

Δ11-1 PURPOSE OF BATTERY The battery supplies current to operate the starting motor and the ignition system when the engine is being started. The battery also supplies current for the lights and other electric equipment when the alternator or generator is not handling the electric load. The amount of current the battery can supply is limited by the *capacity* of the battery. This, in turn, depends on the amount of chemicals it contains. Therefore, the battery is an *electrochemical device.* It produces electricity by converting chemical energy into electric energy.

Δ11-2 BATTERY CONSTRUCTION Several different types of batteries are used with small engines. Many small engines use a 12-volt lead-acid storage battery (Fig. 11-1). It has a series of six cells in the battery case. Each cell has a number of battery plates with separators between them (Fig. 11-2). When battery liquid, called *electrolyte,* is put into each cell, the cell produces an electrical pressure of 2 volts.

The six cells in the battery produce a total of 12 volts. More exactly, each cell has a voltage of 2.1 volts when the

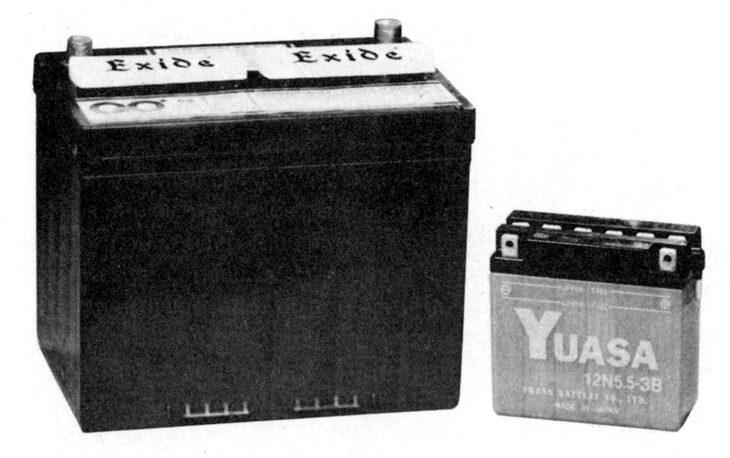

FIG. 11-1 Comparative sizes of batteries used in small-engine electric systems. Left, an automobile battery. Right, a motorcycle, or utility, battery. *(ATW)*

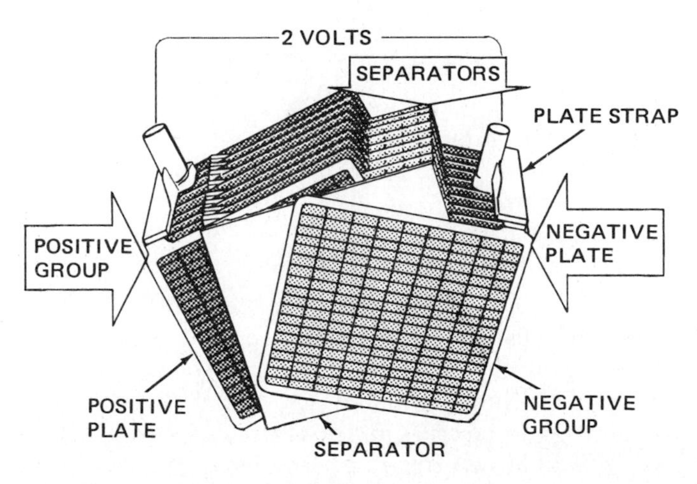

FIG. 11-2 A partly assembled battery element.

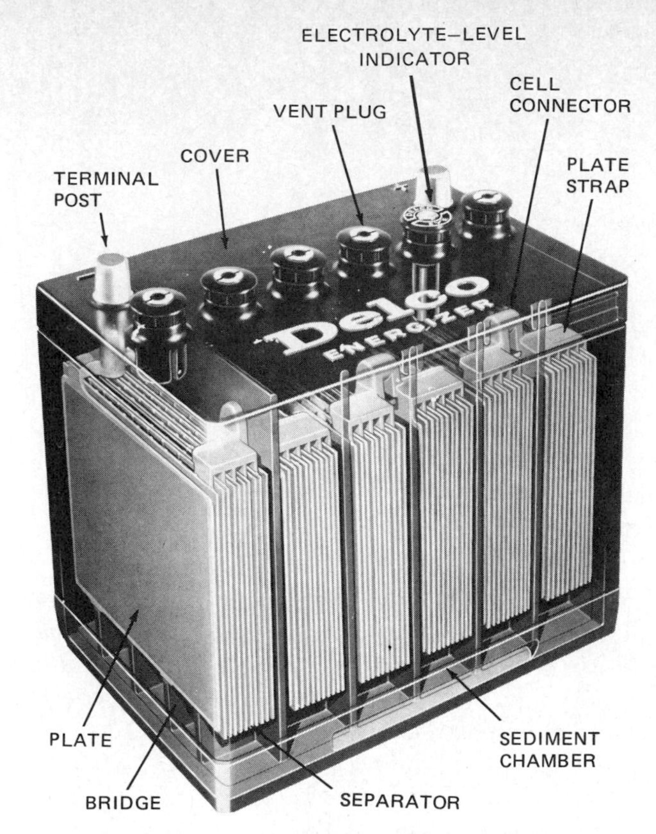

FIG. 11-3 Phantom view of a 12-volt automotive battery, showing the construction of the cells. *(Delco-Remy Division of General Motors Corporation)*

electrolyte has a *specific gravity* (Δ11-3) of 1.250 and its temperature is 80°F [27°C]. If there were only three cells, the result would be a 6-volt battery. Some small-engine starting systems use 6-volt batteries (Δ11-5).

Many batteries have vent plugs on the cover of each cell (Fig. 11-3). These vent plugs let hydrogen and oxygen gas escape from the battery while it is being charged (Fig. 11-4). The vent plugs can be removed so that water can be added when the electrolyte level is low (Δ11-8). Most small batteries have filler caps which do not have vent holes through them. A single *vent tube* allows any gas to escape (Fig. 11-5).

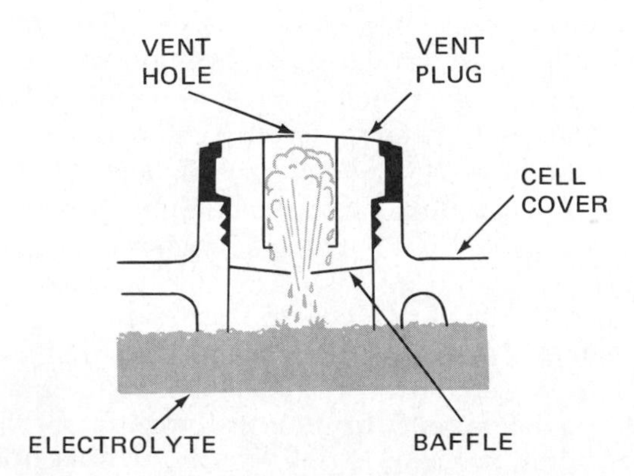

FIG. 11-4 A vent plug in a cell cover has a small vent hole to allow gases to escape. *(Delco-Remy Division of General Motors Corporation)*

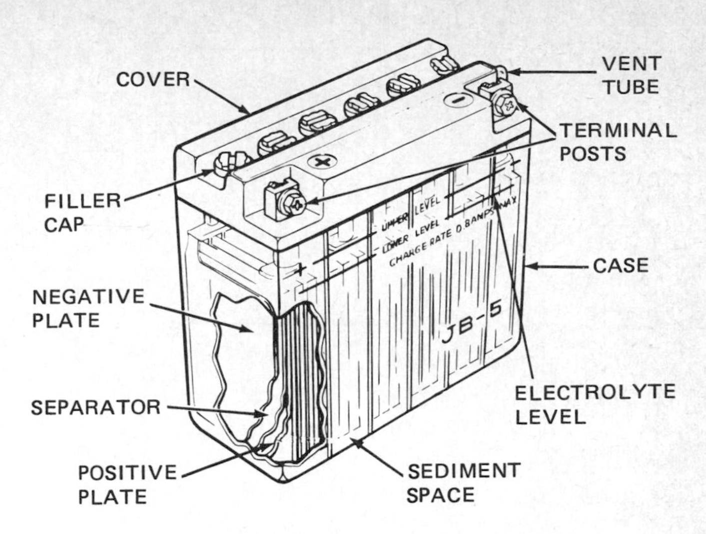

FIG. 11-5 A cutaway battery, which has a single vent tube through which the charging gases escape. *(Lawn Boy Division of Outboard Marine Corporation)*

Some batteries have a sealed top. Because sealed batteries are made differently and never need more water, they do not need a vent plug in each cell. These batteries are often called *maintenance-free batteries.* Some batteries have the terminals on the side instead of on top.

Δ11-3 BATTERY ELECTROLYTE

The battery electrolyte is made up of about 60 percent water and 40 percent sulfuric acid (in a fully charged battery). When electric current is taken out of the battery, the sulfuric acid gradually goes into the battery plates. This causes the electrolyte to become "weaker." As this happens, the battery runs down, or goes "dead." It then has to be recharged. The recharging job requires a *battery charger.*

On the engine, the alternator or generator (Chap. 16) is the battery charger. It pushes an electric current through the battery, restoring the battery to a charged condition. The charging current must flow into the battery in a direction opposite to the direction in which it was taken out.

Δ11-4 BATTERY RATINGS

The amount of current that a battery can deliver depends on the total area and volume of active plate material. It also depends on the amount and strength of the electrolyte. This is the percentage of sulfuric acid in it. In addition, as battery temperature falls, its ability to deliver current decreases (Fig. 11-6). Chemical actions are greatly reduced, and so battery power falls.

Other factors that influence battery capacity include the number of plates per cell, the size of the plates, cell size, and quantity of electrolyte. The ratings most commonly used in referring to battery capacity are described below.

1. Reserve Capacity Reserve capacity is the length of time in minutes that a fully charged battery at 80°F [27°C]

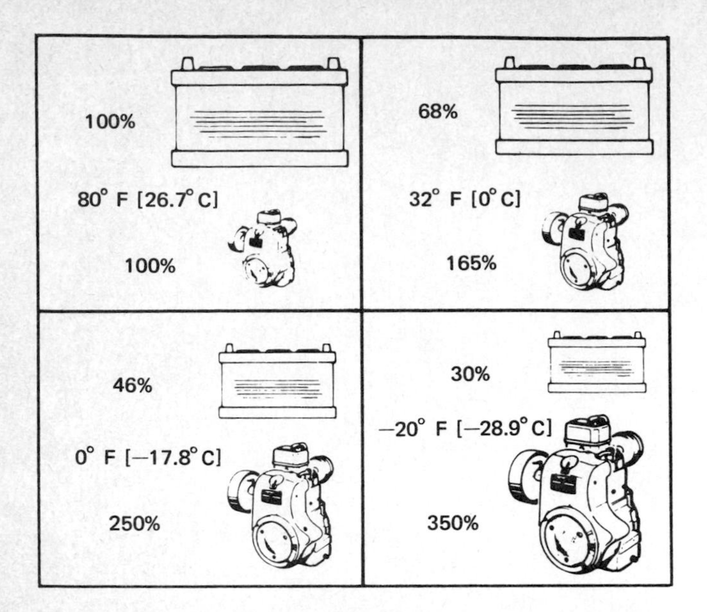

FIG. 11-6 Battery power shrinks while cranking power required increases with falling temperatures. *(Tecumseh Products Company)*

can deliver 25 amperes. A typical rating would be 125 minutes. This figure tells how long a battery can carry the electric operating load when the alternator or generator quits. Some battery manufacturers rate their batteries in ampere-hours. To find the ampere-hour capacity of a battery, multiply the number of amperes by the number of hours. For example, the 125-minute rating means the battery is a 52.1 ampere-hour unit.

$$25 \times \frac{125}{60} = 52.1$$

2. Cold Cranking Rate One of the two cold cranking rates is the number of amperes that a battery can deliver for 30 seconds when it is at 0°F [−18°C] without the cell voltages falling below 1.2 volts. A typical rating for a battery with a reserve capacity of 125 minutes would be 430 amperes. This figure indicates the ability of the battery to crank the engine at low temperatures. The second cold cranking rate is measured at −20°F [−29°C]. In this case, the final voltage is allowed to drop to 1.0 volt per cell. A typical rating for a battery with a reserve capacity of 125 minutes would be 320 amperes.

Other battery ratings include overcharge life units, charge acceptance, and watts.

Δ11-5 A SIX-VOLT BATTERY WITH DIODE

Some small engines use batteries with only three cells. These are 6-volt batteries. Some 6-volt batteries have built-in diodes which change alternating household current (ac) into direct current (dc) for charging the battery.

Direct current is not available from the home electric system. It uses only alternating current. Figure 11-7 shows a battery with a diode which changes the alternating current to direct current. The diode is built into the

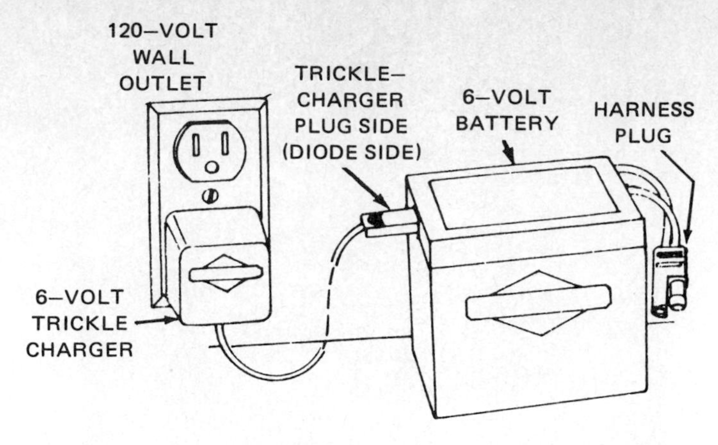

FIG. 11-7 Six-volt battery with built-in diode. The trickle charger is shown plugged into the battery for charging. *(Briggs & Stratton Corporation)*

top of the battery. The trickle charger is a small battery charger of limited capacity. It cuts down the 120 volts of the household electric system to only a few volts. But it is still alternating current. However, the diode converts it into direct current by allowing current to pass through in only one direction (**Δ**10-18).

Δ11-6 NICKEL-CADMIUM BATTERIES

Another type of battery is used to operate the starting motor of some small engines. This is the nickel-cadmium, or ni-cad, battery. It is smaller than a lead-acid battery of the same capacity. Therefore, the ni-cad battery can be used with small equipment that does not have room for a lead-acid battery.

Figure 11-8 shows a ni-cad battery attached to the handle of a lawn mower. Like the lead-acid battery, the ni-cad battery is also rechargeable. In normal use, the ni-cad battery will provide 40 to 60 starts of the engine before the battery must be recharged.

The ni-cad battery is made up of a series of cells, connected in series so their voltages add. A typical ni-cad battery is made up of 10 cells packaged in 5 sticks (Fig. 11-9). A fully charged battery of this type will have a voltage of 12.0 to 12.5 volts.

Each cell consists of positive and negative plates, separators, electrolyte, and cell container. The cells are sealed. The cell electrolyte cannot be checked. The only check is for voltage. When the battery needs recharging, it uses a special battery charger which plugs into the 120-volt household electric system (Fig. 11-10). This charger converts the alternating current into direct current, and supplies a low-voltage, low-amperage, charging current.

CAUTION **Only the trickle charger that comes with the ni-cad battery should be used to charge it. This special charger limits the charging current to 100 milliamps, a very small amount of current. Use of any other battery charger may cause the battery to explode, possibly injuring you or anyone else nearby and destroying the battery.**

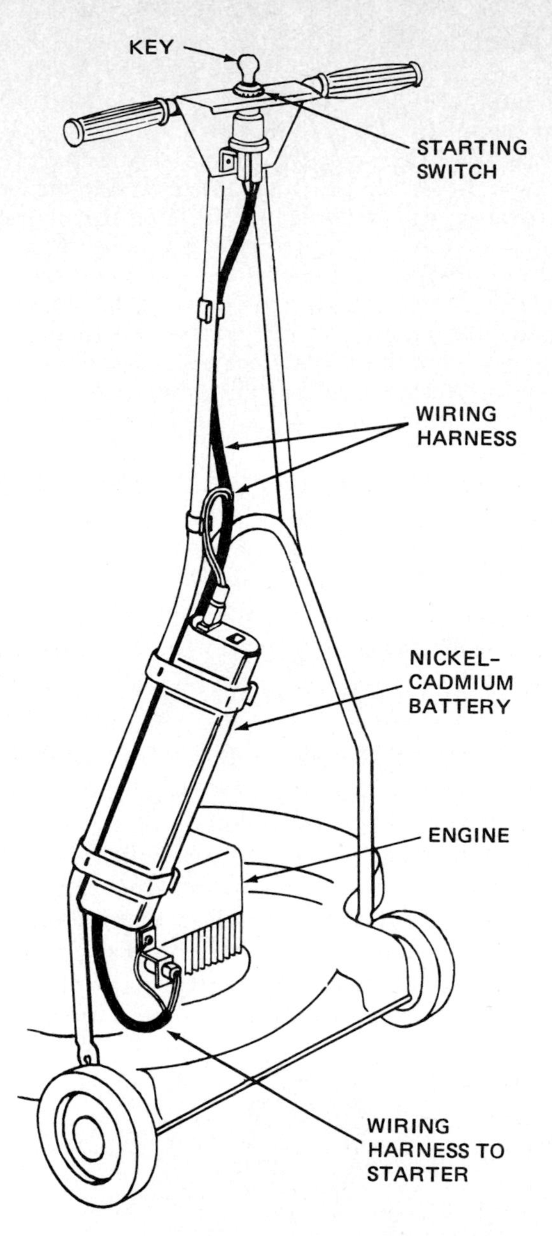

FIG. 11-8 **A lawn mower that uses a nickel-cadmium battery to operate the electric starting motor.** *(Briggs & Stratton Corporation)*

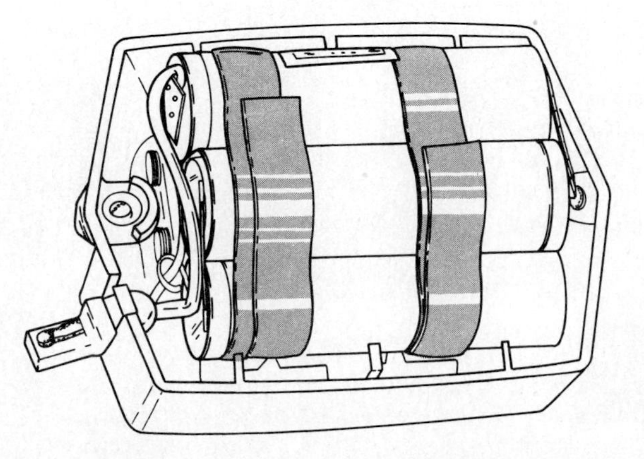

FIG. 11-9 **A typical ni-cad battery, consisting of ten cells packaged in five sticks.** *(Lawn Boy Division of Outboard Marine Corporation)*

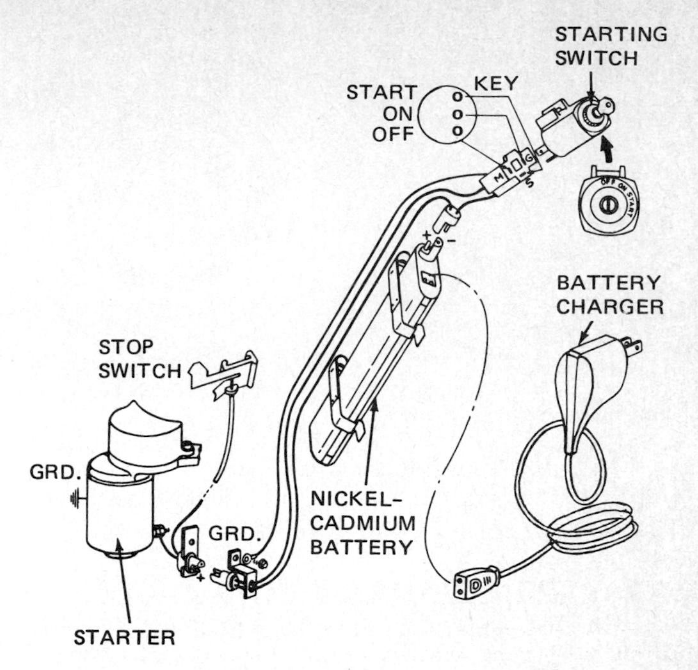

FIG. 11-10 **A starting system using a nickel-cadmium battery which includes a special battery charger.** *(Briggs & Stratton Corporation)*

BATTERY SERVICE

Δ11-7 BATTERY MAINTENANCE Battery failure is one of the more common reasons why a small engine with a battery-operated starter will not start. Usually, battery failure can be avoided if the battery is checked regularly. Here are the steps in proper battery maintenance:

1. Visually inspect the battery.
2. Check the electrolyte level in all cells.
3. Add water if the electrolyte level is low.
4. Clean corrosion from the battery terminals and top.
5. Check the condition of the battery with a tester.
6. Recharge the battery if it is low.

These last five maintenance steps are described in the following sections.

Δ11-8 CHECKING THE ELECTROLYTE LEVEL AND ADDING WATER On vent-cap batteries, the electrolyte level can be checked by removing the caps. Some batteries have a split ring which indicates the electrolyte level (Fig. 11-11). If the level is low, add water. Some batteries have an electrolyte-level indicator (a "Delco Eye"). It gives a visual indication of the electrolyte level (Fig. 11-12). Black means the level is okay. White means the level is low.

CAREFUL Do not add too much water. This can cause electrolyte to leak out and corrode the battery carrier and other metal nearby.

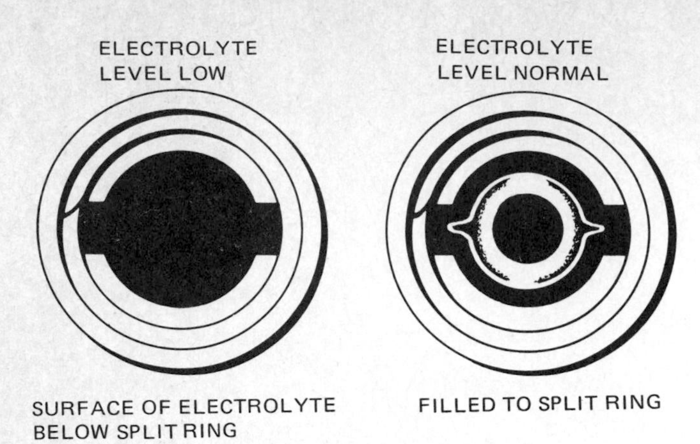

FIG. 11-11 The appearance of the electrolyte and split ring when the electrolyte is too low and when it is correct. *(Delco-Remy Division of General Motors Corporation)*

Δ11-9 CLEANING OFF CORROSION Battery terminals, especially those located on top of the battery, tend to corrode (Fig. 11-13). Corrosion builds up around the terminals and the cable clamps. To remove corrosion and clean the battery, mix 2 tablespoons [30 cc] of baking soda in a quart (liter) of warm water. Brush on the solution (Fig. 11-14). Wait until the foaming stops, and then flush off with water. If corrosion is bad, use a battery terminal cleaner to clean the terminal posts and cable clamps (Fig. 11-15).

CAUTION **The vent caps on filler plugs must be in place before cleaning the battery. Don't wash the top of the vent cap with either the baking-soda solution or the flushing water. Some of the baking soda might get down into the electrolyte through the vent hole in the cap. If this happens, the battery may be permanently damaged.**

Don't be in a hurry for the battery to dry! If you can't wait for the battery to dry by itself, use a throwaway rag or paper towel. Then throw the rag or paper towel into the trash. Never use shop towels for wiping a battery. The shop air hose must never be used to air-dry the battery. The airstream might pull electrolyte out of the cell through the vent cap. Someone nearby could be seriously injured by this spray of battery acid.

Δ11-10 CHECKING THE BATTERY CONDITION There are several methods of testing the battery to find out its condition and state of charge. The hydrometer test and the high-discharge, or battery-capacity, test are the most common. When you work in the shop, your instructor will show you how to use the testers available.

Δ11-11 HYDROMETER TEST The hydrometer measures the specific gravity of the electrolyte. There are two types of hydrometer. One contains a series of plastic balls; the other, a glass float with a marked stem on it (Fig. 11-16). Both are used in the same way. Insert the rubber tube into the battery cell so that the end of the tube is in the electrolyte. Then squeeze and release the bulb (Fig. 11-17). This draws electrolyte up into the glass tube. In the ball-type hydrometer, the number of balls

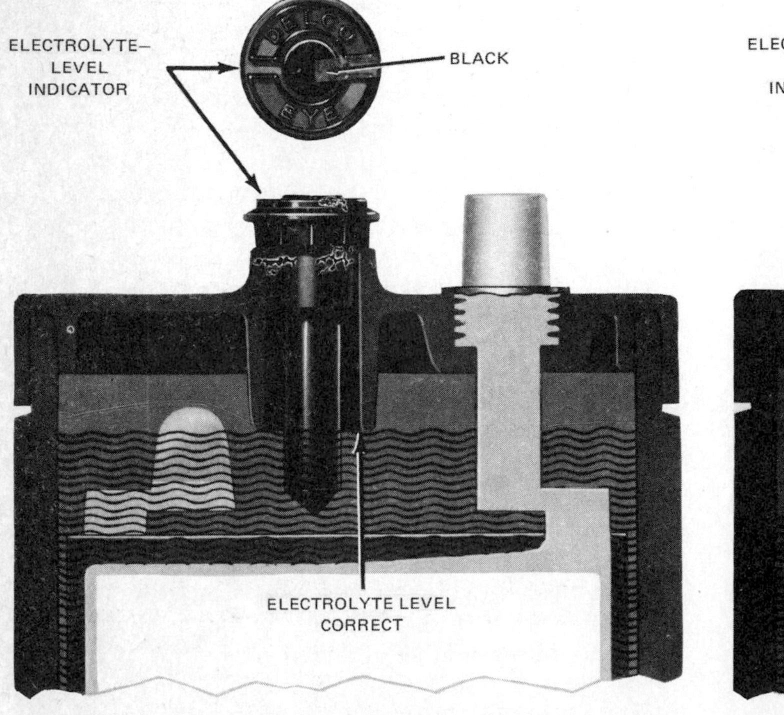

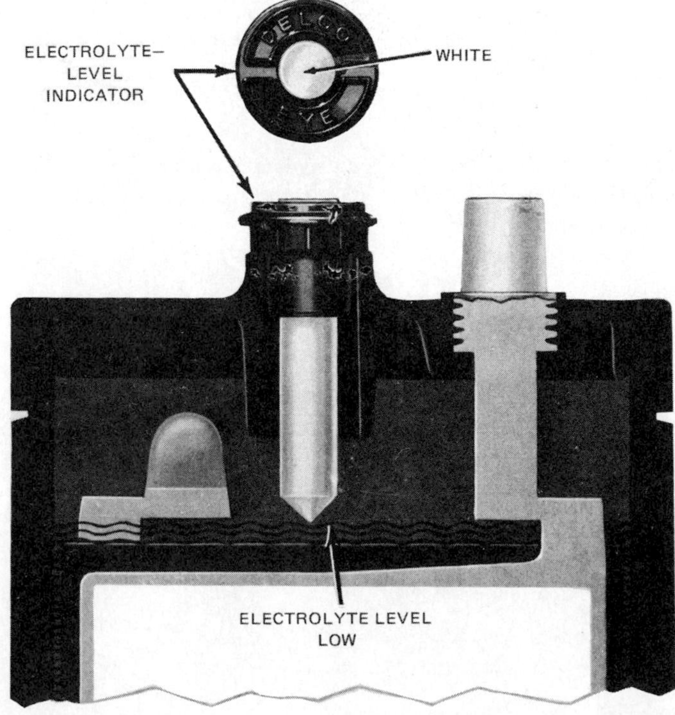

FIG. 11-12 A battery cell with an electrolyte-level indicator. Left, correct electrolyte level. Right, low electrolyte level. *(Delco-Remy Division of General Motors Corporation)*

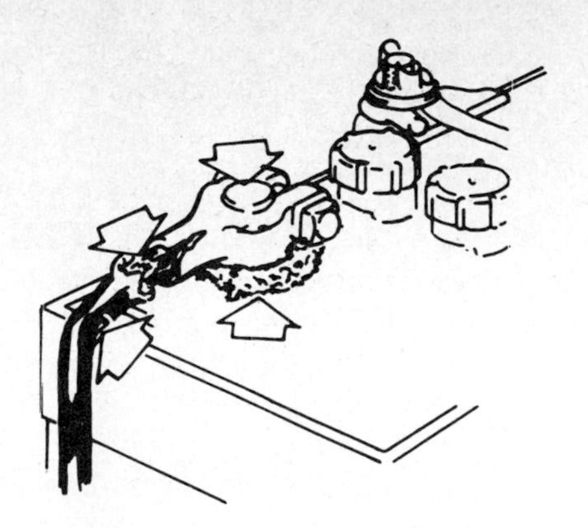

FIG. 11-13 Corroded battery cable and terminal post.

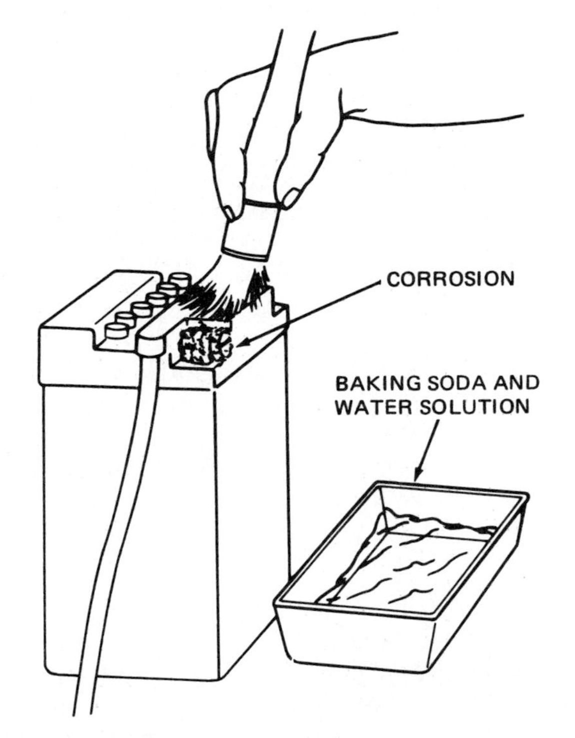

FIG. 11-14 Cleaning corrosion off a battery with a solution of baking soda and water.

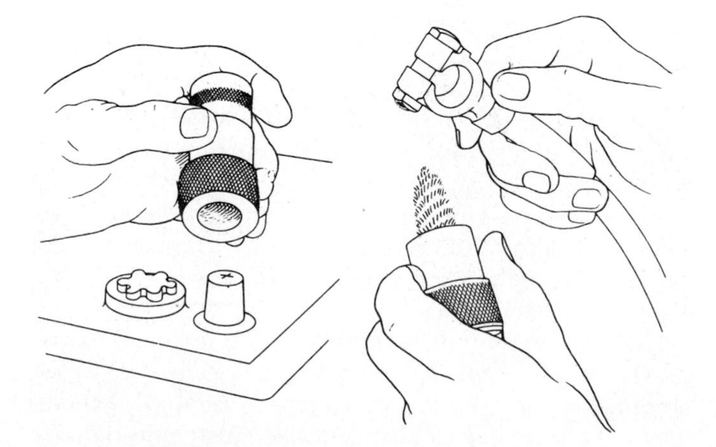

FIG. 11-15 Using a battery-cleaning brush to clean a battery terminal post and cable clamp.

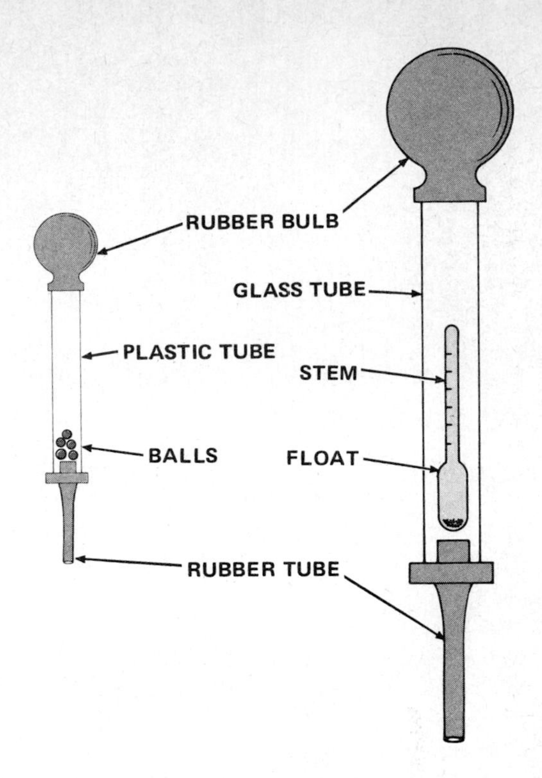

FIG. 11-16 Two types of battery hydrometers.

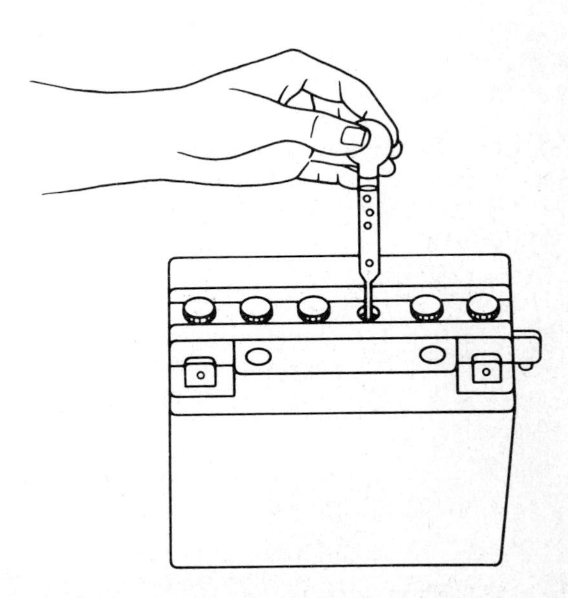

FIG. 11-17 Using a ball-type hydrometer to test the specific gravity of the electrolyte. *(ATW)*

that float indicates the state of charge of the battery cell. If all float, the cell is fully charged. If none floats, the battery cell is run down or dead.

To use the float-type hydrometer (right in Fig. 11-16), put the end of the rubber tube into the electrolyte. Squeeze and release the rubber bulb to draw electrolyte up into the glass tube. Note how far the float stem sticks out above the level of the electrolyte (Fig. 11-18).

Take the reading at eye level (Fig. 11-19). If the reading is between 1.260 and 1.290, the battery is fully charged. If the reading is between 1.200 and 1.230, the battery is

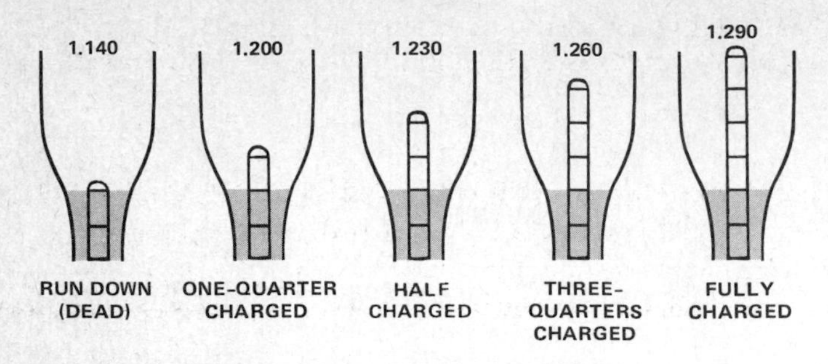

FIG. 11-18 Various specific-gravity readings.

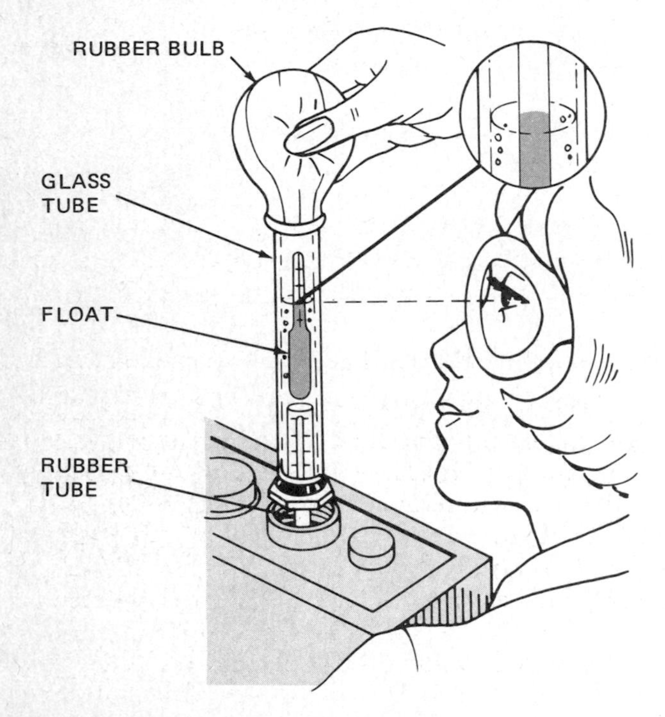

FIG. 11-19 Using a float-type hydrometer to test the specific gravity of the electrolyte. The reading must be taken at eye level.

about half-charged. If the reading is around 1.140, the battery is almost dead and needs a recharge.

NOTE We do not usually refer to the decimal point in speaking of specific gravity. For example, we say the specific gravity is twelve-sixty (1.260) or eleven twenty-five (1.125).

CAUTION **The electrolyte contains sulfuric acid! This is a dangerous acid that can give you serious skin burns. If it gets in your eyes, it can cause blindness. Sulfuric acid will eat holes in clothing, make spots on car paint finishes, and corrode any metal it touches. Be very careful when using the hydrometer to avoid spilling the electrolyte.**

If you get battery acid on your skin, flush it off at once with plenty of water. If you get acid in your eyes, flush your eyes with water over and over again (Fig. 1-11). Then get to a doctor at once! Acid can blind you.

After you have checked the electrolyte in the battery cell, squeeze the bulb to return the electrolyte to the cell from which you withdrew it.

Δ11-12 BATTERY-CAPACITY TEST The battery-capacity test is also called the *high-discharge* test. In this test, the battery is discharged at a high rate for 15 seconds. The discharge rate should be 3 times the ampere-hour rating of the battery. After 15 seconds, battery voltage is checked. If the voltage falls below 9.6 volts, the battery is either discharged or worn out. If worn out, it should be replaced with a new battery. If discharged, it should be recharged and used again.

Δ11-13 CHARGING A BATTERY If a battery is run down but otherwise in good condition, it can be recharged and then put back into operation. In addition to normal operation of the charging system on the engine, there are two ways to charge a battery. These are by using a slow charger (Fig. 11-20) and by using a quick charger (Fig. 11-21).

Certain precautions must be taken, regardless of which type of battery charger is used. However, with the quick charger, you must be careful not to overcharge the battery. This will ruin it. Always follow the instructions supplied with the battery charger.

CAUTION **The gases that form in the tops of battery cells during charging are very explosive. Never smoke or have an open flame around the battery. An explosion could occur that would blow the battery apart and you could be seriously injured.**

Δ11-14 IMPORTANCE OF KEEPING A BATTERY CHARGED A run-down battery will not start the engine. A discharged battery will have a shorter life than a battery that is kept fully charged. Also, a discharged battery will freeze more readily. It will freeze at about 18°F [−8°C]. But it takes −95°F [−35°C] to freeze a fully charged battery. In cold weather, keep the battery charged!

Δ11-15 REMOVING A BATTERY When removing a battery from the vehicle or engine, be careful not to damage the battery. Two types of battery terminals are used: lead posts on top (Fig. 11-3) and side screws (Fig. 11-5). Batteries with the terminals on top can be ruined if too much force is used to loosen the cable clamps attached to the terminals.

There are two types of clamps for top-terminal batteries: nut and bolt (Fig. 11-22) and spring ring (Fig. 11-23). The illustrations show how to loosen them. If a clamp sticks to a terminal, don't pry it off. This could damage the battery cell. Instead, use a clamp puller (Fig. 11-24).

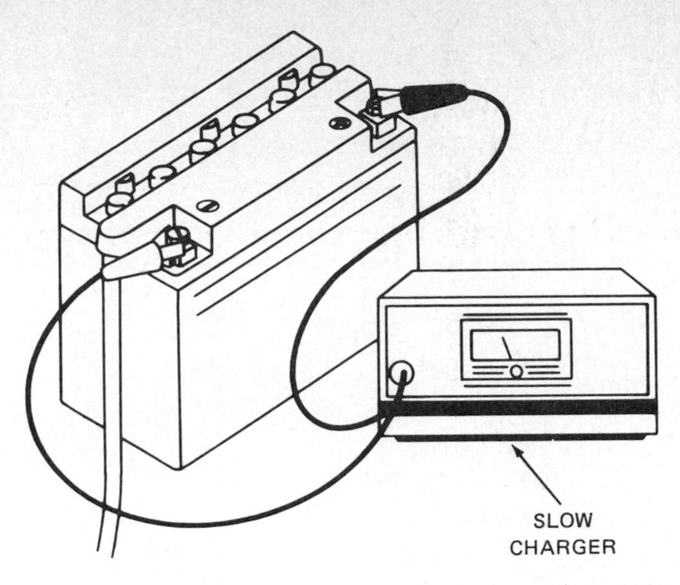

FIG. 11-20 Charging a battery with a slow charger. *(Honda Motor Company, Ltd.)*

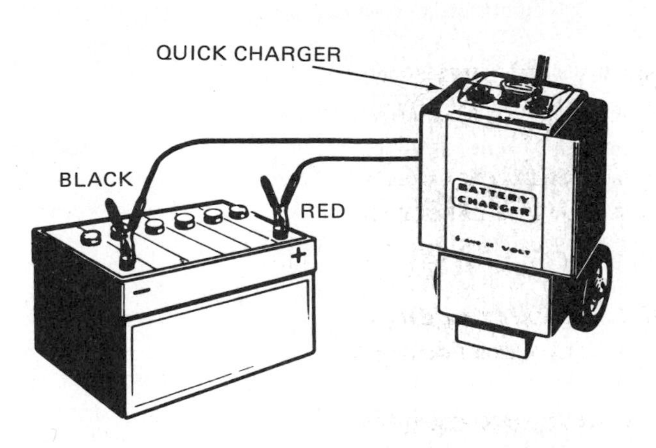

FIG. 11-21 Charging a battery with a fast, or quick, charger.

CAUTION When removing a battery, disconnect the ground battery cable first. This is the uninsulated strap which is fastened to the negative battery terminal and to the frame or engine. If you do not disconnect this cable first, you could accidentally ground the other battery terminal with pliers or puller. This would produce a direct short across the battery, and a very high current would flow. There would be sparks and possible damage to the tool or battery, and you could get hurt.

CAREFUL When installing a battery, do not install it backward. The negative terminal post is smaller than the positive terminal post (Fig. 11-25). Reconnect the positive terminal clamp first, and then the grounded, or negative, clamp. Do not carry a battery by the terminal post. Use a case-type battery carrier.

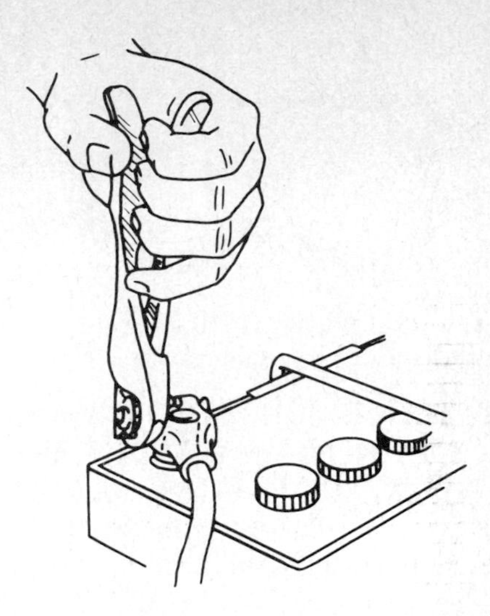

FIG. 11-22 Using battery pliers to loosen the nut-and-bolt type of battery-cable clamp.

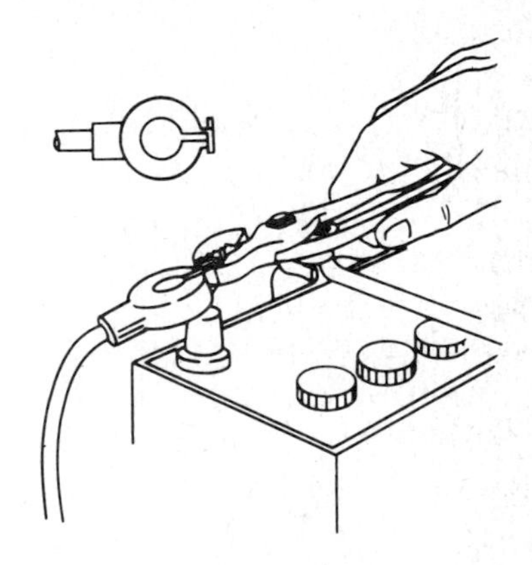

FIG. 11-23 Using pliers to loosen the spring-ring type of cable clamp from a battery post.

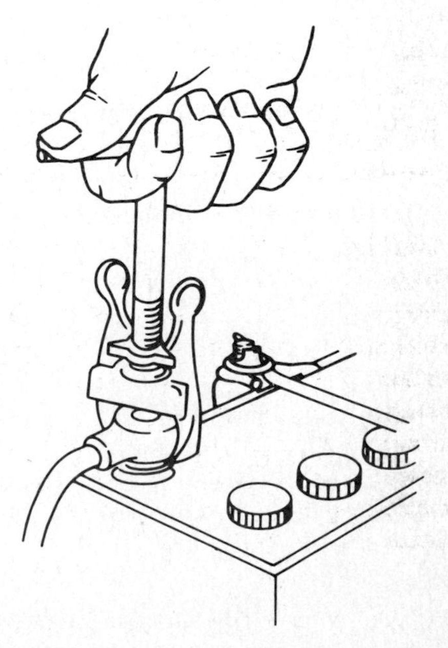

FIG. 11-24 Using a battery-clamp puller to pull the cable from the battery terminal.

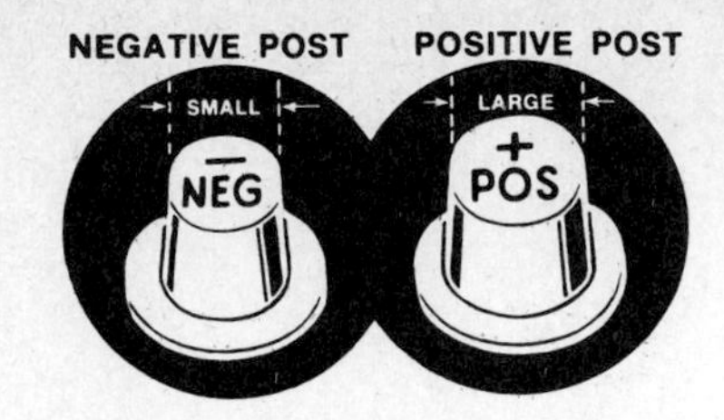

FIG. 11-25 The positive terminal post of the battery is larger than the negative terminal post.

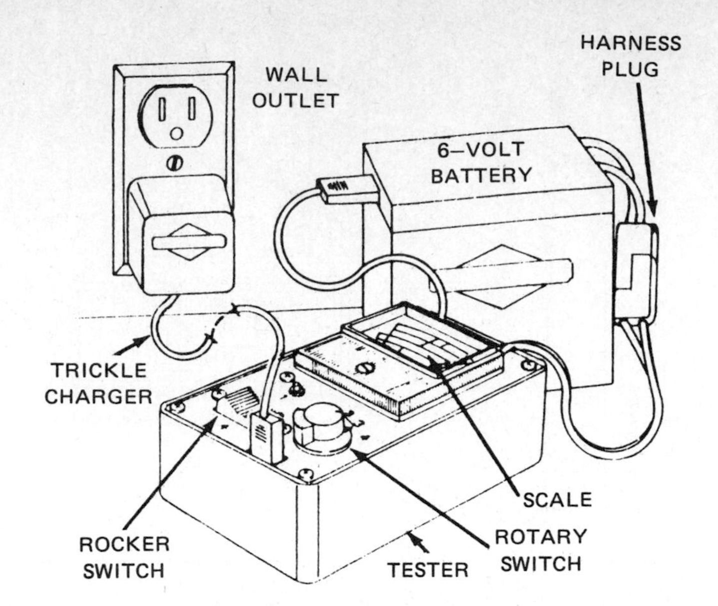

FIG. 11-26 Special tester being used to check the 6-volt battery which has a built-in diode. *(Briggs & Stratton Corporation)*

Δ11-16 CHECKING A 6-VOLT BATTERY WITH A BUILT-IN DIODE This battery is described in Δ11-5 and illustrated in Fig. 11-7. A special tester is required to test the diode and the battery condition (Fig. 11-26). The diode test checks for a shorted or open diode. Battery condition is checked by placing a load on the battery and noting whether the voltage drops too much. If it does, the battery is defective and should be replaced.

CHAPTER 11: REVIEW QUESTIONS

Select the *one* correct, best or most probable answer to each question. You can find the answer in the section indicated at the right of the question.

1. Electric current is produced in the battery by the chemical actions between the (Δ*11-2*)
   a. terminals and cells
   b. plates and terminals
   c. plates and electrolyte
   d. case and cells
2. The ni-cad battery (Δ*11-6*)
   a. cannot be recharged
   b. can be recharged with any battery charger
   c. will not discharge
   d. can be recharged only with a trickle charger
3. A fully charged battery is indicated by a specific-gravity reading of (Δ*11-11*)
   a. 1.140
   b. 1.200
   c. 1.230
   d. 1.265
4. To check the state of charge of the battery, you should (Δ*11-10*)
   a. make a hydrometer test
   b. make a battery-capacity test
   c. both a and b
   d. neither a nor b
5. A battery is being tested by having a battery-capacity test made on it. Mechanic A says the battery voltage should not drop below 9.6 volts while the load is applied for 15 seconds. Mechanic B says the load should be 3 times the ampere-hour capacity of the battery. Who is right? (Δ*11-13*)
   a. A only
   b. B only
   c. both A and B
   d. neither A nor B
6. Reasons for keeping the battery charged include (Δ*11-15*)
   a. a run-down battery cannot start the engine
   b. a run-down battery can go bad more quickly than a fully charged one
   c. a low battery can freeze at low temperatures
   d. all of the above
7. In a fully charged battery, the electrolyte is (Δ*11-3*)
   a. 60 percent water and 40 percent acid
   b. 40 percent water and 60 percent acid
   c. 30 percent water and 70 percent acid
   d. 50 percent water and 50 percent acid
8. Maintenance-free batteries usually (Δ*11-2*)
   a. have no vent plugs
   b. are sealed
   c. never need water
   d. all of the above
9. Two important battery ratings are (Δ*11-4*)
   a. size of plates and separators
   b. reserve capacity and cold cranking rate
   c. reserve capacity and hot cranking rate
   d. size and weight of battery
10. Be careful when working around batteries because
    a. they contain sulfuric acid (Δ*11-11*)
    b. the electrolyte can burn your skin
    c. the electrolyte can blind you if it gets in your eyes
    d. all of the above.

CHAPTER

12

Starting Systems for Small Engines

After studying this chapter, you should be able to:

1. ***List the types of starters used on small engines.***
2. ***Explain the operation of the various types of mechanical starters.***
3. ***Describe the construction and operation of starting motors.***
4. ***Explain the purpose of starter drives and describe how they work***
5. ***Discuss the purpose of the safety interlock.***
6. ***Describe the construction and operation of the mower safety brake.***
7. ***Describe the construction and operation of a starter-generator.***

Δ12-1 TYPES OF STARTERS Both mechanical and electric starters are used on small engines. Mechanical starters are manually operated, or hand-crank, starting systems. They do not use a motor as do electric starters. Mechanical starters are usually used on small engines when weight, cost, or size must be limited. Mechanical starters are generally classed as rope-wind, rope-rewind, windup, and kick-lever types. Electric starters provide the convenience and quick power of motor starting. Although normally used with larger engines, electric starters are being used more and more on small engines. Electric starters are generally classified according to the power source used for operation. One type is operated by connecting the starter motor to 120-volt house current. The other type uses a battery.

MECHANICAL STARTERS

Δ12-2 ROPE-WIND STARTERS The engine which is equipped for rope-wind starting has a pulley attached to the crankshaft. The flange of the pulley is slotted, as shown in Fig. 12-1. The starting rope has a knot on one end and a grip handle on the other. To use the rope, you hook the knot end into the flange slot. Then you wind the rope around the pulley, adjust the choke, make sure the ignition is on, and give the rope a strong pull. This spins the crankshaft to start the engine. It may take more than one windup-and-pull operation to get the engine started.

Δ12-3 ROPE-REWIND STARTERS To avoid having to rewind the rope each time you attempt to start, manufacturers introduced the rope-rewind starter

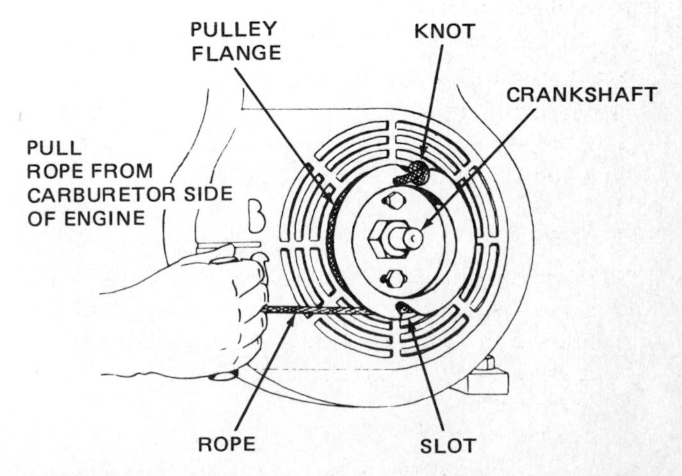

FIG. 12-1 The rope-wind starter is the simplest of all small-engine starters. Wind the rope on the pulley and then pull the handle to spin the engine crankshaft. *(Briggs & Stratton Corporation)*

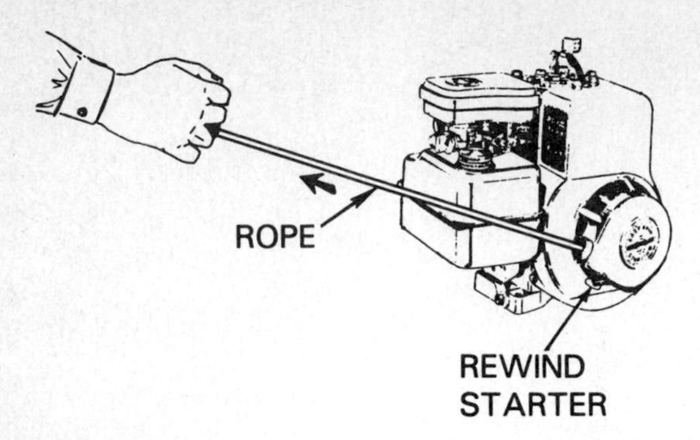

FIG. 12-2 To use the rope-rewind starter, pull out on the handle to crank the engine. Then release the pull on the handle and allow the rope to rewind on the pulley.

(Fig. 12-2). This starter has a rope permanently connected and includes a recoil spring that rewinds the rope on the pulley after each starting attempt. Figure 12-3 shows the basic parts of the rope-rewind starter.

When you pull the rope, the starter pulley is turned. Centrifugal force from movement of the pulley causes the pawls to fly out and lock the pulley to the crankshaft so that it rotates as the rope is pulled out. At the same time, the recoil spring is being wound up. The inside end of this spring is attached to the pulley. The outside end is attached to the starter housing.

After you have pulled the rope all the way out and then released it, the spring has enough tension in it to spin the pulley back in the opposite direction. This rewinds the rope on the pulley. Rewinding takes place whether or not the engine has started. On the rewind cycle, the pawls are ineffective because they do not catch in the teeth on the inside of the crankshaft adapter. At the end of the rewind cycle, the pawls are retracted by the small attaching spring. The starter is then ready for another starting attempt.

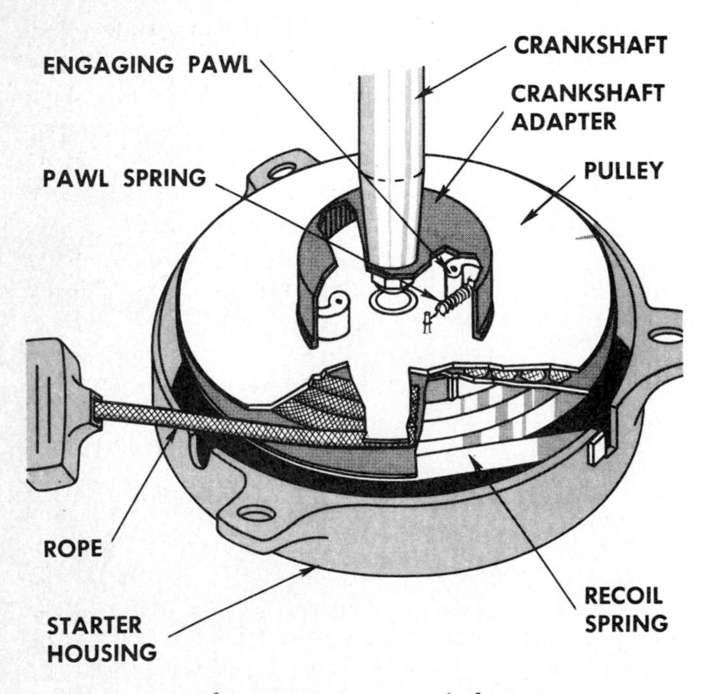

FIG. 12-3 A partly cutaway rope-rewind starter.

Δ12-4 WINDUP STARTERS

The windup starter (Fig. 12-4) is designed to reduce the amount of effort required to start a small engine. With the rope-wind and rope-rewind starters, you must exert a strong pull to spin the crankshaft. The windup starter requires much less force. You simply wind up a spring and then release it. When the spring unwinds, it spins the crankshaft. There are several designs of the windup starter, but each operates in the same manner.

First, set the release lever so it will hold the spring on windup. Next, swing the crank handle out and rotate it to wind up the spring. Some models lock the spring when the crank handle is returned to the running position. The release lever is then moved to allow the spring to unwind and crank the engine. On some models, the spring is automatically released when the crank handle is returned to the running position.

Figure 12-5 shows the operation of a typical windup starter. It includes a crank with a ratchet and a second crank attached to one end of a spring. The other end of the spring is attached to a shaft that is part of the spring-holding mechanism. The shaft lower end is attached to the starter drive. The starter drive has a dog-and-ratchet arrangement inside the flywheel cup.

With the starter control lever set, the holding mechanism is locked in place to hold the inner end of the spring. When the crank handle is turned, the ratchet

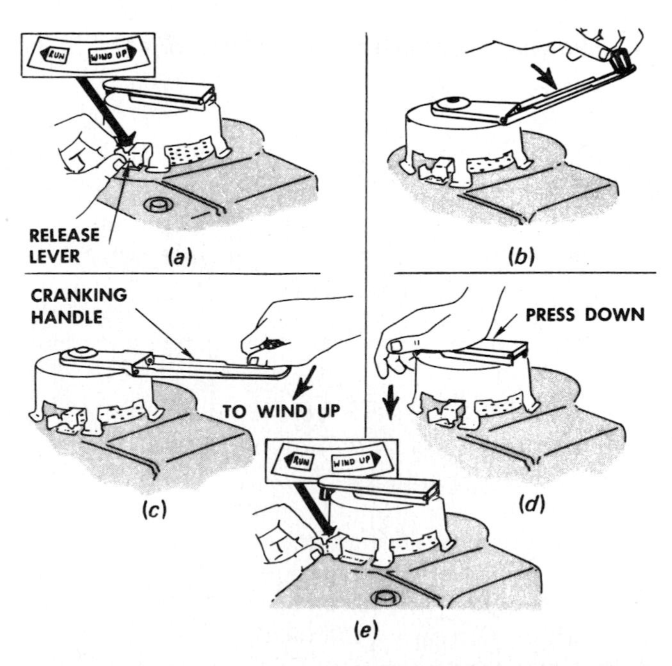

FIG. 12-4 Steps in using a windup starter. *(a)* Lock the spring by moving the control lever to WINDUP. *(b)* Open the crank handle. *(c)* Wind up the recoil spring. *(d)* Fold the handle. *(e)* Release the spring by moving the control lever to RUN.

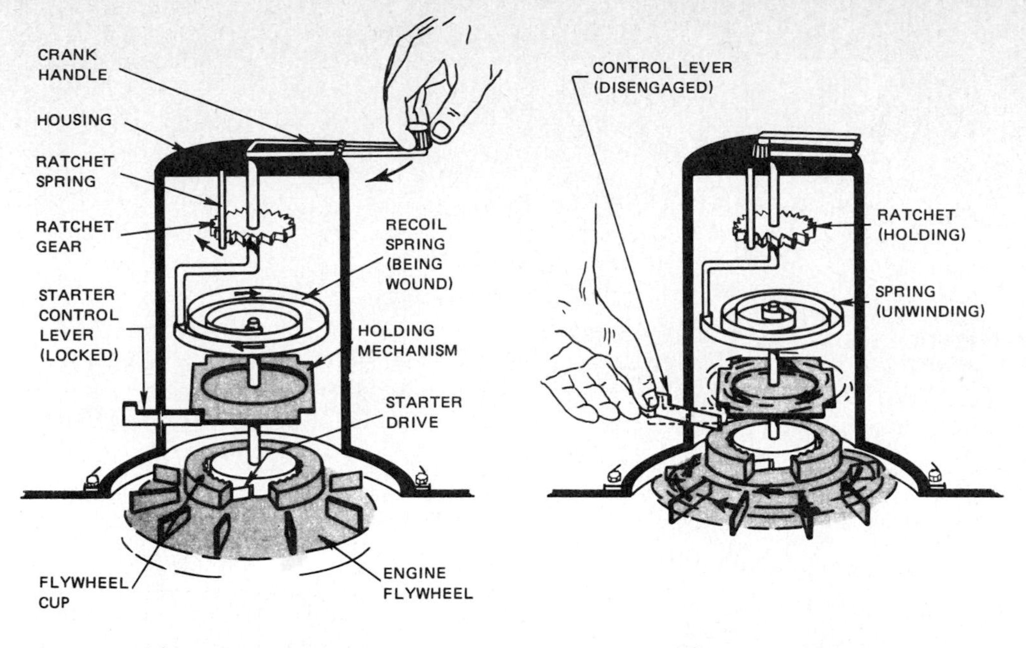

FIG. 12-5 Operation of the windup starter. ***(a)*** **The recoil spring is being wound up. The control lever locks the holding mechanism.** ***(b)*** **The control lever is pushed to unlock the holding mechanism, allowing the spring to unwind and crank the engine.**

gear and crank wind up the spring. The ratchet spring at the top prevents the spring from unwinding. When the spring is completely wound up, the control lever is released so that the spring starts to unwind from the inside out. This engages the ratchets or starter dogs inside the flywheel cup so that the flywheel is rotated to crank the engine.

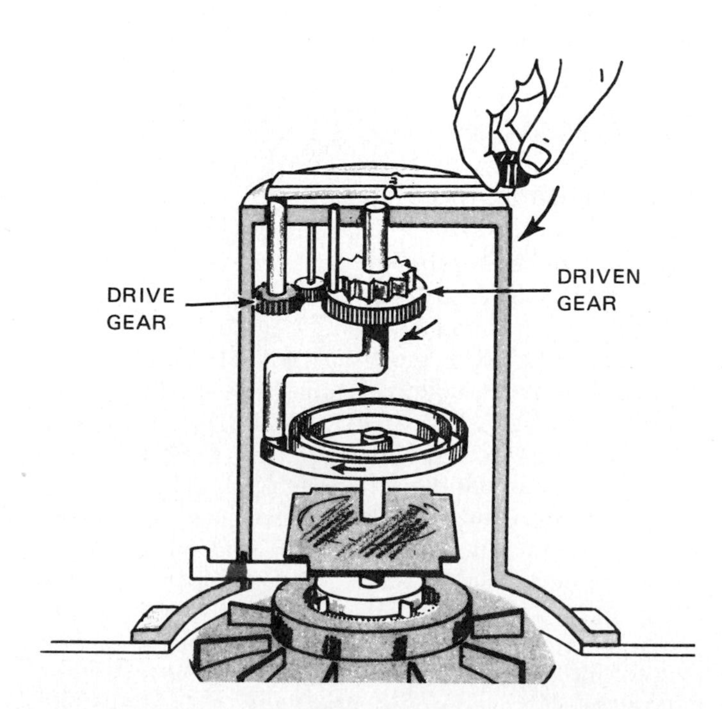

FIG. 12-6 Some windup starters have a gear reduction which makes it easier to turn the crank. However, the crank must be turned more times to wind up the recoil spring.

Some designs include a reduction-gear arrangement (Fig. 12-6). This makes it easier to wind up the spring. Although less force is required to operate the crank, it has to be turned more times to wind up the spring.

Typical windup starters are shown in Fig. 12-7. At the top of the illustration, the major parts are shown as they would appear when removed from the engine. The lower part shows a different windup-starter model disassembled.

Δ12-5 KICK STARTERS

The kick starter uses leg power for its operation and is a popular starter found on minibikes and motorcycles. All kick starters operate in a similar manner. Figure 12-8 shows the gear train for one starter model. When the kick pedal is kicked down by means of leg power, the rotary motion is carried through the gear train to the engine crankshaft, causing it to spin. When the pedal is released, a strong spring returns it to the former raised position. It may take several kicks to start a cold engine.

The gear train in Fig. 12-8 has a considerable increase in gear ratio from the kick-shaft gear to the crankshaft primary pinion gear. This increases the crankshaft revolution rate so that it spins rapidly when the kick-shaft gear rotates. Once the engine is running, the starter is disengaged from the crankshaft. The one-way starter motion is achieved in different ways, depending on the design.

The kick starter shown in Fig. 12-9, illustrates how one-way starter motion is possible. The kick-shaft gear is free to rotate on the kick shaft. A ratchet wheel (also called a gear) can move back and forth on splines of the

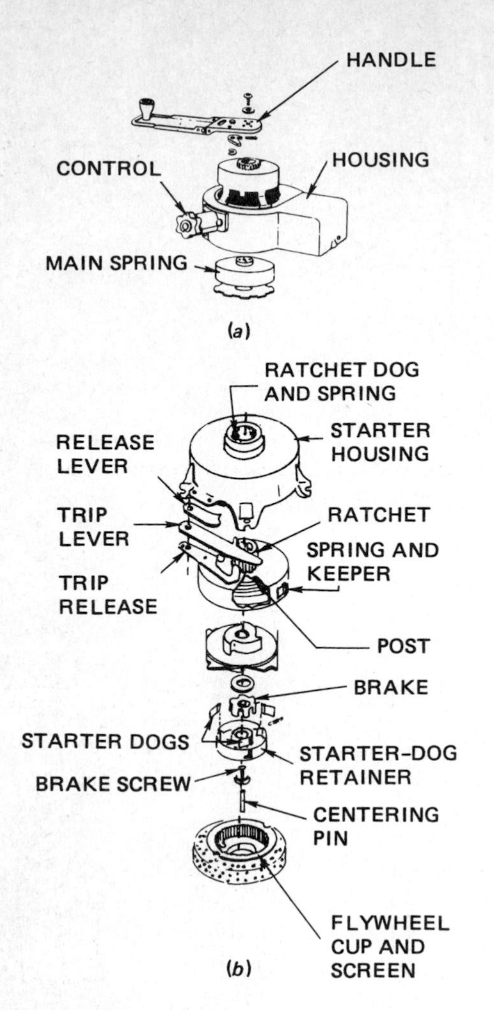

FIG. 12-7 **Assembled and disassembled windup starters. In *(a)*, the starter must be disassembled from the handle end. In *(b)*, the starter is disassembled by first removing the drive mechanism.**

kick shaft. When the engine is running, the arm on the ratchet wheel is resting behind the stopper guide. This holds the ratchet wheel away from the kick-shaft gear.

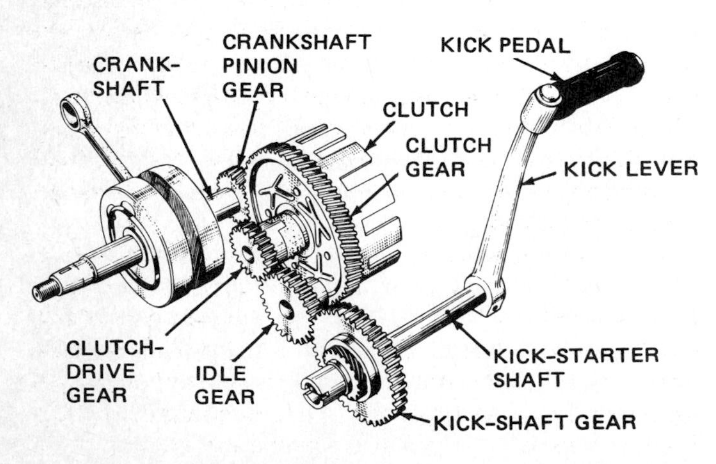

FIG. 12-8 **A kick starter, showing the gear train from the kick lever to the gear on the crankshaft.** *(Suzuki Motor Company)*

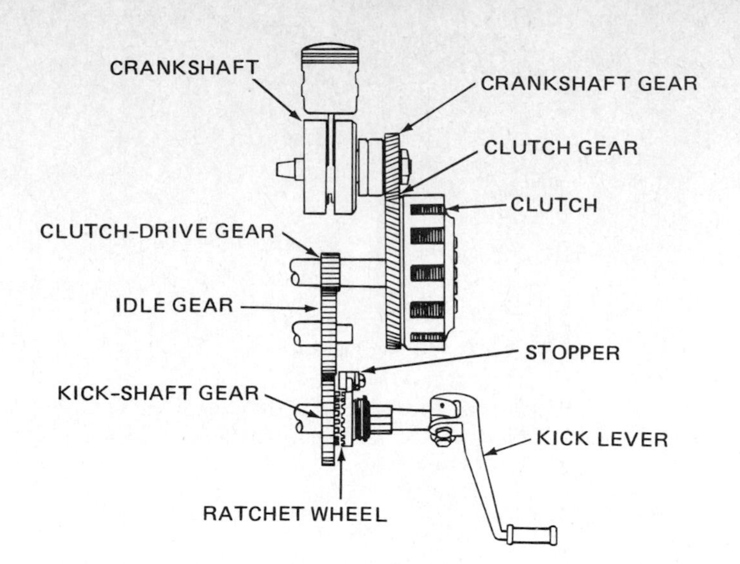

FIG. 12-9 **A kick starter, showing how one-way starter motion is obtained.** *(Suzuki Motor Company)*

However, when the kick pedal is pushed down, the kick shaft turns, forcing the ratchet wheel to turn. The arm on the ratchet wheel moves out from behind the stopper guide. The spring-back of the ratchet wheel now forces the ratchet forward so that the ratchet teeth mesh to engage the kick-shaft gear. Further movement of the kick pedal forces the kick-shaft gear to turn. This transmits motion through the gears, causing the crankshaft to spin.

As the engine starts, it backdrives the kick-shaft gear. However, the ratchet is a one-way device, and so the kick-shaft gear spins without driving the ratchet wheel. The ratchet teeth are disengaged, and no motion is transmitted to the kick-starter shaft.

ELECTRIC STARTING MOTORS

△12-6 ELECTRIC MOTORS In the electric starting motor, an electric motor provides the power to spin the engine crankshaft and start the engine. The battery furnishes the electricity to operate the motor. (Some electric starting motors for small engines use 120-volt house current.) Taking electricity out of the battery runs it down. This means that there must be a charging system to restore the battery's state-of-charge. The electric system for an engine using electric starting is more complicated than when hand cranking is used.

Some small engines use a 12-volt starter-generator (Fig. 12-10), which cranks the engine for starting. It also remains connected to the engine and battery. Then the starter-generator is driven by the engine after it starts to produce an electric current. This electric current charges the battery and also handles any electric loads that might be turned on, such as lights.

NOTE Many small-engine technicians (and manufacturers' service manuals) refer to the motor in the electric

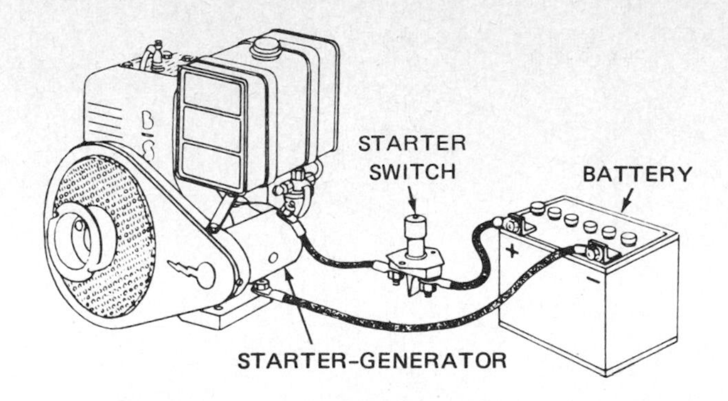

FIG. 12-10 An electric starting system using a starter-generator. After the engine starts, the generator recharges the battery. *(Briggs & Stratton Corporation)*

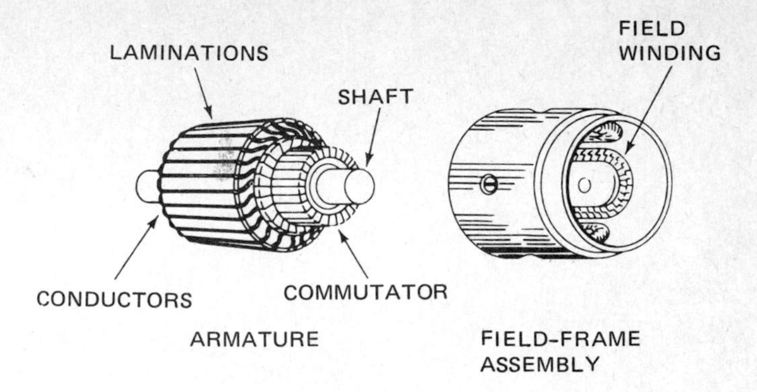

FIG. 12-11 Two major parts of an electric starting motor: the armature and the field-frame assembly. *(Delco-Remy Division of General Motors Corporation)*

starting system as the *starter*. However, to distinguish electric starters from the mechanical starters described earlier, we call them *starting motors*.

Δ12-7 BASIC MOTOR PRINCIPLES

Most electric starting motors are low-voltage direct-current motors. They operate from storage batteries (Fig. 12-10). The two main components of the starting motor are the *armature* and the *field-frame assembly* (Fig. 12-11). The armature has a series of heavy conductors connected to the commutator bars. In the assembly, brushes ride on the commutator (Fig. 12-12). The brushes complete the circuit through the armature conductors.

The field-frame assembly has field windings, also of heavy conductors. In the starting motor, the field windings and the armature windings are connected in series (Fig. 12-12). When the starting motor is connected to the battery, two opposing magnetic fields are produced, one from the armature and the other from the field windings. The opposing fields are so strong that they cause the armature to spin. This spins the engine crankshaft so that the engine starts. While unlike magnetic poles attract, like magnetic poles repel each other (Chap. 10). It is the repelling force on the armature conductors that causes the armature to spin.

Figure 12-13 shows a disassembled starting motor for a small engine. When the starting motor is assembled, the two brushes (upper left) rest on the commutator of the armature. The circuit for this starting motor is shown in Fig. 12-12.

Starting motors for bigger engines, which require greater cranking torque, have four or more field windings and have heavier conductors in the armature. This allows more current to flow through the starting motor so that stronger magnetic fields, and greater cranking power, are produced.

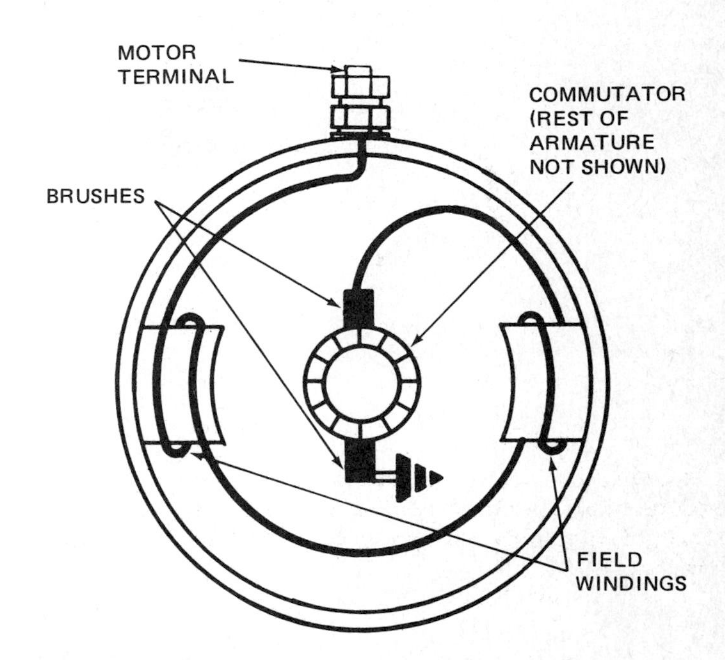

FIG. 12-12 Wiring diagram of a starting motor. The conductors in the armature are not shown.

Δ12-8 DRIVE ARRANGEMENTS

The drive arrangement, to send the cranking power from the armature to the crankshaft, uses a small pinion on the armature shaft. This pinion meshes with teeth cut in the flywheel (Fig. 12-14). The difference in the number of teeth on the pinion and the flywheel produces gear reduction. This means that, typically, the armature must rotate about 15 times to cause the flywheel to rotate once. The armature may revolve about 2000 to 3000 rpm when the starting motor is operated. This causes the flywheel to spin at speeds as high as 300 rpm. This is ample speed for starting the engine.

After the engine starts, it may increase in speed to 3000 rpm or more. If the starting-motor drive pinion remained in mesh with the flywheel, it would be spun at 45,000 rpm because of the 15 : 1 gear ratio. This means that the armature would be spun at this high speed. Centrifugal force would cause the conductors and commutator segments to be thrown out of the armature, ruining it. To prevent such damage, automatic meshing and demeshing devices are used. There are two general types, inertia and overrunning clutch. Both types are described below.

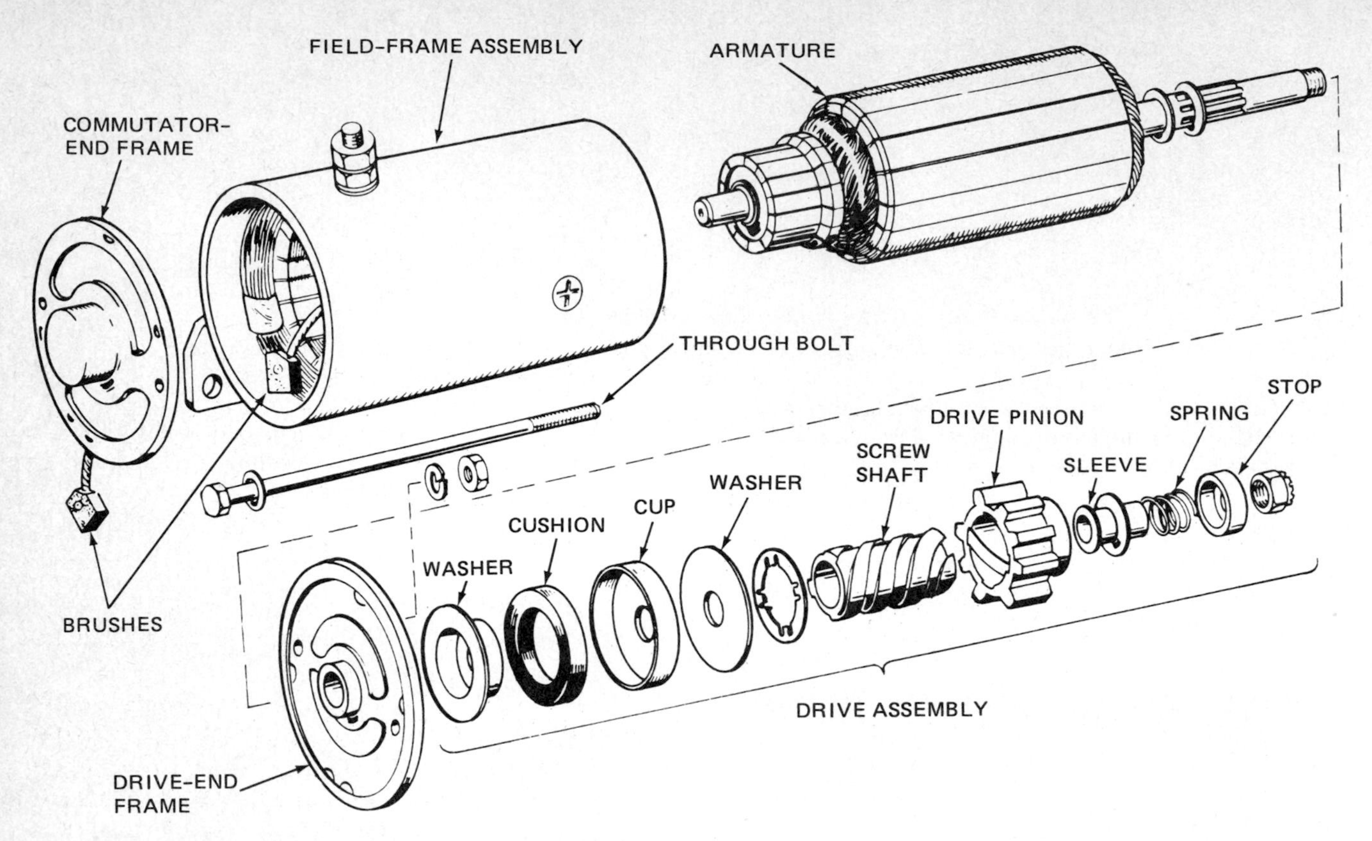

FIG. 12-13 Disassembled rubber-compression-type drive and the starting motor in which it is assembled. *(Onan Corporation)*

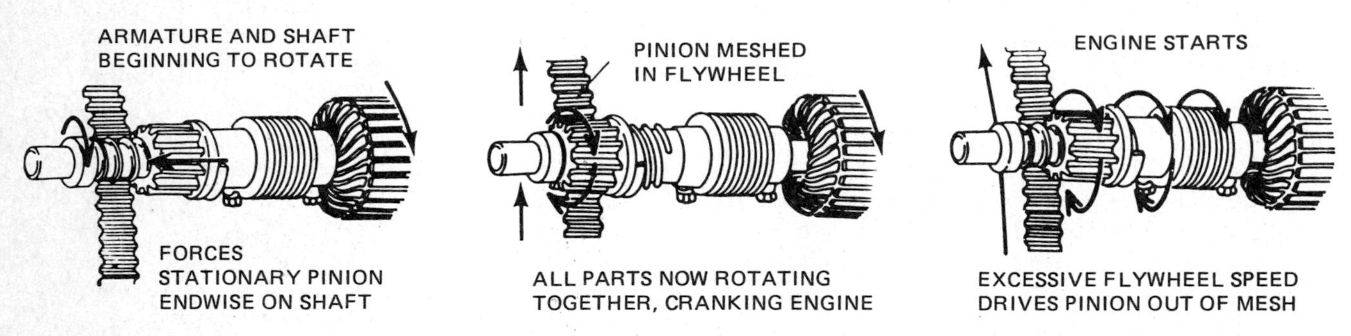

FIG. 12-14 Operation of a Bendix drive. *(Delco-Remy Division of General Motors Corporation)*

Δ12-9 INERTIA DRIVE The inertia drive depends on the drive pinion to produce meshing. *Inertia* is the property that all things resist any change in motion. When the drive pinion is not rotating, it resists any force that attempts to set it into motion. There are several types of inertia drive. One of the most widely used in the *Bendix drive.*

In the Bendix drive (Fig. 12-14), the drive pinion is mounted loosely on a sleeve. The sleeve has screw threads matching internal threads in the pinion. When the starting motor is at rest, the drive pinion is not meshed with the flywheel teeth. As the starting motor switch is closed, the armature begins to rotate. This causes the sleeve to rotate also, since it is fastened to the armature shaft through the heavy spiral Bendix spring.

Inertia prevents the pinion from instantly picking up speed with the sleeve. Therefore, the sleeve turns within the pinion, just as a screw would turn in a nut held stationary. This forces the pinion endways along the sleeve so that the pinion goes into mesh with the flywheel teeth. As the pinion reaches the pinion stop, the endways movement stops. The pinion must now turn with the armature, causing the engine to be cranked. The spiral spring takes up the shock of meshing.

After the engine starts and increases in speed, the flywheel rotates the drive pinion faster than the armature is turning. This causes the pinion to be spun back out of mesh from the flywheel. The pinion turns backward on the sleeve. The screw threads on the pinion and sleeve cause the pinion to be backed out of mesh with the flywheel.

The rubber-compression type of drive (Fig. 12-13) is used on starting motors for outboard engines and on other small engines. It is similar to the compression-

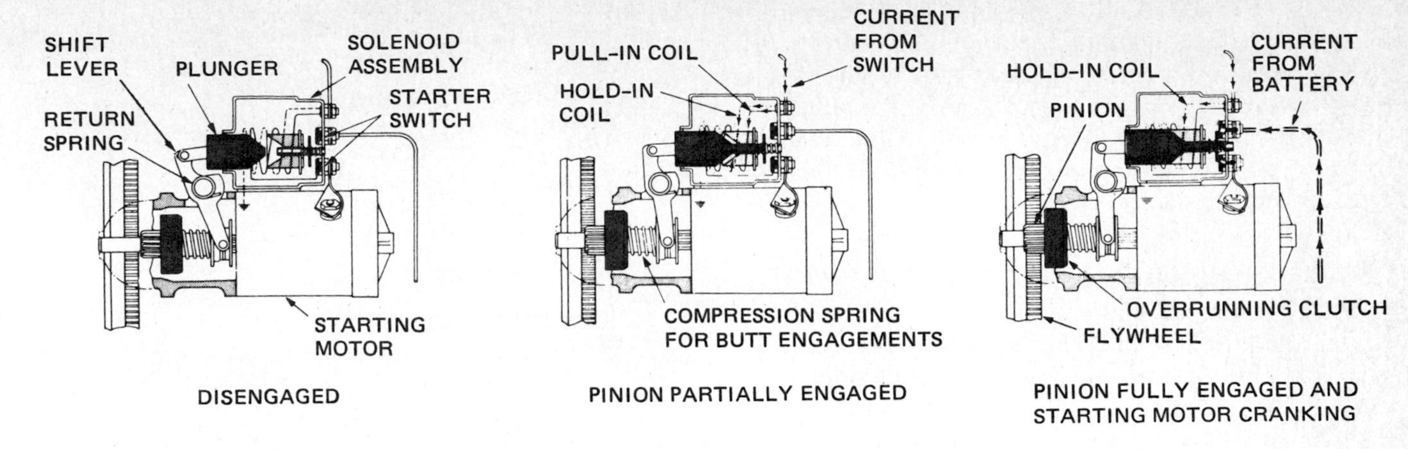

FIG. 12-15 Actions of the solenoid and overrunning clutch as the pinion engages with the flywheel teeth.

spring type. A rubber cushion absorbs the shock as the drive pinion meshes with the flywheel.

▵12-10 OVERRUNNING CLUTCH

More positive meshing and demeshing of the pinion and flywheel teeth is provided by the overrunning-clutch drive. The overrunning clutch uses a shift lever (Fig. 12-15) to slide the pinion along the armature shaft and into, or out of, mesh with the flywheel teeth. The overrunning clutch is designed to transmit driving torque from the starting-motor armature to the flywheel. Then the clutch permits the pinion to overrun (run faster than) the armature after the engine has started.

The overrunning-clutch drive (Fig. 12-16) consists of a shell-and-sleeve assembly, which is splined internally to match splines on the armature shaft. A pinion-and-collar assembly fits loosely into the shell. The collar makes contact with four hardened-steel rollers, which are assembled into notches cut from the shell. These notches taper slightly inward. There is less room in the end away from the rollers than in the end where the rollers are shown. Spring-loaded plungers rest against the rollers.

The shift lever, which causes the clutch assembly to move endways along the armature shaft, is operated either by manual linkage or by a solenoid (Fig. 12-15). When the shift lever is operated, it moves the clutch assembly along the armature shaft until the pinion teeth mesh with the flywheel teeth. If the teeth should butt instead of mesh, the clutch spring compresses and spring-loads the pinion against the flywheel teeth. Then, as soon as the armature begins to rotate, the pinion will mesh.

Full shift-lever travel closes the starting-motor switch contacts, and the armature begins to revolve. This rotates the shell-and-sleeve assembly in a clockwise direction (in the end view of Fig. 12-16). The rollers rotate between the shell and the pinion collar, moving away from their plungers and toward the sections of the notches in the shell, which are smaller. This jams the rollers tightly between the pinion collar and the shell, and the pinion is forced to rotate with the armature and crank the engine. Figure 12-15 illustrates the engaging action in a solenoid-operated starting motor.

When the engine begins to operate, it attempts to drive the starting-motor armature, through the pinion, faster than the armature rotates under its own power. Therefore, the pinion rotates faster than the shell, turning the rollers back toward their plungers, where there is enough room to let them slip freely. The pinion-and-collar assembly can now overrun the shell-and-sleeve assembly and the armature. This gives the armature enough protection for the short time that the operator

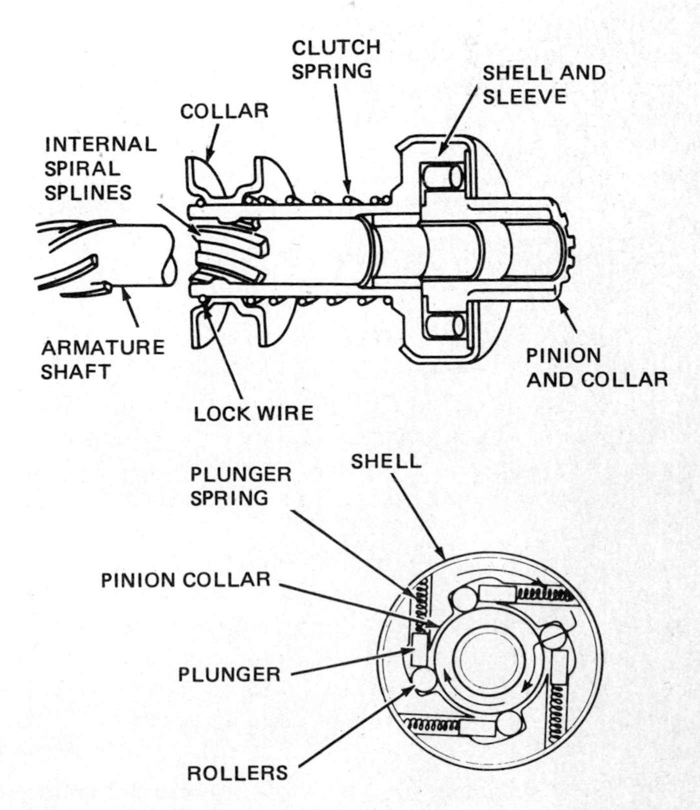

FIG. 12-16 Cutaway and sectional views of an overrunning clutch. *(Delco-Remy Division of General Motors Corporation)*

leaves the ignition switch in the START position after the engine has started (or until the automatic controls take over and open the starting-motor-control circuit). When the force on the shift lever is relieved, the shift-lever spring slides the clutch assembly back along the armature shaft so that the pinion is demeshed. At the same time, the starting-motor switch is opened.

∆12-11 CONTROLLING THE STARTING MOTOR

Starting-motor controls have varied from a simple hand- or foot-operated switch to automatic devices that close the circuit when the accelerator is depressed. The system used today for many small engines has starting contacts in the ignition switch.

A typical key-and-ignition switch for starting-motor control on a small engine is shown in Fig. 12-17. When the ignition key is turned against spring force past the ON position to START, the starting contacts close. This connects the starting-motor solenoid or magnetic switch to the battery. The solenoid or magnetic switch then operates to connect the battery directly to the starting motor. As soon as the engine starts, the operator releases the ignition switch. A spring then returns it from the START to the ON position. This disconnects the starting motor from the battery, and the starting motor stops. However, the ignition remains connected to the battery, so that the engine continues to run.

∆12-12 MAGNETIC SWITCH

The operation of a magnetic switch is based on the fact that a flow of current in a winding creates a magnetic field. The winding is wrapped around a hollow core, and a cylindrical iron plunger is placed partway into this core. When the winding is energized, the resulting magnetic field pulls

FIG. 12-17 A starting system using a key-operated ignition switch which also controls the starting motor. *(Ariens Company)*

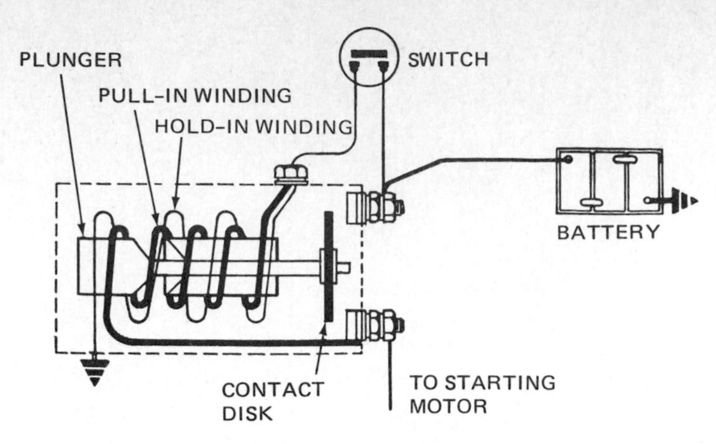

FIG. 12-18 Wiring diagram of a magnetic switch.

the plunger farther into the core (Fig. 12-18). A contact disk is attached to the plunger, and two contacts are placed so that the plunger movement forces the disk against the contacts. This closes the circuit between the battery and the starting motor.

Some magnetic switches have two windings (Fig. 12-18). There is a pull-in winding of a few turns of heavy wire and a hold-in winding of many turns of fine wire. When the control switch is closed, current from the battery flows through both these windings. The current returns to the battery from the hold-in winding directly through ground. The current passing through the pull-in winding must flow through the starting motor before returning through ground to the battery. This hookup may seem unnecessary. But its purpose is to short out (across the main switch contacts) the pull-in winding when the magnetic switch operates, pull in the plunger, and force the contact disk against the two contacts.

Shorting out the pull-in winding lessens the drain on the battery and leaves more energy for cranking. More magnetism is needed to pull the plunger in, so both windings work together to accomplish this. But once the plunger is in, less magnetism is needed to hold it in. Therefore, the pull-in winding is shorted out. Only the hold-in winding operates to hold the plunger in place as long as the control switch remains closed.

∆12-13 SOLENOID SWITCH

On most automobile engines and some small engines, overrunning-clutch starting motors use a solenoid to move the overrunning clutch and close the starting -motor switch. The solenoid is larger than the magnetic switch and is mounted on the starting motor (Fig. 12-15). When the solenoid operates, it first shifts the drive pinion into mesh with the flywheel teeth (Fig. 12-15). Then the solenoid closes the circuit between the battery and the starting motor.

The solenoid has two windings: a pull-in winding and a hold-in winding. The pull-in winding is connected across the starter-switch contacts in the solenoid. Both windings work together to pull the solenoid plunger in and move the overrunning-clutch pinion into mesh. But once mesh is completed, it takes much less magnetism to hold the plunger in. Therefore, the pull-in winding is shorted out as the main contacts are connected by the solenoid disk.

Δ12-14 SAFETY INTERLOCKS

For the operator's safety, manufacturers of equipment such as garden tractors, riding mowers, and snowblowers install a *safety interlock* in the cranking circuit or in the ignition system. These safety interlocks allow the engine to be cranked only when the transmission is in neutral, when the power takeoff is disengaged, or when the operator is sitting on the seat of the vehicle.

Figure 12-19 shows in schematic view an engine that has electric starting and battery ignition (Chap. 13). In the safety-interlock circuit on the vehicle, three electric switches are wired in series. Each switch is spring-loaded to hold it in the open position. When some action takes place—for example, when the operator sits down on the seat—the switch closes. When the proper actions have been taken to close all three switches, turning the key in the ignition switch will activate the starter.

CAUTION If the engine must be cranked with the safety interlock bypassed, make sure that all safety conditions are followed to permit the engine to operate safely. This is to prevent personal injury to you or someone nearby, and damage to the engine or equipment. *Never* return the equipment to the owner with the safety interlock disabled. You could be held liable for any subsequent injuries or property damage.

Δ12-15 MOWER SAFETY BRAKE

Statistics from hospital emergency rooms indicate that more than 70,000 people a year are injured by lawn mowers. Many of these injuries are due to toes which have been caught under the mower while it is operating. In addition, a number of injuries have occurred during an attempted repair while the mower is running. The mowers causing the injuries are rotary lawn mowers, which have the blade spinning horizontally.

To guard against such injuries, the Consumer Products Safety Commission has imposed new standards for lawn-mower construction. The basic requirement is that the mower handle must have a control lever that the operator must hold down while operating the mower. As soon as the operator releases the mower handle, a brake stops the blade within 3 seconds.

There are two types of mower-blade brakes. In one system, the blade is disengaged from the engine and brought to a stop by a brake band. In the other system, the brake is applied to the engine so that the engine and the blade both are brought to a stop. With either system, the brake is actuated the moment the operator releases the mower handle.

One system, which brakes the engine and blade both, is shown in Fig. 12-20. When the operator releases the mower handle and brake control, the brake is applied, shutting down the engine and blade. Then, if the ignition switch is still left on, the engine can be restarted as soon as the operator takes hold of the mower handle and brake control.

Figure 12-20 shows the brake applied. Figure 12-21 shows the brake released. Figure 12-22 shows the wiring system. Note that the engine is killed by shorting out the

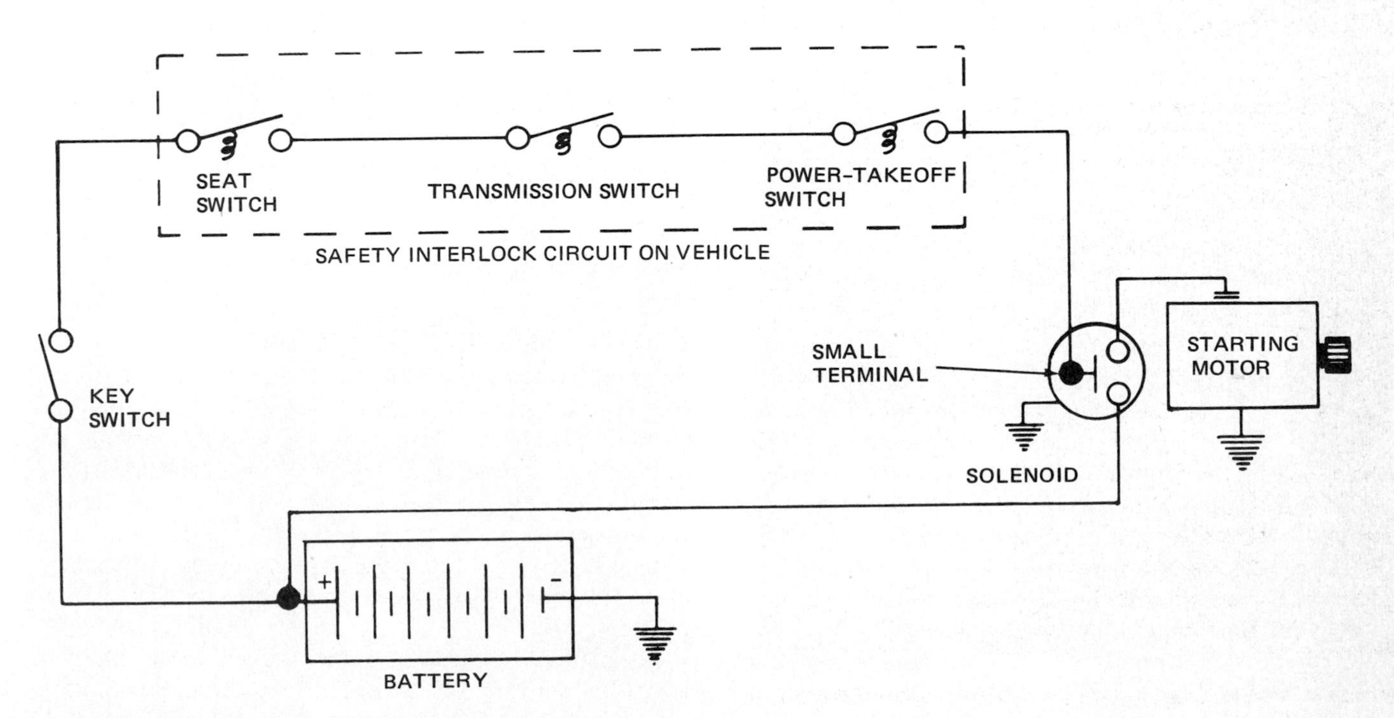

FIG. 12-19 Wiring diagram of a safety-interlock system. *(Kohler Company)*

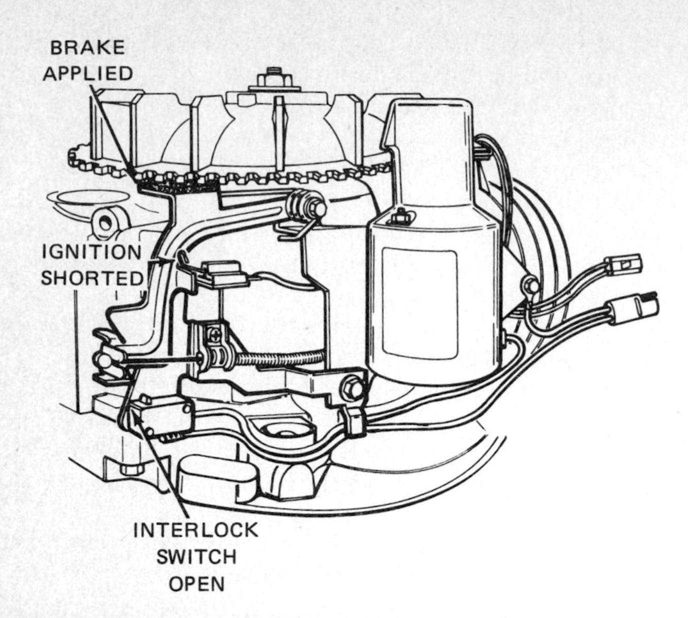

FIG. 12-20 Flywheel brake system with the brake applied, the ignition shorted, and the interlock switch open. *(Tecumseh Products Company)*

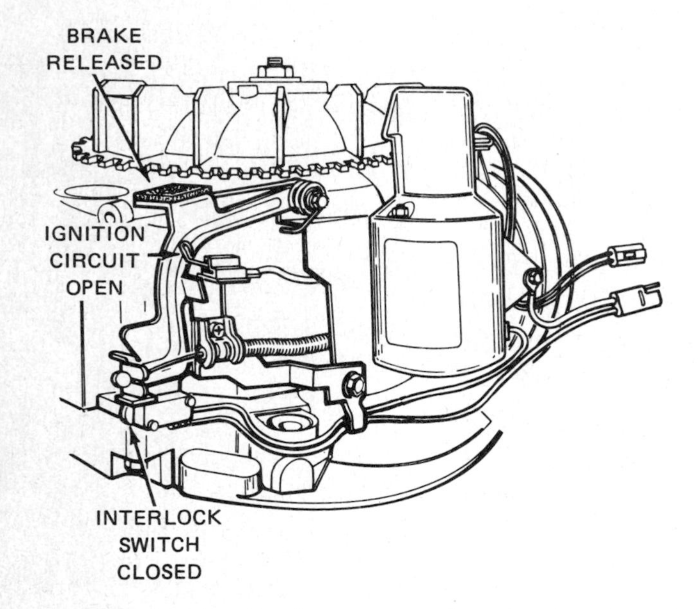

FIG. 12-21 Flywheel brake system with the brake released, the ignition circuit open, and the interlock-switch closed. *(Tecumseh Products Company)*

ignition. At the same time, a torsion spring rotates the brake lever to force the brake pad against the underside of the flywheel and opens the starting-motor interlock switch. When the operator is ready to start mowing again and takes hold of the mower handle, the following happens. The brake pad is pulled away from the flywheel, the ignition short-out switch is opened, and the starter interlock switch is closed. Now, the starting motor is energized, and the engine starts.

The mowers equipped with the automatic braking system are called *compliance* engines. Some operators do not like them and bypass them by taping the control to the mower handle so that the brake will not apply when the mower handle is released. If you find the control

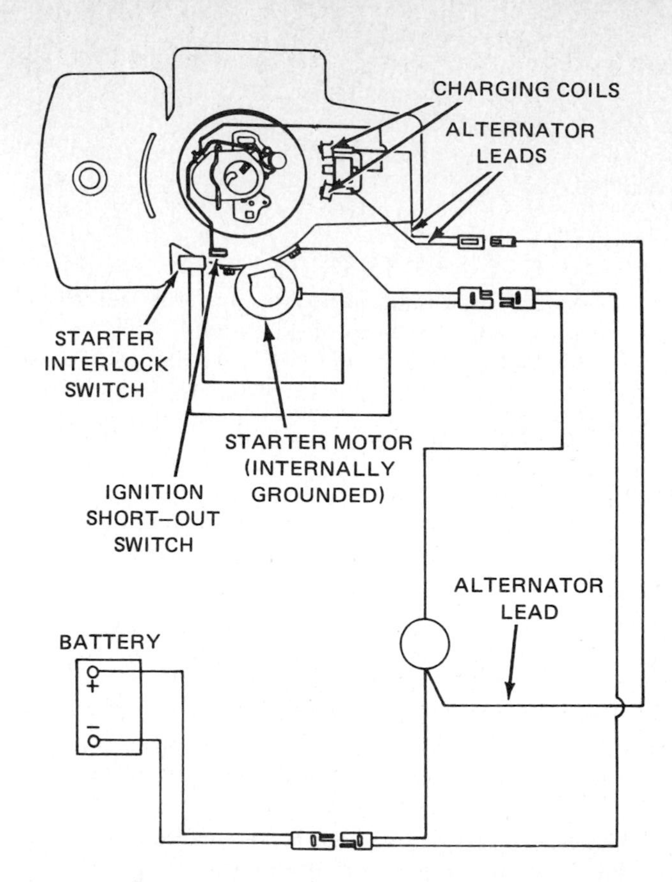

FIG. 12-22 Wiring diagram for a mower safety-brake system. *(Tecumseh Products Company)*

taped down on a mower you are servicing, remove the tape and explain to the operator that the brake is a safety feature. It is there to keep the operator from adding to the total 70,000 people who are hurt by mowers each year.

Figure 12-23 shows a special tester or *Blade Monitor* which accurately measures the time it takes for a blade to stop after the brake system is actuated. This tester also reads top no-load governed speed, idle speed, engine cranking speed, and maximum allowable blade-tip speed.

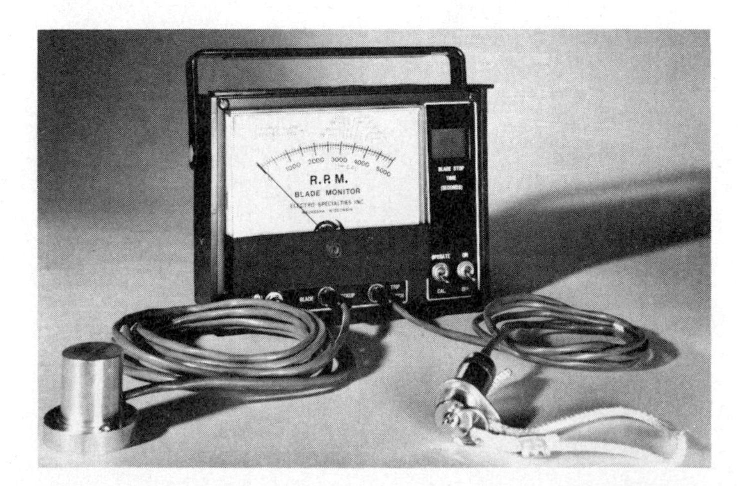

FIG. 12-23 Blade Monitor for measuring the number of seconds it takes for the blade to stop turning after the brake is applied. *(Tecumseh Products Company)*

120-VOLT STARTING MOTORS

Δ12-16 ELECTRIC STARTING Some engines use starting motors that are powered from the 120-volt home or shop wiring outlet. The starting motor is connected by an extension cord to the electric outlet (Fig. 12-24).

CAUTION **When using the 120-volt starting motor, connections should always be made at the 120-volt outlet receptacle, as shown in Fig. 12-24, and never at the starting motor. If you make or break the electric connection at the starting motor, the spark could ignite gasoline vapor from the carburetor and cause a fire. Always make sure the extension cord is in good condition, without frayed insulation. The extension cord should have the third, or ground, lead in it.**

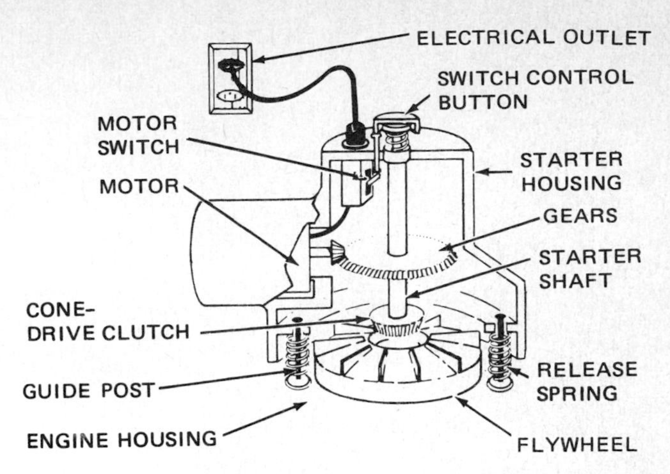

FIG. 12-25 Cone drive clutch, used with a 120-volt starting motor.

Δ12-17 120-VOLT ELECTRIC-STARTING MOTOR DRIVES The 120-volt electric starting motor uses a small electric motor which develops sufficient power to spin the engine crankshaft and start the engine. There are several types of electric starting motor drives. One is very similar to the type used on a battery-powered starting motor, covered earlier in this chapter. The other two types use different drive arrangements that are special for small engines.

FIG. 12-24 Electric starting motor which uses house current to operate. *(Briggs & Stratton Corporation)*

Δ12-18 CONE-DRIVE-TYPE 120-VOLT STARTING MOTOR One type of 120-volt electric starting motor uses a cone-shaped friction-drive clutch (Fig. 12-25). To operate this starting motor, press down on the switch-control button. This connects the electric motor to the 120-volt source so that the motor spins. As soon as it gets up to speed, press down on the starter housing. This engages the cone-shaped drive clutch so that the flywheel and crankshaft rotate.

When the engine starts, release the starter housing and the switch-control button. The release springs lift the starter housing so that the clutch disengages. At the same time the spring under the switch-control button lifts the button so that the starting motor is disconnected from the 120-volt source.

Δ12-19 SPLIT-PULLEY-DRIVE STARTING MOTOR This type, shown in Fig. 12-26, also engages by friction, but it does this automatically. The upper part of Fig. 12-26 shows how the split-pulley drive operates. When the starting motor is not operating, the two halves of the pulley are apart, as shown at the upper left. When the starting motor is operated, the upper half of the pulley comes up to speed along with the motor armature and shaft because the upper half of the pulley is attached to the shaft. The lower half of the pulley, being somewhat free, does not pick up speed instantly. Instead, it lags behind because of inertia. (Inertia is that property that all things have which resists change of position.) Therefore, the lower pulley half is momentarily stationary.

The pin in the motor shaft then pushes against the ramp in the lower pulley half, forcing the pulley half to move upward, as shown to the upper right in Fig. 12-26. Now the sides of the two pulley halves clamp the drive belt so that the drive belt is forced to move. This causes the drive pulley located above the engine flywheel to spin, cranking the engine.

After the engine starts and the motor is disconnected from the 120-volt source, the belt tension applied to the

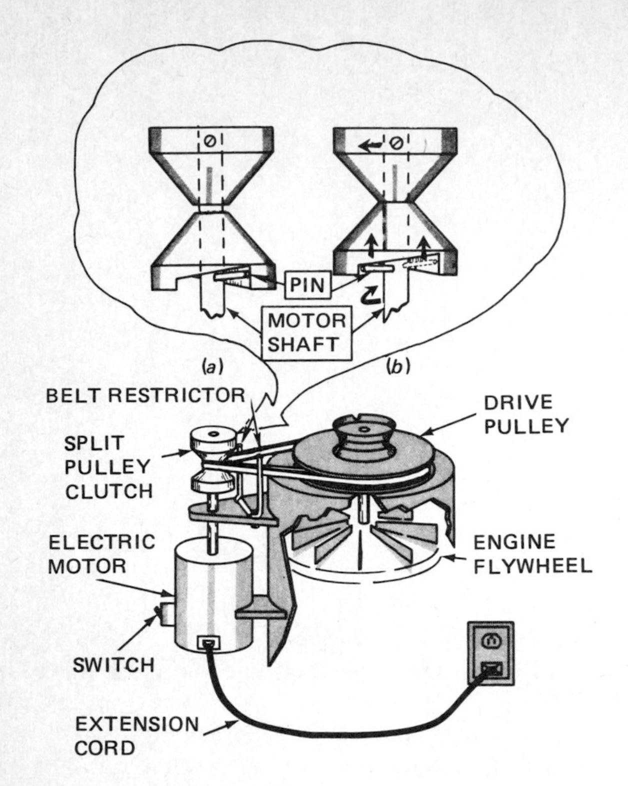

FIG. 12-26 A split-pulley drive, used with a 120-volt starting motor. The top part of the illustration shows the two positions of the lower pulley half. *(a)* Not cranking. *(b)* Cranking.

pulley forces the lower half to continue to move for a moment. This allows the lower half to drop down to the disengaged position, as shown to the upper left in Fig. 12-26.

Δ12-20 BENDIX-DRIVE STARTING MOTOR

This starting motor is similar to the electric starting motors with Bendix drive that are battery-powered. At one time, the Bendix drive was used on many automotive starting motors, but today most use an overrunning clutch. The Bendix drive is used in some small-engine starters because of its simplicity. The Bendix drive was described in Δ12-9.

Δ12-21 STARTER-GENERATOR

In many small-engine installations using battery-type electric starting motors, the starting motor may also be a generator. This makes it a starter-generator assembly. As a starting motor, it cranks the engine to start it. Then when the engine is running, the starter-generator acts as a generator to produce current that puts back into the battery the current taken out by cranking.

Figure 12-27 shows a typical wiring circuit for a starter-generator system. It includes a generator regulator (described later). The starter-generator is connected by a V-belt to the engine (Fig. 12-28), and is continuously connected during both cranking and generating.

The starter-generator has two sets of field windings, one for cranking the engine and the other for producing current. When the starter switch is closed, battery current flows through the starter field windings. These windings are made of heavy copper wire so that a heavy current can flow from the battery through them. This produces a strong magnetic field, which results in a strong cranking force. The armature is spun and the engine crankshaft is rotated so that the engine starts.

Then the operator opens the starter circuit by releasing the starter switch. This opens the starter field windings so starter action is ended. Now, as the engine comes up to speed and drives the starter-generator, the generator begins to produce current. A magnetic field is produced in the generator by the generator field windings, which are made up of relatively light copper wire. These windings are shunted, or connected across, the armature and use up a small part of the current that the armature produces. This creates a magnetic field in which the armature spins. The armature windings, which have

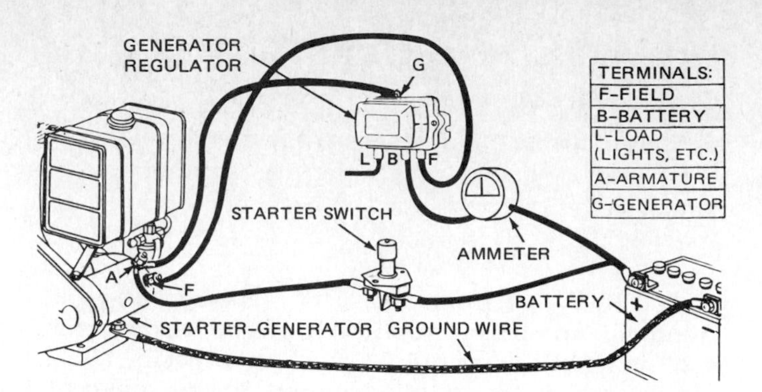

FIG. 12-27 Wiring circuit of a typical starter-generator system. The starter-generator starts the engine, and then generates current to charge the battery. The system includes a regulator to control the generator.

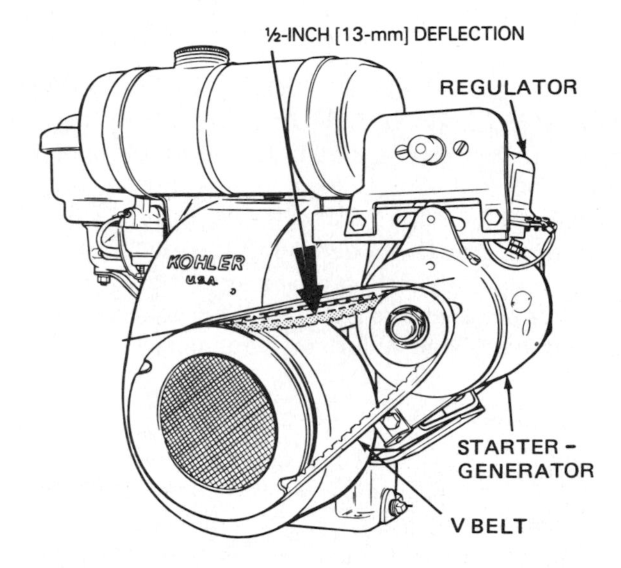

FIG. 12-28 Belt drive for a starter-generator. Proper belt tension is indictated as allowing ½-inch [13-mm] deflection. It is adjusted by moving the starter-generator toward or away from the engine. *(Kohler Company)*

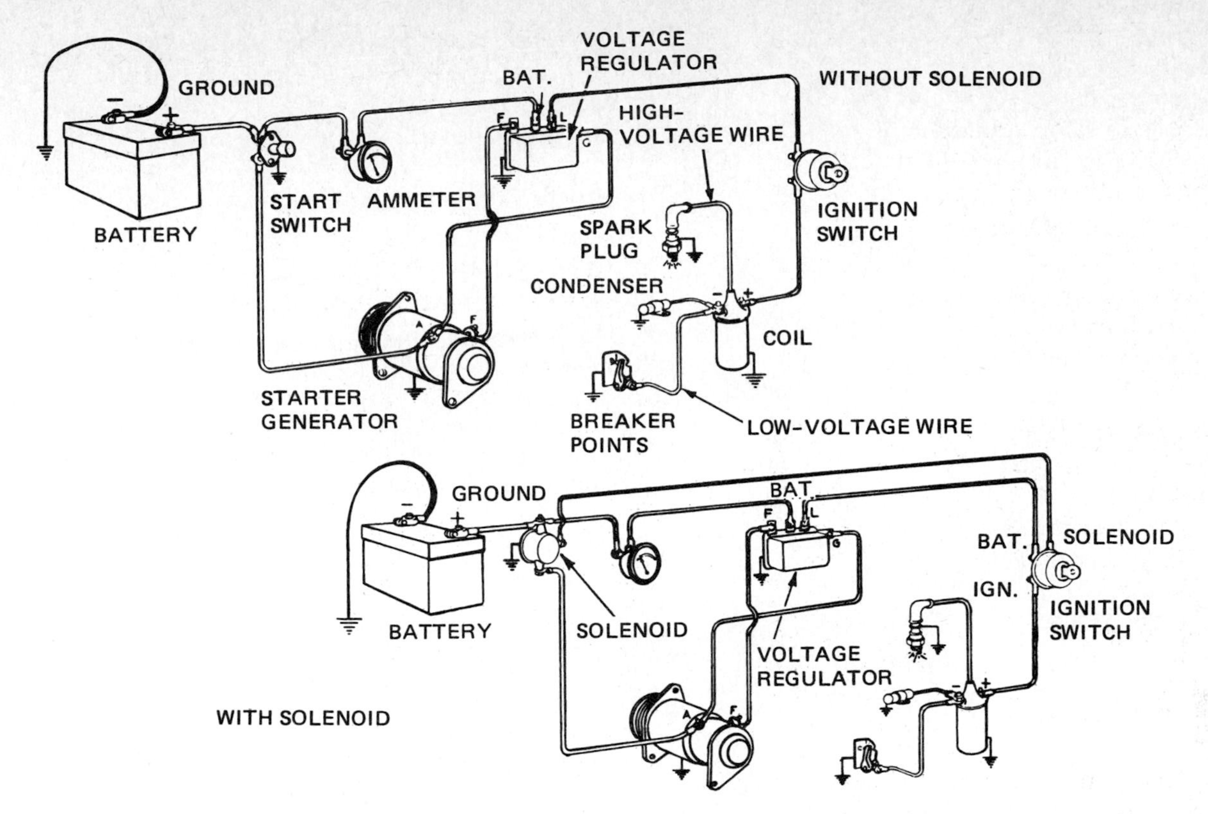

FIG. 12-29 Wiring diagrams for two types of starter-generator systems. Top, without a solenoid. Bottom, with a solenoid. *(Kohler Company)*

been serving as starter windings during the starting cycle, now begin to serve as current producers.

Figure 12-29 shows schematically two variations of the basic starter-generating system, one using a starter solenoid. The purpose of the solenoid is to make it possible for the starter switch to be located some distance away from the battery and starter. This reduces the length of heavy cable needed to complete the circuit between the battery and starter. Only a light wire is needed between the switch and solenoid, because the solenoid needs only a small amount of current to make it work. (The bigger the current flow, the heavier the wire needed to carry it.)

When the switch is turned to SOLENOID for starting, the solenoid is connected to the battery and it produces a magnetic field. This magnetic field pulls in an iron plunger which forces heavy contacts to close. These heavy contacts make a direct connection from the battery to the starting motor so that the engine is cranked. The system shown in Figure 12-29 includes an ignition system. This is a battery-ignition system which is described in Chap. 14. The generator and regulator are described in Chap. 16.

CHAPTER 12: REVIEW QUESTIONS

Select the *one* correct, best, or most probable answer to each question. You can find the answer in the section indicated at the right of the question.

1. The two types of mechanical starters using ropes are (**Δ*12-2 and 12-3***)
   a. rope-wind and rope-kick
   b. rope-rewind and rope-wind
   c. rope-kick and rope-rewind
   d. kick-rewind and rewind
2. The simpler of the two rope-type mechanical starters is the (**Δ*12-2 and 12-3***)
   a. kick starter
   b. rope-rewind
   c. windup
   d. rope-wind
3. To use the windup mechanical starter, you (**Δ*12-4***)
   a. set the release lever
   b. turn a crank
   c. release the spring
   d. all of the above
4. Mechanic A says kick starters are used in motorcycles. Mechanic B says they are used in minibikes. Who is right? (**Δ*12-5***)
   a. mechanic A
   b. mechanic B
   c. both A and B
   d. neither A nor B
5. Most electric starters for small engines are (**Δ*12-1***)
   a. 120-volt units
   b. ac units
   c. battery-operated
   d. parallel-wound

6. The two basic parts of a battery-operated starting motor are the (**Δ***12-7*)
   a. brushes and commutator
   b. armature and field-frame assembly
   c. Bendix drive and overrunning clutch
   d. field windings and brushes
7. The two basic types of drive used with battery-operated starters are (**Δ***12-9 and 12-10*)
   a. Bendix drive and overrunning-clutch drive
   b. solenoid-operated and mechanical
   c. Bendix drive and rubber-compression drive
   d. inertia drive and Bendix drive
8. The safety-interlock system allows engine cranking only when the (**Δ***12-14*)
   a. transmission is in neutral
   b. power takeoff is disengaged
   c. operator is sitting on the seat
   d. all of the above
9. The purpose of the mower safety brake is to
   a. prevent mower runaway (**Δ***12-15*)
   b. prevent blade runaway
   c. stop the blade in 3 seconds after the control lever is released
   d. slow the engine when it begins to overheat
10. The starter-generator is (**Δ***12-21*)
    a. a 120-volt unit
    b. used only on automotive engines
    c. a combined starting motor and generator
    d. an ac unit

CHAPTER 13

Starting-Systems Service

After studying this chapter, you should be able to:

1. ***List the steps in trouble diagnosis of mechanical starters.***
2. ***Service each type of mechanical starter.***
3. ***Explain how to diagnose trouble in and service electric starting motors.***
4. ***List the possible causes when the starting motor does not operate and when it operates slowly but the engine does not start.***
5. ***Rebuild and test a battery-powered starting motor.***
6. ***Test the starting-motor drive.***

Δ13-1 MECHANICAL-STARTER TROUBLE DIAGNOSIS

The following points should be considered whenever you begin to solve a starting problem:

1. Wear safety goggles and follow all safety cautions when working on mechanical starters.
2. Look for obvious troubles first, such as a broken rewind spring, a weak battery, poor battery connections, or a defective starter drive.
3. Try to pinpoint the trouble before removing the starter.
4. On electric starting motors, disconnect the battery ground cable before removing the starting motor.
5. Disassemble the starting motor only as far as necessary to correct the trouble, unless you are rebuilding the starting motor.
6. Test the starting motor before reinstalling it.

When diagnosing a trouble in mechanical starters, first determine if the starter is really the problem. Generally, if the mechanical starter is able to crank the engine, the problem is elsewhere—possibly in the ignition or fuel system. However, most small engines require normal cranking speed to generate ignition voltage. Be sure that the starter is cranking the engine at the proper speed.

The chart on page 138 lists various mechanical-starter troubles and the checks or corrections to be made for each condition. The chart does not list all the problems that could happen, nor the obvious ones, such as a broken pull rope or handle.

CAUTION Some types of mechanical starters have powerful springs wound up inside. Use caution when disassembling these starters. Improper disassembly could allow the spring to unexpectedly unwind with great force. To prevent eye injury, wear safety goggles when working with the starter spring.

Δ13-2 ROPE-WIND STARTERS

This type of starter operates by winding several turns of rope around the engine flywheel pulley and then pulling it quickly to crank the engine. The only parts usually needing service on this type starter are the rope and handle. You can buy a new rope and handle assembly, or either part if only one goes bad. The new rope should be the same size and length as the old one. Most engines use a 3/16-inch [5-mm] nylon braided rope about 5 feet [1.5 m] long. The rope should be long enough to wrap around the flywheel pulley up to five times plus an additional 1 foot [305 mm] to make knots at the two ends.

Two kinds of knots can be used to hold the rope in the handle at one end and on the pulley at the other. The procedure for making the knots is illustrated in Figs. 13-1 and 13-2. Some handles use a pin tied into the knot to retain it in the handle (Fig. 13-3). Before making any knots

MECHANICAL-STARTER TROUBLE-DIAGNOSIS CHART

Complaint	Check or correction
1. Starter spins but engine does not crank	a. Check engagement of starter pawls, rachet, or dogs
2. Engine cranks but starter pull has no feel of tension (rope-wind and rope-rewind starter)	a. Check engine compression
3. Rope does not rewind on starter (rope-rewind starter)	a. Check operation of rewind spring
4. Starter slips while cranking engine	a. Check starter pawls, dogs, or rachet teeth b. Check flywheel teeth c. Check for dirt or grease in drive mechanism
5. Starter does not wind up (windup starter)	a. Check windup spring
6. Starter does not crank engine fast enough	a. Check that starter engages and does not slip b. Check starter spring (windup starter or kick starter) c. Check for engine drag
7. Engine cranks properly but does not start	a. Check ignition system for spark b. Check fuel system for fuel, water in fuel, line obstructions, and overchoking c. Check engine for low compression

in the rope, singe both ends with a match to prevent the rope from unraveling. This also assures that the knots will not slip and loosen after they are tied. In servicing any rope-pull starter, replace the rope if it is frayed or ragged.

Δ13-3 ROPE-REWIND STARTERS

The rope-rewind starter uses a pull rope with a strong spring to rewind the rope after it is pulled out for cranking. Several designs of this type of starter are in use. All work on the same general principle but require different servicing procedures. These include replacing the rope, replacing the spring, and repairing the drive mechanism.

Δ13-4 REPLACING THE ROPE

Remove the starter from the engine. Usually the starter is attached by three or four screws. On some engines, you must remove

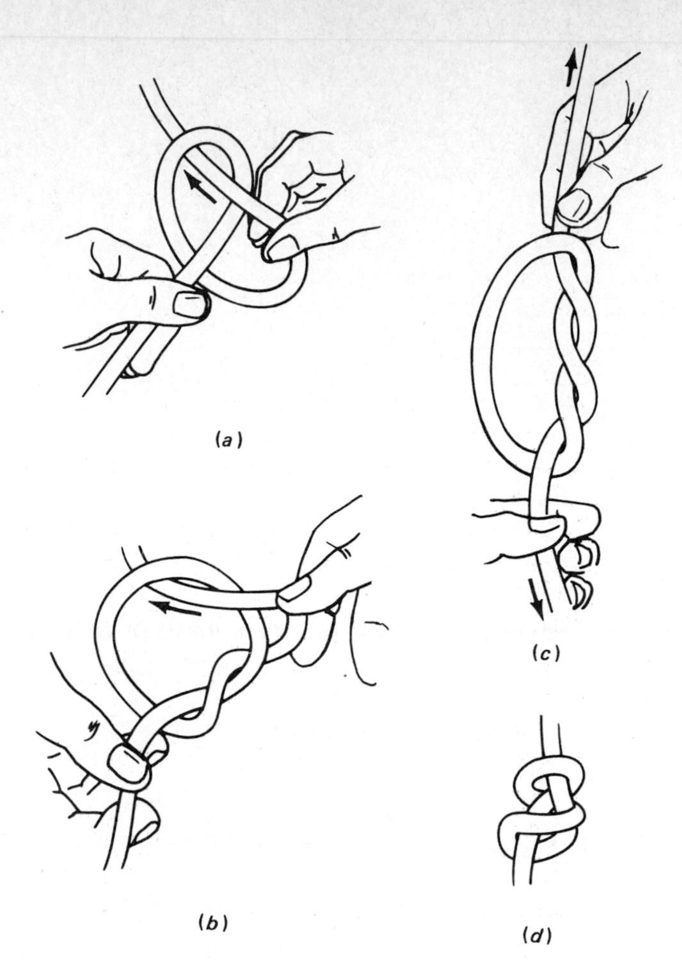

FIG. 13-1 How to make a double-overhand knot.

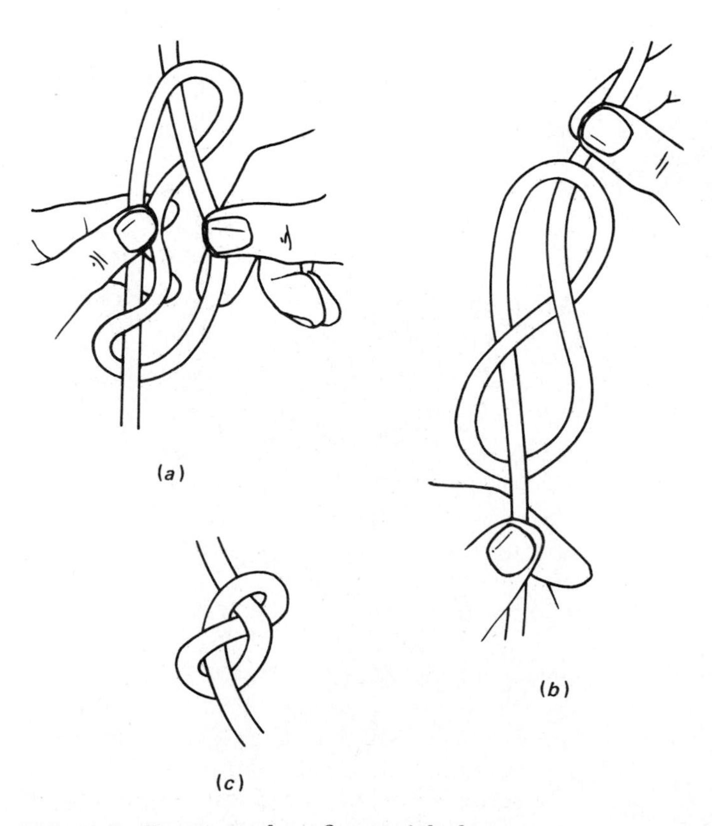

FIG. 13-2 How to make a figure-eight knot.

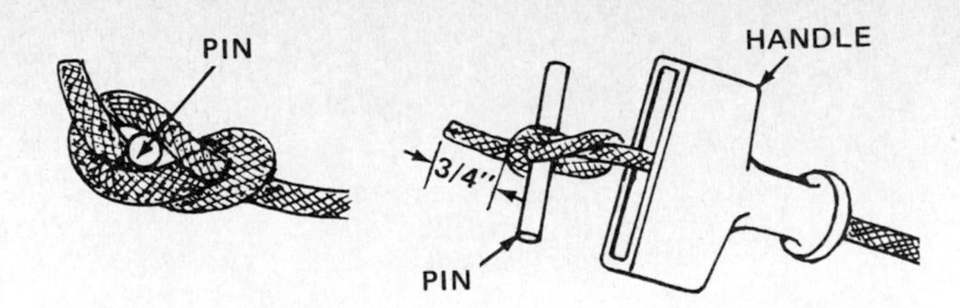

FIG. 13-3 **Handle retained by a pin tied into the knot.** *(Briggs & Stratton Corporation)*

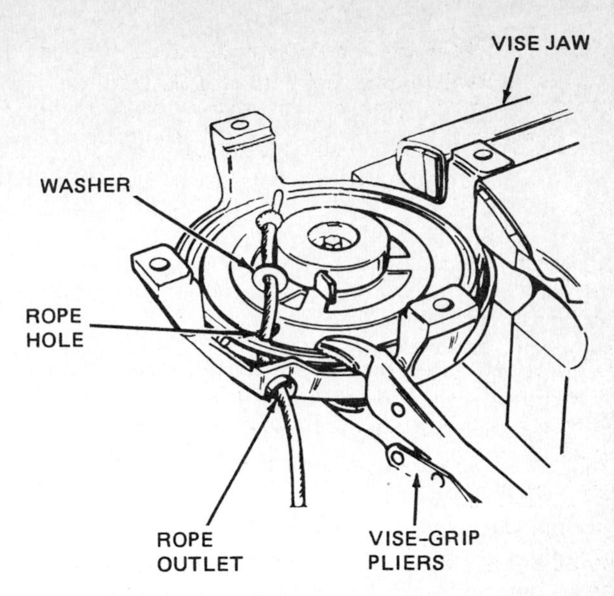

FIG. 13-5 **Holding the pulley with Vise-Grip pliers.** *(Tecumseh Products Company)*

a top cover or other piece to get to the starter. You can work on the starter if it is laid on a workbench, but you will find it easier to work on if you clamp it upside down in a vise (Fig. 13-4). If the knot in the rope is visible, you can replace the rope without disassembling the starter. If the knot is not visible, you will have to disassemble the starter. With the starter disassembled, always check the spring. When the rope breaks, it sometimes bends the spring.

On the visible-knot starter (Fig. 13-4), pull the rope all the way out against the tension of the recoil spring. The pulley must be held in place so the spring does not unwind when the new rope is installed. You can hold the pulley with Vise-Grip pliers (Fig. 13-5), or you can use a wrench and square stick (Fig. 13-6). When the wood-stick rewind method is used, the end of the wrench is wired to the housing, as shown in the inset in Fig. 13-6.

If the spring has no tension when the rope is pulled out, the spring has unwound. You can wind it up again with a wrench and square stick or with a screwdriver as shown in Figs. 13-6 and 13-7.

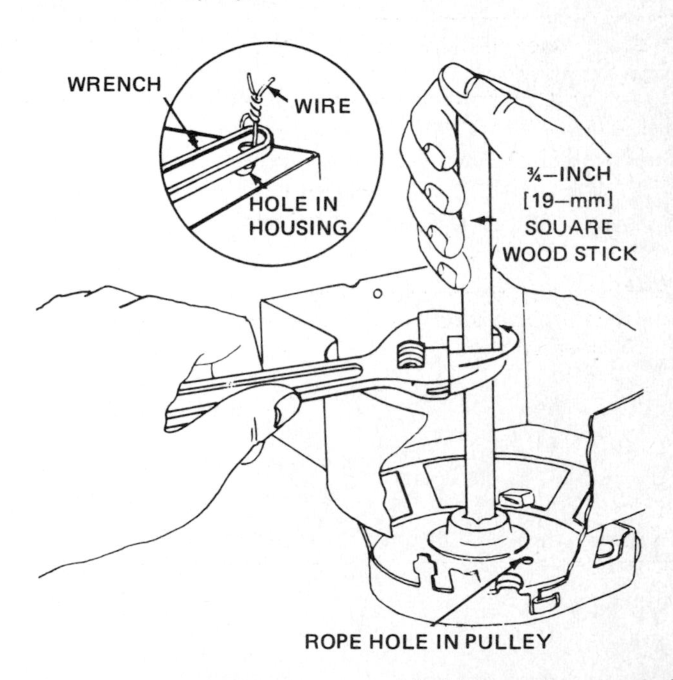

FIG. 13-6 **On some starters you can hold the pulley, or wind up the recoil spring, with a square piece of wood and a wrench.** *(Briggs & Stratton Corporation)*

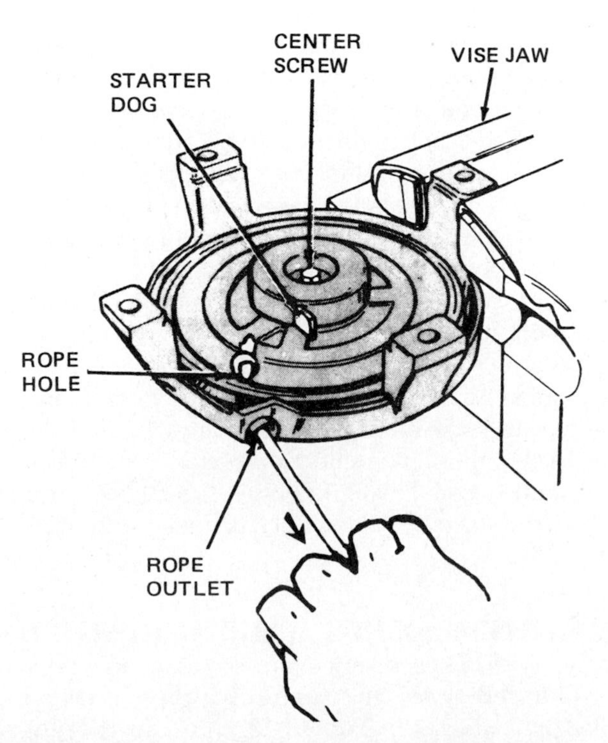

FIG. 13-4 **To begin the repair of the visible-knot type of rope-rewind starter, pull the rope all the way out.**

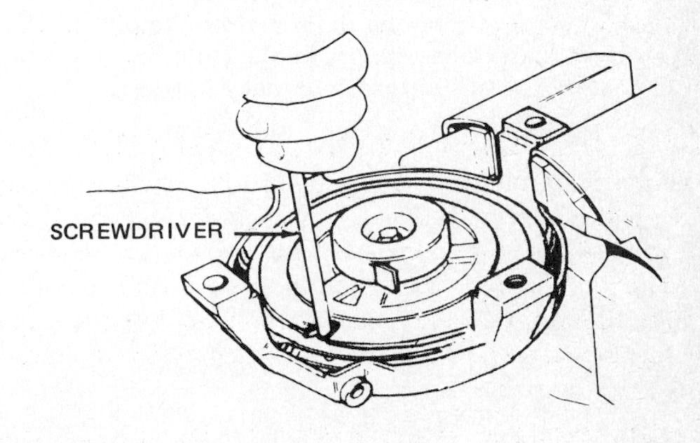

FIG. 13-7 **On some starters, you can wind up the recoil spring with a screwdriver.**

Whenever you wind up a starter spring, always make sure you wind it in the correct direction. The general rule is to wind the spring up tight and then back it off one complete turn. If the spring cannot be wound up, it is probably broken or detached. How to service the spring is described later.

The new rope should be the same size and length as the old. Singe both ends of the rope with the flame of a match. This prevents the rope from fraying and also prevents the knot from loosening. Thread the rope through the rope eyelet in the housing and then through the hole in the pulley, as shown in Figs. 13-8 and 13-9. If the holes line up, threading the rope is not difficult. But when the holes do not line up, threading is made easier by using a hooked wire through the rope end, as shown in Fig. 13-8.

On this starter, be sure that the rope passes on the inside of the guide lug. After the rope is in the pulley, tie a knot in the rope end. On some starters, a thin washer with a ¾-inch [19-mm] outside diameter and a ¼-inch [6.35-mm] hole is placed on the rope before the knot is tied (Fig. 13-5). When the knot is tied, pull the rope so the knot is near the hole and does not interfere with anything when the pulley rotates. Now, with the rope pulled taut, release the Vise-Grip pliers or the wrench holding the wood stick. Slowly let the rope wind up on the pulley. Then attach the handle to the other rope end, as explained earlier for rope-wind starters.

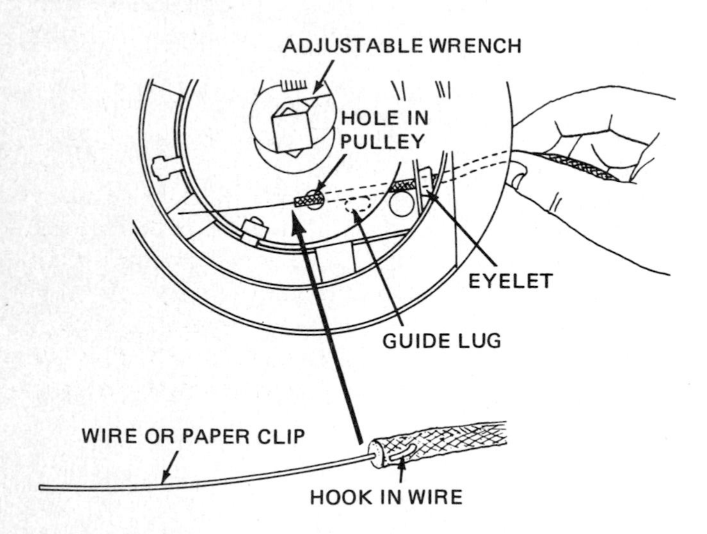

FIG. 13-8 **When rethreading the rope into the pulley hole, you may find that a piece of wire hooked into the end of the rope, as shown at the bottom, will make it easier.**

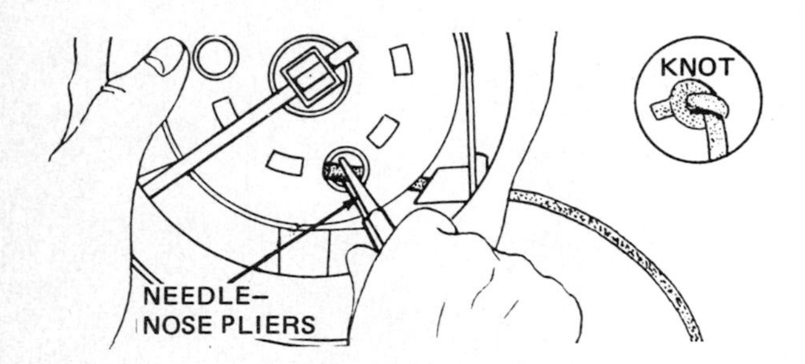

FIG. 13-9 **Installing the rope on a starter that does not have a guide lug.** *(Briggs & Stratton Corporation)*

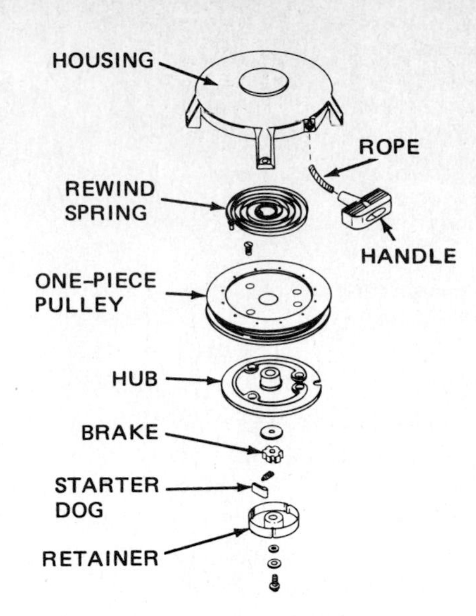

FIG. 13-10 **Disassembled rope-rewind starter with a one-piece pulley.** *(Tecumseh Products Company)*

On the hidden-knot starter, the knot is not visible, and the starter must be disassembled to replace the rope. Disconnect the handle from the end of the rope and untie the knot. Hold the pulley with a cloth or a gloved hand and allow it to turn slowly. This removes the tension on the spring. Then remove the starter drive from the pulley assembly. This requires disassembly of the starter-drive mechanism (Δ13-6). There are two types of pulleys: the one-piece and the two-piece.

The one-piece pulley is shown disassembled in Fig. 13-10. When the starter has a one-piece pulley, remove the pulley, leaving the spring in the housing. Then the rope can be replaced on the pulley, as described earlier.

Figure 13-11 shows a rope-rewind starter that uses a two-piece pulley. When the starter has a two-piece pulley, remove one side of the pulley, leaving the spring under the other half in the housing. You will now be able to remove the rope and install a new rope.

Reinstall the pulley after winding the rope on it. The drive mechanism is assembled as explained later. Rewind the spring and secure the pulley in place with Vise-Grip pliers or with a wrench and wood stick. Assemble the handle on the rope end so that the pulley cannot unwind. Then release the Vise-Grip pliers or wood stick. Reinstall the starter on the engine, making sure that the starter is properly aligned. Some starters have an alignment pin which is inserted into a hole in the crankshaft to secure alignment. If the alignment is not correct, the starter will not work properly and will wear out rapidly.

Δ13-5 REPLACING THE SPRING There are two types of recoil springs: the removable type and the packaged type. One removable type is shown installed in a starter in Fig. 13-12. You can recognize it because you can see the spring end on the outside. The packaged type is a stronger spring and comes pre-oiled

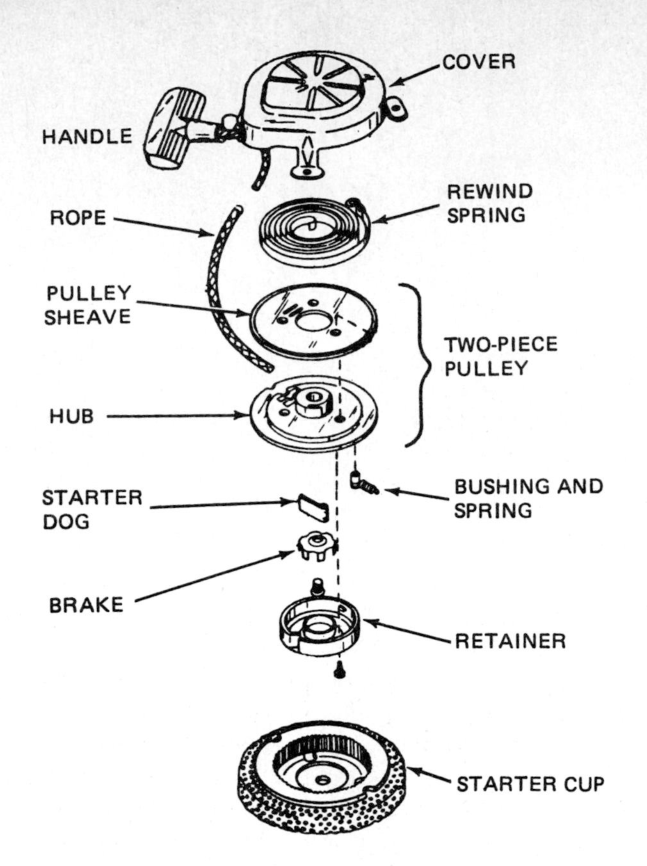

FIG. 13-11 Disassembled rope-rewind starter with a two-piece pulley.

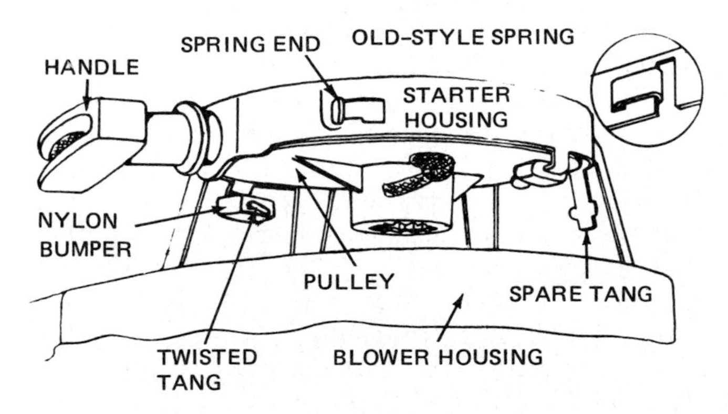

FIG. 13-12 A removable type of recoil spring. The end of this spring can be seen from outside the starter.

and compressed inside a retainer housing. The retainer housing may be of the permanent type and is installed in the starter along with the spring. On others, the housing is discarded when the spring is installed in the starter.

If the spring is of the removable type accessible from outside the starter housing, remove the starter and clamp it in a vise. Remove the rope from the handle, and with a gloved hand or cloth, slowly let the pulley rotate to release the spring tension. Pull the spring out of the housing as far as possible with pliers, as shown in Fig. 13-13. Use a cloth or wear gloves to protect your hands. Remove the pulley. If the pulley is held in place by bumper tangs as shown in Fig. 13-13, bend up one tang to remove the pulley. If the pulley is held in place by the starter-drive mechanism, remove it first. Starter-drive service is described in Δ13-6.

Disconnect the spring from the pulley, as shown in Fig. 13-14. If you are going to use the spring again, straighten it, as shown in Fig. 13-15. Wear gloves to protect your hands.

Attach the replacement spring to the pulley. Install the pulley inside the starter housing with the spring extending out, as shown in Fig. 13-13. Now with gloves or cloth protection for your hands, guide the spring into the housing. Let the spring wind up, using either of the methods shown in Figs. 13-6 and 13-7. Be sure the spring end securely locks into the side of the housing at the end of the rewind. Attach the rope to the pulley and install the handle. Then install the starter on the engine and check its operation.

If the spring is of the packaged type (Fig. 13-16), the retainer housing may be of the temporary type or it may be permanent. If it is temporary, the spring is slid from the temporary housing into the permanent housing on the starter. If the retainer housing is of the permanent type, the old housing is removed from the starter and the new housing with spring is installed.

If the spring is of the semicoiled removable type, it should be removed from its package as shown in Fig. 13-17. The spring can then be wound into the housing on the starter.

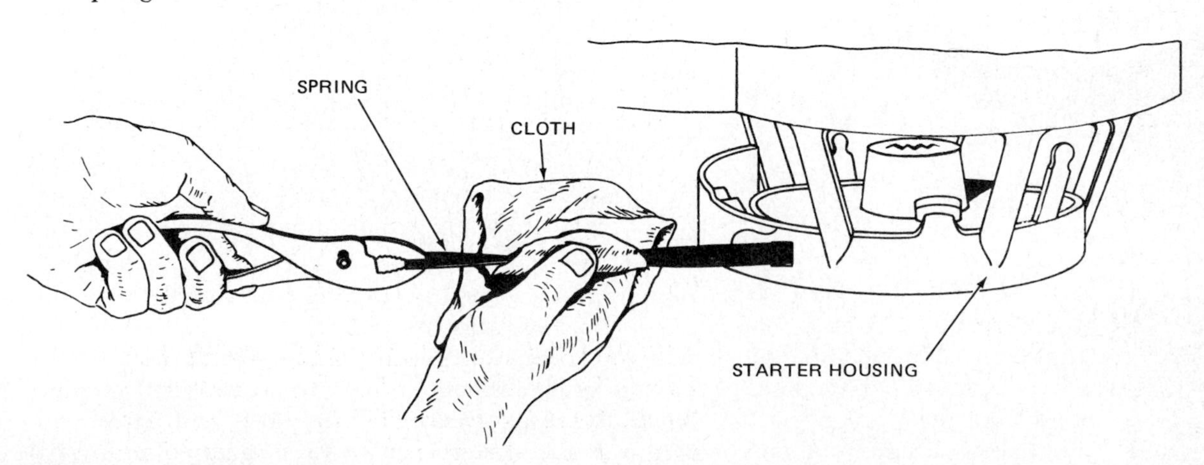

FIG. 13-13 Use pliers and a cloth or glove on your hand to pull the spring out of the starter housing.

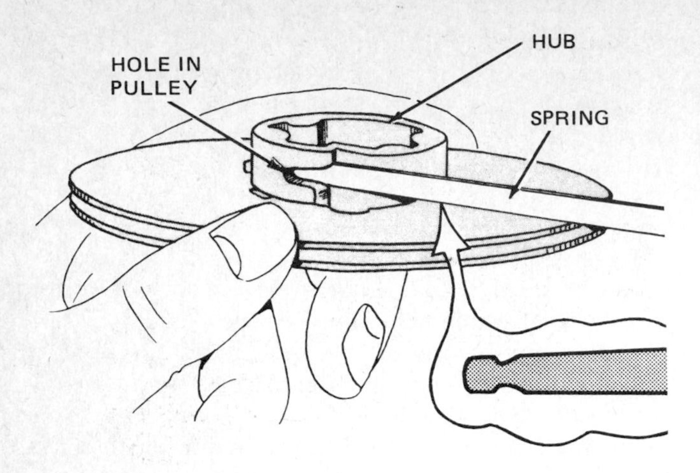

FIG. 13-14 Unhooking the spring from the pulley.

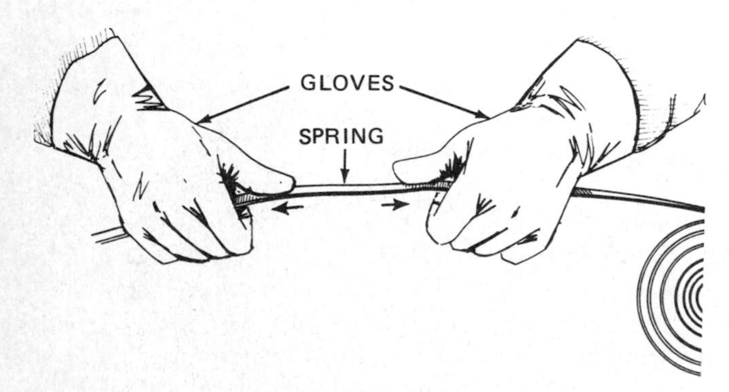

FIG. 13-15 Straightening the spring to provide more tension. Wear gloves when handling the spring.

FIG. 13-16 A packaged type of recoil spring. Most replacement recoil springs are enclosed like this in a housing for safe handling.

Δ13-6 DRIVE-MECHANISM SERVICE

All rope-rewind starters are similar in construction. The main difference is in the ratchet mechanism that causes the pulley to engage the crankshaft for cranking and then releases when the engine starts. There are four general types (Fig. 13-18), based on the method used for

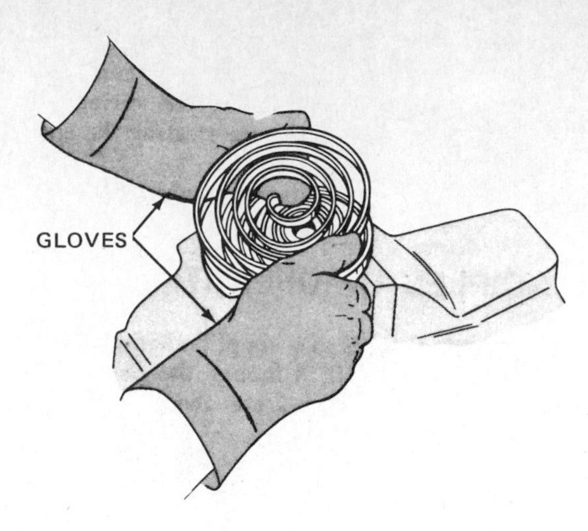

FIG. 13-17 A removable or semicoiled type of recoil spring.

flywheel engagement. These methods are pawls, dogs, friction shoes, and steel balls.

To service the drive mechanism, remove the starter and clamp it in a vise (Fig. 13-4). Pull the rope and feel the action of the drive mechanism. When the rope is pulled, a properly working drive mechanism should quickly engage the flywheel adapter. If service is required, remove the drive mechanism (Fig. 13-19). Disassemble it.

Note very carefully what parts go where when you take the mechanism apart. Then you will know exactly how everything goes together when you reassemble it. Be especially careful to put the engaging mechanism back in exactly the same way you found it. If you get it in wrong, the starter will not work. Lubricate the parts lightly with graphite or multipurpose grease on reassembly.

Check for proper operation after reassembly. Then install the starter on the engine. Recheck the starter for proper operation.

Δ13-7 VERTICAL-PULL STARTERS

The rewind starter (Fig. 13-20) is used on some outboard engines and lawn-mower engines. It is a vertical-pull starter using a Bendix drive. To service the starter, remove the handle and let the rope slowly recoil to remove the spring tension. Then unscrew the starter-mounting bolt, and remove the starter from the engine. Hold the starter together while removing it to prevent releasing the spring.

To replace the rope on this starter, take off the retainer clip on the end of the Bendix drive and slide off the plastic starter pinion (Fig. 13-20). The pinion gear mounts on the worm gear with the grooved side toward the pulley. Remove the screws that hold the pulley plate to the pulley and disassemble the pulley (Fig. 13-21). Obtain the new replacement rope and singe the ends. Remove the old rope and install one end of the new rope in the pulley. Install the pulley plate on the pulley and tighten the screws. Hold the starter so the worm gear points toward you, and wind the rope clockwise on the pulley.

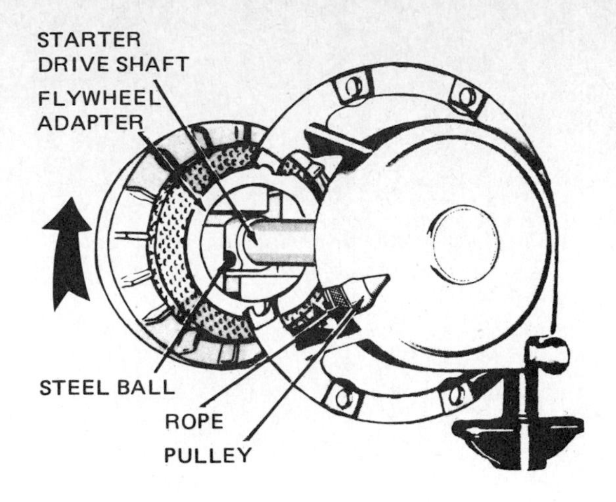

(*a*) WEDGING STEEL BALLS

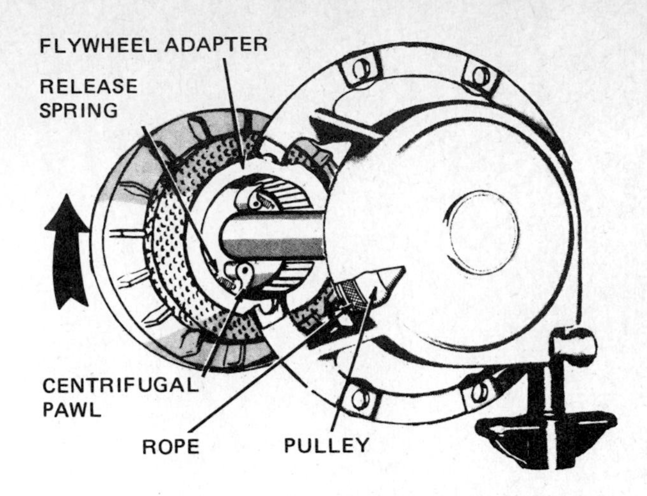

(*b*) CENTRIFUGALLY ACTUATED PAWLS

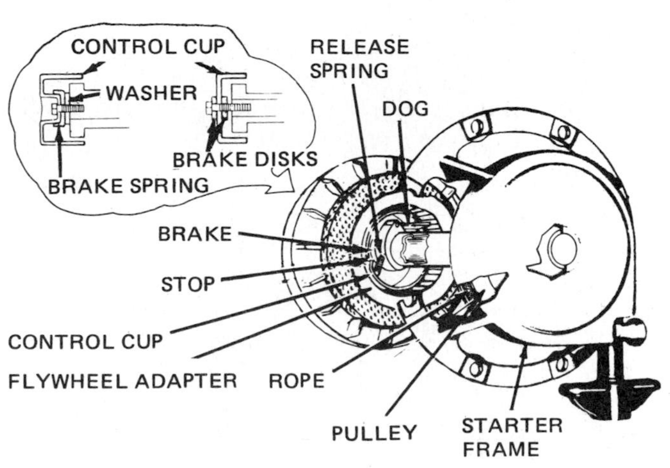

(*c*) CAM-OPERATED DOG

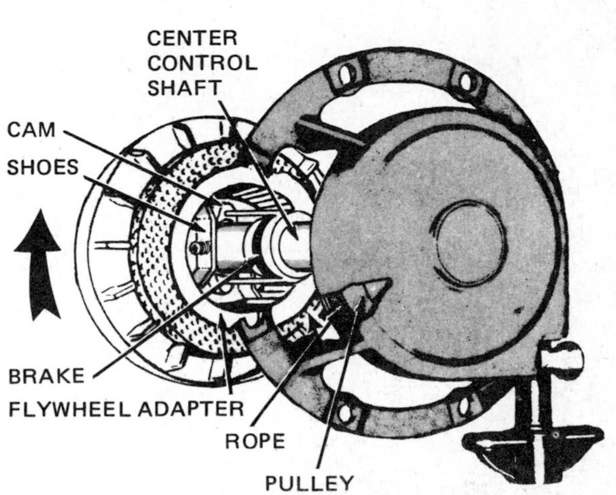

(*d*) CAM-OPERATED SHOES

FIG. 13-18 Various types of drive mechanisms for rope-rewind starters.

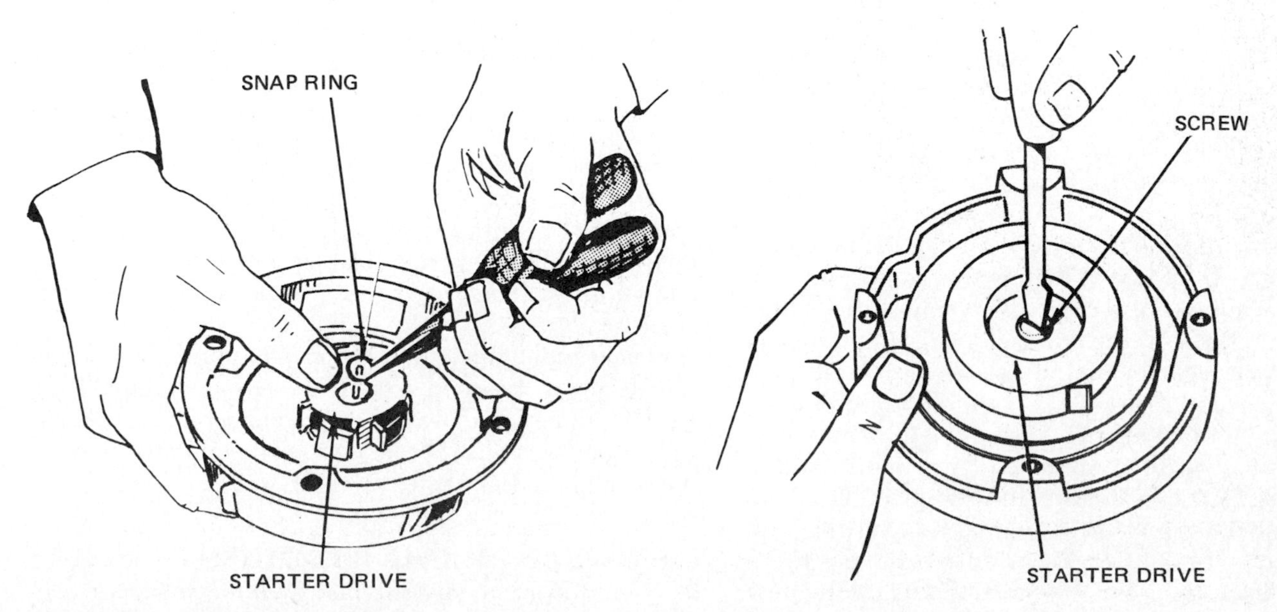

FIG. 13-19 *Left:* On some starters, remove the drive mechanism by removing a snap ring. *Right:* On others, remove a screw.

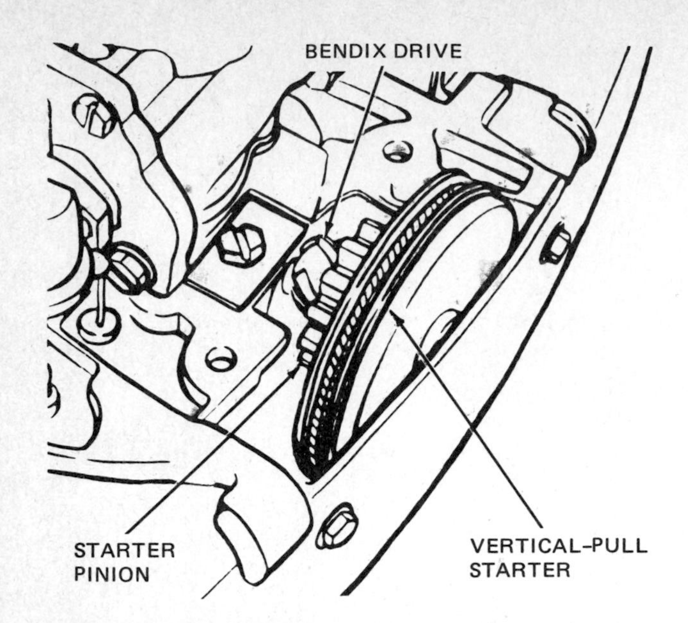

FIG. 13-20 **Vertical-pull rope-rewind starter using a Bendix drive.** *(Evinrude Motors Division of Outboard Marine Corporation)*

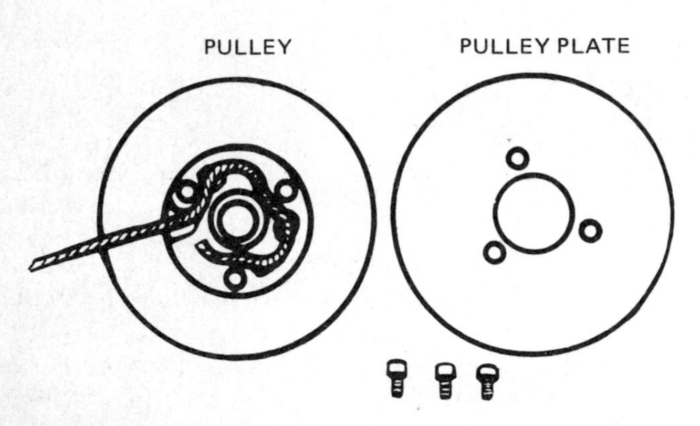

FIG. 13-21 **Position of the rope in the pulley.** *(Evinrude Motors Division of Outboard Marine Corporation)*

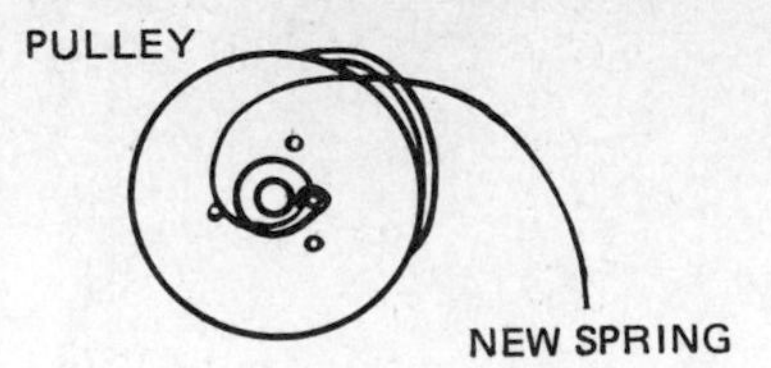

FIG. 13-22 **Installing a new starter spring.** *(Evinrude Motors Division of Outboard Marine Corporation)*

FIG. 13-23 **Winding the new spring into the cover.** *(Evinrude Motors Division of Outboard Marine Corporation)*

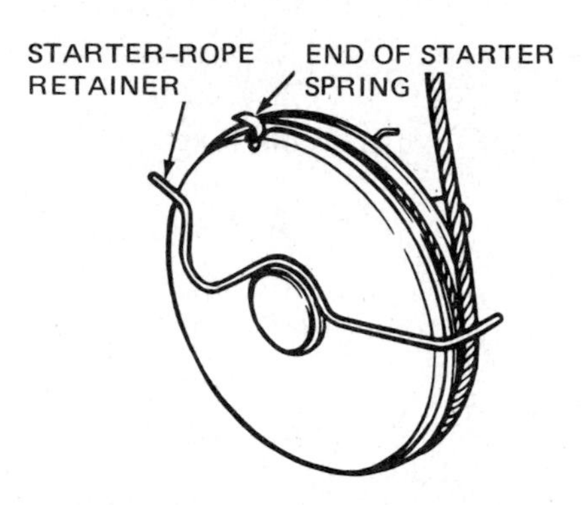

FIG. 13-24 **When installed, the end of the starter spring hooks over the edge of the cover slot.** *(Evinrude Motors Division of Outboard Marine Corporation)*

To install a new starter spring, remove the outer pulley cup and old spring. Curl the end of the new spring on the outside of the starter pulley (Fig. 13-22). Position the cover over the pulley so that the spring is guided through the slot in the cover. Clamp the starter in a vise using wooden blocks to prevent damage to the worm-gear shaft. Wind the rope on the pulley. Then pull the rope to turn the pulley (Fig. 13-23). Repeat this as necessary to draw the spring into the cover. The end of the starter spring should hook on the cover slot (Fig. 13-24). Install the starter-rope retainer on the outside of the pulley.

If the starter could be pulled but it did not crank the engine, the pinion spring may be sprung or damaged. A distorted spring will not grip the pinion properly, and so the pinion will fail to engage the flywheel. The two prongs of the pinion spring should be ¼ inch [6.35 mm] or less apart when the spring is off the pinion. If the spring needs replacement, use care to stretch the new one only wide enough to allow it to snap into the groove of the pinion.

Place the starter assembly on the engine with the end of the starter spring up and in the cutout made for it (Fig. 13-25). One prong of the starter-pinion spring should be above the mounting plate edge and the other prong below it. The starter-rope retainer will lock into place as shown in Fig. 13-26 when the starter is positioned correctly.

Install and tighten the starter mounting bolt. Completely wind the rope around the starter pulley. Then turn the pulley an extra one to one and a half more turns for proper tension. Thread the rope through the handle, and fasten the handle to the rope.

△13-8 SERVICING WINDUP STARTERS

The windup starter is also called an *impulse starter, self-starter, ratchet starter, speed starter,* and other names. It uses a ratchet and pawl, or dog, mechanism to wind up

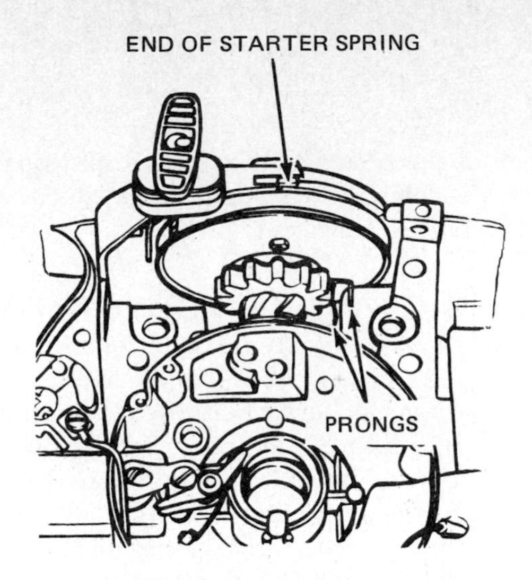

FIG. 13-25 Proper position of the end of the starter spring. *(Evinrude Motors Division of Outboard Marine Corporation)*

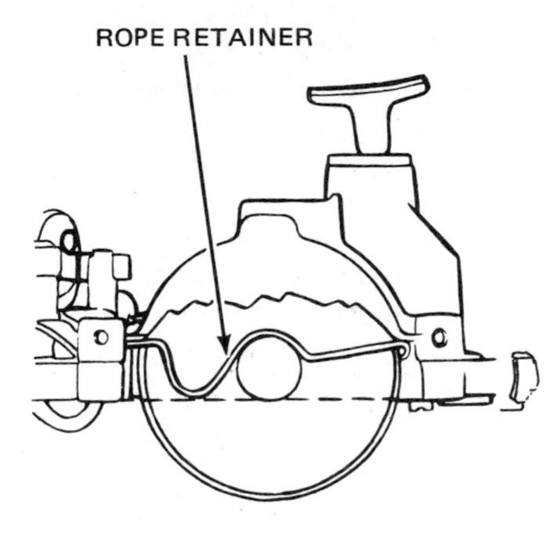

FIG. 13-26 Starter-rope retainer will lock in place when the starter is properly positioned. *(Evinrude Motors Division of Outboard Marine Corporation)*

and hold the tension of an internal spring. When the spring is released, it overcomes engine compression and friction to crank the engine. The ratchet is usually rotated directly through gearing by a windup handle. The spring is released by an external lever or control knob. Two types of windup starters are shown in Fig. 13-27.

Before beginning to work on a windup starter, always make sure the starter spring is fully unwound. Wind the spring several turns with the handle. Then trip the release lever or operate the control knob and let the spring unwind. If the spring does not unwind, on some models you can release the spring tension by taking off the handle (Fig. 13-28). Hold the handle with one hand, and remove the Phillips-head screw holding the handle to the starter.

> ***CAUTION*** **Never attempt to work on a windup starter without releasing the spring tension. The spring is very strong and could cause serious injury if it should pop out of the starter during disassembly.**

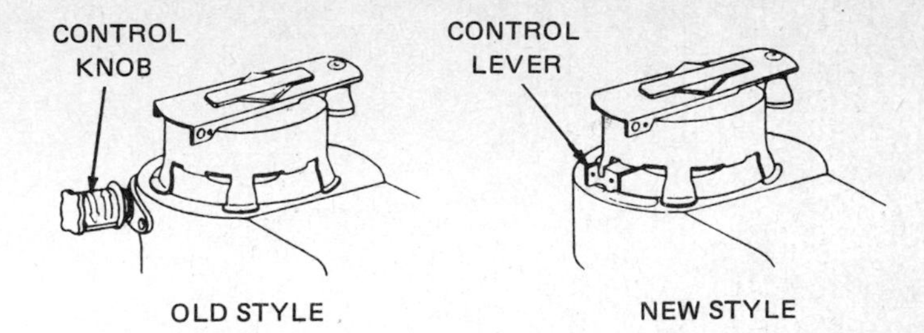

FIG. 13-27 Left, old-style windup starter with control knob. Right, newer-style starter with control lever. *(Briggs & Stratton Corporation)*

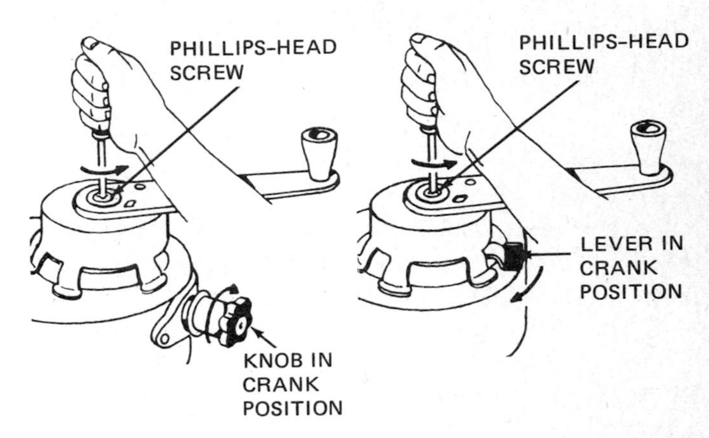

FIG. 13-28 Releasing the windup spring by removing the Phillips-head screw. Hold the handle with one hand while removing the screw.

If the starter does not crank the engine, the spring is broken or the starter clutch balls are not engaging. Put the control knob or release lever in position for START, and crank the handle 10 turns clockwise. While turning the crank handle, watch the starter clutch ratchet (Fig. 13-29). If it does not move, the spring is broken. If the ratchet moves, it is the ratchet that needs servicing.

Some windup starters are disassembled from the drive end. Others are disassembled from the handle end. If the handle is riveted or welded to the shaft, remove the drive mechanism first. The drive mechanism and retainer screws are similar to those on rope-rewind starters, but heavier. If necessary, remove the drive mechanism for repair. Remove the mainspring assembly, with its retainer.

> ***CAUTION*** **Do not remove the spring from the retainer unless the service manual for the engine you are working on specifically says to do it and tells you how. These springs are very strong and can injure you if they are released improperly!**

When disassembling the drive mechanism, observe carefully the location and position of all parts so you can correctly put the mechanism back together. Inspect the housing and operating parts for cracks, worn gear teeth or ratchet, and broken or bent pawls. Watch for small parts such as springs, washers, and spacers. Clean all

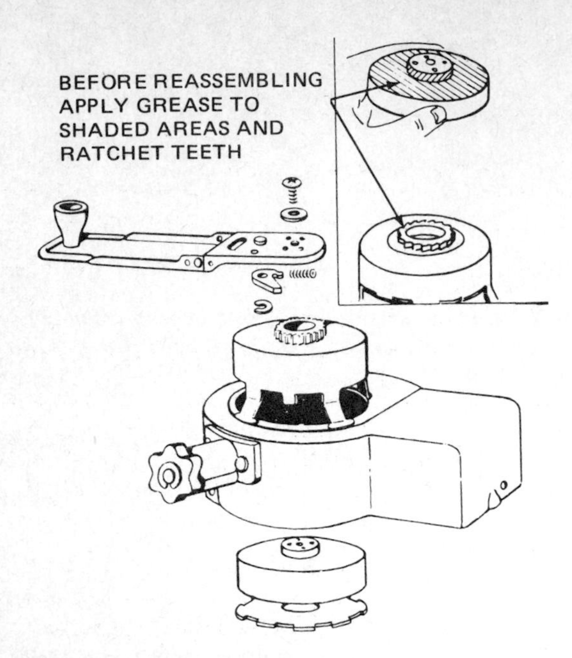

FIG. 13-29 Partly disassembled windup starter. *(Briggs & Stratton Corporation)*

parts and lubricate all moving surfaces with engine oil.

If you install a new spring, be sure to destroy the old spring so that no one will get hurt tampering with it. The best way is to heat the old spring with a torch. This removes the temper and tension in the spring.

On reassembly, make sure you put all parts back into their proper places. Reinstall the starter on the engine, and check it for proper operation.

∆13-9 SERVICING ELECTRIC STARTING MOTORS There are two general types of electric starting motors. These are low-voltage dc (6-volt and 12-volt) motors and 120-volt motors that operate on house current. A 6-volt starting motor operates from a 6-volt battery. A 12-volt starting motor operates from a 12-volt battery. To keep the battery charged, the engine has a charging system that includes a generator or an alternator (Chap. 16).

∆13-10 12-VOLT STARTING-MOTOR TROUBLE DIAGNOSIS The most common complaint about the starting motor is that the engine cranks very slowly, or does not crank at all. You can usually pinpoint the problem by turning on the lights (if provided). Then operate (or attempt to operate) the starting motor. (Many engines equipped with a 12-volt starting motor also have lights.) Under normal conditions, the battery voltage drops and the lights dim slightly as the engine is cranked. However, if trouble exists, the lights may do one of the following:

1. Stay bright with no cranking action
2. Dim slightly with no cranking action
3. Dim considerably with no cranking action
4. Go out with no cranking action
5. Not come on at all

The chart on page 147 lists various improper operating conditions, their possible causes, and the checks or corrections to be made for each condition.

∆13-11 CHECKING 12-VOLT STARTING MOTORS Two general types of drives are used with battery-powered starting motors for small engines. One type is the various mechanical drives. They include the Bendix drive (Fig. 12-14), the overrunning-clutch drive (Fig. 12-15), and the rubber-compression drive (Fig. 12-13). The other type is a belt drive (Fig. 13-30). It has an overrunning clutch (a *belt clutch*), either in the pulley on the starting motor or in the crankshaft pulley. When the starting motor is turning, the clutch locks so the belt turns the crankshaft. But as soon as the engine starts and attempts to backdrive the starting motor, the clutch unlocks. This prevents the engine from turning the starting motor at a high speed, which would damage it.

Regardless of type, no periodic service is required on a starting-motor drive. However, the belt-drive system must have the drive belt checked and adjusted periodically. To do this, remove the belt guard. Check the condition of the belt and its deflection (Fig. 13-30). Various conditions of the drive belt that require replacement are shown in Fig. 9-25.

Most electric starting motors are prelubricated and require no further lubrication during normal service. About the only trouble that can occur inside the starting motor results from worn brushes or commutator. Sometimes starting motors are damaged by prolonged cranking that causes overheating. The small-engine starting motor should not be used for longer than about 10 seconds at a time. If the engine does not start within this time, stop cranking. Allow the starting motor to cool for 1 minute or longer before attempting to restart.

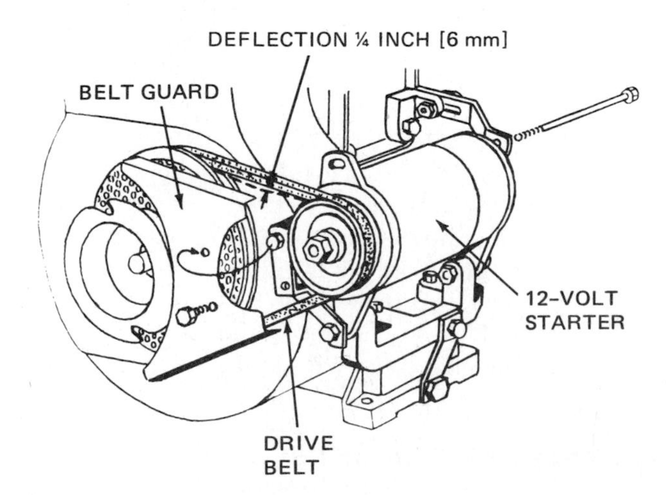

FIG. 13-30 A 12-volt starting motor that uses a drive belt. *(Briggs & Stratton Corporation)*

12-VOLT STARTING-MOTOR TROUBLE-DIAGNOSIS CHART

Complaint	Possible cause	Check or correction
1. No cranking, no lights	a. Battery dead	Recharge or replace battery
	b. Open circuit	Clean and tighten connections; replace wiring
2. No cranking, lights go out	a. Poor connection, probably at battery	Clean cable clamp and terminal; tighten clamp
3. No cranking, lights dim slightly	a. Pinion (Bendix) not engaging	Clean pinion and sleeve; replace damaged parts
	b. Excessive resistance or open circuit in starter	Clean commutator; replace brushes; repair poor connections
4. No cranking, lights dim heavily	a. Battery weak	Check and recharge or replace battery
	b. Very low temperature	Make sure battery is fully charged and engine, wiring circuit, and starter are OK
	c. Pinion (Bendix) jammed	Free pinion
	d. Frozen shaft bearing, direct short in starter	Repair starter
5. No cranking, lights stay bright	a. Open circuit in switch	Check switch contacts and connections
	b. Open in control circuit	Check solenoid, ignition switch, and connections
	c. Open circuit in starter	Check commutator, brushes, and connections
6. Engine cranks slowly but does not start	a. Battery run down	Check and recharge or replace battery
	b. Very low temperature	Make sure battery is fully charged and engine, wiring circuit, and starter are OK
	c. Starter defective	Test starter
	d. Undersized battery cables	Install cables of adequate size
	e. Mechanical trouble in engine	Check engine
	f. Operator may have run battery down	
7. Engine cranks at normal speed but does not start	a. Ignition system defective	Check ignition system
	b. Fuel system defective	Check fuel pump, line, choke, and carburetor
	c. Air leaks in intake manifold or carburetor	Tighten mounting; replace gasket as needed
	d. Engine defective	Check compression, valve timing, etc.
8. Solenoid plunger chatters	a. Hold-in winding of solenoid open	Replace solenoid
	b. Weak battery	Charge battery
9. Pinion disengages slowly after starting	a. Sticky solenoid plunger	Clean and free plunger
	b. Overruning clutch sticks on armature shaft	Clean armature shaft and clutch sleeve
	c. Overrunning clutch defective	Replace clutch
	d. Shift-lever return spring weak	Install new spring

Checking the Electric Connections When an engine cranks slowly, or fails to crank, and low temperature is not the cause, check the battery. Battery service is covered in Chap. 11. Then look at the cables and connections. If cables are frayed and connections are bad, not enough current can get through to produce normal cranking.

Checking the Starting-Motor Control On systems using a magnetic switch or a solenoid, bypass the switch or solenoid by connecting a heavy jumper wire to the two terminals. If the starting motor works, the trouble is in the switch or solenoid.

Inspecting the Starting Motor If everything outside of the starting motor looks satisfactory, then check the brushes and commutator inside the starting motor. Some starting motors have a cover band which can be removed (Fig. 13-31). On other starting motors (Fig. 12-13), the end frame or cap assembly must be removed. This is done by

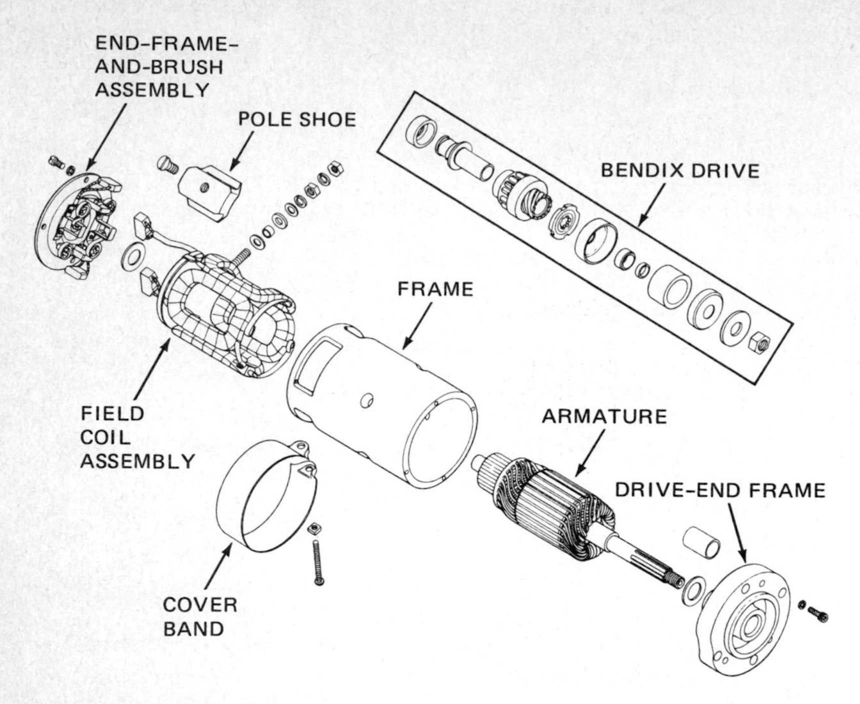

FIG. 13-31 **Disassembled electric starting motor with a Bendix drive.** *(Kohler Company)*

removing the two through bolts so that the cap can be partially slipped off. To remove the cap completely, take the insulated brush out of the brush holder. This is done by lifting the spring and pulling the brush out.

Inspecting Brushes and Commutator If the brushes (Fig. 13-32) are worn down to less than half the length of a new brush, they must be replaced. If the commutator is worn or rough, clean it with fine sandpaper or a brush-seating stone (Fig. 13-33).

New brushes can be installed in most starting motors. The grounded brush lead may be riveted to the end cap. The rivet must be drilled out so the new brush lead can be riveted into place. The insulated brush is connected to the field lead and must be unsoldered before removal. Then the new brush lead is soldered to the field lead. Rosin-core solder must be used when soldering electric connections.

According to some manufacturers' recommendations, cleaning of the commutator and replacement of brushes are the only services that should be attempted. If there are other troubles in the starting motor, replace it with a new or rebuilt unit. Complete disassembly of some starting motors is not recommended.

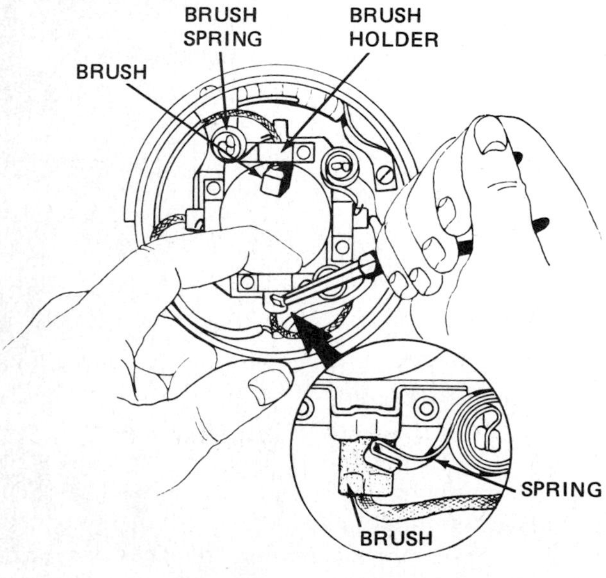

FIG. 13-32 Checking the brushes for wear. *(Briggs & Stratton Corporation)*

△13-12 STARTING-MOTOR SERVICE

When the starting motor is disassembled, the armature, field-frame assembly, and brush holders should be checked with an ohmmeter or test light for shorts, opens, and grounds (Fig. 13-34). If the armature is in good condition, place it in the growler and check for shorts (Fig. 13-35).

If the commutator on the armature is worn, burned, or out-of-round, place the armature in an armature lathe (Fig. 13-36). Then operate the lathe to take a light cut from around the commutator. Remove only enough metal to clean the commutator and restore it to a round condition. Some manufacturers recommend that the mica between the commutator bars should be undercut. Your instructor will show you how to perform these operations in the shop.

When you are working on a small-engine starting motor and have removed only the end cap to check the commutator and brushes, lubricate the bushing and armature shaft before putting the end cap back on. Coat the

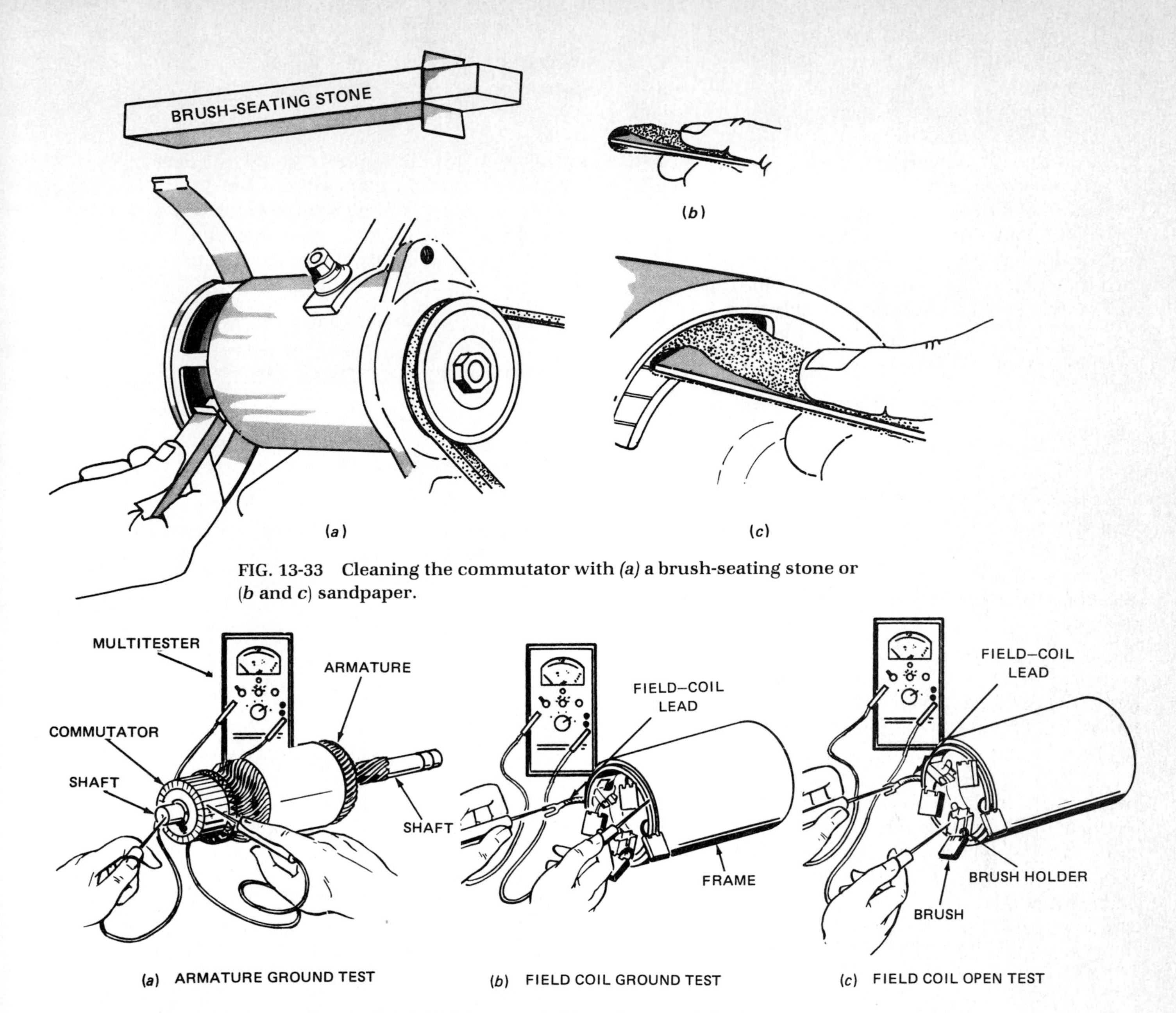

FIG. 13-33 Cleaning the commutator with *(a)* a brush-seating stone or (*b* and *c*) sandpaper.

FIG. 13-34 Checking the armature and field-frame assembly using a multitester.

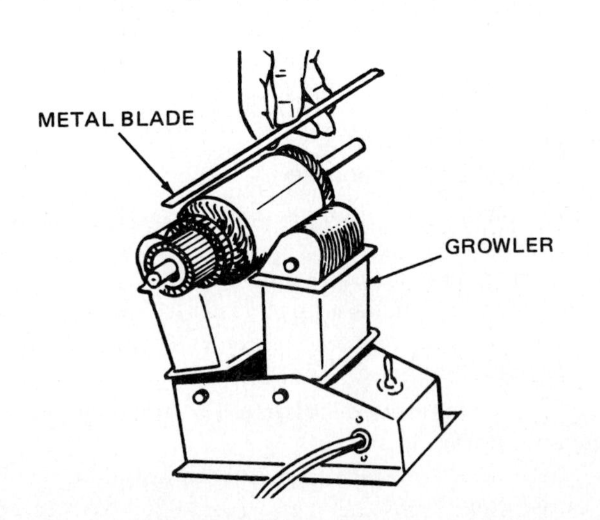

FIG. 13-35 Testing an armature for short circuits on a growler.

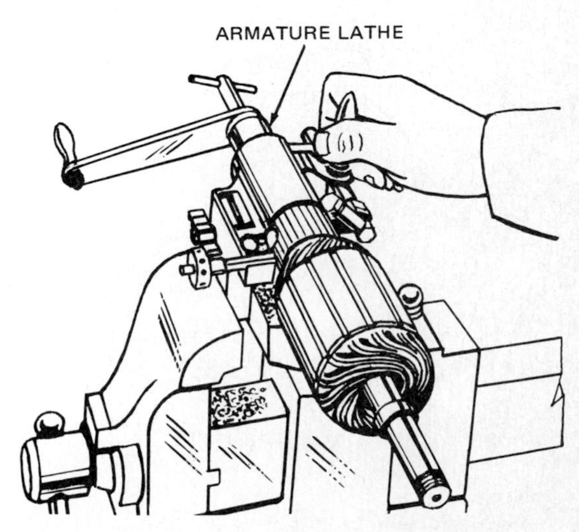

FIG. 13-36 Using a vise-mounted armature lathe to refinish the commutator bars.

bushing in the end cap and the end of the armature shaft lightly with SAE10 oil before installing the end cap. Do not put on too much oil or it will get onto the commutator and brushes and cause trouble.

When installing the end cap, lift the brushes to allow the commutator to pass under them. This can be done by pulling up on the brush springs with needle-nose pliers (Fig. 13-32). After the cap is partly on, let the springs down so the brushes will rest on the commutator. Then the end cap can be pushed on into place and the through bolts installed to complete the assembly.

Δ13-13 CHECKING THE BENDIX DRIVE

To check the Bendix drive, remove the starting motor from the engine. If the drive pinion or splined sleeve is damaged, the drive must be replaced. Do not lubricate the Bendix drive. This can cause it to stick.

While the starting motor is off, check the ring gear on the engine flywheel. If the teeth are battered or broken, the ring gear should be replaced. One method of replacing the ring gear is shown in Fig. 13-37. This ring gear is attached with rivets on the original assembly. The rivets must be drilled out. To do this, mark the centers of the rivets with a center punch. Then drill out the rivets with a 3/16-inch drill. Clean the holes after drilling out the rivets. Then attach the new ring gear with the four screws and locknuts supplied in the ring-gear kit.

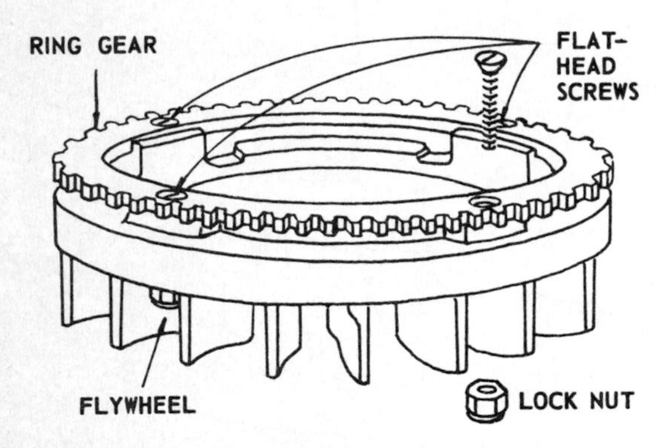

FIG. 13-37 Method of replacing ring gear in one model of a small engine. The old ring is removed by drilling out the attaching rivets. Then the new ring gear is attached with screws and locknuts. *(Briggs & Stratton Corporation)*

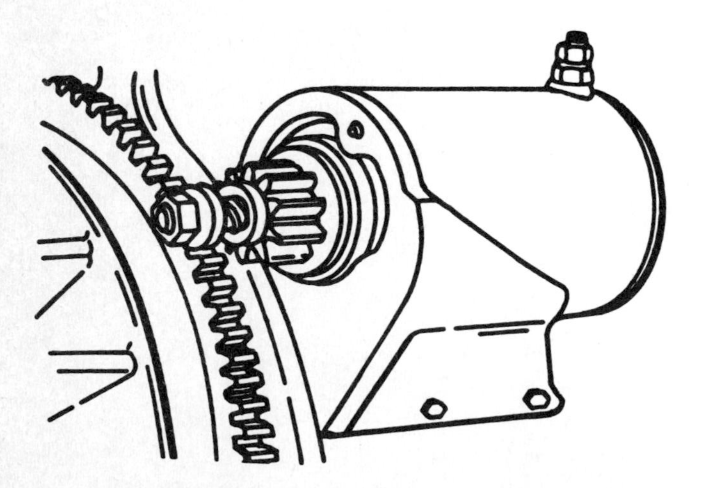

FIG. 13-38 Relationship of ring gear to pinion gear of a Bendix drive. Incorrect alignment will cause poor meshing and rapid tooth wear. *(Kohler Company)*

When reinstalling the starting motor on the engine, use the original mounting bolts and lock washers. These bolts provide the proper alignment of the drive gear to the ring gear (Fig. 13-38). Incorrect alignment can cause gear clash and damage to the gears.

CHAPTER 13: REVIEW QUESTIONS

Select the *one* correct, best, or most probable answer to each question. You can find the answer in the section indicated at the right of each question.

1. In the rope-wind starter, the only parts usually needing service are the (**Δ*13-2***)
   a. rope and spring
   b. rope and drive mechanism
   c. rope and handle
   d. kick plate and handle
2. General services for the rope-rewind starters include replacing the rope and (**Δ*13-3***)
   a. replacing the handle and the spring
   b. replacing the spring and repairing the drive mechanism
   c. repairing the windup and installing a handle
   d. tying a new knot in the rope and replacing the spring
3. The first step in replacing the rope in a rope-rewind starter is to (**Δ*13-4***)
   a. remove the starter from the engine
   b. untie the rope
   c. singe the ends of the new rope with a match
   d. cut out the old rope
4. The two types of recoil springs for the rope-rewind starter are (**Δ*13-5***)
   a. straight and leaf
   b. coil and recoil
   c. removable and riveted
   d. removable and packaged
5. Before starting to work on a windup starter, first make sure the (**Δ*13-8***)
   a. spring is completely wound up
   b. spring is completely unwound
   c. ratchet is released
   d. spring is removed
6. The most common problem with battery-operated starting motors is that (**Δ*13-10***)
   a. the motor overspeeds while cranking
   b. the lights dim
   c. the engine cranks slowly or not at all
   d. the battery is overcharged

7. There are no lights and no cranking when a start is attempted. Mechanic A says the trouble is a dead battery. Mechanic B says there are bad connections. Who is right? (**Δ*13-10***)
   a. mechanic A
   b. mechanic B
   c. both A and B
   d. neither A nor B

8. Lights burn normally when turned on, but go out when starting is attempted. Mechanic A says the trouble is a low battery. Mechanic B says the trouble is a bad connection. Who is right? (**Δ*13-10***)
   a. mechanic A
   b. mechanic B
   c. both A and B
   d. neither A nor B

CHAPTER

14

Ignition Systems

After studying this chapter, you should be able to:

1. ***List the three general types of small-engine ignition systems.***
2. ***Explain the purpose of the ignition system and describe the basic components in each system.***
3. ***Describe the construction and operation of the magneto-ignition system.***
4. ***Explain the difference between a magneto-ignition system with breaker points and a breakerless magneto-ignition system.***
5. ***Explain the difference between the breakerless magneto-ignition system and the capacitor-discharge-ignition system.***

Δ14-1 TYPES OF IGNITION SYSTEMS A variety of ignition systems have been used with small spark-ignition engines. Today, almost all ignition systems can be classified as one of three different types. These are *magneto ignition, battery ignition,* and *capacitor-discharge* (CD) *ignition.* Although they differ in construction and operation, they all perform the same job. That is to produce and distribute high-voltage surges to the spark plugs at the correct time. Each type is described in the following sections.

Δ14-2 PURPOSE OF IGNITION SYSTEM The ignition system (Fig. 14-1) supplies electric sparks that ignite, or set fire to, the compressed air-fuel mixture near the end of the compression stroke. The mixture burns, producing a high pressure that pushes the piston down the cylinder (Δ4-2).

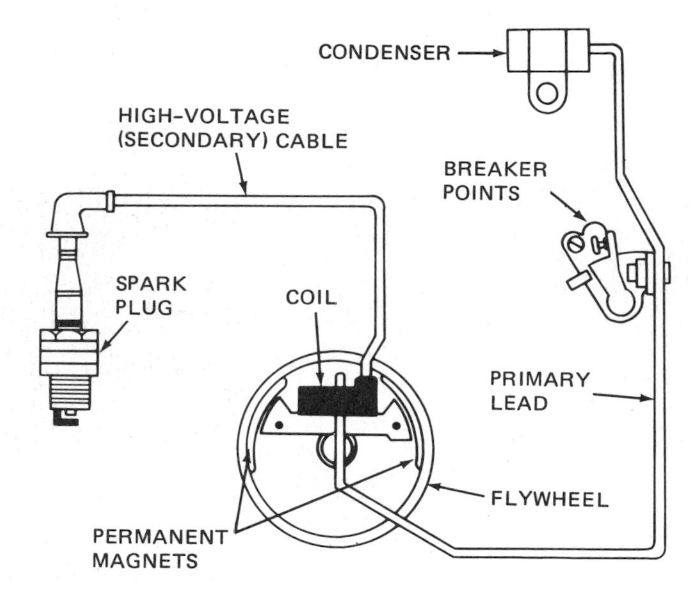

FIG. 14-1 Typical flywheel-magneto-ignition system. *(Kohler Company)*

The ignition system usually has an advance mechanism to vary the *ignition timing* (Δ14-16). This is the instant that the spark occurs. When the engine is idling, the spark occurs just before the piston reaches TDC on the compression stroke. But at higher speeds, the spark is advanced (moved ahead) so that it occurs earlier. This gives the mixture enough time to start burning so that it will produce maximum pressure when the piston reaches TDC and starts down on the power stroke.

When the throttle is only partly open, less mixture can get into the cylinder. There is less mixture to compress, and it will burn slower after it is ignited. Some engines have another mechanism in the ignition system to advance the spark under these conditions. The mechanism is usually called a *vacuum-advance mechanism.* It advances the spark so that the mixture has sufficient time to burn.

MAGNETO-IGNITION SYSTEMS

Δ14-3 TYPES OF MAGNETOS A *magneto* is an engine-driven device that generates its own primary current, transforms that current into high-voltage surges, and delivers them to the spark plug at the proper time. Some magnetos are built into the engine and use the flywheel as part of the magneto (Fig. 14-1). These are called *flywheel magnetos.* Larger engines may have a separate self-contained magneto (a *unit magneto*) installed on the outside of the engine.

Magnetos may also be classified according to the type of switch used to allow and stop the flow of current in the primary circuit (Δ14-5). If a mechanical switch such as breaker points (also called *contact points*) is used, the assembly is called a *breaker-point flywheel magneto.* When the breaker points are replaced with a solid-state electronic switch such as a transistor (Δ10-19), the system is a *breakerless flywheel magneto* (Fig. 14-2).

Regardless of type, the magneto supplies the ignition voltage. No other source of electricity (such as a battery, generator, or alternator) is required. One advantage to magneto ignition is that the ignition voltage is a function of engine speed. The higher the engine speed, the higher the voltage available from the magneto to fire the spark plug.

Δ14-4 OPERATION OF BREAKER-POINT FLYWHEEL MAGNETO In a flywheel magneto, movement of a magnetic field past a stationary coil of wire induces a voltage in the coil (Fig. 14-3). As a result, current will flow through the electric circuit of which the coil is a part.

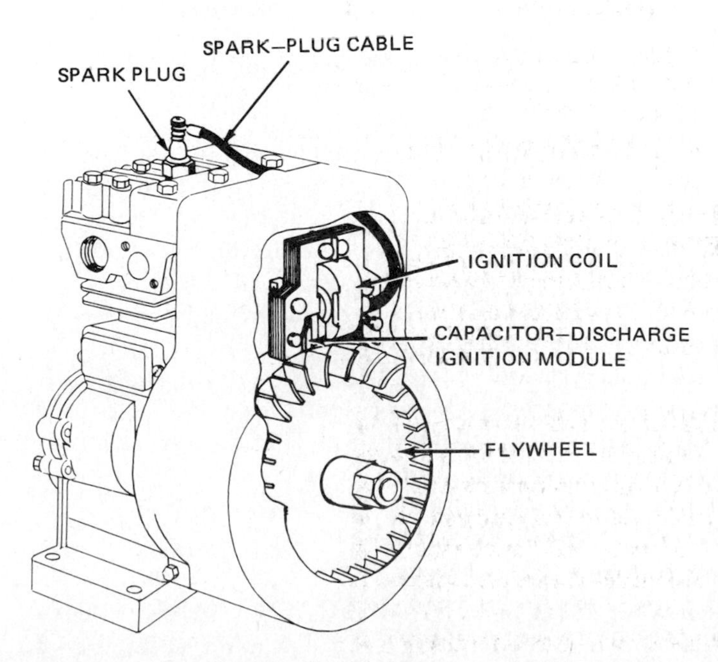

FIG. 14-2 A breakerless flywheel-magneto-ignition system. *(Briggs & Stratton Corporation)*

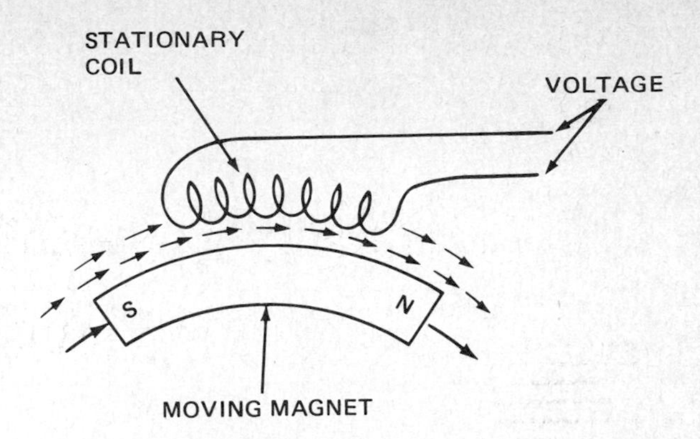

FIG. 14-3 When a magnet moves past a stationary coil of wire, a voltage is induced in the coil.

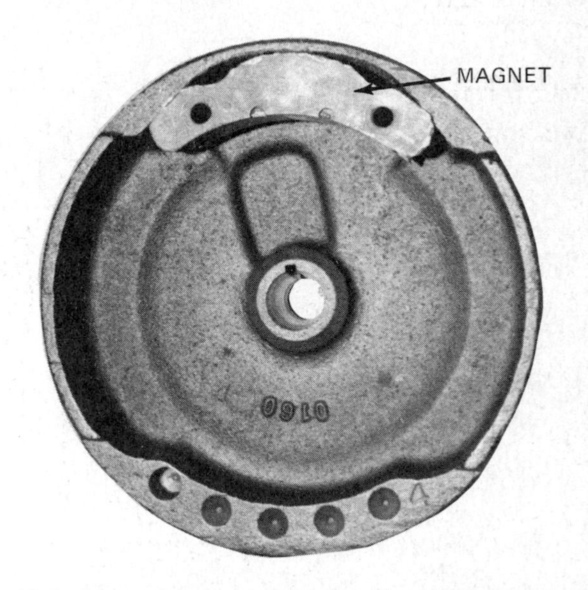

FIG. 14-4 Magnet mounted in the rim of the engine flywheel. *(Clinton Engine Corporation)*

Figure 14-3 shows a stationary coil positioned above a moving magnet. The permanent magnet is attached to the inside of the flywheel (Fig. 14-4). A magnetic field surrounds the magnet (Chap. 10). As the flywheel turns, the magnet moves past the coil (Fig. 14-5). The lines of force around the magnet pass through the windings of the coil. As the lines of force cut the coil windings, a voltage is induced in the coil. This is called *electromagnetic induction.* If the coil is in a complete electric circuit, the voltage will force a current to flow.

Δ14-5 PRIMARY AND SECONDARY CIRCUITS All ignition systems have two basic circuits. These are a low-voltage primary circuit (Fig. 14-6) and a high-voltage secondary circuit (Fig. 14-7). In the breaker-point flywheel magneto, the primary circuit includes the primary winding in the ignition coil, the breaker points and condenser, the on-off ("ignition kill") switch, the engine frame, and the primary wiring. The secondary circuit includes the secondary winding in the ignition coil, the spark-plug cable, and the spark plug.

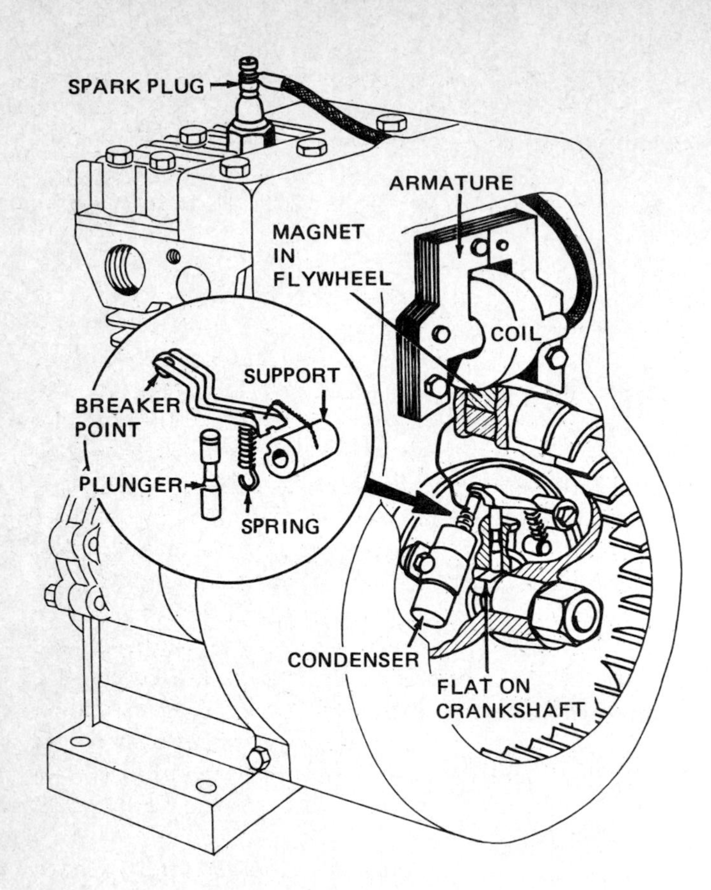

FIG. 14-5 Location and construction of the magneto. *(Briggs & Stratton Corporation)*

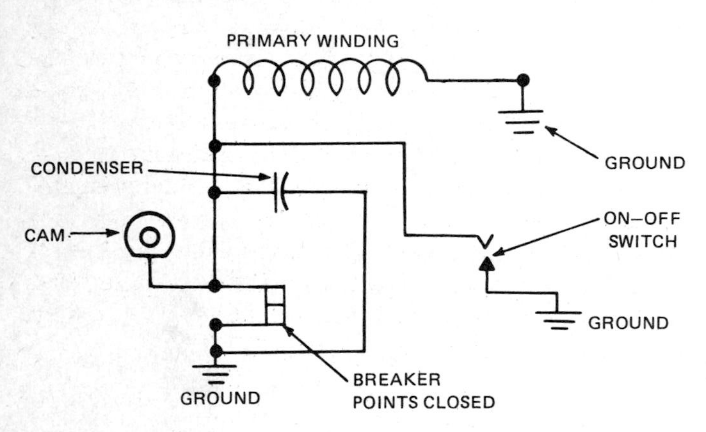

FIG. 14-6 Wiring diagram of a magneto-ignition system with current flowing through the primary circuit. *(Lawn Boy Division of Outboard Marine Corporation)*

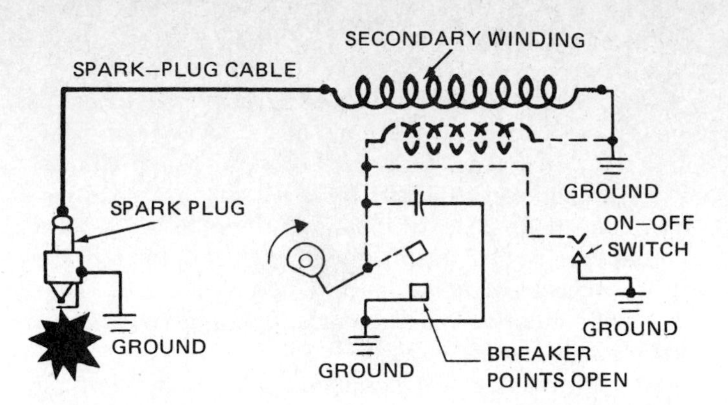

FIG. 14-7 Wiring diagram of a magneto-ignition system with the breaker points open. Current has stopped flowing in the primary circuit. When this happens, a high-voltage surge is induced in the secondary circuit which produces a spark at the spark-plug gap in the cylinder. *(Lawn Boy Division of Outboard Marine Corporation)*

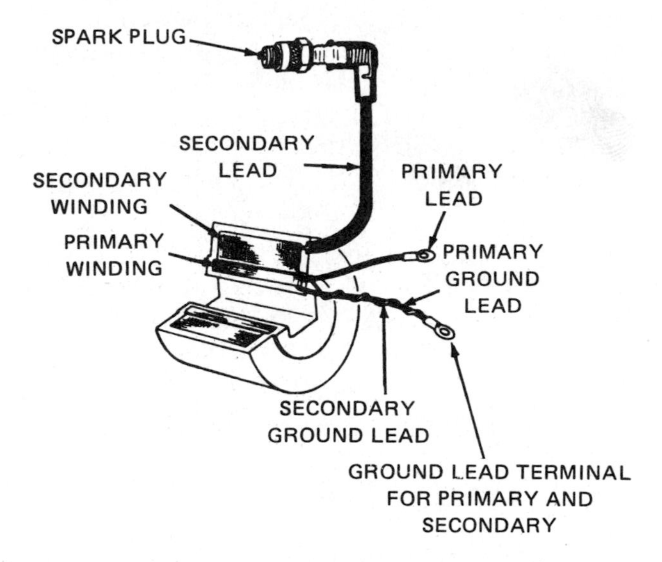

FIG. 14-8 Construction of one type of small-engine ignition coil. *(Tecumseh Products Company)*

Δ14-6 IGNITION COIL

The ignition coil (Fig. 14-8) changes the low voltage in the primary circuit into the high voltage needed by the secondary circuit to fire the spark plug. The ignition coil has two windings. These are a primary winding of a relatively few turns of heavy wire and a secondary winding of many turns of fine wire. The windings are wound around a laminated iron core, which conducts magnetic lines of force better than air.

On some engines, the coil is mounted on the engine, inside the flywheel (Fig. 14-1). Other engines have the coil outside the flywheel (Figs. 14-5 and 14-9).

When the breaker points are closed, current flows in the primary winding of the coil (Fig. 14-9*a*). This causes a magnetic field to build up around the coil. When the breaker points open and break the circuit (Fig. 14-9*b*), the current flow stops and the magnetic field collapses.

As the collapsing lines of force cut through the secondary winding, a very high voltage is induced in it. The secondary circuit is complete, although the spark-plug gap prevents low-voltage current flow. But when the secondary voltage builds up high enough, a spark jumps across the gap at the spark plug (Fig. 14-9*c*). The heat from the arc ignites the air-fuel mixture, and the power stroke follows.

Δ14-7 BREAKER POINTS AND CONDENSER

Also included in the primary circuit is a set of *breaker points* (Fig. 14-10). One point is stationary. The other is mounted on a movable lever, or arm. This allows movement of the movable point to open and close the

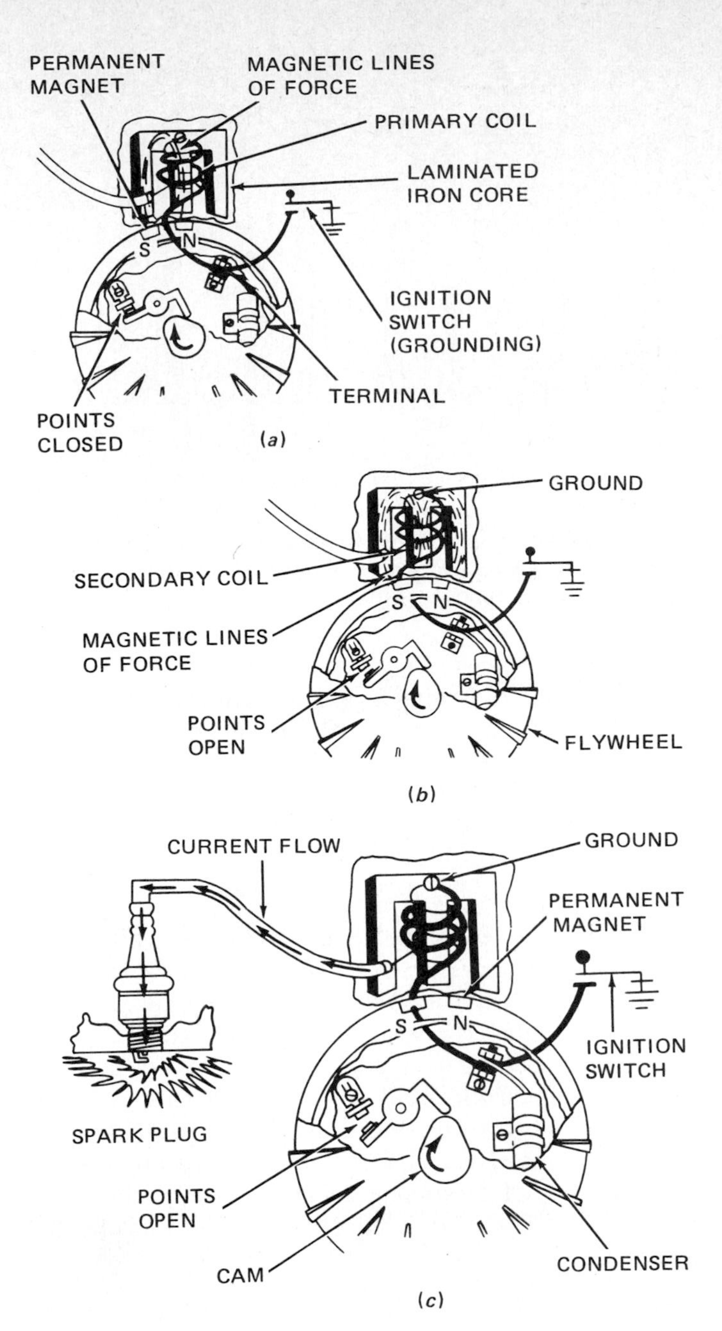

FIG. 14-9 Operation of the flywheel magneto *(a)* as energy builds up in the primary winding and *(b)* at the instant the points separate. *(c)* The induced high voltage produces the spark.

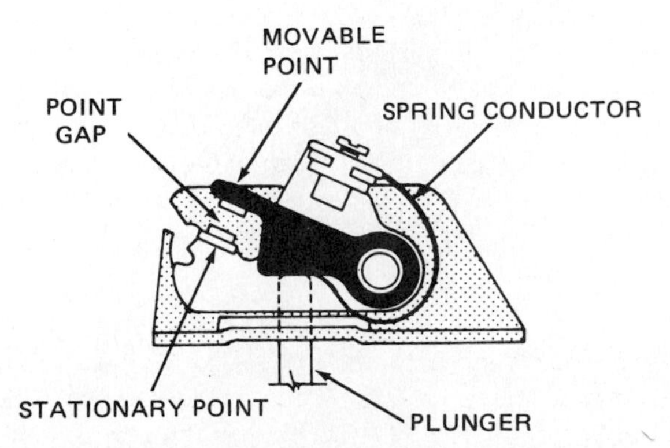

FIG. 14-10 Set of breaker points for a small engine. *(Kohler Company)*

primary circuit. Various arrangements are used to open and close the points. However, most use a cam either on the crankshaft or on the camshaft. The arrangement depends on the number of cylinders in the engine, and whether it operates on the two-stroke or four-stroke cycle.

A single-cylinder two-cycle engine requires a spark every crankshaft revolution. This can be provided by using a cam on the crankshaft to open the points. A single-cylinder four-cycle engine needs a spark every other crankshaft revolution. The camshaft makes one revolution for every two of the crankshaft. Therefore, the breaker points in an single-cylinder four-cycle engine are often opened by a cam on the camshaft. It turns at one-half the speed of the crankshaft. However, other arrangements are also used by engine manufacturers.

Figure 14-5 shows a flywheel magneto in which the movable-point arm rests on a plunger that rides on a cam on the crankshaft. The cam is round, except for a flat spot (Fig. 14-6). When the crankshaft and cam rotate, the breaker points remain closed while the plunger is riding on the flat spot. Then the flat spot moves out from under the plunger, and the round part of the cam moves in under the plunger (Fig. 14-7). This pushes the plunger up, opening the points (Fig. 14-10).

The ignition spark occurs at the spark plug at almost the instant the points open (Fig. 14-9). When the points begin to open, a voltage of about 250 volts is induced in the primary circuit. This voltage is sometimes called the *ignition-system intermediate voltage.* It causes a small momentary arc to form across the breaker points. The arc forms when the current tries to flow across the points as they open to break the primary current. This arc could damage the points and shorten their useful life.

To reduce the arc, a *condenser* (Fig. 14-11), or *capacitor,* is connected across the contact points. The condenser has two purposes. It brings the primary curent to a quick stop, which helps collapse the magnetic field. The condenser also reduces arcing across the points. This ensures longer point life.

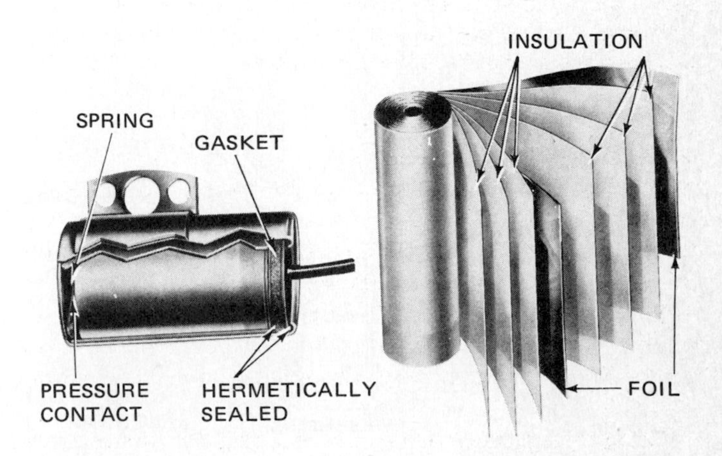

FIG. 14-11 Condenser (capacitor) assembled and with the winding partly unwound.

Δ14-8 SPARK PLUGS

The spark plug is a metal shell in which a porcelain insulator is fastened (Fig. 14-12). An electrode extends through the center of the insulator. The metal shell has a short electrode attached to one side. This outer electrode is bent inward to produce the proper gap between it and the center electrode.

Some spark plugs have a built-in resistor which forms part of the center electrode. The resistor reduces television and radio interference (static) from the high-voltage surges in the ignition secondary circuit. This interference is called *radio-frequency interference,* or RFI.

Δ14-9 SPARK-PLUG HEAT RANGE AND REACH

The heat range of the spark plug determines how hot it will get. Plug temperature depends on how far the heat must travel from the center electrode to the much cooler cylinder head (Fig. 14-13). If the heat path is long, the plug will run hot. If the heat path is short, the plug will run much cooler. However, if the plug runs too cool, sooty carbon will deposit on the insulator around the center electrode. This could soon build up enough to short out the plug. Then the high-voltage surges would arc across the carbon instead of producing a spark across the spark-plug gap. Using a hotter plug will burn the carbon away, or prevent it from forming.

Spark-plug reach is the distance from the seat of the shell to its lower edge (Fig. 14-12). If the reach is too long, the spark plug will protrude too far into the combustion chamber. This could interfere with the turbulence of the air-fuel mixture and reduce combustion action. Also, a piston could hit the end and damage the engine. However, the plug must reach into the combustion chamber far enough so that the spark plug will be properly positioned in the combustion chamber.

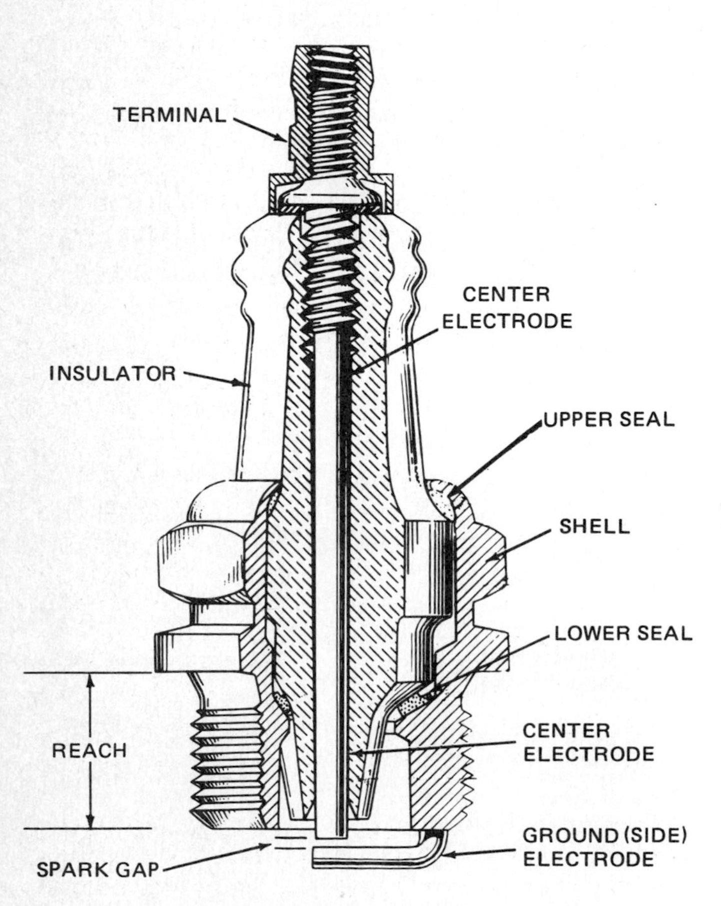

FIG. 14-12 Construction of a spark plug.

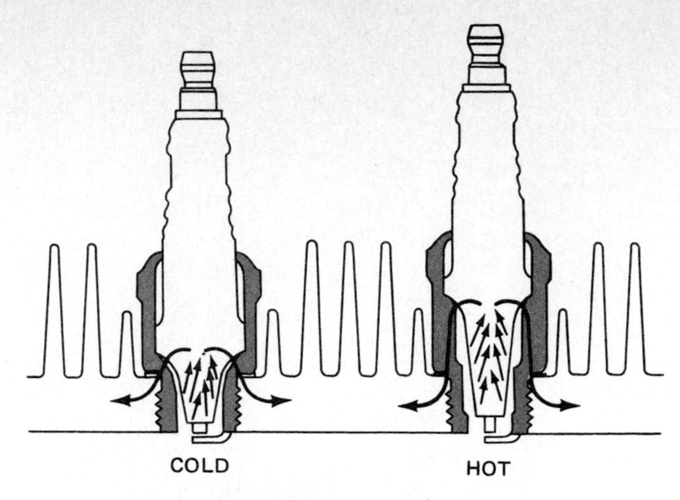

FIG. 14-13 The heat range of spark plugs. The longer the heat path (indicated by the arrows), the hotter the plug runs. *(Champion Spark Plug Company)*

Δ14-10 SURFACE-GAP SPARK PLUGS

Some small engines equipped with capacitor-discharge ignition (**Δ**14-18) use *surface-gap spark plugs* (Fig. 14-14). This type of plug has a ceramic insulator (instead of an air gap) in the space between the lower end of the shell and the center electrode. This is called a *surface gap.* The spark travels from the center electrode across the insulator to the ground electrode, which is formed by the entire lower end of the shell. In a running engine, the spark normally rotates around the surface gap to a different location.

FIG. 14-14 A surface-gap spark plug. *(Champion Spark Plug Company)*

These spark plugs are very "cold." Therefore, the heat-range values of other spark plugs do not apply. In addition, the surface-gap spark plug is almost free of preignition.

Δ14-11 MAGNETO-IGNITION SWITCH

An ON-OFF switch (also called *stop switch*, or *kill switch*) is used in many magneto-ignition systems to stop the engine (Fig. 14-7). When this switch is closed, it grounds the breaker-point end of the primary winding. Now current continues to flow in the primary winding, and the opening of the points does not interrupt it. As a result, there is no collapse of the magnetic field so no secondary spark occurs. Therefore, the engine stops.

The engine can also be stopped by a *grounding blade* located near the spark plug (Fig. 14-15). The blade can be bent to ground the insulated terminal of the spark plug. When the blade is touching the spark-plug terminal, the secondary current flows through the blade. No spark occurs at the spark-plug gap.

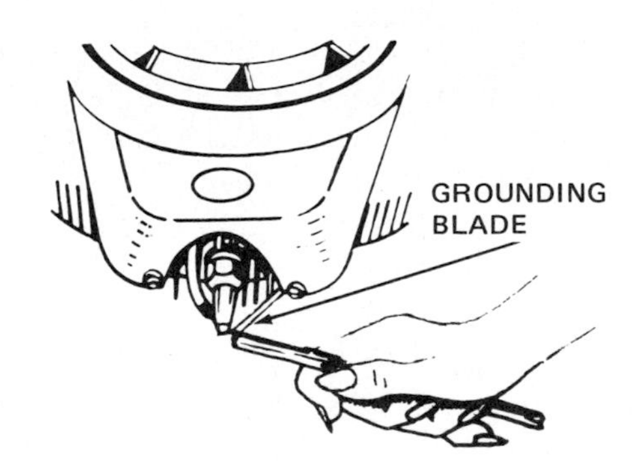

FIG. 14-15 Grounding blade near the spark plug used to stop the engine.

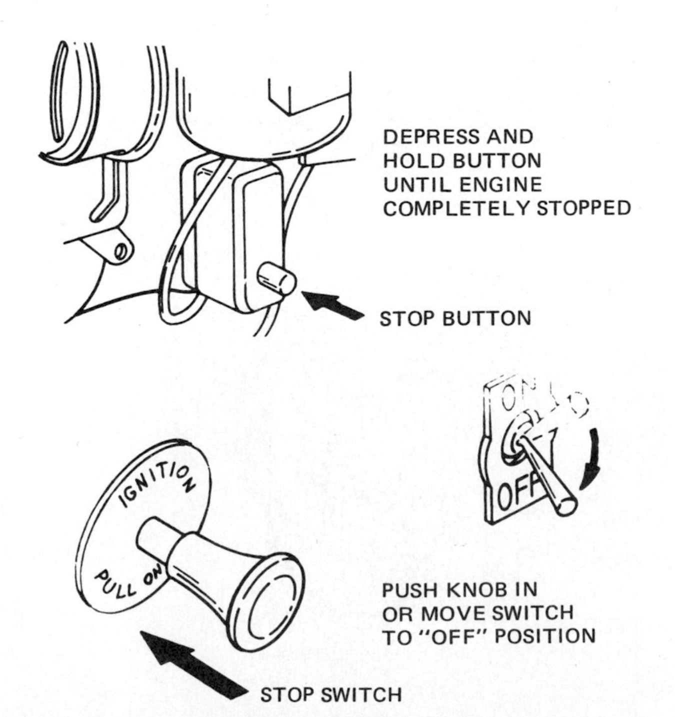

FIG. 14-16 A stop button or a stop switch is used on some engines to shut down the engine. *(Kohler Company)*

Figure 14-16 shows other types of ignition switches. All work by grinding the primary winding of the ignition coil. Then, even though current continues to flow in it, the magneto cannot produce a secondary voltage.

Δ14-12 EXTERNAL MAGNETO

Some engines have an externally mounted *unit magneto* (Fig. 14-17). The operation of the unit magneto is basically the same as that of the flywheel magneto described earlier. However, the unit magneto is constructed differently.

In the unit magento (Fig. 14-17), the rotor is driven through an *impulse coupling* (Δ14-13). As the rotor spins, it produces a magnetic field in the laminated iron frame on which the coil primary and secondary windings are wound. Each half turn of the rotor causes a complete reversal of the magnetic field. This causes the lines of force to continually build up and collapse through the windings. As a result, a current flow is induced in the primary winding as long as the breaker points are closed.

When the current flow is at its greatest, the breaker points are opened by the cam on the end of the rotor shaft (Fig. 14-18*b*). This stops the flow of current, and so the magnetic lines of force collapse rapidly. As the lines of force cut through both the primary and secondary windings, a high voltage is induced in the secondary winding. This induced voltage is high enough to force the spark to jump the spark-plug gap.

The condenser does the same job as in the flywheel magneto (Δ14-7). However, by adding a rotor and cap in the secondary circuit, this type of magneto can be used on multicylinder engines. Figure 14-19 shows how the ignition rotor and cap act as a rotary switch. As the rotor turns, it sends the high-voltage surges from the secondary winding of the coil to each spark plug at the proper time.

Δ14-13 IMPULSE COUPLING

The rotor in the unit magneto (Δ14-12) is driven by an *impulse coupling* (Fig. 14-18). It improves engine starting by doing two things. First, the impulse coupling retards ignition timing for better engine starting during cranking. Second, it flips the magneto rotor ahead at the proper moment so that the rotor spins very rapidly for a part turn. This produces a stronger spark because the faster the rotor spins, the higher the voltage induced in the secondary winding.

A spring in the impulse coupling provides the impulse action. During cranking, spring tension increases during the first half turn of the coupling. As it is turned further, the wound-up spring suddenly releases, spinning the rotor ahead.

After the engine starts, centrifugal force causes the impulse coupling to unlock so that it does not function. Now the magneto rotor turns steadily in time with the engine.

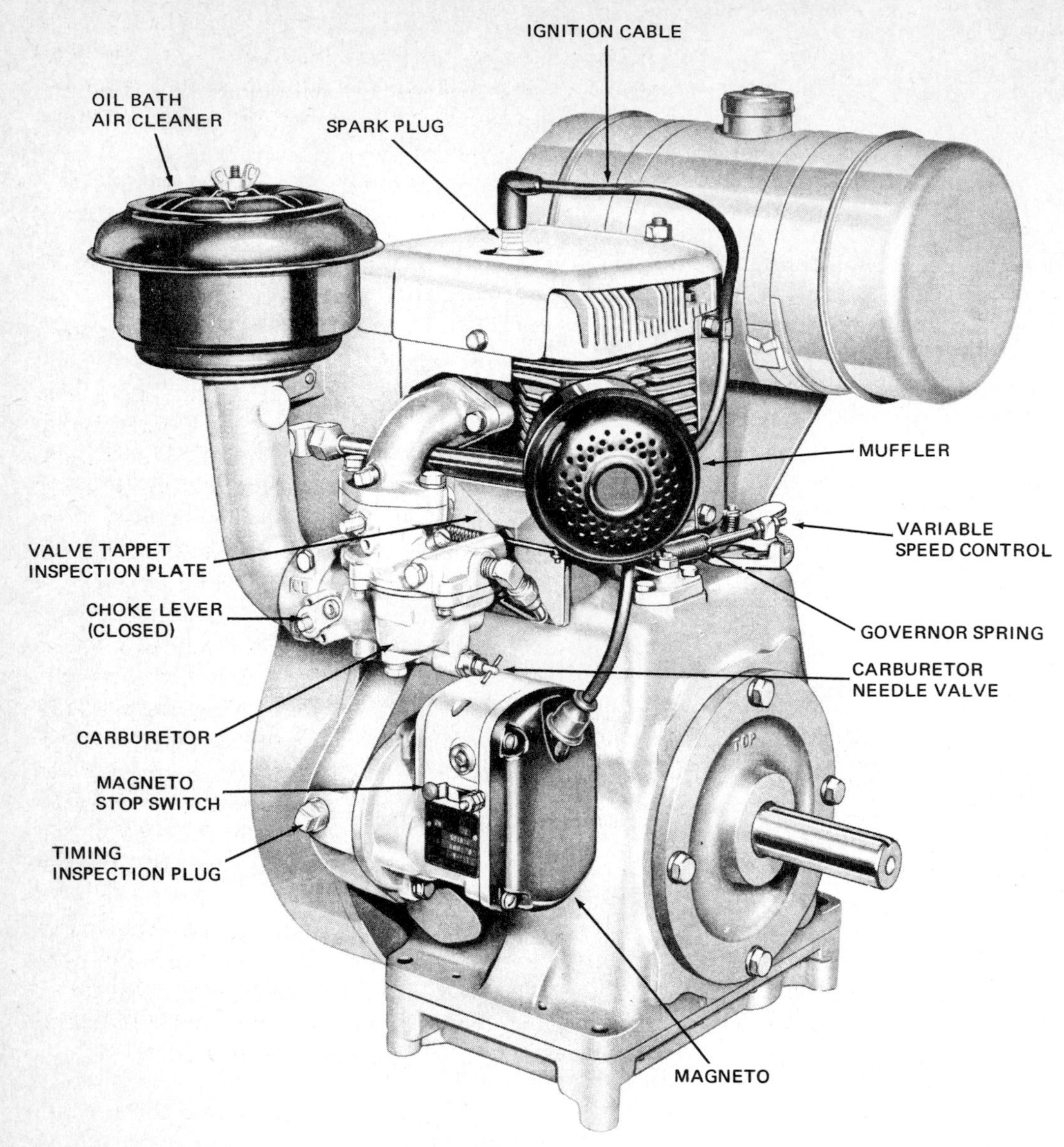

FIG. 14-17 An externally mounted magneto on a one-cylinder four-cycle engine. *(Teledyne Wisconsin Motor)*

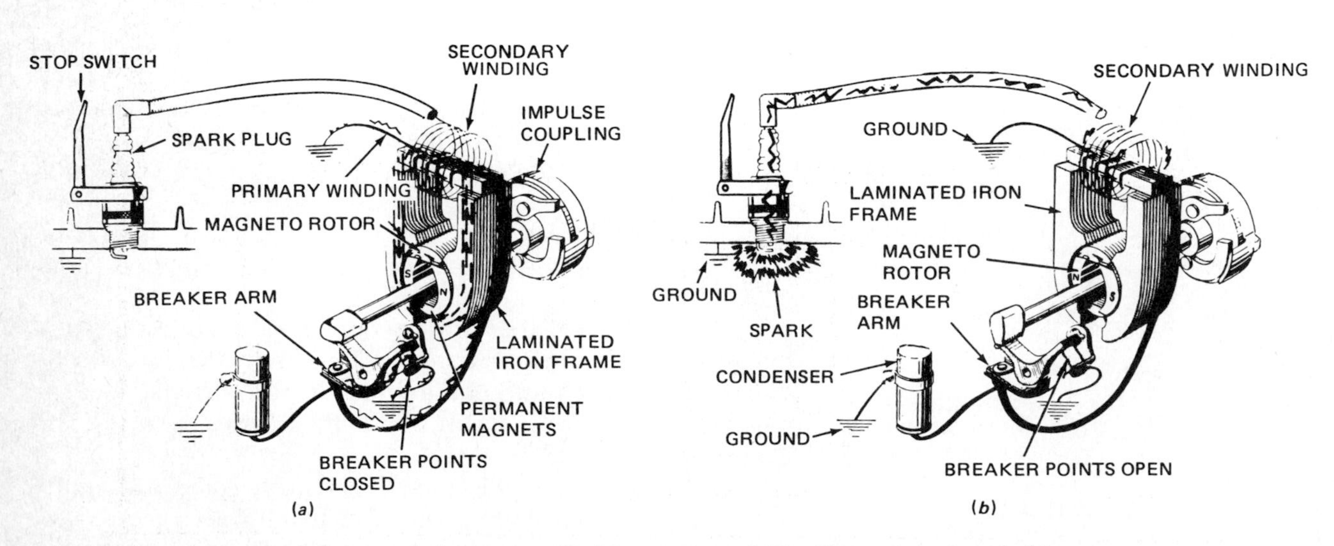

FIG. 14-18 Operation of the external magneto-ignition system. *(a)* Breaker points closed. *(b)* Breaker points open.

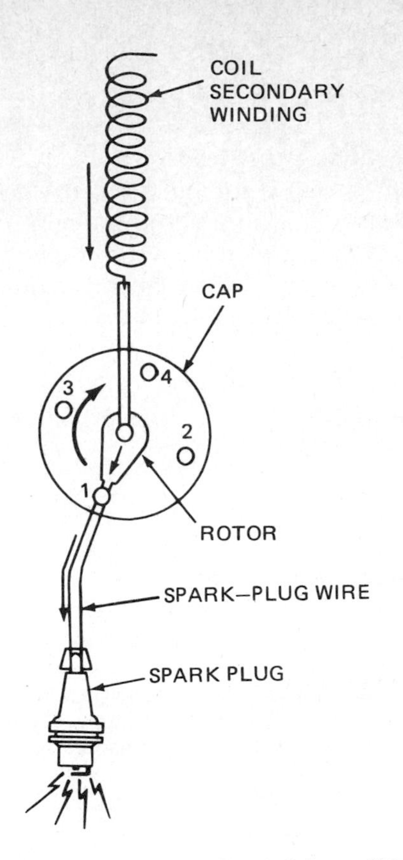

FIG. 14-19 The secondary winding of the coil is connected through the rotor in the distributor to the spark plug.

BATTERY-IGNITION SYSTEMS

Δ14-14 OPERATION OF BATTERY-IGNITION SYSTEM The battery-ignition system (Fig. 14-20) differs in three ways from the magneto-ignition system (Δ14-4). First, in the battery-ignition system, the primary current is supplied by a battery. No flywheel magneto is used. Second, the ignition switch must be closed for the battery-ignition system to work. Third, a separate ignition coil will be mounted on the engine.

Figure 14-21 shows a two-cylinder engine with a battery-ignition system. Two spark-plug cables are attached to the top of the coil. As in the ignition coil in the magneto, this coil also has a primary winding and a secondary winding. However, in the battery-ignition system, the primary winding in the coil is connected to the secondary winding (Fig. 14-20).

Operation of the battery-ignition system on a small engine is shown in Figs. 14-22 and 14-23. Rotation of a cam causes the breaker points to open and close. When the points are closed, the primary winding of the ignition coil is connected to the battery. This allows current to flow through the primary winding and build up a magnetic field (Fig. 14-22).

When the cam rotates so that the lobe on the cam opens the points (Fig. 14-23), the current stops flowing in the primary winding. The magnetic field collapses, inducing a high voltage in the secondary winding. This causes a high-voltage surge of current to flow from the coil, through the spark-plug cable, to the spark plug. As the current jumps the gap at the spark plug, heat from the arc ignites the compressed air-fuel mixture in the engine cylinder. Combustion begins, and the power stroke follows.

A condenser is connected across the points to aid in the collapse of the magnetic field (Fig. 14-23). The condenser also helps control arcing of the points as they begin to separate. These are the same jobs, performed in the same way, as in the magneto-ignition system (Δ14-7).

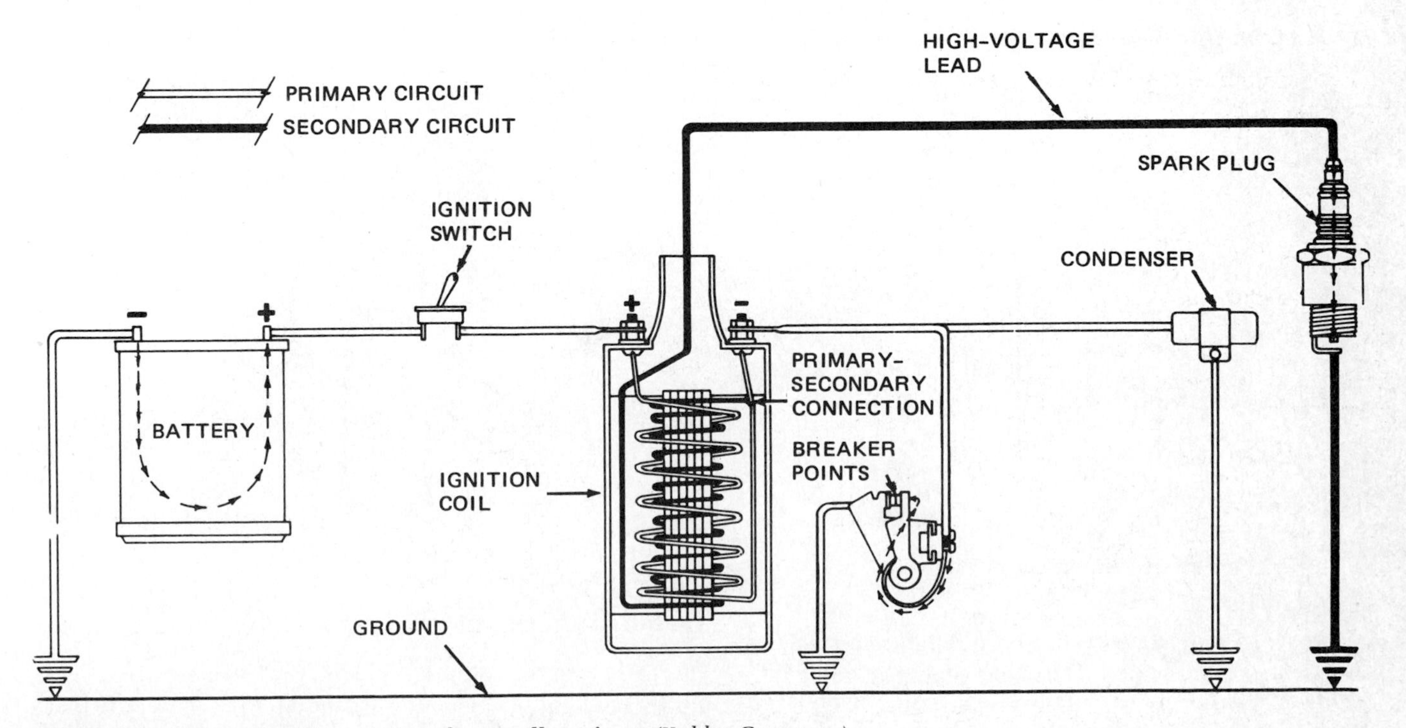

FIG. 14-20 Battery-ignition system for small engines. *(Kohler Company)*

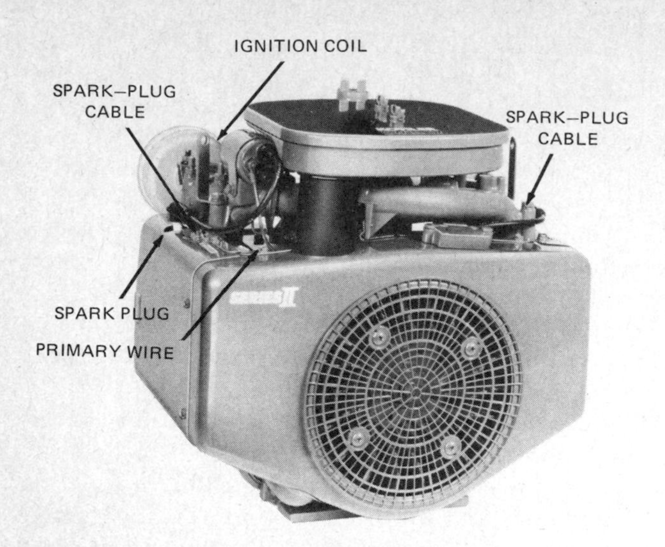

FIG. 14-21 Components of a battery-ignition system on a two-cylinder four-cycle engine. *(Kohler Company)*

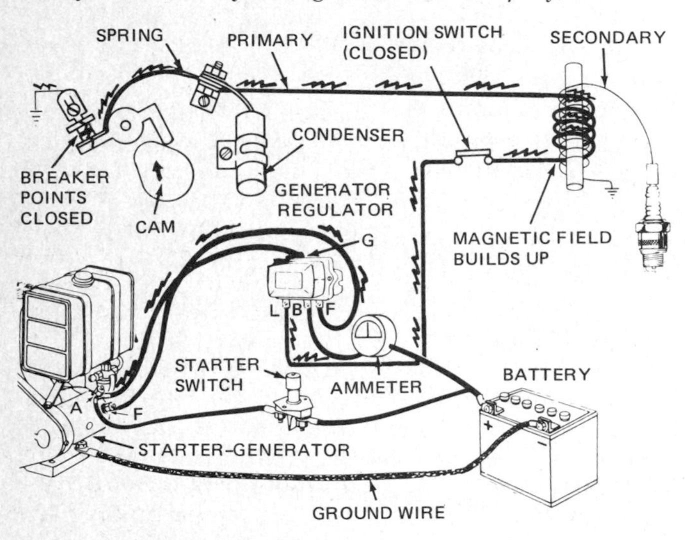

FIG. 14-22 Schematic diagram of a battery-ignition system with the breaker points closed and current flowing from the battery through the primary winding of the ignition coil.

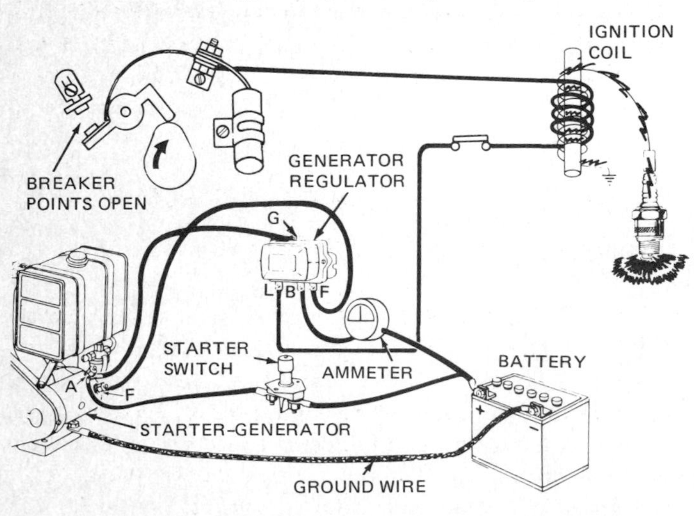

FIG. 14-23 Schematic diagram of a battery-ignition system with the breaker points open. The collapsing magnetic field in the ignition coil produces a high voltage in the secondary winding of the coil. This causes a spark to occur at the spark plug.

Δ14-15 TIMERS AND DISTRIBUTORS

The battery-ignition system is used on most multicylinder spark-ignition engines. On some single-cylinder four-cycle engines, the breaker points are located in a separate housing called the *timer* (Fig. 14-24). The ignition cam (Fig. 14-25) is located on the end of the timer shaft, which is gear-driven from the engine crankshaft.

Multicylinder engines with battery ignition usually have an *ignition distributor* (Fig. 14-26). In the distributor,

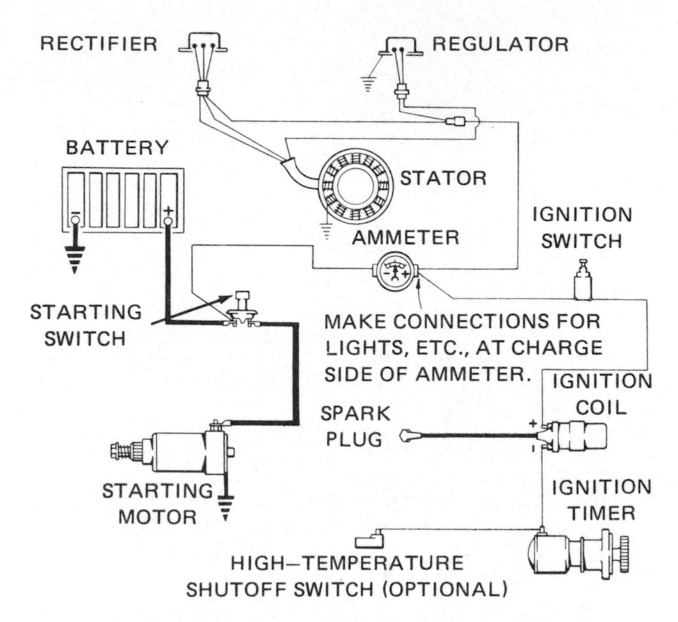

FIG. 14-24 Ignition timer used on a one-cylinder four-cycle engine that has battery ignition. *(Teledyne Wisconsin Motor)*

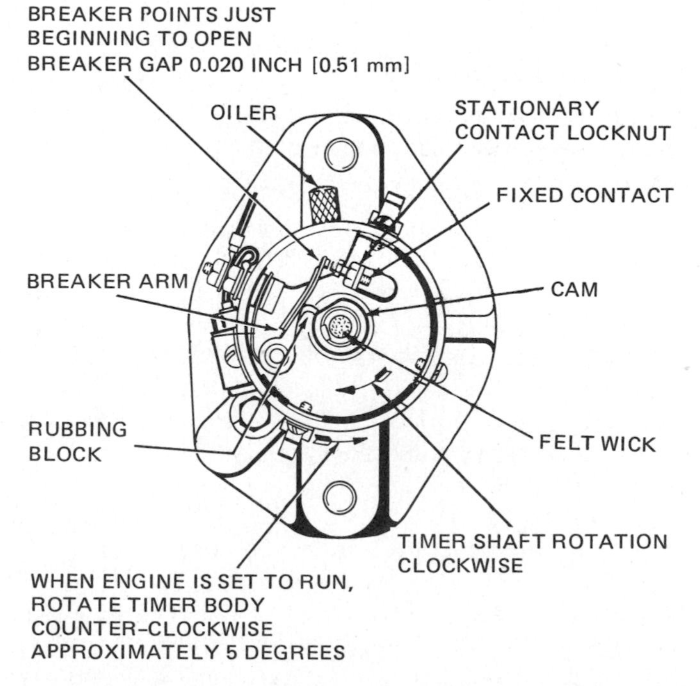

FIG. 14-25 As the rubbing block on the breaker arm moves off the flat spot of the cam, the points open. This provides the ignition spark for a one-cylinder four-cycle engine. *(Teledyne Wisconsin Motor)*

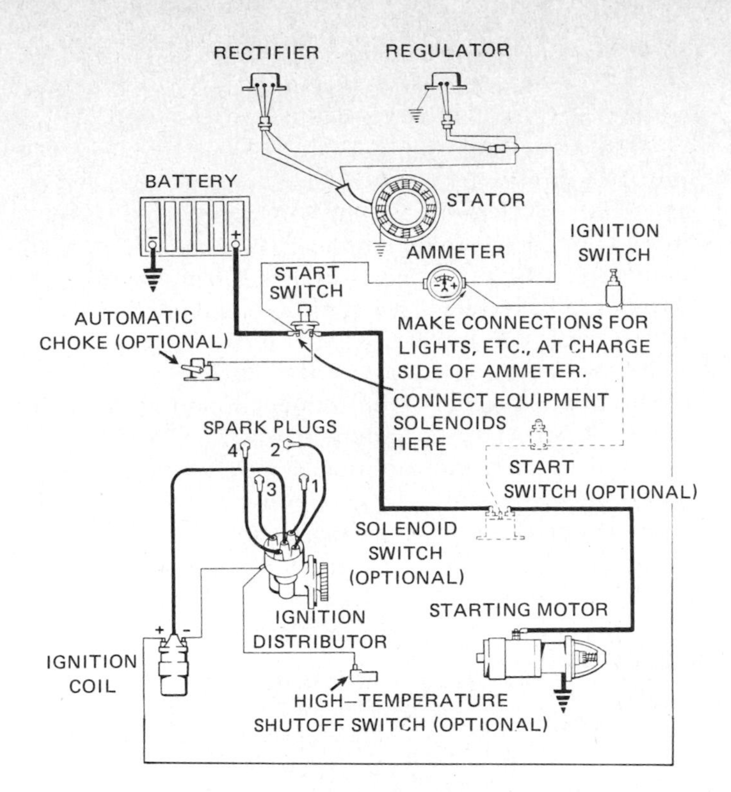

FIG. 14-26 Battery-ignition system which includes an ignition distributor for a small four-cylinder four-cycle engine. *(Teledyne Wisconsin Motor)*

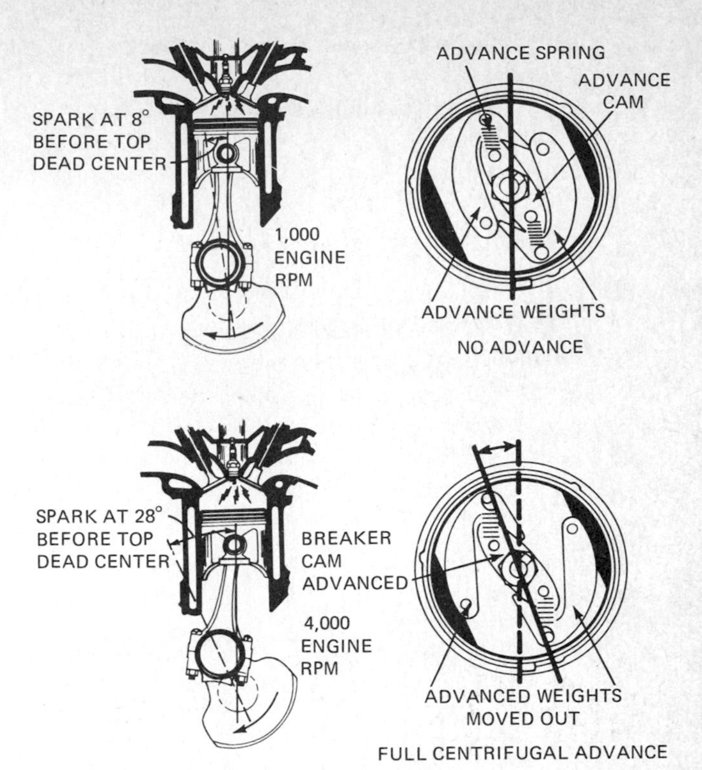

FIG. 14-27 Centrifugal-advance mechanism in the no-advance and full-advance positions. In the example shown, the ignition is timed at 8° before TDC on idle. There is no centrifugal advance at 1000 engine rpm. But there is a total advance of 28° (20° centrifugal advance plus 8° due to original timing) at 4000 engine rpm.

the cam usually has the same number of lobes as the engine has cylinders. This provides a spark for each cylinder. A rotor and cap then distribute the sparks at the proper time through the secondary wiring to the spark plugs (Fig. 14-19).

In some ignition systems, the breaker cam has only one-half as many lobes as engine cylinders. In these systems, two sets of breaker points are arranged to close and open alternately. This produces the same effect as described above.

Δ14-16 IGNITION TIMING

Ignition timing refers to the delivery of the spark from the coil to the spark plug at the proper time for the power stroke. The "time" that the spark occurs is relative to the position of the piston in the cylinder. Therefore, ignition timing is normally given as degrees before or after the TDC position of the piston. Some engine manufacturers specify ignition timing with the piston a short distance from the TDC position. The measurement of this distance is given in thousandths of an inch, or in millimeters.

Some small engines have *fixed timing.* The points are opened at the best time for the normal operating speed of the engine. With fixed timing, the engine is always timed at its maximum advance. This system is widely used with small single-cylinder engines which normally operate at a fixed speed. However, in engines operated at varying speed, fixed timing is not satisfactory. The timing must vary, according to the speed of the engine.

Most ignition distributors include a *centrifugal-advance* (or *mechanical-advance) mechanism.* It automatically advances the spark by pushing the breaker cam ahead as engine speed increases. When the engine is idling, each spark is timed to occur at the spark-plug gap just as the piston approaches TDC on the compression stroke (Fig. 14-27). At any speed faster than idle, the spark timing is advanced so that the spark occurs earlier in the cycle. This gives the mixture more time to burn.

CAPACITOR-DISCHARGE-IGNITION SYSTEMS

Δ14-17 BREAKERLESS IGNITION SYSTEMS

All the ignition systems discussed so far in this chapter have used breaker points to make and break the primary circuit. However, breaker points bounce at high speeds, burn away from excessive arcing, and change the ignition timing (Δ14-16) as wear occurs on the point faces and movable-arm rubbing block. For these reasons, many small engines now have some type of maintenance-free *breakerless* (also called *electronic, solid-state,* or *semiconductor)* ignition system.

In the breakerless ignition system, no breaker points are used. The primary circuit is opened and closed by a solid-state electronic switch which has no moving parts. However, additional electronic components and circuits (Δ10-20) are needed to control the operation of the switch. The most widely used type of breakerless ignition system for small engines is the capacitor-discharge-ignition (CDI) system (Δ14-18).

Δ14-18 OPERATION OF CAPACITOR-DISCHARGE IGNITION

Any ignition system has two basic circuits. These are the primary circuit and the secondary circuit (Δ14-5). After the high voltage builds up in the secondary circuit, the actions in firing the spark plug are the same in all ignition systems. However, the primary-circuit actions in CDI systems differ from those described earlier.

A capacitor-discharge-ignition system (Fig. 14-28) stores its primary energy in a large capacitor. (In other ignition systems, the primary energy is stored in a magnetic field around the ignition coil.) Then, when a spark is needed at the spark plug, the capacitor is discharged. This sends a surge of current flowing through the primary winding of the ignition coil. As the magnetic field builds up, a high voltage is induced in the secondary winding. When the voltage gets high enough, the spark jumps across the gap at the spark plug.

Notice that the operation of the capacitor-discharge-ignition system differs from other ignition systems. In them, the spark at the plug occurs when the primary circuit is opened and the magnetic field collapses (Fig. 14-23). In the capacitor-discharge-ignition system, the spark occurs when the primary circuit is completed

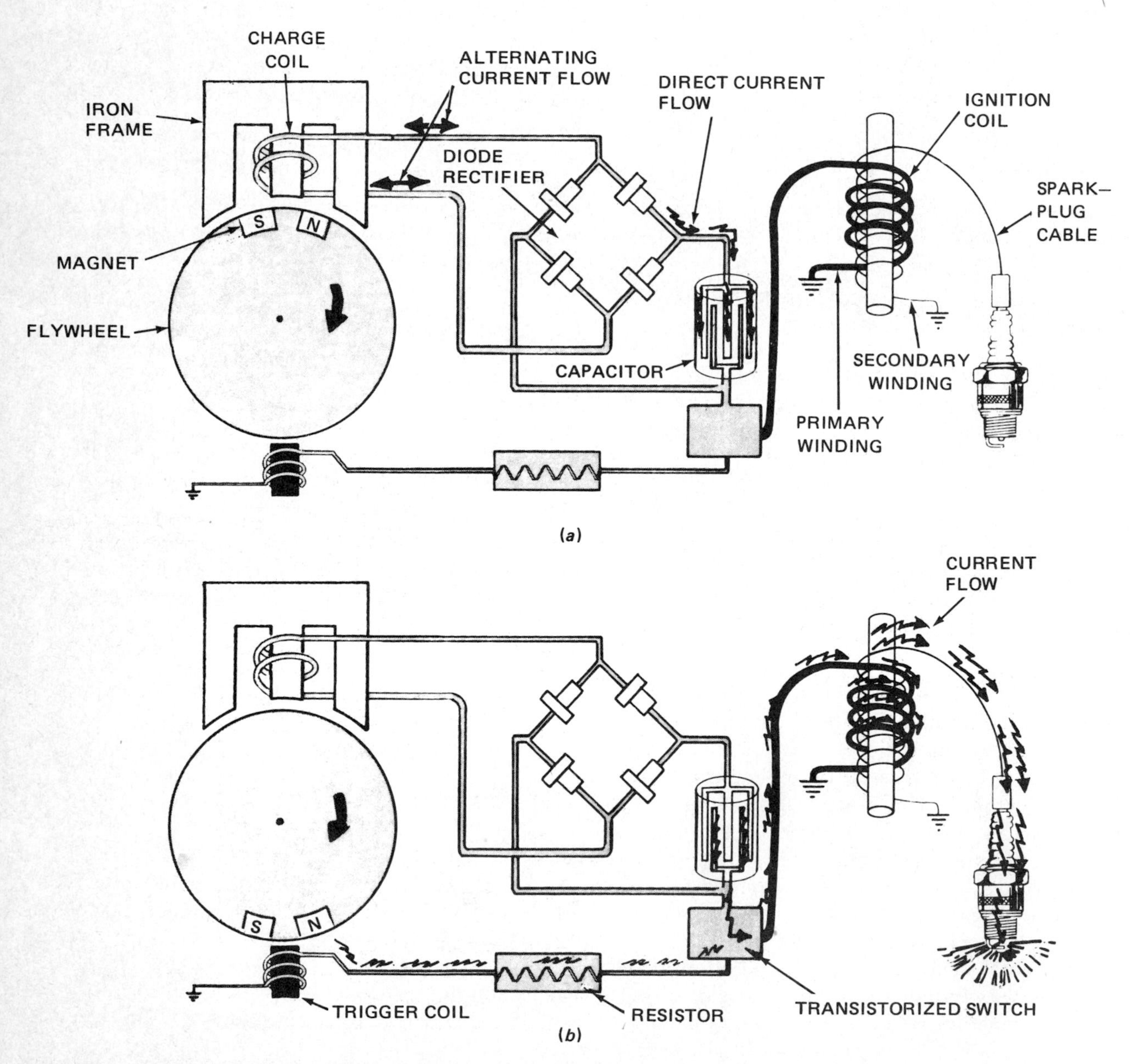

FIG. 14-28 **A capacitor-discharge-ignition system.** *(a)* Electric energy is being stored in the capacitor. *(b)* The capacitor discharges through the primary winding of the coil, which induces a high voltage in the secondary winding to fire the spark plug.

(through the transistorized switch). Now the capacitor discharges through the primary winding of the coil. The resulting magnetic field continues to build up until the voltage in the secondary winding is high enough to fire the spark plug.

Δ14-19 TYPICAL CDI SYSTEM

Figure 14-2 shows a small engine with one type of capacitor-discharge-ignition system. Figure 14-29 shows the ignition module removed from the engine. Almost all CDI systems work in the same general way. However, many different arrangements of the components have been used. An engine with CDI that has a battery and a magneto-alternator system is shown in Fig. 14-30. The charge coil that charges the capacitor is part of the alternator stator. The stator also includes windings which produce low-voltage ac. This low-voltage ac passes through the rectifier-regulator, which changes the current to dc to charge the battery. (Charging systems are covered in detail in Chap. 16.)

A CDI system for rotary-lawn-mower engines is shown in Fig. 14-31. As each magnet in the flywheel rotates past the charge coil, a voltage of about 200 volts ac is induced in it. Input diode D1 converts the ac to dc, which now charges the capacitor. The magnet continues rotating past the trigger coil, where a low-voltage signal is produced. This signal causes an electronic switch [also called a *silicon-controlled rectifier* (SCR), or *thyristor*] to close. Now the capacitor discharges. The current flows through the SCR and then through the primary winding of the ignition coil (*pulse transformer* in Fig. 14-31).

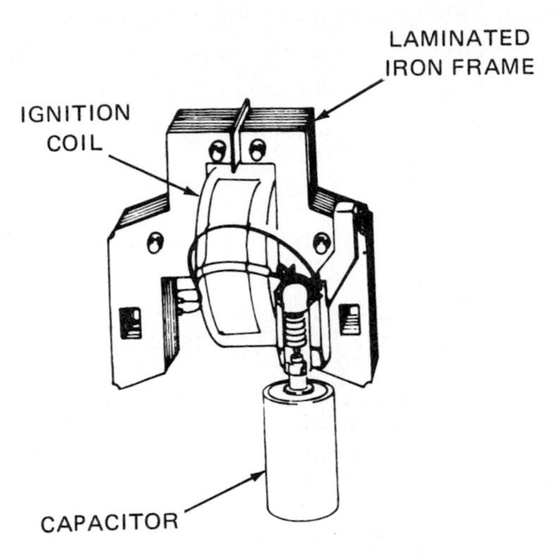

FIG. 14-29 A capacitor-discharge-ignition module removed from the engine. *(Briggs & Stratton Corporation)*

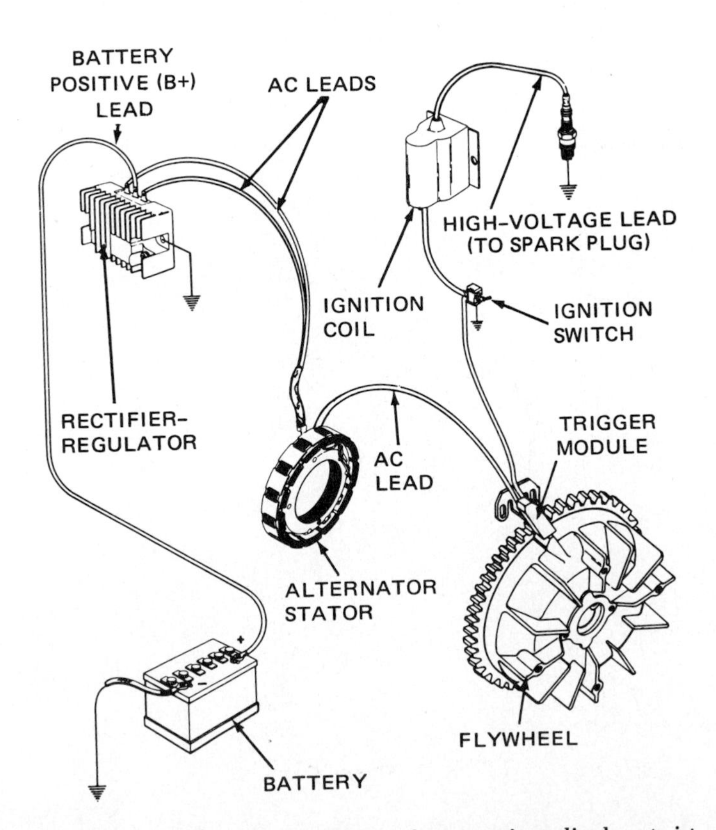

FIG. 14-30 Schematic diagram of a capacitor-discharge-ignition system on an engine that has a battery and a magneto alternator.

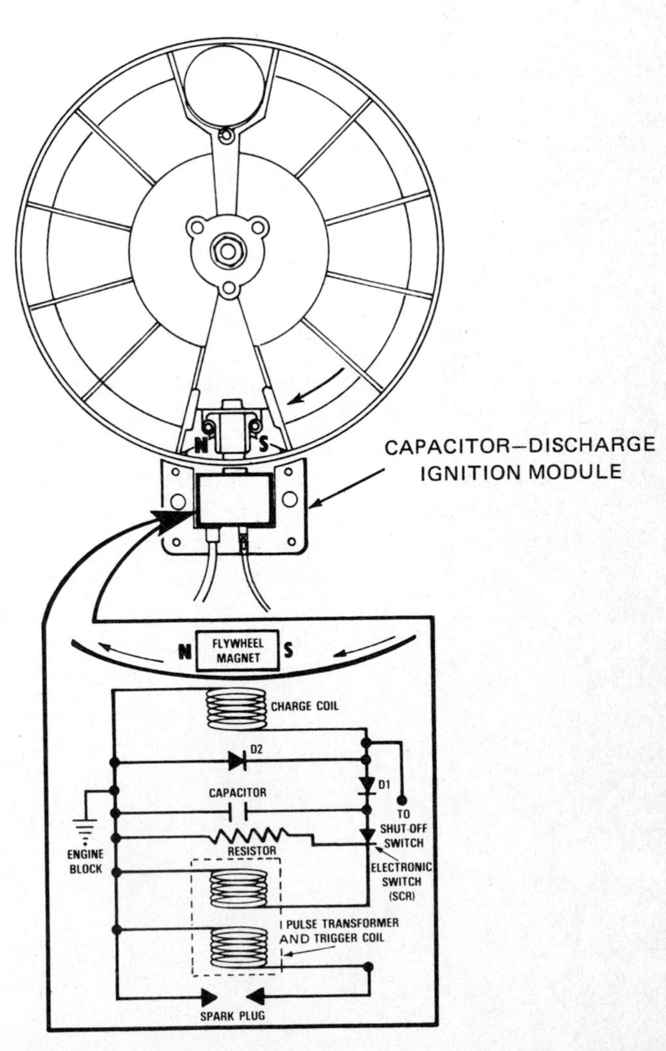

FIG. 14-31 Wiring diagram of a capacitor-discharge-ignition system for a rotary-lawn-mower engine. *(Tecumseh Products Company)*

As the current flows through the primary winding, the magnetic field builds up. This induces a voltage of up to 30,000 volts in the secondary winding. When the voltage gets high enough, the high-voltage surge flows through the spark-plug wire to the spark plug. There the current creates an arc by jumping across the spark-plug electrodes. This ignites the compressed air-fuel mixture.

CHAPTER 14: REVIEW QUESTIONS

Select the *one* correct, best, or most probable answer to each question. You can find the answers in the section indicated at the right of each question.

1. In the ignition system, the low-voltage circuit is the (**Δ*14-5***)
   a. primary circuit
   b. intermediate circuit
   c. secondary circuit
   d. none of the above
2. The ignition coil has (**Δ*14-6***)
   a. one winding
   b. two windings
   c. three windings
   d. four windings
3. In a battery-ignition system, the battery is connected to the primary winding of the ignition coil through the (**Δ*14-14***)
   a. secondary winding
   b. condenser
   c. spark plug
   d. primary wiring
4. In the breaker-point flywheel magneto, the purpose of the condenser is to help prevent excessive arcing of the points and to produce (**Δ*14-7***)
   a. quick collapse of the magnetic field
   b. slow collapse of the magnetic field
   c. high voltage across the breaker points
   d. minimum primary voltage
5. When the ignition points open, (**Δ*14-6***)
   a. primary current flow stops
   b. the magnetic field collapses
   c. the spark plug fires
   d. all of the above
6. To stop a running engine that has a flywheel magneto, the magneto-ignition switch should be (**Δ*14-11***)
   a. moved to the open position
   b. disconnected from the battery
   c. moved to the closed position
   d. connected to the battery
7. Mechanic A says that single-cylinder air-cooled engines may use magneto ignition or battery ignition. Mechanic B says that these engines may use almost any type of breakerless ignition or a magneto-alternator system. Who is right? (**Δ*14-1, 14-17, 14-19***)
   a. mechanic A
   b. mechanic B
   c. both A and B
   d. neither A nor B
8. In which ignition system are the actions in the secondary circuit different from in the other systems? (**Δ*14-18***)
   a. capacitor-discharge-ignition system
   b. breakerless flywheel magneto
   c. unit magneto
   d. none of the above
9. In which ignition system does the spark at the plug occur as the magnetic field is building up? (**Δ*14-18***)
   a. capacitor-discharge-ignition system
   b. breakerless flywheel magneto
   c. unit magneto
   d. none of the above
10. The device in the distributor that pushes the breaker cam ahead as engine speed increases is the (**Δ*14-16***)
    a. vacuum-advance mechanism
    b. centrifugal-advance mechanism
    c. trigger coil
    d. charge coil

Ignition-Systems Service

After studying this chapter, you should be able to:

1. ***Diagnose trouble in the ignition system.***
2. ***Make a spark test and interpret the results.***
3. ***Replace and adjust breaker points.***
4. ***Static-time the ignition.***
5. ***Time the ignition using a timing light.***

Δ15-1 IGNITION-SYSTEM TROUBLE DIAGNOSIS The ignition system is a basic part of the engine. Whatever happens in the ignition system affects engine operation. If the engine fails to start even though it cranks normally, or if the engine misses, lacks power, overheats, backfires, or pings, the trouble could be caused by a faulty ignition system.

Normal ignition-system operation requires all components to be in good condition and properly adjusted. In addition, the spark must be properly timed. When the ignition system is suspected of causing trouble, and especially if the engine will not start, make a *spark test*. The procedure is described in Δ15-2.

The chart below lists the possible troubles in the battery-ignition system and the checks or corrections to be made for each condition. Engine troubles are listed in the engine trouble-diagnosis chart in Chap. 19.

Δ15-2 MAKING A SPARK TEST To make a spark test, disconnect the high-voltage lead from the spark plug. Pull back the rubber boot on the spark plug to expose the metal clip, or put a bolt into the boot to contact the metal clip.

BATTERY-IGNITION-SYSTEM TROUBLE-DIAGNOSIS CHART

Complaint	Possible cause	Check or correction
1. Engine cranks normally but will not start	a. Open primary circuit	Check connections, coil, breaker points, and switch for open
	b. Coil primary grounded	Replace coil
	c. Points not opening	Adjust
	d. Points burned	Clean or replace
	e. Out of time	Check and adjust timing
	f. Condensor defective	Replace
	g. Coil secondary open or grounded	Replace coil
	h. High-voltage leakage	Check coil head, distributor cap, rotor, and leads
	i. Spark plugs fouled	Clean and adjust or replace
	j. Defects in electronic control unit or pickup-coil circuit	Replace defective part
	k. Fuel system faulty	Check
	l. Engine faulty	Repair engine
2. Engine runs but misses—one cylinder	a. Defective spark plug	Clean or replace
	b. Distributor cap or lead defective	Replace
	c. Engine defects such as stuck valve or defective rings, piston, or gasket	Repair engine

(Chart Continued on Next Page)

BATTERY-IGNITION-SYSTEM TROUBLE-DIAGNOSIS CHART *(CONTINUED)*

Complaint	Possible cause	Check or correction
3. Engine runs but misses—different cylinders	a. Points dirty, worn, or out of adjustment	Clean, replace or adjust as necessary
	b. Condenser defective	Replace
	c. Advance mechanism defective	Repair or replace distributor
	d. Defective high-voltage wiring	Replace
	e. Defective (weak) coil	Replace
	f. Bad connections	Clean and tighten connections
	g. High-voltage leakage	Check coil head, distributor cap, rotor, and leads
	h. Defective spark plugs	Clean adjust, or replace
	i. Defective fuel system	Check
	j. Defects in engine such as loss of compression or faulty valve action	Repair engine
4. Engine lacks power	a. Timing off	Retime ignition
	b. Exhaust system clogged	Clear
	c. Excessive load resistance	Check load
	d. Heavy engine oil	Use correct oil
	e. Wrong fuel	Use correct fuel
	f. Engine overheats	See item 5
	g. Other defects listed under item 3	
5. Engine overheats	a. Late ignition timing	Retime ignition
	b. Lack of coolant or other trouble in cooling system	Add coolant, or repair cooling system
	c. Late valve timing or other engine conditions	Repair engine
6. Engine backfires	a. Ignition timing off	Retime ignition
	b. Ignition cross-firing	Check high-voltage wiring, cap, and rotor for leakage points
	c. Spark plugs of wrong heat range	Install correct plugs
	d. Engine overheating	See item 5
	e. Fuel system not supplying proper air-fuel ratio	Repair fuel system
	f. Engine defects such as hot valves or carbon	Repair engine
7. Engine detonates or pings	a. Improper timing	Retime engine
	b. Advance mechanisms faulty	Rebuild or replace distributor
	c. Points out of adjustment	Readjust
	d. Distributor bearing worn or shaft bent	Rebuild or replace distributor
	e. Spark plugs of wrong heat range	Replace with correct plugs
	f. Low-octane fuel	Use fuel of proper octane
	g. Conditions listed under item 6	
8. Burned or oxidized breaker points	a. Excessive resistance in condenser circuit	Tighten condenser mounting and connection; replace condenser if bad
	b. High voltage	Readjust voltage regulator
	c. Excessive dwell, too little gap	Reset breaker points
	d. Weak spring tension	Adjust breaker spring tension
	e. Oil or crankcase vapors entering distributor	Check engine PCV system; avoid overlubricating distributor
9. Spark plug defective	a. Cracked insulator	Install new plug
	b. Spark plug sooty	Install hotter plug; correct condition in fuel system or engine causing high fuel consumption
	c. Spark plug white or gray, with blistered insulator	Install cooler plug

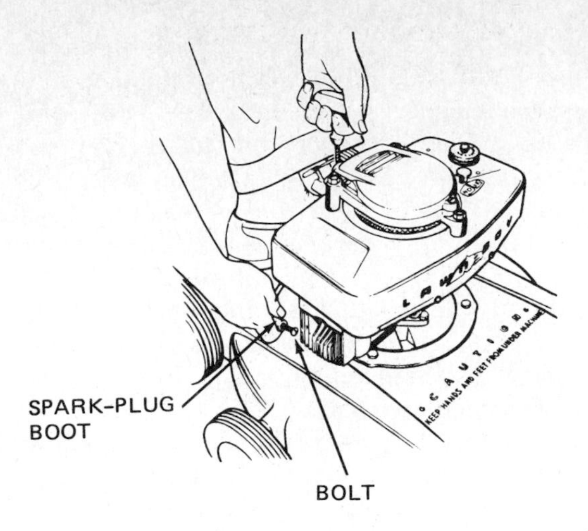

FIG. 15-1 **Making a spark test.** *(Lawn Boy Division of Outboard Marine Corporation)*

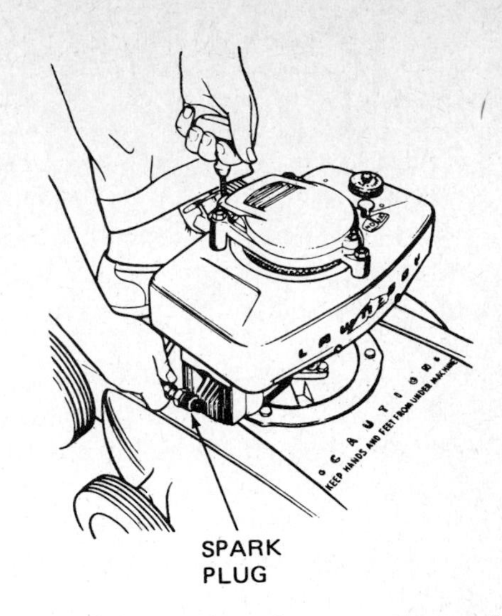

FIG. 15-2 **Checking for a spark across the spark-plug gap.** *(Lawn Boy Division of Outboard Marine Corporation)*

Hold the metal clip or bolt about 3/16 inch [5 mm] from the cylinder head and crank the engine (Fig. 15-1). If a strong spark jumps to the cylinder head as the engine is cranked, the ignition system is probably working properly. If no spark occurs, the ignition system is probably at fault and should be checked.

CAUTION **Electronic ignition systems can deliver a very high voltage. Proceed with care when making the spark test and working on the ignition system. The ignition switch must be *off* and the battery (if used) *disconnected.* Wear special rubber gloves or use insulated pliers to hold the plug wire while making the spark test. The high voltage that some ignition systems deliver can give you a painful and dangerous shock!**

If a spark jumps from the bolt or clip to the cylinder head during the spark test, then the spark plug may be defective. Remove and examine it. If no defects are seen, reconnect the spark-plug lead. Then lay or hold the spark-plug shell against the cylinder head and crank the engine (Fig. 15-2).

Watch for a spark at the plug gap. If no spark jumps the gap as the engine is cranked, the plug is probably at fault. Examine it closely for cracks, black sooty deposits on the insulator or electrodes, burned electrodes, or a wide gap. Any of these could prevent a good spark. Spark-plug service is covered in **Δ**15-4.

Some manufacturers recommend that the spark test be made using a *spark tester* (Fig. 15-3). Remove the lead from the spark plug and connect the lead to the spark tester. Clamp the tester to a good ground on the engine. Then crank the engine. Bright, hot sparks jumping the gap indicate the ignition system is performing satisfactorily.

Another method of testing the spark is to use the *spark-gap tester* (Fig. 15-4). To use the tester, remove the spark-plug lead and connect it to the tester. Ground the electrode of the tester that gives a 0.166-inch [4-mm] gap.

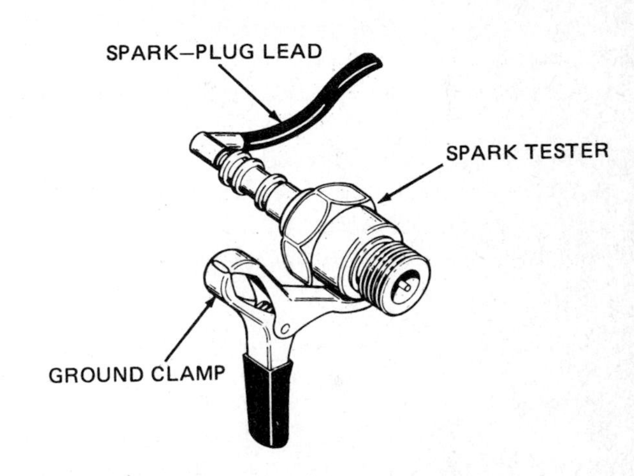

FIG. 15-3 **Using a spark tester to make a spark test.** *(Tecumseh Products Company)*

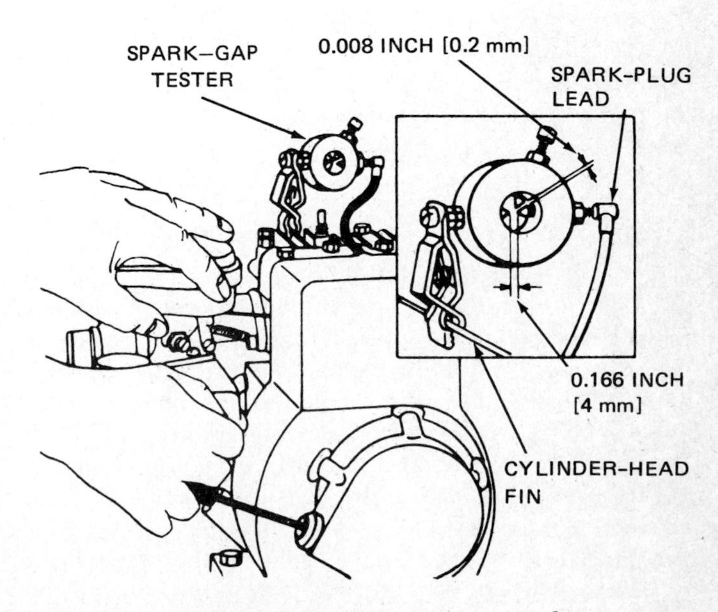

FIG. 15-4 **Checking the spark with a spark-gap tester.** *(Briggs & Stratton Corporation)*

Then crank the engine. If a good spark jumps the gap, the ignition system is working properly.

This tester may also be used to make an *in-line,* or *operational, test.* If the engine runs but misses, insert the tester between the spark-plug lead and the spark plug. Run the engine, and watch the spark at the tester gap. If the spark is not regular and steady, a spark miss is occurring. This indicates trouble in the ignition system.

Δ15-3 SPARK-TEST RESULTS

If the spark test indicates that the ignition system is not the cause of the problem, then check the fuel system, the cooling system, and the engine itself.

When no spark occurs during the spark test, check for the following:

1. Defective insulation or poor connections in the wiring.
2. A grounded ignition switch which prevents the opening of the primary circuit so that primary current is not interrupted.
3. A shorted condenser, which has the same effect as a grounded ignition switch.
4. A magneto coil that is shorted, open, or grounded.
5. A loss of magnetism in permanent magnets in the flywheel or in the unit-magneto rotor.
6. Breaker points that are dirty, worn, or out of adjustment.

If the engine backfires or kicks back while cranking, the breaker-point gap may be too wide. This causes ignition to occur too early in the compression stroke. When this happens, the resulting increase in pressure forces the piston back down before it can reach TDC. An early spark will also cause detonation while the engine is running.

To correct these conditions, reset the ignition timing. On some magnetos, the only adjustment is to change the breaker-point opening. On others, the breaker plate is moved to change the timing. The methods for timing various small-engine ignition systems are described in later sections.

Δ15-4 SPARK-PLUG SERVICE

Worn, fouled, or improperly gapped spark plugs often cause engine misfire and poor operation. Spark plugs with improper heat range (**Δ**14-9) will foul and wear rapidly.

Spark plugs should be cleaned and adjusted periodically. Before removing the spark plug, clean the area around the seat. One method is to loosen the plug slightly and then blow out around the seat with compressed air. This prevents any dirt and debris from falling into the engine cylinder as the plug is removed.

Examine the spark plugs as soon as they have been removed. The deposits on the firing end indicate the general condition of the piston rings, valves, carburetor, and ignition system. Figure 15-5 shows various spark-plug conditions and their causes.

Many shops have spark-plug cleaners that clean a used plug by sandblasting the firing end (the electrodes and the porcelain insulator) with an abrasive. However, some small-engine manufacturers do not recommend this. The abrasive may stick to the plug, enter the cylinder, and cause excessive wear and damage to the piston, rings, cylinder, and bearings. These manufacturers recommend installing a new plug if the old plug cannot be gapped and reinstalled. Some manufacturers also caution against filing the center electrode flat on a worn spark plug. Installing a new plug is recommended.

To check the gap between the electrodes, use a wire thickness gauge (Fig. 15-6). The gap can be adjusted by bending the ground electrode. Then the plug can be reinstalled if it is in otherwise good condition. Always select and install the correct spark plug for the engine. But first inspect and gap the plugs to the manufacturer's specifications.

If a spark plug fouls frequently, check for the following conditions:

1. Carburetor set too rich
2. Choke not opening fully
3. Poor grade of gasoline
4. Clogged exhaust system
5. Incorrect spark plug
6. Crankcase breather plugged
7. Crankcase-oil level too high
8. Engine using excessive oil

FLYWHEEL-MAGNETO SERVICE

Δ15-5 REMOVING ENGINE FLYWHEEL

On many small engines with a flywheel magneto, the flywheel must be removed before you can service and adjust the breaker points. Servicing of breaker points that are mounted in an external box is described in later sections.

> ***CAUTION*** **To prevent accidental starting, always disconnect the spark-plug wire from the spark plug before beginning work on an engine. Then attach a jumper wire to the end of the spark-plug wire to ground it.**

If the engine has a rope-rewind or a windup starter, remove the starter assembly (Chap. 13). On some engines with an electric starter-generator, the flywheel will have a stub shaft on which the drive pulley mounts. This stub shaft must be removed. Then the flywheel shroud must be removed so you can get at the flywheel.

Next, remove the flywheel-attaching nut by turning it in the direction opposite to normal crankshaft rotation. Use a flywheel holder (Fig. 15-7) to prevent the flywheel from turning. Figure 15-8 shows another type of flywheel holder. It can be used on some engines with rope, rewind, or windup starters.

NORMAL

Brown to grayish tan color and slight electrode wear. Correct heat range for engine and operating conditions.

RECOMMENDATION: Service and reinstall. Replace if over recommended mileage.

WORN

Center electrode worn away too much to be filed flat.

RECOMMENDATION: Replace with new spark plugs of proper heat range.

PREIGNITION

Improper heat range, incorrect ignition timing, lean air-fuel mixture, or hot spots in combustion chamber has caused melted electrodes. Center electrode generally melts first and ground electrode follows. Normally, insulators are white, but may be dirty due to misfiring or flying debris in combustion chamber.

RECOMMENDATION: Check for correct plug heat range, overadvanced ignition timing, lean air-fuel mixture, clogged cooling system, leaking intake manifold, and lack of lubrication.

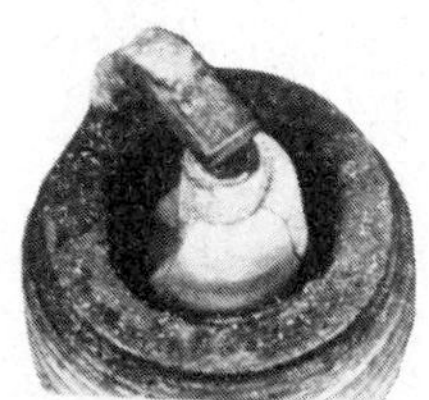

DETONATION

Insulator has cracked and broken away as a result of the shock waves created by detonation.

RECOMMENDATION: Cause of detonation must be found and corrected. Check for use of low-octane fuel, improper air-fuel mixture, incorrect ignition timing, overheating, and increased octane requirement in the engine.

MECHANICAL DAMAGE

Something such as a foreign object in cylinder, or the piston, has struck the ground electrode. It has been forced into the center electrode, which has bent, and broken off the insulator.

RECOMMENDATION: Check that plug has the correct reach for the engine, that the gap was set properly, and that no foreign object remains in cylinder.

GAP BRIDGED

Deposits in the fuel have formed a bridge between the electrodes, eliminating the air gap and grounding the plug.

RECOMMENDATION: Sometimes plug can be serviced and reinstalled. Check for excess additives in fuel.

CARBON DEPOSITS

Dry soot, frequently caused by use of spark plug with incorrect heat range.

RECOMMENDATION: Carbon deposits indicate rich mixture or weak ignition. Check for clogged air cleaner, high float level, sticky choke or worn contact points. Hotter plugs will provide additional fouling protection.

OIL DEPOSITS

Oily coating.

RECOMMENDATION: Caused by poor oil control. Oil is leaking past worn valve guides or piston rings into the combustion chamber. Hotter spark plug may temporarily relieve problem, but correct the cause with necessary repairs.

SPLASHED DEPOSITS

Spotted deposits. Occurs shortly after long-delayed tuneup. After a long period of misfiring, deposits may be loosened when normal combustion temperatures are restored by tuneup. During a high-speed run, these materials shed off the piston and head and are thrown against the hot insulator.

RECOMMENDATION: Clean and service the plugs properly and reinstall.

ASH DEPOSITS

Poor oil control, use of improper oil, or use of improper additives in fuel or oil has caused an accumulation of ash which completely covers the electrodes.

RECOMMENDATION: Eliminate source of ash. Install new spark plugs.

OVERHEATED

Blistered, white insulator, eroded electrodes and absence of deposits.

RECOMMENDATION: Check for correct plug heat range, overadvanced ignition timing, low coolant level or restricted flow, lean air-fuel mixture, leaking intake manifold, sticking valves, and if car is driven at high speeds most of the time.

HIGH-TEMPERATURE GLAZING

Insulator has yellowish, varnish-like color. Indicates combustion chamber temperatures have risen suddenly during hard, fast acceleration. Normal deposits do not get a chance to blow off. Instead, they melt to form a conductive coating.

RECOMMENDATION: If condition recurs, use plug type one step colder.

FIG. 15-5 Various spark-plug conditions and their causes. *(Champion Spark Plug Company)*

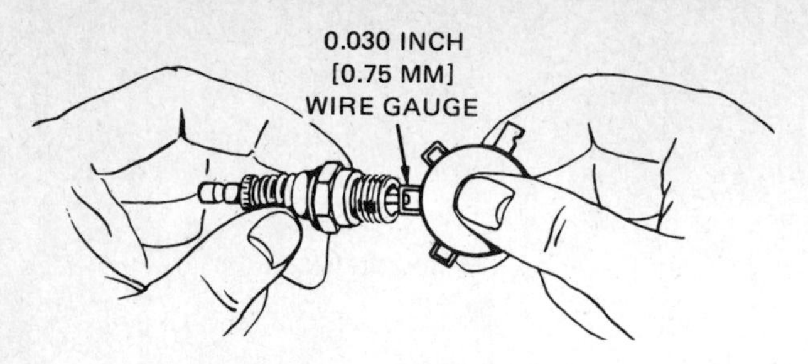

FIG. 15-6 Checking spark-plug gap with a wire thickness gauge. *(Briggs & Stratton Corporation)*

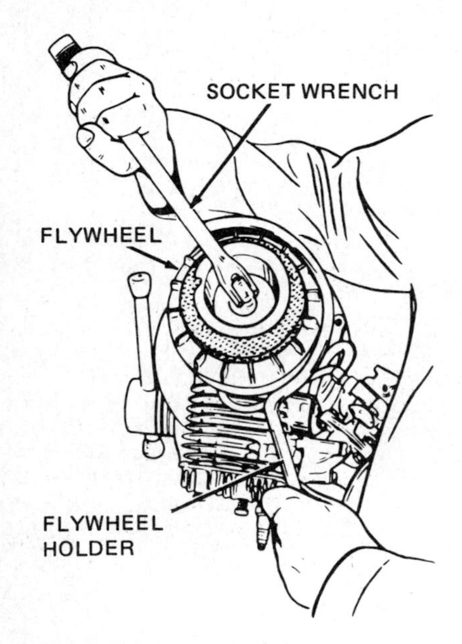

FIG. 15-7 Removing the flywheel-attaching nut while holding the flywheel with a flywheel holder. *(Tecumseh Products Company)*

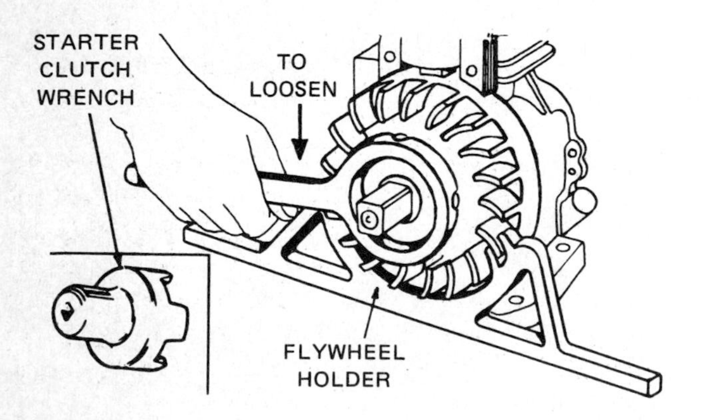

FIG. 15-8 A bench-mounted flywheel holder. *(Briggs & Stratton Corporation)*

Remove the flywheel from the crankshaft by using a flywheel puller (Figs. 15-9 and 15-10) or a knockoff puller (Fig. 15-11). If the shaft is tapered, you can install a nut on the threads. Turn the nut onto the shaft so that the threads on the end of the shaft are almost exposed. Then strike the nut with a soft hammer (Fig. 15-12).

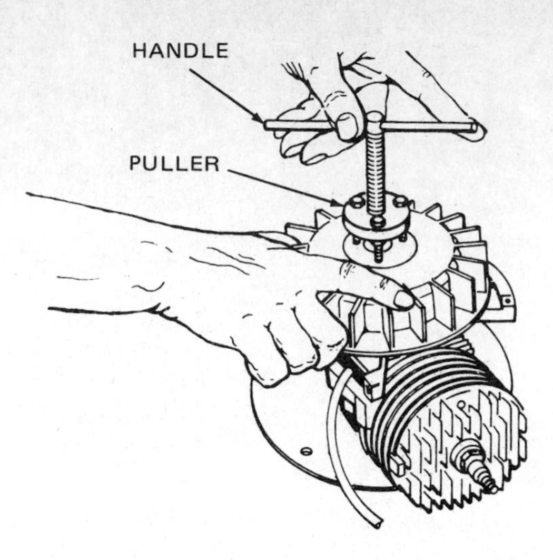

FIG. 15-9 Using a pressure-screw puller to remove the flywheel from the shaft. *(Tecumseh Products Company)*

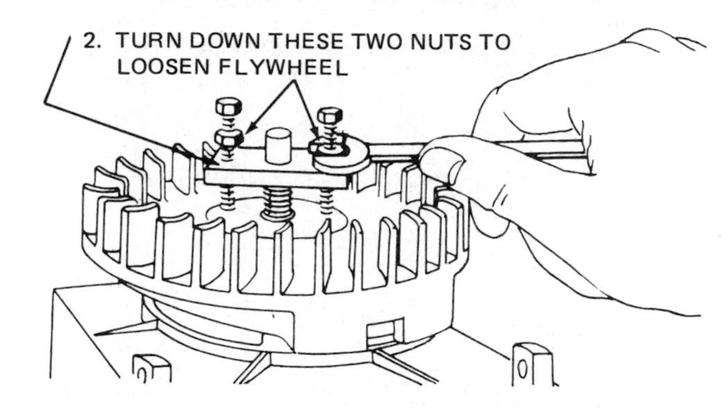

FIG. 15-10 Using self-tapping screws to remove the flywheel from the crankshaft. *(Briggs & Stratton Corporation)*

NOTE Some manufacturers caution against trying to remove the flywheel with a jaw-type puller attached to the outer rim. This will break the flywheel.

When removing the flywheel, do not drop it. This could cause the permanent magnets to loose most of their magnetism. Heat from a welder or cutting torch will also take the magnetism out of magnets.

On some engines, the breaker points are protected by a dust cover (Fig. 15-13). Remove this cover carefully. If it is bent during removal, it will not seal on reassembly. Then a new cover will be required. Note that the place where the leads come out from under the cover is sealed. This seal is made with nonhardening No. 2 Permatex or a similar sealer as the cover is installed. A poor seal allows dirt to get on the breaker points.

Δ15-6 INSPECTING FLYWHEEL AND KEY

After the flywheel is removed, inspect its condition. Many air-cooled engines have a fan with blades or vanes cast into the flywheel (Fig. 15-12). If any of these are

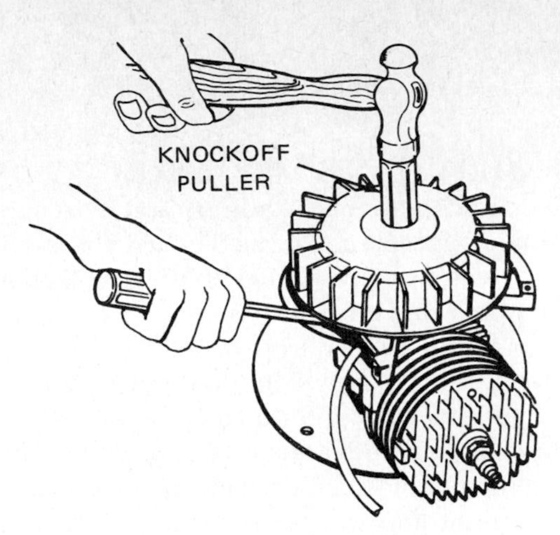

FIG. 15-11 A knockoff puller is used on some engines to remove the flywheel from the crankshaft. *(Tecumseh Products Company)*

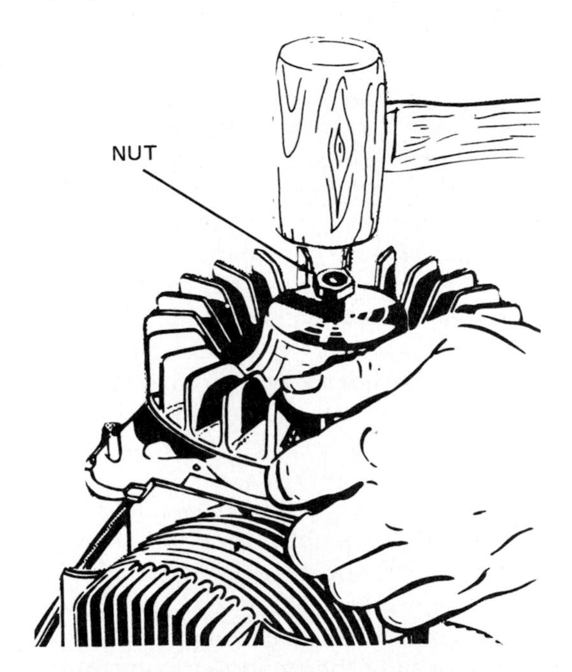

FIG. 15-12 Thread the nut onto the shaft and then strike the nut with a soft hammer to remove the flywheel from a tapered shaft. *(Outboard Marine Corporation)*

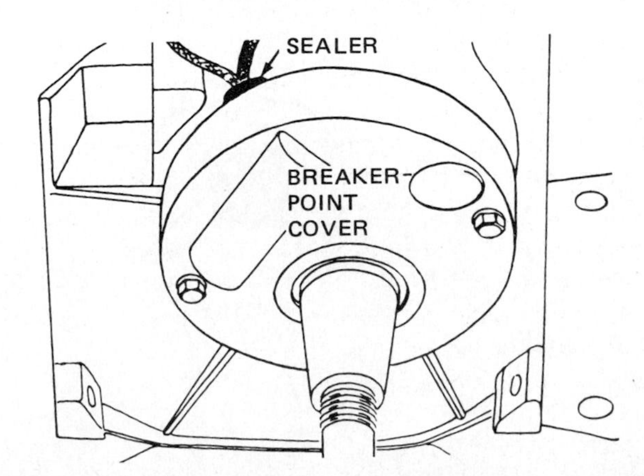

FIG. 15-13 On some engines, the breaker points are protected by a dust cover, or breaker-point cover. *(Briggs & Stratton Corporation)*

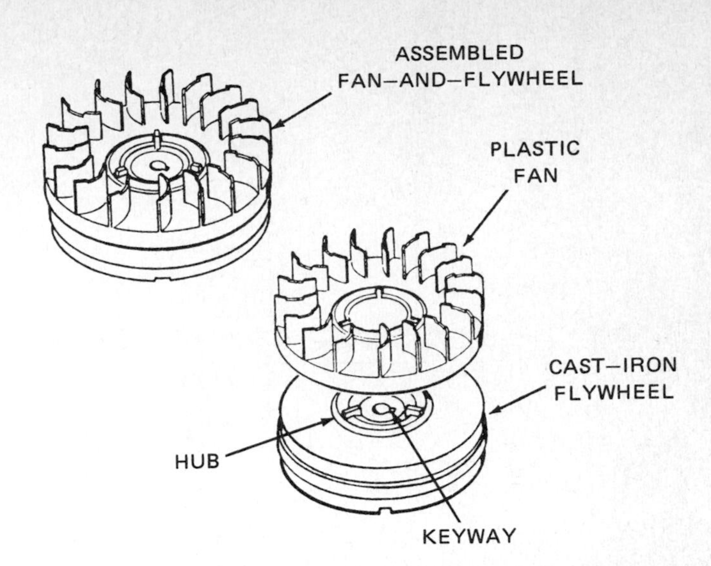

FIG. 15-14 A plastic fan that attaches to the cast-iron flywheel. *(Tecumseh Products Company)*

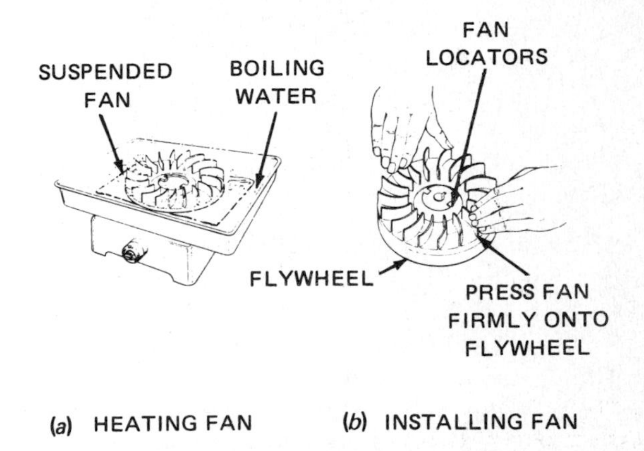

FIG. 15-15 Installing a plastic fan on a cast-iron flywheel. *(Tecumseh Products Company)*

broken, the engine will be out of balance. This will cause excessive vibration and rapid wear. It may also cause the crankshaft to break.

Some flywheels have a plastic fan that attaches to the cast-iron flywheel (Fig. 15-14). If the fan is damaged, tap the outside edge of the fan until it separates from the flywheel. To install the new plastic fan, suspend the fan in a pan of boiling water (Fig. 15-15*a*). Then install the fan on the flywheel, making sure that the locators on the fan are properly positioned on the flywheel hub (Fig. 15-15*b*).

The magnets in the flywheel seldom lose their magnetism. A quick check of the magnets' strength can be made by placing the flywheel upside down on a wood surface. Hold a screwdriver by the end of the handle, with the blade down (Fig. 15-16). Move the blade to within ¾ inch [19 mm] of each magnet. A good magnet should be strong enough to pull the screwdriver blade against the magnet.

If the magnets are weak, they should be replaced or remagnetized in a special magnetizer. Weak alnico magnets cannot be recharged and must be replaced. Never

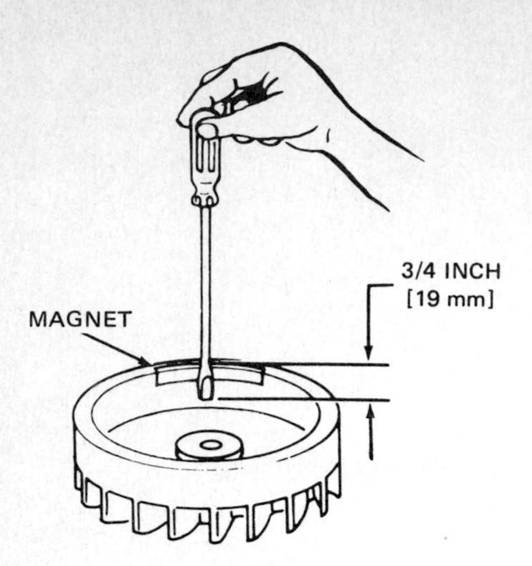

FIG. 15-16 Making a quick check of the strength of the flywheel magnet. *(Tecumseh Products Company)*

store flywheels in nested piles. This can cause the magnets to loose their strength.

> **CAREFUL** Welding on an engine with a flywheel magneto can cause the magnets to lose strength. Magnets can also be weakened by heating, dropping, or receiving other hard blows.

A pin or key is used to locate the flywheel in its proper position on the crankshaft. This must be correct for proper ignition timing. The key is often made of soft metal which easily shears (Fig. 15-17), especially if the crankshaft receives a sudden heavy blow. This allows the flywheel to shift slightly, which makes the spark occur out-of-time. As a result, the engine stops. It may not start again, or it may be difficult to start.

Anytime the flywheel is removed, inspect the condition of the keyway in the flywheel hub (Fig. 15-14). Then inspect the key and the keyway in the crankshaft. The keyways should not be damaged or enlarged. If the key is sheared or partially sheared (Fig. 15-17), replace it with a new soft-metal key. Do not use a steel key.

△15-7 SERVICING BREAKER POINTS

Breaker-point condition and adjustment have a great affect on engine operation. Points that are burned or have any type of coating or oxidation on them will allow little or no current to pass through. As a result, the engine may not run, or it may run poorly.

The size of the gap between the points when they are fully open affects ignition timing. If the gap is too wide, the ignition timing is advanced. The spark will occur too early. Detonation or kickback may occur during cranking. If the point gap is too close, ignition timing is retarded. This causes the engine to lose power and overheat.

Cleaning Points After the breaker points are exposed by removal of the flywheel and the dust cover (where present), examine them for oxidation or pits. If they are only slightly burned or pitted, they can be cleaned with an ignition file (Fig. 15-18). It is not necessary to file the points until they are smooth. Just remove the worst of the high spots. Blow out all dust after cleaning the contacts. Pull a strip of clean bond paper between the points (with points closed) to remove the last traces of filings. Then open the points to withdraw the paper.

Never use emery cloth or sandpaper to clean the points. Particles of emery or sand will embed in the points and cause erratic operation and possible point burning.

Installing New Points If the points are badly burned, worn, or pitted, they should be replaced. Severe burning of the points could be due to a defective condenser, improper adjustment, or oil on the contact surfaces. Check for these conditions before replacing the points. Various methods of attaching the stationary point and breaker

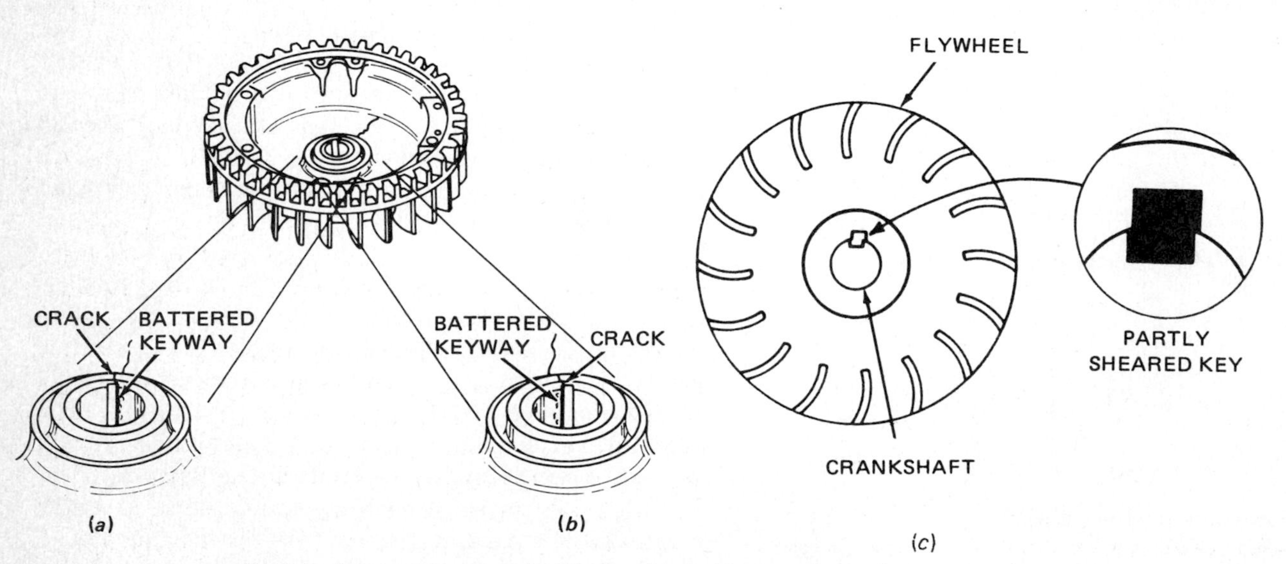

FIG. 15-17 Inspecting the key and keyway in the flywheel. *(Tecumseh Products Company)*

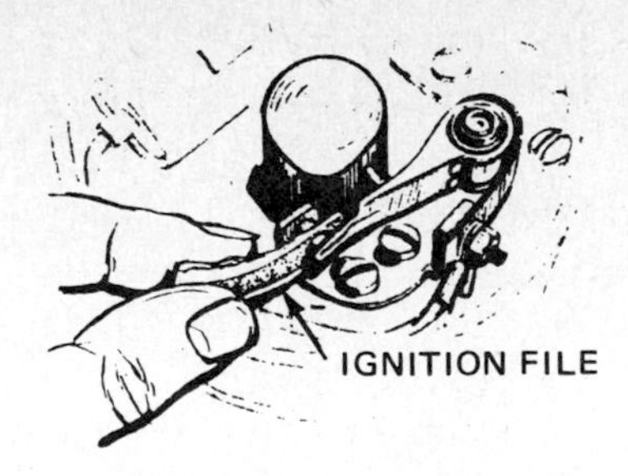

FIG. 15-18 Cleaning the breaker points with an ignition file.

arm are used. On one type, the point assembly is removed by removing the condenser wire from the breaker-point clip and then loosening the adjusting lock screw so the assembly can be slipped off. On the type shown in Fig. 15-19, the breaker arm is removed by loosening the screw holding the post in position. The stationary contact is on the condenser and is removed along with the condenser by loosening the condenser-clamp screw.

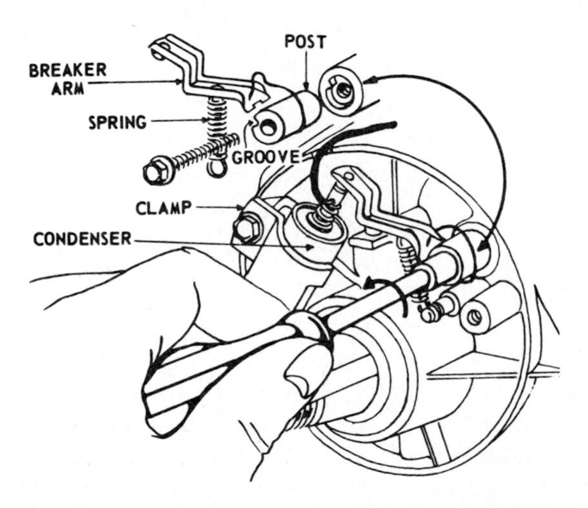

FIG. 15-19 This breaker-point assembly is removed by loosening the screw holding the post. The condenser, with the stationary point, is removed separately by loosening the condenser-clamp screw. *(Briggs & Stratton Corporation)*

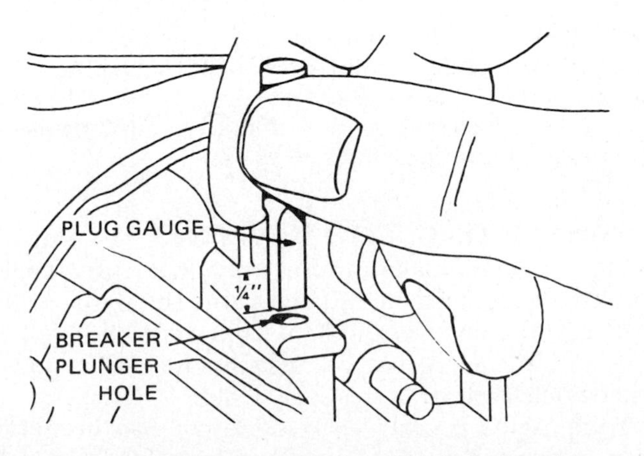

FIG. 15-20 A plug gauge is used to check the breaker-point plunger hole for wear. *(Briggs & Stratton Corporation)*

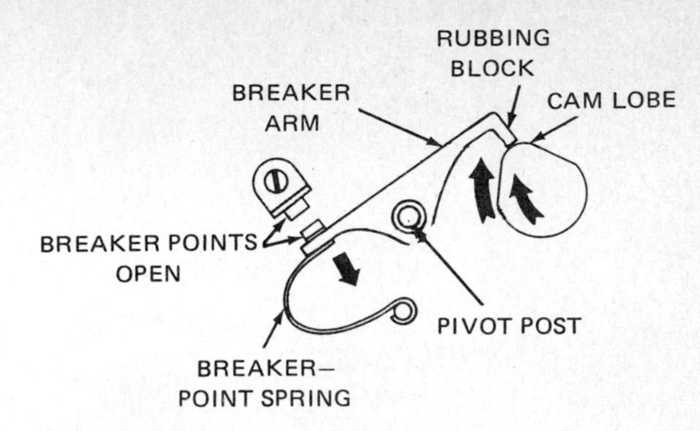

FIG. 15-21 When inspecting the condition of the breaker points, always check for wear of the rubbing block.

On engines which use a pushrod or plunger to operate the breaker arm, check the plunger and plunger hole whenever the points are removed. A plug gauge is used to check the hole for wear (Fig. 15-20). If the hole is enlarged, it will allow oil to enter the breaker-point compartment. There, the oil will get on the points and cause them to burn. If the hole is worn, it must be reamed out and a special bushing installed. The plunger should be checked for wear. If the plunger is worn and too short, replace it with a new plunger. Be sure the plunger is installed correctly.

On the type of breaker arm that uses a rubbing block (Fig. 15-21), check for rubbing-block wear. As the rubbing block wears, it changes the breaker-point opening and also can change the timing.

Examine the breaker cam for wear. Some cams on two-cycle engines are integral with the crankshaft. If the cam is badly worn, the crankshaft must be replaced. On other engines, the cam is a separate collar locked to the crankshaft by a key. On these, a worn cam can be replaced.

Check for a leaky crankshaft seal. It will allow oil to get on the breaker points so that they burn rapidly. A defective seal should be replaced.

After removing the old points, install the new points. Then check the point opening and adjust it as necessary. Finally, check the ignition timing.

Adjusting Points To adjust the breaker points, first make sure that they are properly aligned. Figure 15-22 shows proper point alignment. Usually, the points are properly aligned and no adjustment is required. However, if new points are misaligned, adjust them by slightly bending the bracket supporting the stationary point.

To adjust the point opening, turn the crankshaft in the direction of normal rotation until the cam opens the points to the widest position. Then use the specified thickness gauge to measure the gap between the points (Fig. 15-23). Make sure the thickness gauge is clean so you do not get oil or dirt on the points.

Adjustments are made in different ways according to the method of point attachment. On some magnetos, the

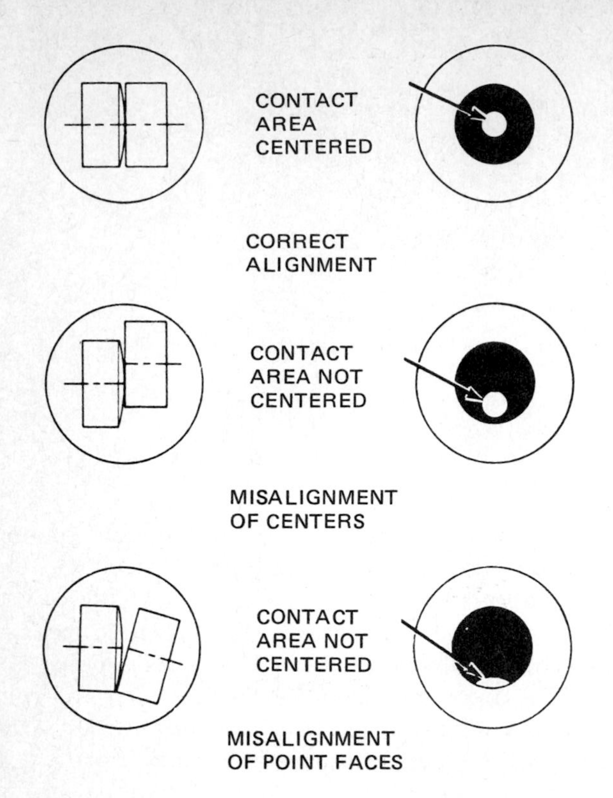

FIG. 15-22 Correct and incorrect alignment of breaker points.

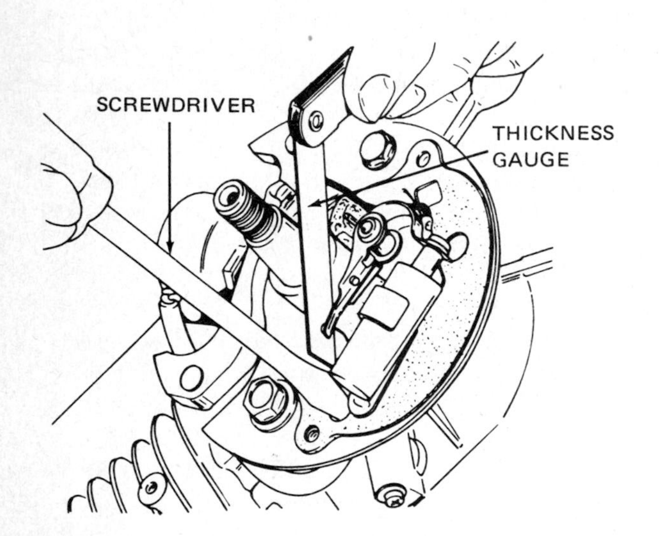

FIG. 15-23 Adjusting the gap between the breaker points in a flywheel magneto. *(Tecumseh Products Company)*

adjustment is made as shown in Fig. 15-23. The lock screw holding the bracket on which the stationary point is mounted is loosened slightly. Then a screwdriver is inserted into the slot and twisted to move the stationary point the correct amount to get the proper point opening. Then the lock screw is tightened.

Another point arrangement is shown in Fig. 15-24. With this arrangement, the stationary point is mounted on the end of the condenser. To make the adjustment, the condenser-clamp screw is loosened slightly. Then the condenser is shifted one way or the other with the screwdriver to get the proper point opening.

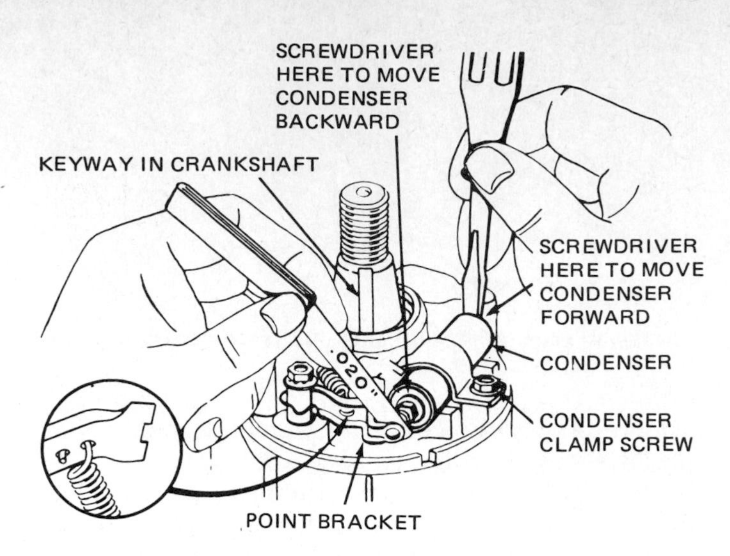

FIG. 15-24 Shifting the condenser one way or the other to adjust the opening between the breaker points. *(Briggs & Stratton Corporation)*

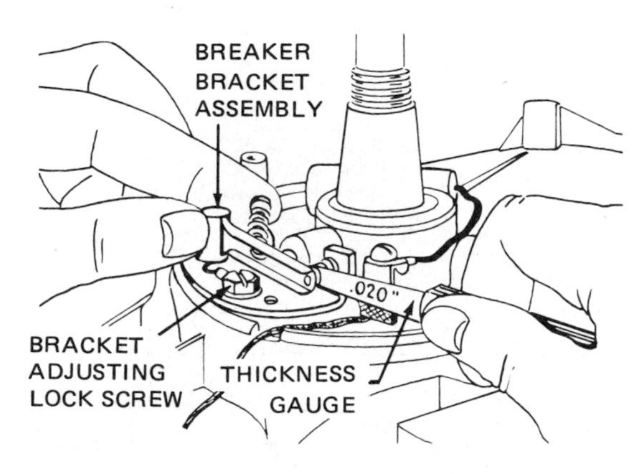

FIG. 15-25 Adjusting the point opening by loosening the lock screw and shifting the bracket on which the stationary point is mounted.

In the arrangement shown in Fig. 15-25, the stationary point is mounted on a bracket. The lock screw must be loosened to shift the bracket as necessary to get the proper point opening. Then the lock screw is tightened.

NOTE A kit can be installed in many engines to convert from breaker-point to solid-state electronic ignition (Fig. 14-31). This eliminates all future problems with breaker points. The electronic module should last as long as the engine, without requiring any service.

Δ15-8 TIMING THE IGNITION

To time the ignition means to make an adjustment that will cause the spark to occur at exactly the right time before the piston reaches TDC on compression. This starts the ignition process at the correct moment so that maximum power will be delivered on the power stroke.

If the timing is early, the engine will backfire while cranking and detonate while running. If the timing is late, the power stroke will be weak. Ignition will not start

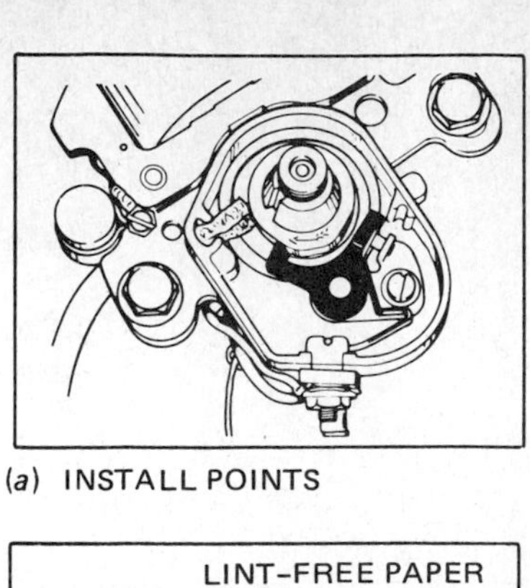

(*a*) INSTALL POINTS

POINT-ALIGNING TOOL

(*b*) ALIGN POINTS

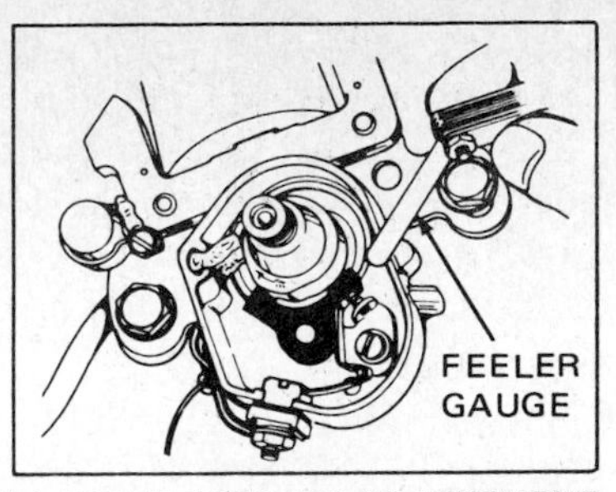

(*c*) POINT OPENING ADJUSTMENT

(*d*) CLEAN POINTS

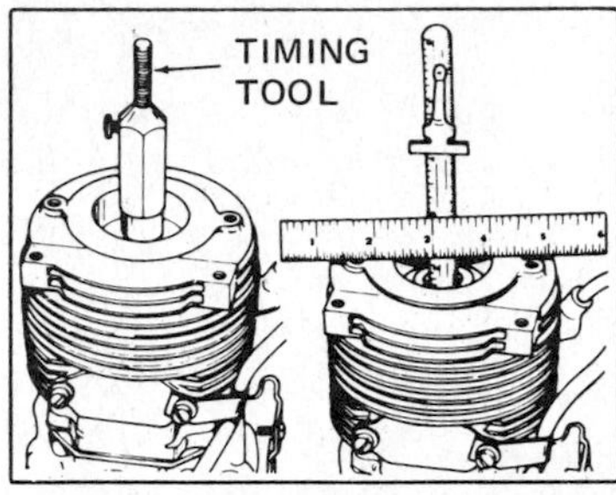

(*e*) INSTALL TIMING TOOL OR RULE

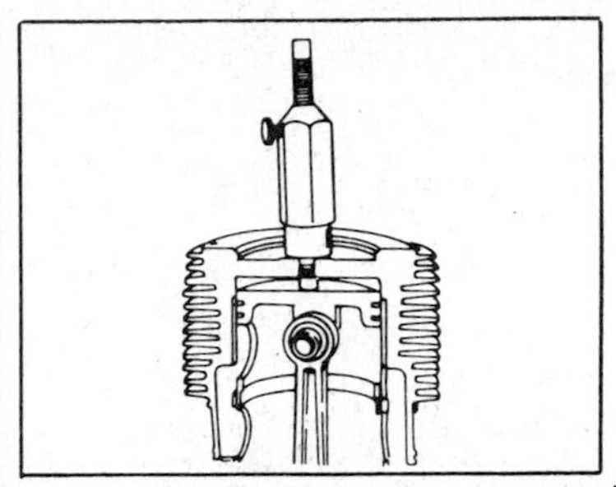

(*f*) FIND TDC (TOP DEAD CENTER)

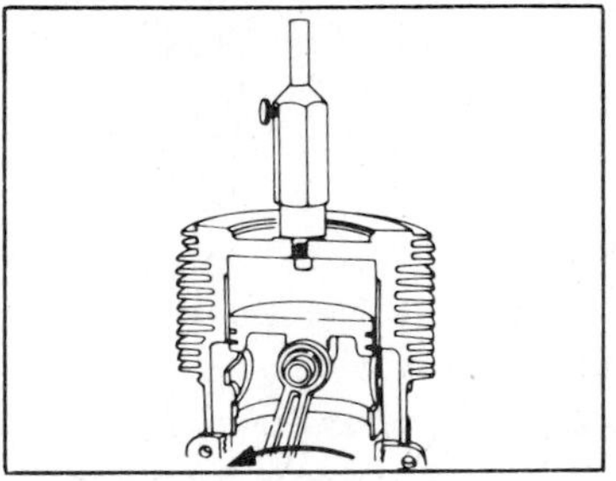

(*g*) BACK OFF ROTATION (OPPOSITE NORMAL RUNNING ROTATION)

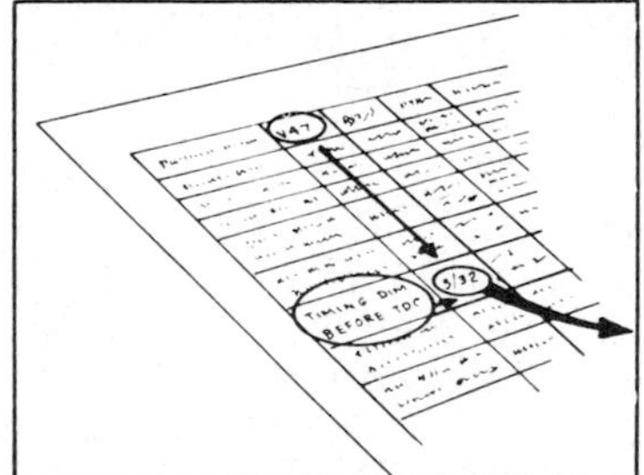

(*h*) FIND BTDC (TIMING DIMENSION SPECIFICATIONS)

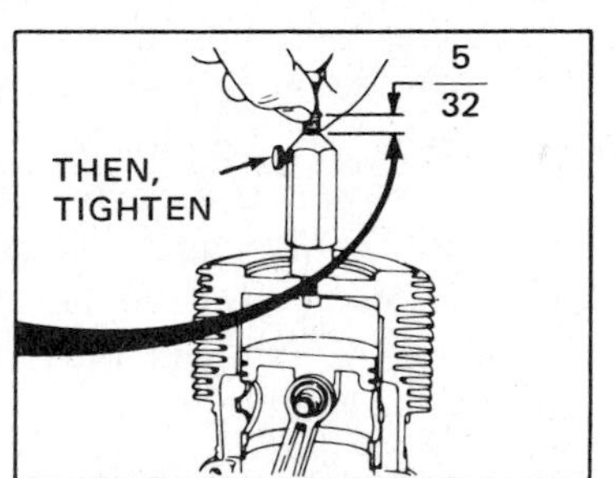

(*i*) APPLY DIMENSION TO TOOL

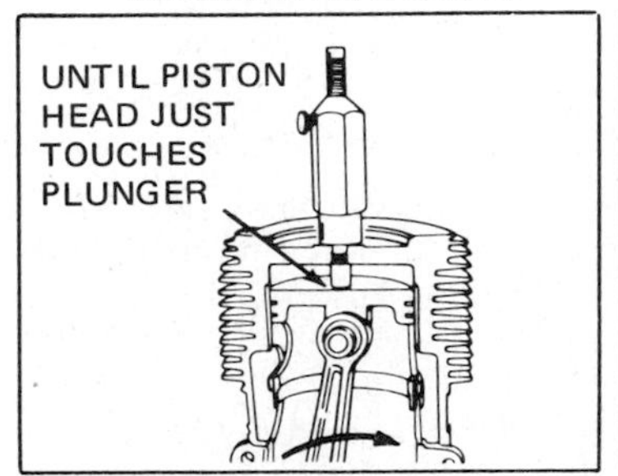

(*j*) BRING UP ON STROKE (NORMAL RUNNING ROTATION)

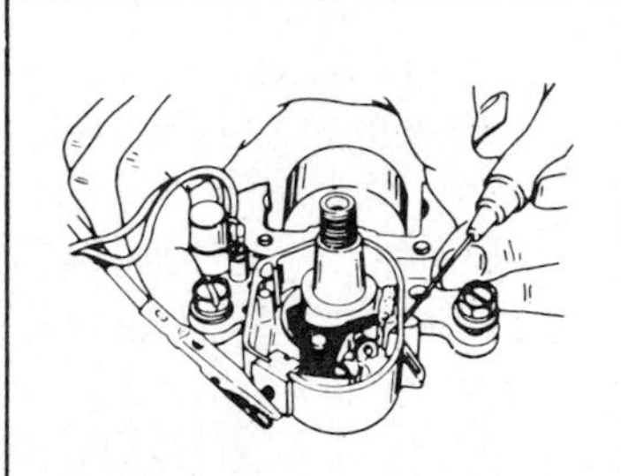

(*k*) INSTALL TIMING LIGHT (OR USE CELLOPHANE)

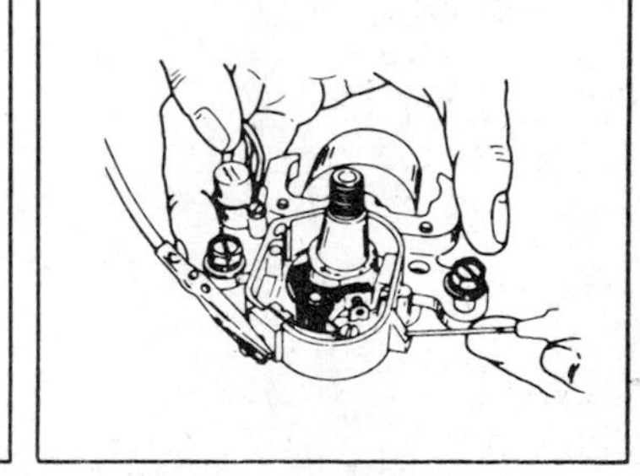

(*l*) ROTATE STATOR UNTIL POINTS JUST OPEN

FIG. 15-26 Steps in timing a two-cycle engine. *(Tecumseh Products Company)*

until after the piston has begun to move down on the power stroke.

Various methods of timing the ignition are used. On many engines, the ignition can be timed either with the engine not running *(static timing)* or with the engine running. Both procedures are described in the following sections.

Δ15-9 STATIC TIMING

The sequence in Fig. 15-26 shows the static ignition-timing procedure for many small two-cycle engines. After installing the points (*a*), align them using the special tool shown to bend the stationary-point support (*b*). Then measure the point opening and adjust it as required (*c*).

Next, clean the points with lint-free paper (*d*), and use a timing tool or rule to locate the TDC position of the piston, as shown at (*e*) and (*f*). Back off the piston by turning the crankshaft backward (*g*).

Find the timing dimensions in the manufacturer's specifications, and adjust the tool to that dimension. Then tighten the thumb screw to lock the dimension, as shown in (*h*) and (*i*).

Next, slowly rotate the crankshaft in the normal running direction until the piston touches the bottom of the tool (*j*). Then connect a test light across the breaker points (*k*). With the stator hold-down screws loosened slightly, shift the stator as shown in (*l*) until the points just open. When this occurs, the light goes out. Tighten the hold-down screws.

As an alternative, you can use a strip of thin cellophane between the points. It will fall out as the points separate. After completing the timing, install lead, cover, flywheel, and lower housing.

Some engines have timing marks on the armature mounting bracket and flywheel (Fig. 15-27). To time the ignition on these engines, remove the flywheel, set the points to the proper opening, install the flywheel, and run the nut on finger-tight. Then rotate the flywheel in the running direction until the points are just opening.

Next, take off the flywheel, being careful not to move the crankshaft. Note the positions of the arrows. If they do not align, slightly loosen the mounting screws holding the armature bracket to the engine cylinder. Slip the flywheel back on the crankshaft, using the key to get correct alignment. Install the flywheel nut finger-tight.

Move the armature and bracket assembly to align the arrows. Remove the flywheel and tighten the armature-bracket mounting screws. Install the flywheel, with the key, and tighten the nut to the specified torque. Finally, set the armature air gap to the proper specifications (Fig. 15-28). Loosen the two armature-attaching screws, and move the armature up or down as necessary.

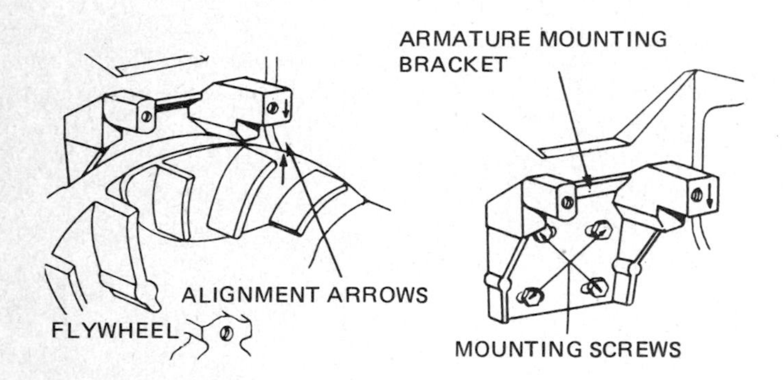

FIG. 15-27 The timing marks or arrows on the armature mounting bracket and flywheel. *(Briggs & Stratton Corporation)*

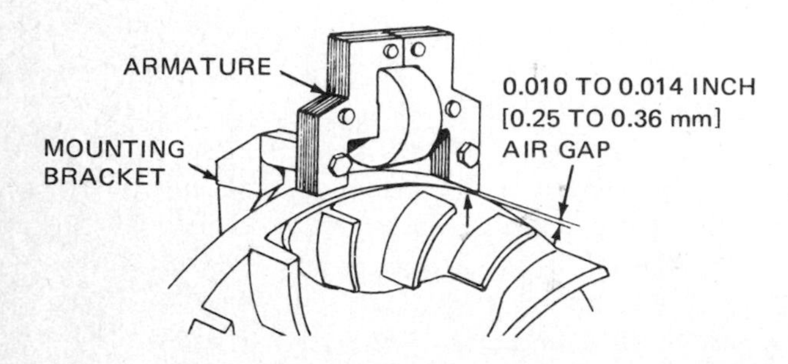

FIG. 15-28 Adjust the armature gap, if necessary, by shifting the armature up or down. *(Briggs & Stratton Corporation)*

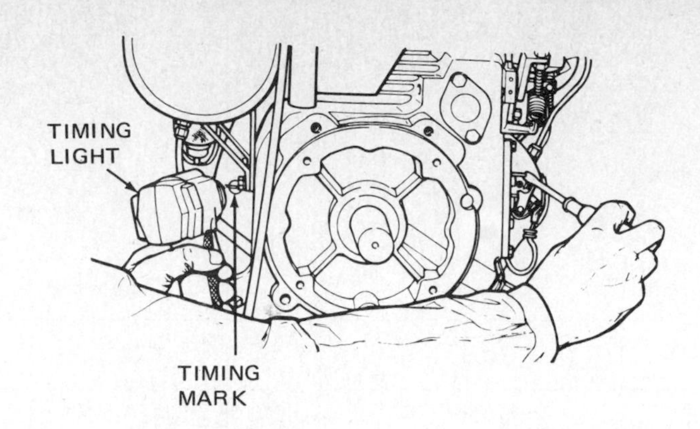

FIG. 15-29 Using a timing light to check the ignition timing with the engine running. *(Kohler Company)*

A simple way to set the proper air gap is to put a postcard between the armature and the flywheel. Set the armature down against the postcard. Then tighten the attaching screws and remove the postcard.

Δ15-10 USING THE TIMING LIGHT

Figure 15-29 shows how to time a running engine by using a timing light. The timing light is connected to the spark-plug cable. Each time the plug fires, a momentary flash of light is produced by the timing light. During the timing operation, the light is directed to a hole in the engine case through which the flywheel can be seen.

The flywheel has a timing mark on it that should align with a mark on the case when the timing is correct. If the marks do not align, loosen the point-opening adjusting screw. Then shift the breaker plate with a screwdriver (Fig. 15-29). When the marks align, tighten the breaker-plate screw.

Ignition timing is checked and set with the enging running at a specified speed. For the engine shown in Fig. 15-29, the specified speed is 1200 to 1800 revolutions per minute (rpm).

Δ15-11 CHECKING COIL AND CONDENSER

In addition to the flywheel and breaker points, the magneto coil and condenser (Fig. 15-30) should also be checked. The condenser can be checked for capacity, insulation resistance, shorts or grounds, and series resistance. These tests are made with the condenser removed from the ignition system or with the condenser lead disconnected. A condenser tester is used to check the condenser. Replace the condenser if it is defective or if its condition is questionable.

Inspect the coil for cracks in the insulation or other damage. Check that the electric leads are intact, especially where they enter the coil. Check the operation of the coil on a coil tester. Replace the coil if it fails to meet specifications. If the laminations in the armature are damaged, replace the armature. Some coils are permanently attached to the laminations. On these, replace the assembly.

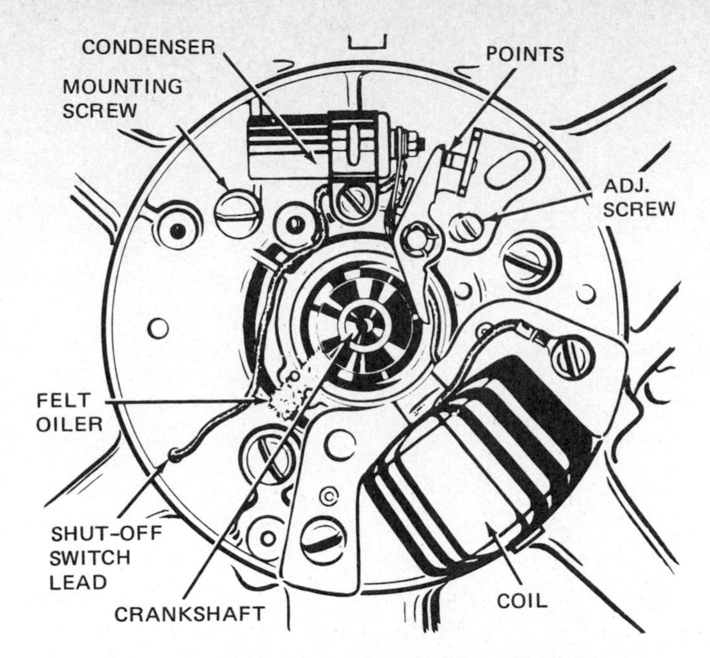

FIG. 15-30 Top view of the magneto, showing the breaker points, coil, and condenser. *(Lawn Boy Division of Outboard Marine Corporation)*

CHAPTER 15: REVIEW QUESTIONS

Select the *one* correct, best, or most probable answer to each question. You can find the answers in the section indicated at the right of each question.

1. The spark test (**Δ*15-2***)
   a. checks the fuel system
   b. starts the engine
   c. checks the ignition system
   d. stops the engine
2. The reason for disconnecting and grounding the spark-plug wire when working on an engine is to
   a. make sure the engine starts (**Δ*15-5***)
   b. make sure the engine does not start
   c. check the ignition system
   d. check the magnets
3. To adjust the spark-plug gap (**Δ*15-4***)
   a. bend the center electrode
   b. move the center electrode down
   c. replace the outer electrode
   d. bend the ground electrode
4. A quick check of the strength of the magnets in the flywheel can be made with (**Δ*15-6***)
   a. a solenoid
   b. a battery
   c. a screwdriver
   d. a spark tester
5. Breaker points can be cleaned with (**Δ*15-7***)
   a. an ignition file
   b. sandpaper
   c. emery cloth
   d. solvent
6. New points are properly aligned by (**Δ*15-7***)
   a. bending the movable point
   b. bending the stationary point
   c. replacing the cam
   d. replacing the crankshaft
7. Ignition static timing is performed with (**Δ*15-8***)
   a. the engine running
   b. a timing light
   c. the engine not running
   d. none of the above
8. When using a timing light, the timing is checked and set (**Δ*15-10***)
   a. with the engine running
   b. by aligning marks on the case and flywheel
   c. both a and b
   d. neither a nor b
9. The condenser can be checked for (**Δ*15-11***)
   a. capacity
   b. grounds
   c. resistance
   d. all of the above
10. When an engine cranks but will not start, the first step is to (**Δ*15-11***)
    a. check the timing
    b. make a spark test
    c. file the breaker points
    d. replace the condenser

CHAPTER

16

Charging Systems

After studying this chapter, you should be able to:

1. ***Explain why a charging system is needed.***
2. ***Describe the difference between a generator and an alternator.***
3. ***Describe the construction and operation of a generator.***
4. ***Explain how generators are regulated.***
5. ***Describe the construction and operation of the flywheel alternator.***
6. ***Explain the operation of the diode rectifier.***

Δ16-1 PURPOSE OF CHARGING SYSTEM

Most small engines that have electric starting motors use electric energy from a battery to power the starting motor. When an electric current is taken out of the battery, the chemical energy in the battery must be restored or a run-down battery will result.

An engine-driven *alternator* or *generator* does this job. It converts mechanical energy from the engine into electric energy. When this electric energy is sent to the battery, it restores the chemical energy which can be released later as electricity. Batteries and battery service are described in Chap. 11.

Δ16-2 GENERATORS AND ALTERNATORS

A *generator* is a device that converts mechanical energy into electric energy. There are two basic types: the direct-current (dc) generator (Fig. 16-1) and the alternating-current (ac) generator. The dc generator is usually just called a *generator*. The ac generator is called an *alternator*. The components of a charging system using a flywheel alternator are shown in Fig. 14-30.

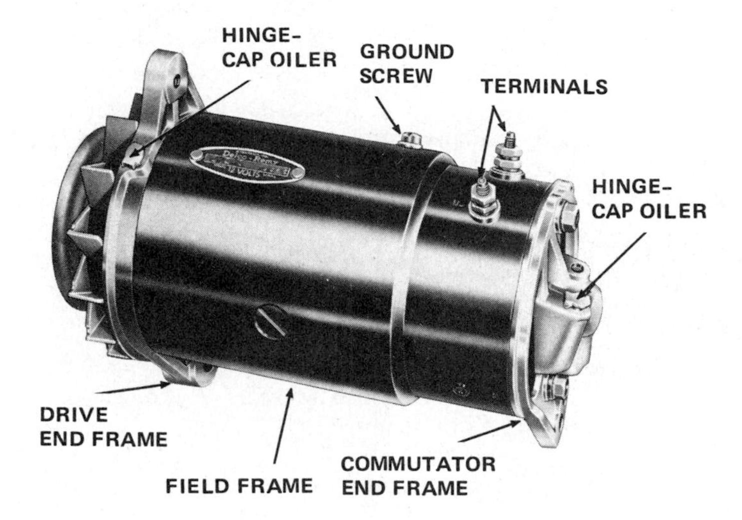

FIG. 16-1 A dc generator. *(Delco-Remy Division of General Motors Corporation)*

Generators and alternators are two different devices that do the same job: producing electric current. In both, alternating current is induced in the conductors. However, the generator uses brushes and a commutator to change the alternating current to direct current. Most alternators use diodes (Δ10-18) as a *rectifier,* which is a device that converts alternating current to direct current. In both charging systems, a *regulator* is often used to prevent overcharging of the battery.

Some generators and alternators are built into the flywheel. Others are separate units, which may be combined with the electric starting motor into the *starter-generator* (Δ12-21). This unit is usually found on engines in the 4- to 15-hp range. It is too big for smaller engines and has insufficient cranking torque for larger engines.

DC GENERATORS

Δ16-3 GENERATOR PRINCIPLES

The generator converts mechanical energy into electric energy by electromagnetic induction (Δ14-4). This is the inducing of a voltage in a conductor when it is moved across a magnetic field.

Mechanical motion is delivered to the generator by the engine. This motion causes conductors to cut through a magnetic field. The field is produced by the *field windings* in the field-frame assembly (Fig. 16-2). The cutting of the magnetic field causes electric current to flow in the conductors. They are assembled into a rotating assembly called the *armature* (Fig. 16-2).

In the armature, the conductors are connected to copper bars in the *commutator*. As the armature rotates, the current induced in the conductors flows through the copper bars and *brushes* to the generator terminals (Fig. 16-3). The brushes are blocks of carbon which rest against the commutator to form a continuous electric circuit. Each brush is usually connected to the circuit by a short lead which attaches to the back end of the brush. The brush lead is then connected to ground, or to one end of the field winding (Fig. 16-3).

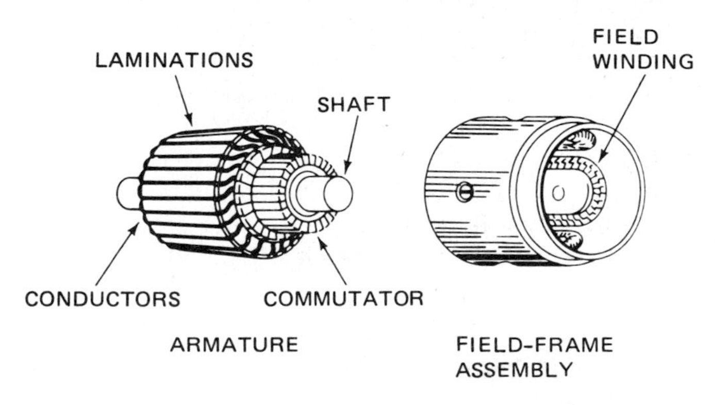

FIG. 16-2 Generator armature and field-frame assembly.

Δ16-4 DRIVING THE GENERATOR

The generator or starter-generator is usually driven by a V-belt from a pulley on the engine crankshaft (Fig. 12-28). On liquid-cooled engines, the belt may also drive the water pump and fan (Chap. 9).

Δ16-5 GENERATOR REGULATORS

As engine speed increases, the speed of the armature in the generator also increases. This increases the voltage and current produced by the generator. Excessive output would damage the generator and other equipment in the electric system. To prevent this, the charging system includes some type of *generator regulator* (Fig. 16-4).

The regulator used with the generator and the starter-generator usually has two units (Fig. 16-5). These are a *cutout relay* (Δ16-7) and a *current-voltage regulator* (Δ16-6). Three unit regulators are used in dc charging systems with heavier electric loads. These have a separate voltage regulator and current regulator.

Δ16-6 CURRENT-VOLTAGE REGULATOR

The current-voltage regulator has a pair of contact points in parallel with a resistance (Fig. 16-6). When the current or voltage starts to go too high, the points open. This puts the resistence in series with the field windings, thereby cutting down the field current and magnetism. As a result, any further rise in generator voltage and current output is prevented.

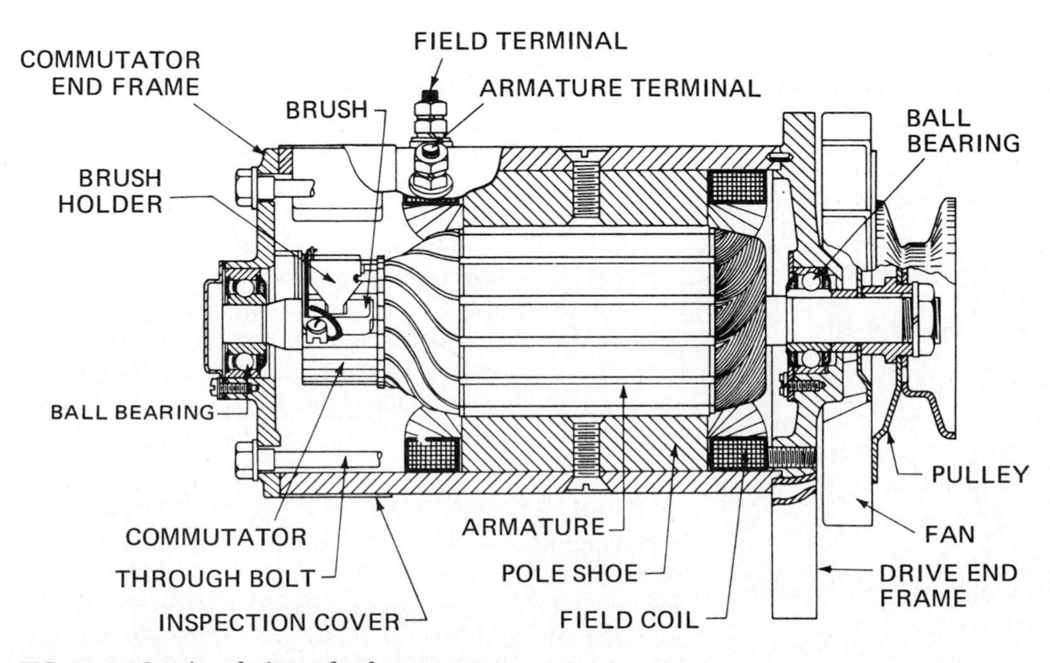

FIG. 16-3 Sectional view of a dc generator.

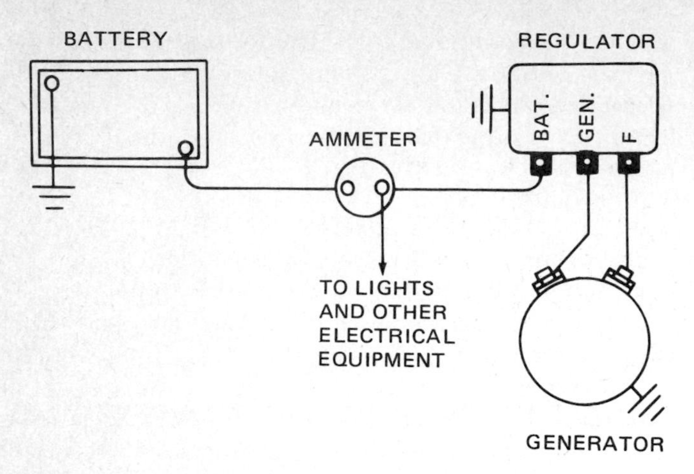

FIG. 16-4 A dc-charging-circuit wiring diagram.

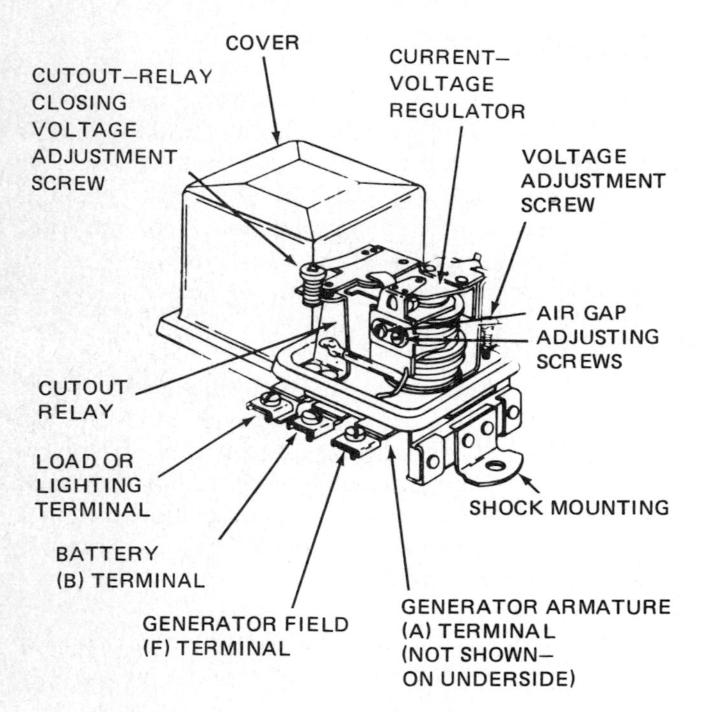

FIG. 16-5 Current-voltage regulator for a dc generator, with the cover removed. *(Kohler Company)*

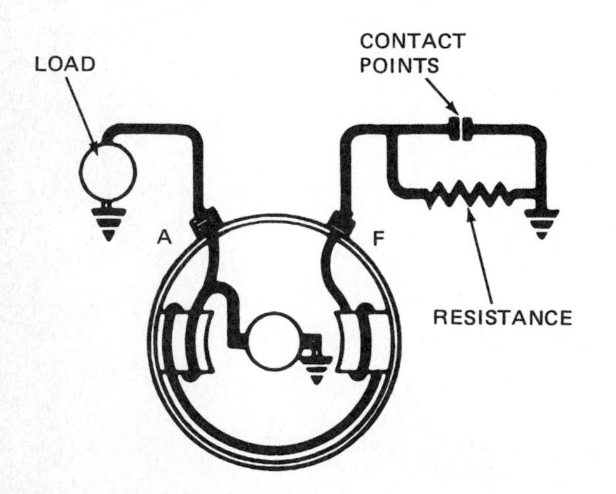

FIG. 16-6 Wiring diagram for a dc generator with an externally grounded field circuit. *(Delco-Remy Division of General Motors Corporation)*

Δ16-7 CUTOUT RELAY Figure 16-7 shows the schematic wiring diagram for a charging system using a dc generator and a two-unit regulator. The cutout relay is shown with its points open. This indicates that the generator is either not turning or turning too slowly to provide a charging voltage.

The cutout relay closes the circuit to the battery when the generator is turning fast enough to produce a charging voltage. This allows the generator to charge the battery. When the engine slows, the charging voltage falls. As soon as the voltage drops so low that a reverse current begins to flow from the battery through the generator, the cutout relay opens the circuit. This prevents the battery from discharging through the generator.

FLYWHEEL ALTERNATORS

Δ16-8 ALTERNATOR PRINCIPLES In the alternator, the conductors are stationary, and a magnetic field is moved through them. Alternating current is induced in the conductors as the north and south poles of the magnet move past (Fig. 16-8).

An alternator has no commutator (**Δ**16-2) to change the alternating current to direct current. Therefore, the electric load must use ac, or some type of *rectifier* is required. In addition, alternators used for charging the battery or operating other dc electric equipment usually have a *regulator*. It prevents excessively high voltage, which could cause damage and overcharge the battery.

Δ16-9 CHARGING COIL Small engines with a 12-volt electric starting motor (Fig. 16-9) must have some arrangement to recharge the battery. One method is to mount an alternator *charging coil* on the engine. As the flywheel rotates, magnets in it pass the charging coil, inducing an alternating current in it. These are the same magnets that are part of the ignition system. A single charging coil is shown in Fig. 16-9. However, the more coils that an alternator has, the higher the current that it can produce.

The current induced in the charging coil is changed to direct current (rectified) by a diode located in the wiring harness. No voltage regulator is used. Therefore, voltage varies with engine speed. On one engine, the dc voltage ranges from 8.0 volts at 2500 rpm to 11.5 volts at 3600 rpm. The ac voltage at the diode ranges from 18.0 volts at 2500 rpm to 26 volts at 3600 rpm. Maximum current is 3 amperes.

Δ16-10 FLYWHEEL ALTERNATOR The construction and operation of the flywheel magneto was covered in Chap. 14. To make it into a flywheel alternator (also called a *flywheel generator*), only the addition of a separate charging coil (or stator assembly) is required.

The basic parts of a typical flywheel alternator are shown in Fig. 14-30. They include a permanent-field

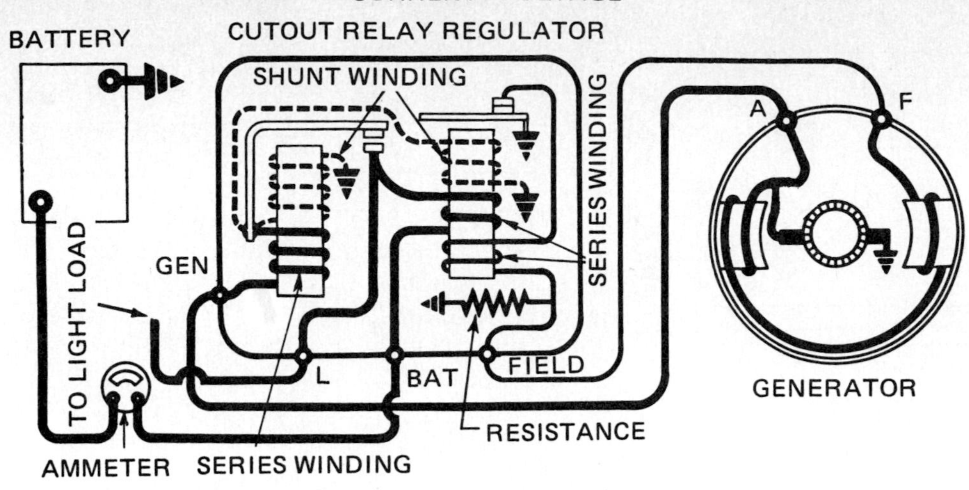

FIG. 16-7 Wiring diagram for a charging system using a dc generator and a two-unit regulator.

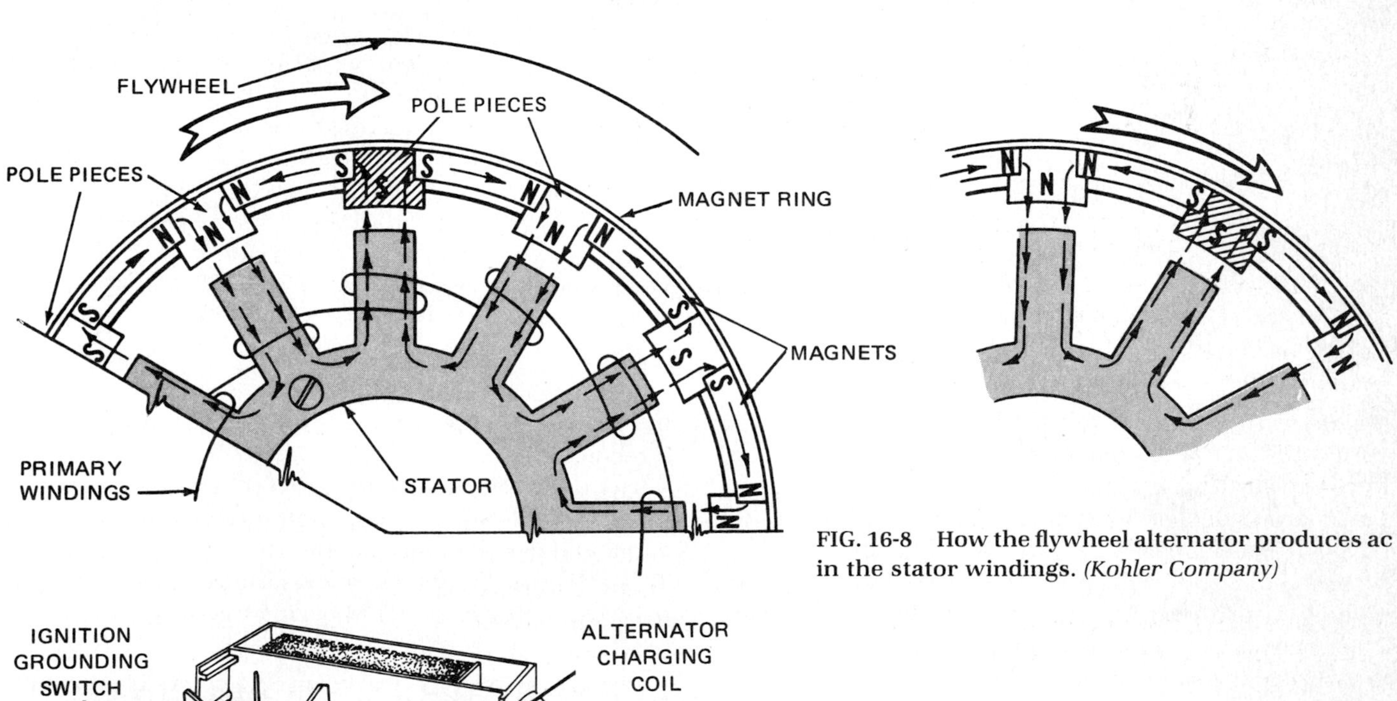

FIG. 16-8 How the flywheel alternator produces ac in the stator windings. *(Kohler Company)*

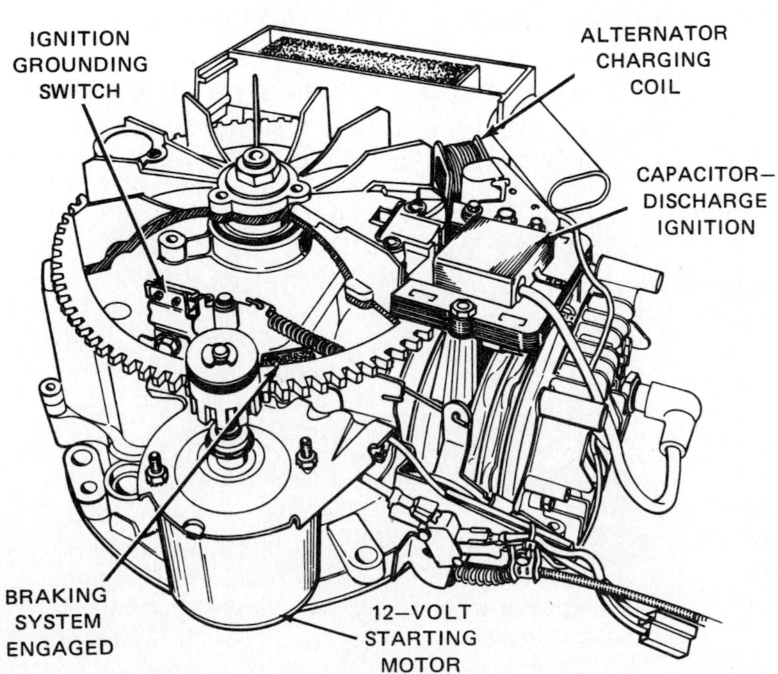

FIG. 16-9 A charging coil to charge the battery on an engine with an electric starting motor. *(Tecumseh Products Company)*

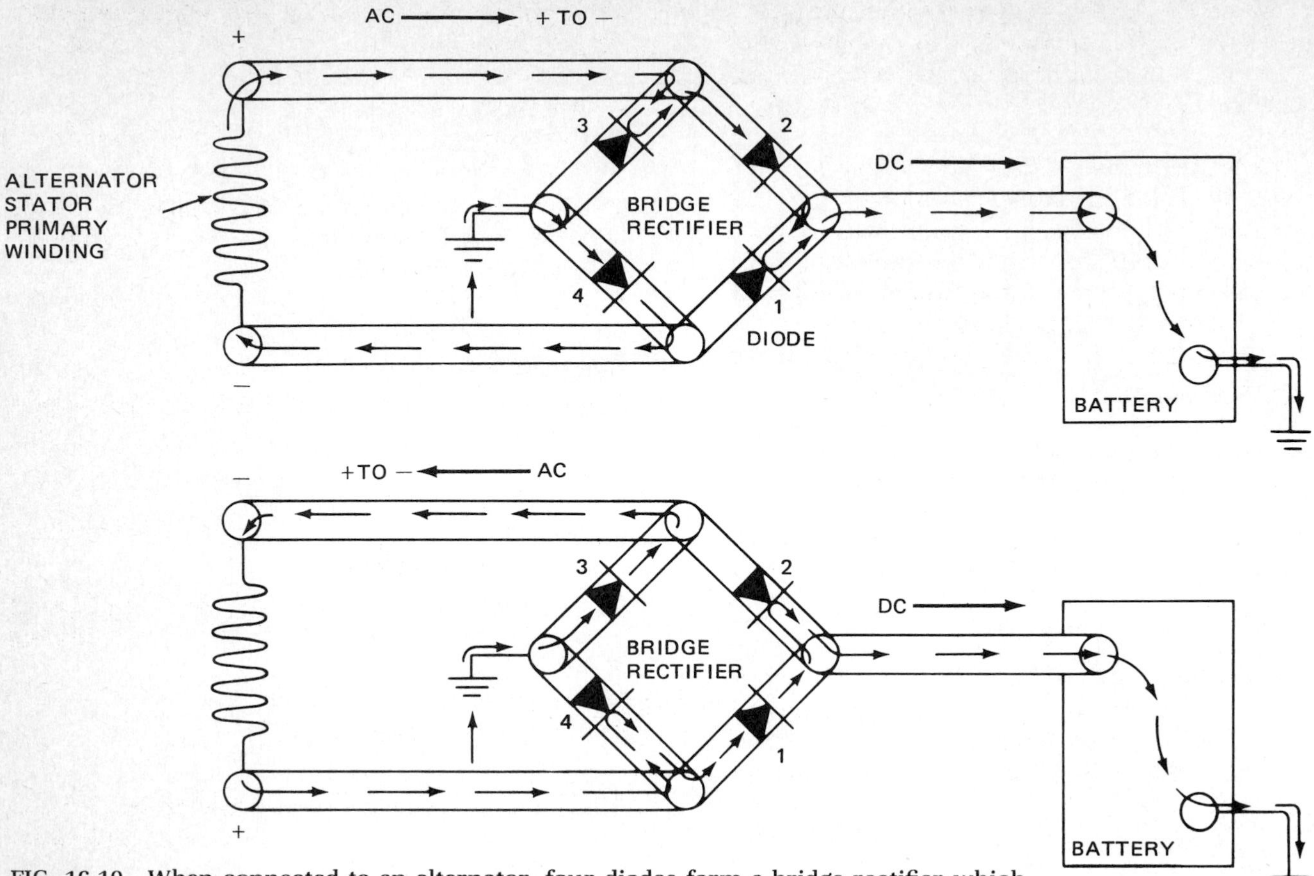

FIG. 16-10 When connected to an alternator, four diodes form a bridge rectifier which rectifies the alternating current and changes it to direct current to charge the battery. *(Kohler Company)*

magnet ring (usually on the flywheel), an alternator stator assembly, and a rectifier-regulator. However, many flywheel-alternator systems do not have a regulator. These are called *nonregulated systems.* Others do not have a rectifier.

Figure 16-8 shows the operation of the flywheel alternator. The magnet ring mounted on the flywheel is made up of permanent magnets. When the magnets spin past the coils on the *stator* (an assembly of *sta*tionary conduc*tors*), an alternating current is induced in the windings. This current may be used to operate ac light bulbs. But for charging the battery, the alternating current must be rectified (**Δ**16-11).

Δ16-11 DIODE RECTIFIERS

The charging coil has a diode in the wiring harness (**Δ**16-9). However, the diode allows only current from one-half the voltage alternations to pass through. To improve the charging-system efficiency, four diodes can be connected to form a *bridge rectifier.* It changes the positive and the negative alternations into a continuous flow of direct current.

Figure 16-10 shows how a bridge rectifier using four diodes — numbered 1 to 4 — changes alternating current to direct current. In the top illustration, current from the stator primary winding flows through diodes 2 and 4 and on to the battery. However, diodes 1 and 3 do not allow the current to pass through. It is flowing in the wrong direction for them.

The lower illustration in Fig. 16-10 shows the actions when the direction of the current reverses. Now diodes 1 and 3 will pass the current. But diodes 2 and 4 will not. This action of the diodes allows the rectifier to continually change ac to dc for charging the battery.

Δ16-12 DUAL-CIRCUIT ALTERNATORS

The dual-circuit alternator is basically two separate alternator systems (Fig. 16-11). One is a nonregulated 5-am-

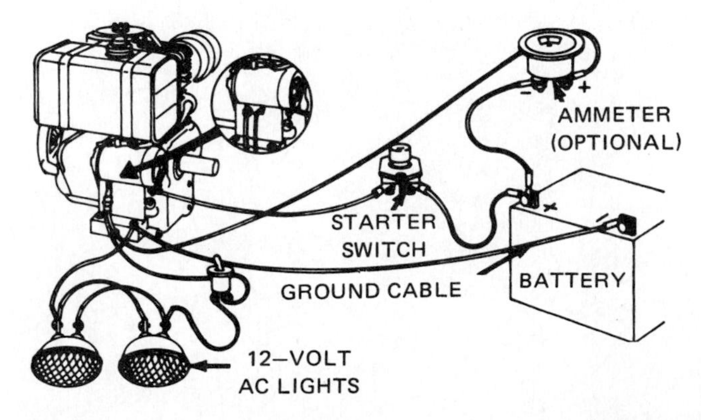

FIG. 16-11 Wiring diagram for a dual-circuit alternator. *(Briggs & Stratton Corporation)*

pere ac circuit. It is used for operating two 12-volt ac headlights. The other circuit provides up to 3 amperes of rectified direct current for charging the battery. However, the alternator is completely separate from the engine starting system.

A single ring of magnets inside the flywheel supplies the magnetic field for both alternators. Since the two are electrically independent, use of the lights does not reduce the charge current flowing to the battery. Also, the lights work as long as the engine is running, even if the battery is disconnected or removed. However, the brightness of the lights changes with engine speed. The dual-circuit alternator uses less than 0.2 hp.

CHAPTER 16: REVIEW QUESTIONS

Select the *one* correct, best, or most probable answer to each question. You can find the answers in the section indicated at the right of each question.

1. In the alternator (**Δ*16-8***)
   a. conductors move through the magnetic field
   b. the magnetic field is stationary
   c. both the conductors and magnetic field are stationary
   d. the magnetic field moves through stationary conductors
2. In the dc generator (**Δ*16-3***)
   a. conductors move through a magnetic field
   b. the magnetic field is in motion
   c. both the conductors and magnetic field are stationary
   d. the magnetic field moves through stationary conductors
3. Generators are usually driven by (**Δ*16-4***)
   a. a set of gears
   b. a belt from the crankshaft pulley
   c. a chain from the crankshaft sprocket
   d. the camshaft
4. Most generators are regulated by (**Δ*16-6***)
   a. controlling the speed
   b. adding resistance in the charging circuit
   c. adding resistance in the field circuit
   d. removing conductors from the armature
5. The regulator used with the starter-generator usually has (**Δ*16-5***)
   a. one unit
   b. two units
   c. three units
   d. four units
6. To change alternating current produced by the alternator to direct current (**Δ*16-11***)
   a. transistors are used
   b. diodes are used
   c. a battery is used
   d. a generator is used
7. In the flywheel alternator, the magnetic field is produced by (**Δ*16-8***)
   a. an electromagnet
   b. the charging coil
   c. the stator
   d. permanent magnets
8. In a nonregulated ac charging system, the highest voltage is (**Δ*16-9***)
   a. the ac voltage
   b. the dc voltage
   c. the battery voltage
   d. the cell voltage
9. When four diodes are used as a bridge rectifier, the battery receives (**Δ*16-11***)
   a. no charging current
   b. one-half of the charging current
   c. all of the charging current
   d. none of the above
10. When a dual-circuit alternator is used on an engine in a riding mower, the headlights receive current from (**Δ*16-12***)
    a. the ac circuit
    b. the dc circuit
    c. both the ac and dc circuits
    d. the battery

CHAPTER 17

Charging-Systems Service

After studying this chapter, you should be able to:

1. ***Discuss the service cautions for dc charging systems.***

2. ***List the possible troubles in the dc charging system.***

3. ***Describe how to service the starter-generator.***

4. ***Discuss the service cautions for ac charging systems.***

5. ***Explain how to diagnose and repair troubles in the flywheel alternator.***

GENERATOR SERVICE

Δ17-1 SERVICE CAUTIONS FOR DC CHARGING SYSTEMS Certain service cautions must be followed while working on dc charging systems. This will prevent damage to the charging system and other electric equipment on the engine. The service cautions include:

1. Be sure the correct regulator is being used. The wrong regulator in a charging system will usually lead to trouble. For example, if a regulator has the wrong polarity, the current will flow through the regulator contacts in the wrong direction. The contact points will then wear away rapidly, and defective regulator action will result.
2. Never close the cutout relay or circuit-breaker contact points by hand with the regulator connected to the battery. This may damage the contact points. On many regulators, enough current could flow to cause the points to weld together. This would result in damage to the generator and regulator.
3. The regulator must be at operating temperature and in the normal operating position when electrical tests are made. Some manufacturers recommend use of a special thermometer to check the temperature to make sure that it is correct when the electric settings are checked.
4. Cycle the generator after every voltage adjustment by slowing the engine to idle and then bringing it back to speed before checking the setting. A variable resistance in the generator field circuit will accomplish the same results.
5. Reinstall the regulator cover after every voltage regulator adjustment. Then check the settings.
6. Make the meter connections according to the markings on the regulator terminals (*not* according to the wiring circuits shown in this book).
7. Never connect a radio bypass condenser from the field terminal of the generator or regulator to ground. This will burn the regulator contacts.
8. Repolarize the generator after every check or adjustment (Fig. 17-1).
9. Never operate the generator on open circuit, with the circuit to the battery or other electric units disconnected.

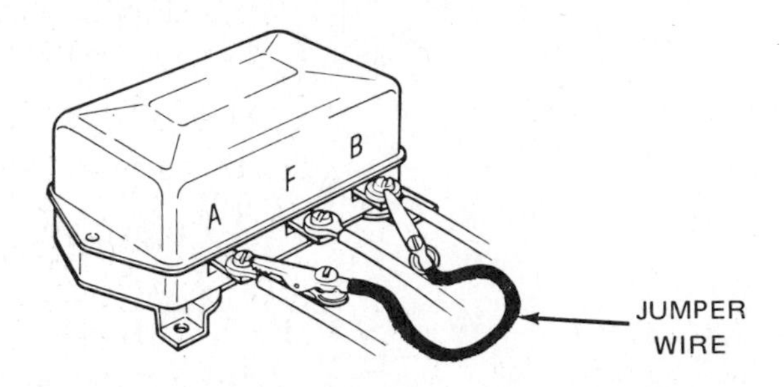

FIG. 17-1 To polarize a generator, momentarily connect a jumper wire between the A and B terminals.

▵17-2 DC-CHARGING-SYSTEM TROUBLE DIAGNOSIS

The typical dc charging system includes a battery, generator, regulator, and connecting wiring (Fig. 16-4). Normally, the regulator limits the generator output so that it produces enough current at the proper voltage to keep the battery fully charged. At the same time, the regulator acts to prevent excessively high voltage which would overcharge the battery.

When the owner complains about a weak battery or generator trouble, tests must be made to determine if there is trouble. If there is, then you must check the battery, generator, regulator, and wiring.

Four general complaints about the dc charging system are:

1. High charging rate
2. Low charging rate
3. No ammeter reading
4. Charging indicator light inoperative

In addition, a generator may become noisy because of mechanical trouble.

The chart below lists the dc-charging-system troubles and the checks or corrections to be made for each condition.

▵17-3 POLARITY OF DC GENERATORS

After any wire has been reconnected to the dc generator, it must be *polarized.* This assures that its output will be in the right direction to charge the battery. Otherwise, output in the reverse direction could damage the generator, regulator, and battery.

To polarize the generator, momentarily connect a jumper wire from the positive battery terminal to the armature terminal of the generator. Another method of making the same connections is to connect the jumper wire between the A and B terminals of the regulator (Fig. 17-1). This is the procedure for most generators. Refer to the manufacturer's service manual on other generators.

CAREFUL Never try to polarize an alternator. This could damage the ac charging system.

▵17-4 GENERATOR SERVICE

Troubles that can be traced to the generator include no output, unsteady or low output, excessive output, and excessive noise. Most generator troubles can be corrected by tightening the drive belt (Fig. 12-28), replacing the brushes, repairing obviously bad connections, and adding lubrication. Other defects, such as reconditioning the commutator, replacing field windings, and replacing bearings, may require complete rebuilding. Sometimes it is more economical to replace a defective generator than to rebuild it.

DC-CHARGING-SYSTEM TROUBLE-DIAGNOSIS CHART

Complaint	Possible cause	Check or correction
1. High charging rate	a. High voltage-regulator setting	Reduce setting
	b. High temperature	Reduce voltage-regulator setting; reduce battery specific gravity
	c. Defective battery	Replace battery
	d. Generator field windings grounded or shorted	Repair or replace generator
	e. Short, ground, or open in regulator	Repair or replace regulator
2. Low charging rate	a. Defective wires or connections	Replace wires; clean and tighten connections
	b. Low voltage-regulator setting	Adjust or replace regulator
	c. Dirty regulator contact points	Clean and replace points or regulator
	d. Open in regulator	Repair or replace generator
	e. Defective generator	Repair or replace generator
	f. Cutout relay not closing	Adjust, repair, or replace regulator
	g. Drive belt loose	Tighten or replace if necessary
3. No ammeter reading	a. Ammeter defective	Replace
	b. No generator output	See item 2
4. Charging indicator-light failure	a. Bulb burned out	Replace bulb
	b. Bad connection	Repair connection
	c. No generator output	See item 2
5. Noisy generator	a. Generator mounting loose	Tighten
	b. Pulley or fan loose	Tighten
	c. Worn brushes or bearings	Repair or replace generator

DC-GENERATOR TROUBLE-DIAGNOSIS CHART

Complaint	Possible cause	Check or correction
1. No output	a. Sticking brushes	Free; replace brushes and springs as necessary
	b. Gummy or dirty commutator	Clean; turn commutator, and undercut mica if needed
	c. Burned commutator	Service commutator, and check the current-regulator setting
	d. Loose connections or broken leads	Tighten or solder connections; replace leads
	e. Grounded armature	Check with test light; repair or replace
	f. Open armature	Repair or replace
	g. Shorted armature	Check on growler; repair or replace
	h. Grounded field	Check with test light; repair or replace
	i. Open field	Check with test light; repair or replace
	j. Shorted field	Check with ammeter; repair or replace
	k. Grounded terminal	Replace insulation or terminal as needed
	l. Broken drive belt	Replace belt
2. Unsteady or low output	a. Loose or worn drive belt	Tighten or replace
	b. Sticking brushes	Free; replace brushes and springs as needed
	c. Low brush-spring tension	Retension or replace springs
	d. Dirty, gummy, or burned commutator	Clean; turn commutator, and undercut mica as needed; check armature for opens
	e. Out-of-round, dirty, worn, or rough commutator	Clean; turn commutator, and undercut mica as needed
	f. Partial short, ground, or open in armature	Repair or replace armature
	g. Partial, short, ground, or open in field	Repair or replace fields
3. Excessive output	a. Grounded field circuit—type with field grounded in regulator	Check with test light; repair or replace
	b. Shorted field circuit—type with field grounded in generator	Test with ammeter; repair or replace
4. Noisy generator	a. Loose mounting or pulley	Tighten
	b. Worn or dirty bearings	Clean or replace
	c. Improperly seated brushes	Seat brushes properly

The chart above lists possible troubles in a dc generator, along with the checks and corrections to be made for each condition. Refer to this chart when you have a specific generator problem.

STARTER-GENERATOR SERVICE

Δ17-5 SERVICING THE STARTER-GENERATOR

The operation of the starter-generator (Fig. 17-2) is described in Δ12-21. Trouble-diagnosis and servicing procedures are similar to those for the dc generator and charging system described earlier. However, because the starter-generator serves a dual function, it should be checked at more frequent intervals.

Following are recommended service procedures for the starter-generator.

Brushes Brushes should be checked after every 200 hours of engine operation and replaced if worn. On the unit shown in Fig. 17-2, this requires removal of the two through bolts and the commutator end frame. Some starter-generators have windows in the brush end of the frame. These windows allow inspection of the brushes after the cover band is removed.

The brushes should be in good condition and make total face contact with the commutator. If they are worn down to less than one-half of their original length, replace them. After new brushes are installed, brush-spring tension should be checked with a spring scale. If the tension is not correct, or if the springs are blue or burned, replace them.

Commutator The commutator should be examined for glaze or dirt. It can be cleaned with #00 sandpaper (Fig. 17-3). If possible, place the armature in a lathe. Then hold the sandpaper against the commutator while it is rotating. On starter-generators with a cover band or windows in the end frame, stick the sandpaper through the window (Fig. 13-33) while the engine is driving the starter-generator. If the commutator is rough, is out-of-round, or

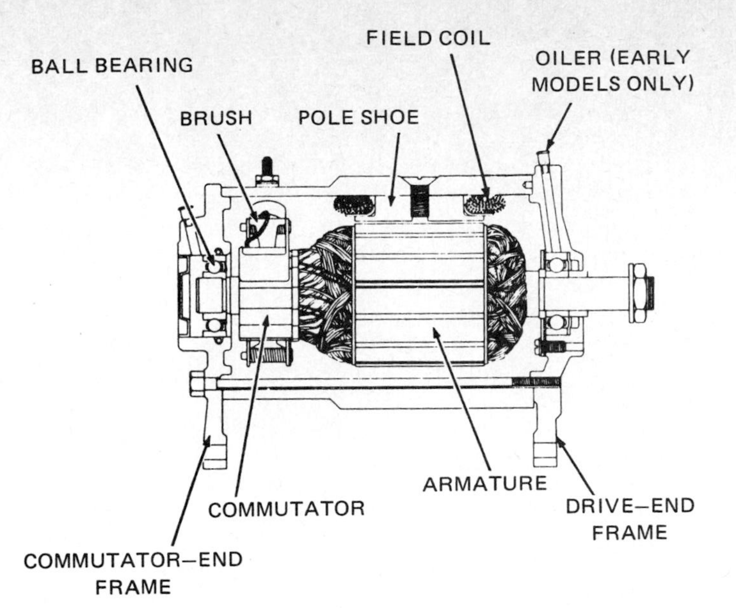

FIG. 17-2 Sectional view of a starter-generator. *(Kohler Company)*

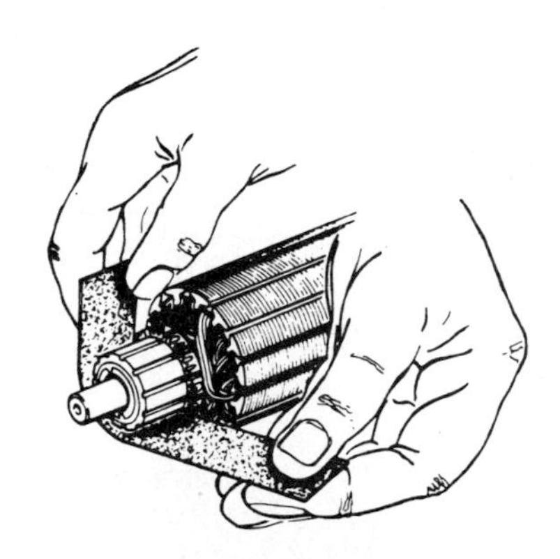

FIG. 17-3 A dirty commutator can be cleaned with sandpaper. *(Tecumseh Products Company)*

has high mica, the commutator should be turned down in a lathe (Fig. 13-36). Then the mica should be undercut (Fig. 17-4).

Never use emery cloth to clean the commutator. Particles of emery can become embedded in the commutator and cause rapid brush wear.

Drive Belt Check that the drive belt is in good condition and adjusted to the proper tension (Fig. 13-30). Low belt tension allows the belt to slip so that poor cranking and low generator output result. Belt slippage causes rapid belt wear. Various conditions of the belt that require replacement are shown in Fig. 9-25.

On some engines, belt tension (or *deflection*) is correct if you can push the belt in ½ inch [13 mm] halfway between the pulleys. Adjust the belt by loosening the starter-generator lock bolt, and swing the unit in or out to get the correct tension. Then tighten the bolt.

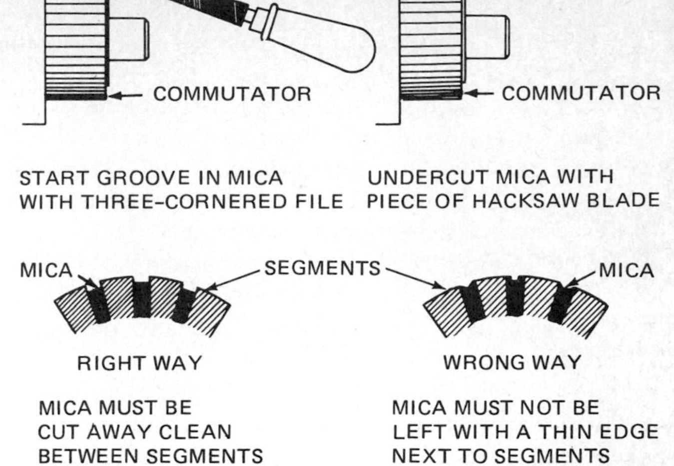

FIG. 17-4 Undercutting the mica on the commutator.

Lubrication Some starter-generators support the armature on sealed ball bearings which never require lubrication. Others should be lubricated periodically as specified by the manufacturer.

△17-6 STARTER-GENERATOR REGULATOR SERVICE

Figure 16-5 shows a regulator for a starter-generator. This unit has a combined current regulator and voltage regulator. The electric settings of the cutout relay and regulator are adjusted by turning screws. Refer to the manufacturer's service manual for the adjusting procedures and specifications.

FLYWHEEL-ALTERNATOR SERVICE

△17-7 SERVICE CAUTIONS FOR AC CHARGING SYSTEMS

Certain service cautions must be followed while working on alternator-equipped (ac) charging systems. This will prevent damage to the charging system and other electric equipment on the engine. These service cautions include:

1. Keep all electric connections clean and tight, and keep wires intact. High resistance due to poor battery connections will cause excessively high charging rates which will damage the battery and cause it to use too much water.
2. All regulator checks and adjustments must be made under specified conditions.
3. Do not connect the battery backward. Reversed battery connections may damage the rectifier, wiring, or other components of the charging system.
4. If booster batteries are used for starting the engine, they must be connected properly to prevent damage to the electric system.

5. Anything that affects the battery or regulator affects voltage regulation.
6. Do not use batteries of higher-than-system voltage either for boosting a battery of lower voltage or for starting. For example, do not use a 12-volt battery to jump-start an engine that uses a 6-volt battery.
7. Alternators must not be operated on open circuit with the field winding energized. High voltages will result, causing possible rectifier failure. Make sure all connections are tight.
8. Do not attempt to polarize the alternator. No polarization is required. Any attempt to polarize may result in damage to the alternator, regulator, or electric circuits.
9. The field circuit must not be grounded at any point. Grounding of the field will damage the regulator. If the field lead from the regulator to the alternator is grounded, the battery will charge excessively.
10. Grounding the alternator output terminal may damage the alternator and/or circuit components even when the system is not in operation. A short circuit between the stator leads to the rectifier will show a discharge on the ammeter, and an undercharged battery will result.

Δ17-8 FLYWHEEL-ALTERNATOR SERVICE

A typical flywheel-alternator charging system is shown in Fig. 17-5. The flywheel has a series of magnets which whirl past the stationary coils (the stator coils) located around either the outside or the inside of the flywheel. The flywheel alternator seldom requires service because it has no separate moving parts.

Described below are procedures for checking one type of charging system that includes a rectifier-regulator (Fig. 17-6). Many different types of charging systems using an alternator are found today. For procedures and specifications, refer to the manufacturer's service manual. Incorrect testing procedures can damage equipment.

Two basic conditions require checking of the flywheel-alternator charging system. These are:

1. No charging current to battery
2. Battery overcharging

No Charging Current to Battery If no charging current is going to the battery and the battery is weak or discharged (check battery to be sure), disconnect the wire from the battery positive (B+) terminal. Connect a dc voltmeter from this lead to the case of the rectifier-regulator (Fig. 17-7). Run the engine near full throttle, and read the voltage. If it is above 14 volts, the charging system is satisfactory. The trouble may be in the ammeter or connections in the circuit.

If the voltage is less than 14 volts but greater than zero, there probably is some defect in the rectifier-regulator. You can try another rectifier-regulator to determine if

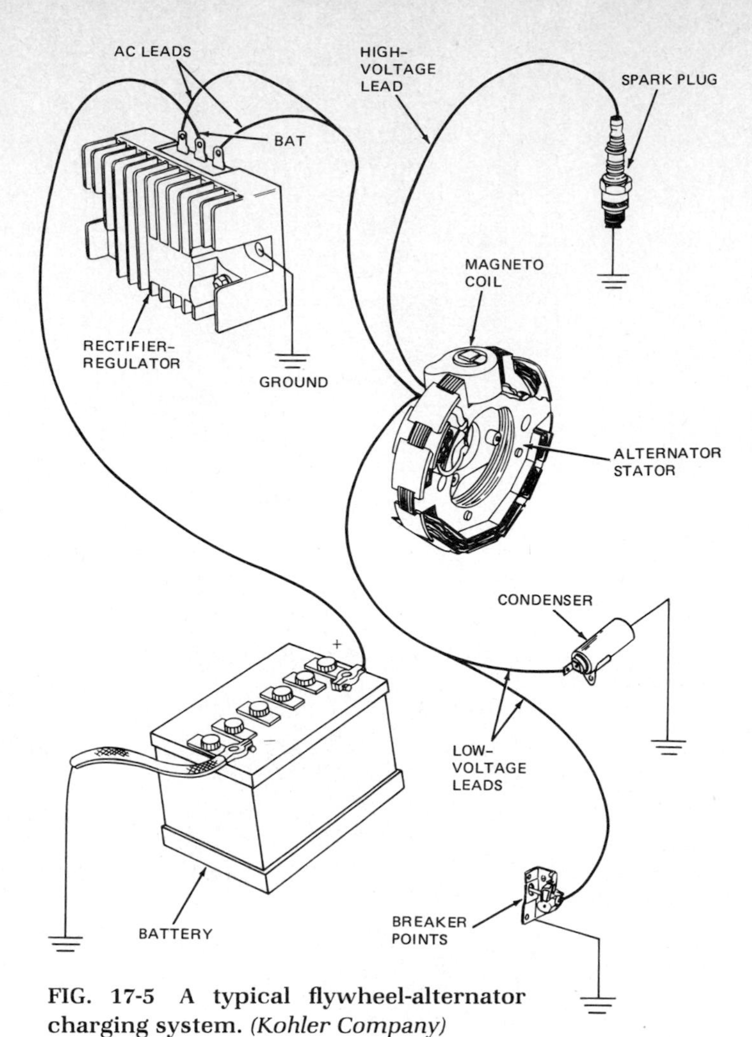

FIG. 17-5 A typical flywheel-alternator charging system. *(Kohler Company)*

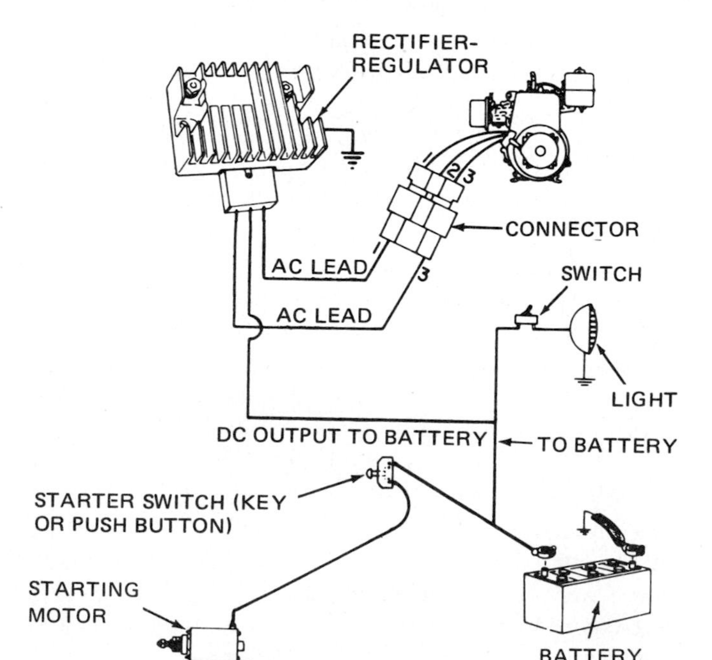

FIG. 17-6 Wiring diagram for a 7-amp flywheel-alternator charging system that includes a rectifier-regulator. *(Tecumseh Products Company)*

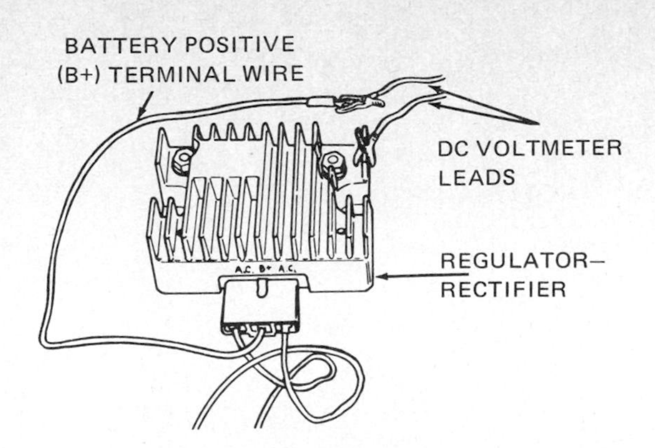

FIG. 17-7 Connecting voltmeter leads to check system voltage. *(Tecumseh Products Company)*

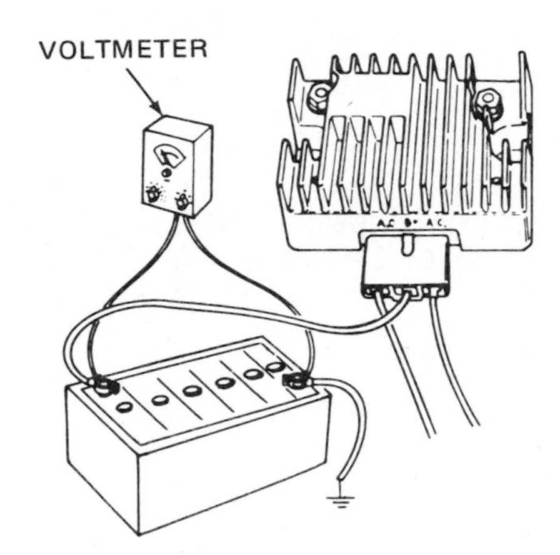

FIG. 17-8 Checking voltage at the battery. *(Tecumseh Products Company)*

this corrects the trouble, or you can check further. If there is no voltage at all, then the trouble can be in either the stator or the rectifier-regulator.

Isolate the trouble by reconnecting the battery. Then check the voltage between the two battery terminals (Fig. 17-8), with the engine operating near full throttle. If the voltage is 13.8 volts or higher, turn on a load such as the lights to reduce the voltage below 13.6 volts. If the charging rate increases, the system is in good condition. If the charging rate does not increase, the stator or rectifier-regulator is at fault and a further check must be made.

If the system has no ammeter, connect a test ammeter into the circuit when making the above test.

Disconnect the plug from the rectifier-regulator. Then test the ac voltage (use an alternating-current voltmeter) as shown in Fig. 17-9, with the engine running near full throttle. If the voltage reads less than 20 volts, the stator is defective. If the voltage reads more than 20 volts, the rectifier-regulator is defective.

If tests indicate the rectifier-regulator is defective, replace it. If tests show the stator is at fault, the trouble probably is due to an open or ground. The coils must be replaced. This requires partial disassembly of the engine. There also is the possibility that the flywheel magnets

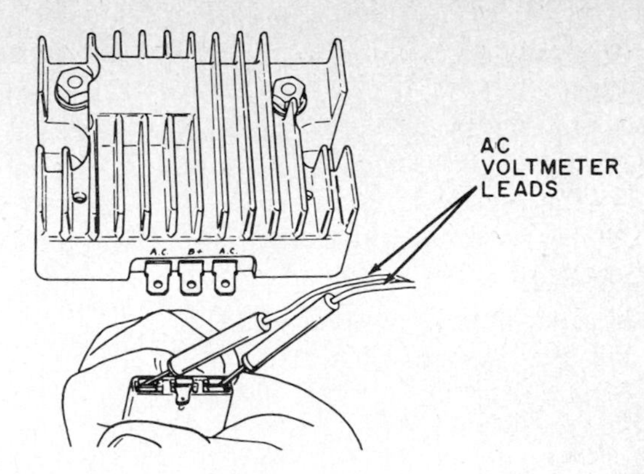

FIG. 17-9 Checking voltage of the stator winding with an ac voltmeter. *(Tecumseh Products Company)*

have weakened. However, this is rare. The magnets can be tested with a screwdriver as shown in Fig. 15-16.

Procedures vary for the disassembly of different engines. Refer to the shop manual for the engine being serviced. General disassembly procedures for small engines are covered in later chapters.

Battery Overcharging This condition is usually caused by a defective rectifier-regulator. You can check with a dc voltmeter connected as shown in Fig. 17-8, while the engine runs at nearly full throttle. If the voltage goes over 14.7 volts, the regulator is defective. Replace the rectifier-regulator.

If the voltage remains under 14.7 volts, the charging system is operating properly. However, there may be some trouble in the battery—such as a shorted cell—which is causing the charging rate to remain high. Battery testing is described in Chap. 11.

CHAPTER 17: REVIEW QUESTIONS

Select the *one* correct, best, or most probable answer to each question. You can find the answers in the section indicated at the right of each question.

1. Technician A says that a dc generator overcharging a battery can be caused by a defective voltage regulator. Technician B says that the trouble can be caused by a defect in the generator field circuit. Who is right? (**Δ**17-2)
   a. technician A
   b. technician B
   c. both A and B
   d. neither A nor B
2. Failure to correctly polarize a generator can cause (**Δ**17-3)
   a. battery overcharge
   b. generator damage
   c. burned-out lights
   d. excessive gasoline usage

3. Most starter-generator regulator adjustments are made by (**Δ***17-6*)
   a. replacing parts
   b. connecting wires
   c. turning screws
   d. replacing the commutator
4. Alternators (**Δ***17-7*)
   a. should be polarized periodically
   b. should never be polarized
   c. are dc units
   d. are high-voltage units
5. The flywheel alternator (**Δ***17-8*)
   a. requires periodic adjustment of the magnets
   b. never requires a rectifier
   c. never requires a regulator
   d. has no separate moving parts

ENGINE SERVICE

Part 5 of this book describes operation, maintenance, trouble diagnosis, and servicing of small spark-ignition and diesel engines. Included in Part 5 are chapters on servicing small two- and four-cycle spark-ignition engines and small four-cycle diesel engines. These include procedures for servicing engine valves and valve trains, pistons and related parts, crankshafts, cylinders, and heads. The servicing procedures discussed are aimed at correcting **specific engine troubles,** *which is one method of engine service.*

Another method of engine service is called **engine rebuilding.** *It involves bringing in old, worn engines and rebuilding them completely. All worn parts are repaired or replaced. Only those old parts that are still in good condition are reused. Still another way to service an engine is to purchase a* **short block** *from the manufacturer. It has the major parts already assembled into it.*

There are five chapters in Part 5. They are:

CHAPTER

18

Operating and Maintaining Spark-Ignition Engines

After studying this chapter, you should be able to:

1. ***Start, operate, and stop a small engine.***
2. ***Perform the daily and regular maintenance procedures required by a small engine.***
3. ***Clean a small engine.***
4. ***Prepare a small engine for winter storage.***
5. ***Perform an engine tuneup.***

Δ18-1 STARTING A SPARK-IGNITION ENGINE If the engine uses a rope-wind or rope-rewind starter, make sure the equipment is level. This reduces the possibility of tipping over the machine when you pull the handle. You can guard against tipping by placing one foot on the machine. Figure 18-1 shows how this is done to start a rotary lawn mower.

CAUTION **Never place your foot under the lawn mower where the rotating blade might hit it. You could be seriously injured. New horizontal-blade lawn mowers have safety brakes that quickly stop the blade when the mower handle is released (Δ12-15). Do not bypass the safety brake by taping down the control.**

Never start and operate the engine in a closed place, such as a garage with the doors closed. The engine can produce enough carbon monoxide in only 3 minutes to kill you!

FIG. 18-1 Make sure that the mower is level and that you have it under control by holding it or by having a foot on it before pulling the handle on the starter rope.

If the equipment has brakes, apply them when starting the engine. If it has a clutch, disengage it if possible, so the machine will not move when the engine starts.

Many engines have a shutoff valve between the fuel tank and the carburetor (Fig. 18-2). If the valve has been closed, open it.

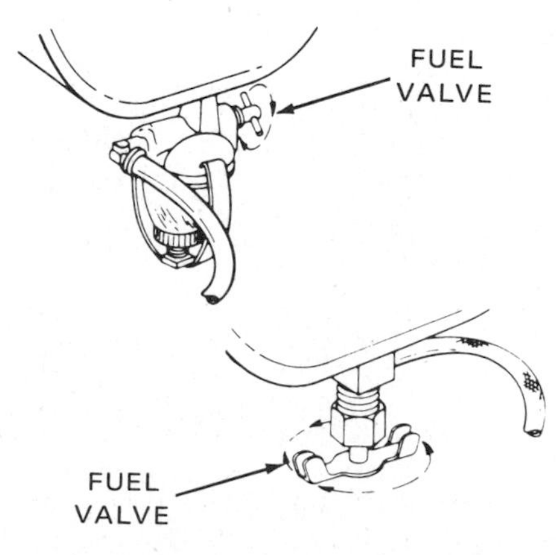

FIG. 18-2 To start the engine, first turn the fuel valve to the ON position. *(Kohler Company)*

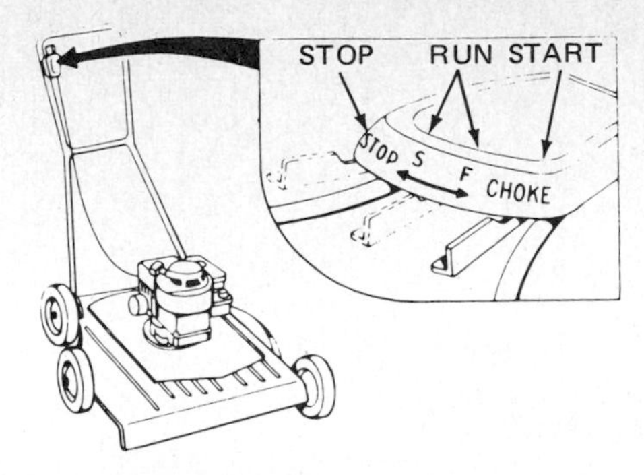

FIG. 18-3 Lawn mower with a single control lever for the choke and throttle. *(Briggs & Stratton Corporation)*

Close the choke valve or prime the engine. Chokes and primers are described in Chaps. 7 and 8. Some manufacturers recommend cranking the engine a few times with the ignition off to allow gasoline to get into the carburetor. Then the engine will start more quickly after the ignition is turned on.

On riding equipment, always operate the controls from the driver's seat. Then if anything goes wrong or the machine suddenly starts moving, you can quickly stop the engine. Never start an engine until you know how to stop it.

Adjust the throttle to the recommended opening for starting. Some engines have a single control for choke and throttle (Fig. 18-3). On these engines, move the lever to the choke position. After the engine starts, move the lever back to the open-throttle position.

Turn on the ignition and crank the engine. If you are using a rope-wind or rope-rewind (recoil) starter, pull the handle until the engine reaches the compression stroke (Fig. 18-4). Then rewind the rope so you can give it a strong, steady pull for starting.

CAUTION **When starting a chain saw (Fig. 18-5), put it on the ground or brace it. You must have full control of the chain saw when you crank it. If you don't, you could lose control of the saw and the blade could severely cut you.**

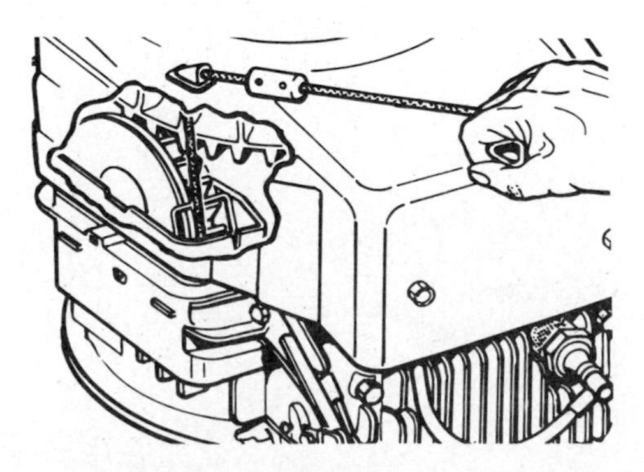

FIG. 18-4 To crank an engine with a rope-wind or rope-rewind (recoil) starter, pull the handle until the engine reaches a compression stroke. Then rewind the rope. *(Tecumseh Products Company)*

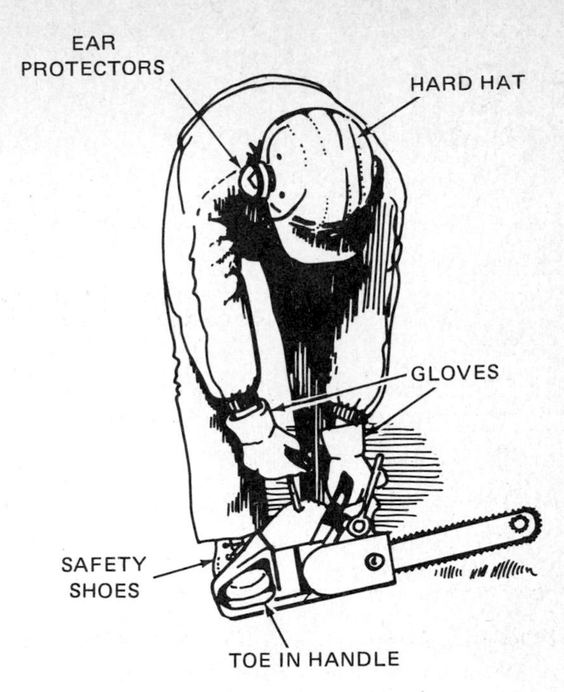

FIG. 18-5 When starting a chain saw, place it on level ground with your foot through the handle to brace it before pulling the starter handle.

Figure 18-6 shows how to start an engine with an electric starting motor. First, close the ignition switch. Then hold the starting switch closed and allow the engine to crank for up to 10 seconds. If the engine does not start, release the starting switch. Then open the choke valve part way and try again. The engine may have flooded (taken in too much gasoline). Avoid cranking for long periods of time. This can damage the starting motor.

After the engine has started, run it at fast idle for a couple minutes. This allows the engine to warm up. Never run a cold engine at high speed or try to take full power from it. The engine must warm up first. As the engine warms up, gradually open the choke.

If the engine does not start promptly, refer to the trouble-diagnosis chart in Chap. 19.

∆18-2 OPERATING A SMALL ENGINE

Overloading and overspeeding are the two most common causes of small-engine trouble and short engine life. Overspeeding the engine by improperly adjusting the governor will shorten engine life. It will also cause the engine to fail from the excessive speed. High speed can cause the operating parts of the equipment to spin faster than the designed speed, with damaging results.

For example, the *tip speed* of the blade in a rotary lawn mower must never exced 19,000 feet [5701 m] per minute. If the engine overspeeds, the high speed might cause the blade to explode. Parts would fly off and could seriously injure anyone nearby.

Overloading an engine can cause overheating and rapid wear of engine parts. When using a mower under

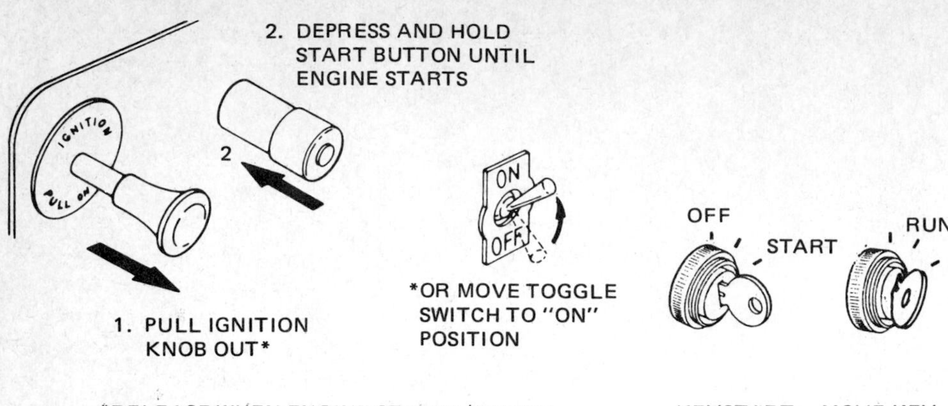

FIG. 18-6 Procedures for starting an engine with an electric starting motor. *(Kohler Company)*

heavy load, such as cutting tall wet grass, cut a narrow swath. Move the mower forward slowly.

For new engines, read and follow the *break-in* and operating instructions in the owner's manual and on the nameplate, decals, and labels on the engine. Rebuilt engines may require the use of a *break-in oil.* Follow the recommendations in the manufacturer's service manual, and the instructions that accompany some replacement parts.

> ***CAUTION*** **Before inspecting or working on the drive end of an engine or power equipment (for example, the side with the blade attached), always make sure the engine cannot start. Turn the ignition off. Then disconnect the spark-plug wire, and connect a jumper wire from its terminal end to a good ground on the engine.**

Δ18-3 STOPPING A SMALL ENGINE

To stop an engine properly, first remove any load. Next, reduce the engine speed to idle, and allow the engine to run for about a minute. Then shut off the engine. This may be done by shorting out the spark plug with a grounding blade or by returning the ignition key switch or toggle switch to the OFF position (Fig. 18-6). A spark-ignition engine may continue to run, or "diesel," after the ignition is turned off. If this happens, stop the engine by closing the choke valve and opening wide the throttle valve.

After the engine stops, close the fuel shutoff valve (Fig. 18-2), if used on the engine. This relieves any pressure on the carburetor diaphragms or float, and prevents fuel leaks.

If the engine will not be used for up to a month, drain the fuel tank and carburetor. This prevents formation of gum which could clog carburetor passages. If the engine will not be used for a longer time, prepare the engine the same as for winter storage (Δ18-10).

Δ18-4 MAINTENANCE PROCEDURES

Proper maintenance prolongs engine life. Many of the most commonly performed maintenance steps are listed below. How to perform some of these steps is described in other sections.

Cleaning the Air Cleaner How to clean and service various types of air cleaners is covered in Δ8-1.

Cleaning the Fuel Strainer or Filter How to clean and service fuel strainers and filters is covered in Δ8-2.

Cleaning Crankcase Breathers How to clean and service the crankcase breather is covered in Δ8-3.

Checking Crankcase-Oil Level (Four-Cycle Engine) Oil level in the crankcase of four-cycle engines should be checked after every 5 hours of engine operation. Some engines have a dipstick, which is attached to the fill plug or cap (Fig. 18-7). After the engine has cooled, remove the dipstick, wipe it dry, reinsert it, and then remove it again. The level of oil on the dipstick shows the

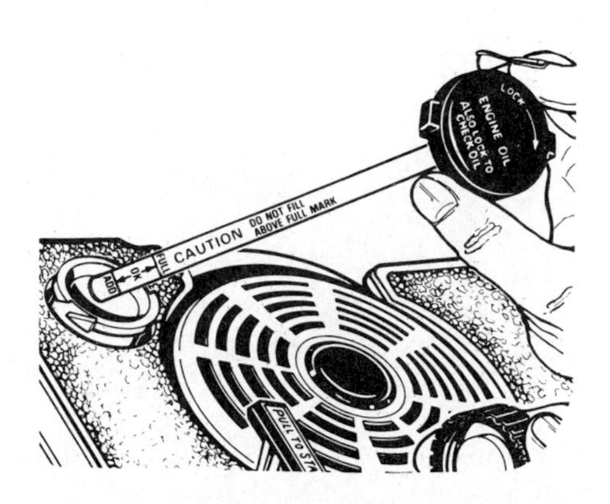

FIG. 18-7 Checking crankcase-oil level in a four-cycle engine. *(Tecumseh Products Company)*

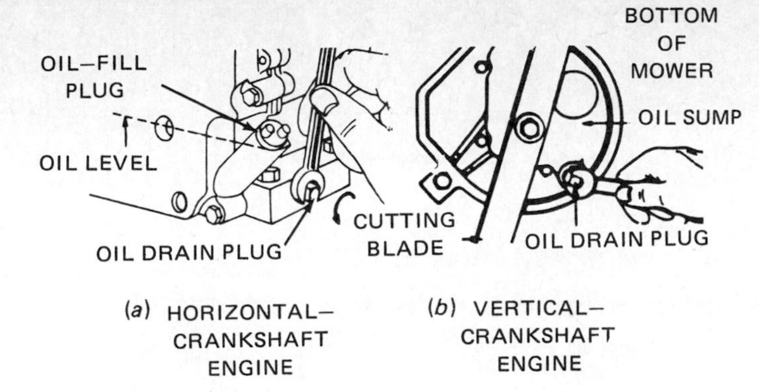

FIG. 18-8 Typical locations of the oil-fill plug on four-cycle engines. *(Briggs & Stratton Corporation)*

level of oil in the crankcase. On other engines, the crankcase-oil level is correct when oil overflows from the fill-plug opening (Fig. 18-8*a*).

Δ18-5 CHANGING OIL AND OIL FILTER

Four-cycle engines require periodic changing of the engine oil and filter.

Changing Oil Change the lubricating oil in the engine after the first 5 hours of operation of a new or rebuilt engine. Thereafter, change the oil after every 25 hours of operation. While the engine is warm, remove the oil drain plug (Fig. 18-8). Allow the oil to drain into a container and then dispose of the oil properly.

On some engines, the oil is drained by removing the drain plug in the bottom of the engine (Fig. 18-8*b*). This requires tipping the engine to remove the drain plug. Always empty the fuel tank first to prevent gasoline spills. Then keep the spark plug or muffler side up. To completely drain all oil from the crankcase, be sure the engine is level while the oil is draining. On some engines, the oil may also be drained through the oil-fill hole, or through the extended oil-fill tube.

If the oil-fill plug has not been removed, clean the area around it, and then remove it. With the engine level, add the specified quantity of engine oil (Fig. 18-9). Be sure that the oil is the proper grade and viscosity. Always follow the recommendations in the manufacturer's service manual, or on the decal or label on the engine. Never overfill the engine crankcase.

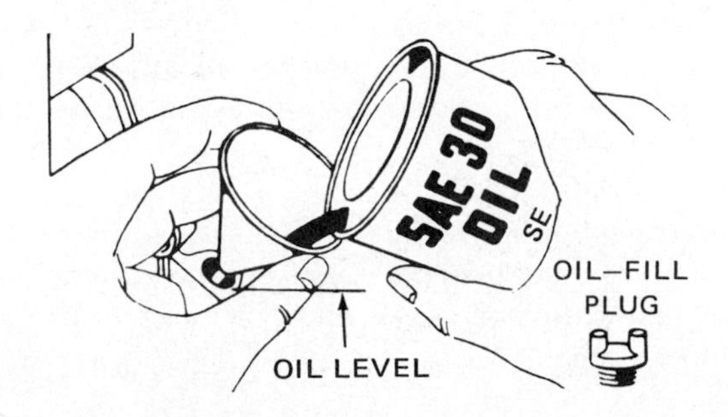

FIG. 18-9 Adding oil to the crankcase of a four-cycle engine. *(Briggs & Stratton Corporation)*

Changing Oil Filter The engine oil filter should be changed with every other oil change (each 100 hours of engine operation). If the engine is operating in extremely dirty conditions, the filter should be changed more frequently. Two types of filter elements are used, the spin-on type (Fig. 6-20) and the replaceable cartridge type (Fig. 6-22). Sometimes air baffles, shrouds, or other parts must be removed to provide access to the oil filter.

Place a container or shop cloths under the filter to catch any oil that may spill during removal. To remove a spin-on filter (Fig. 18-10), install a strap wrench around the filter. Then turn the wrench counterclockwise to remove the filter. Be sure the old gasket does not remain on the engine. Clean the filter adapter, lubricate the gasket on the new filter, and install the new filter. Tighten it only hand-tight. Installation instructions and tightening

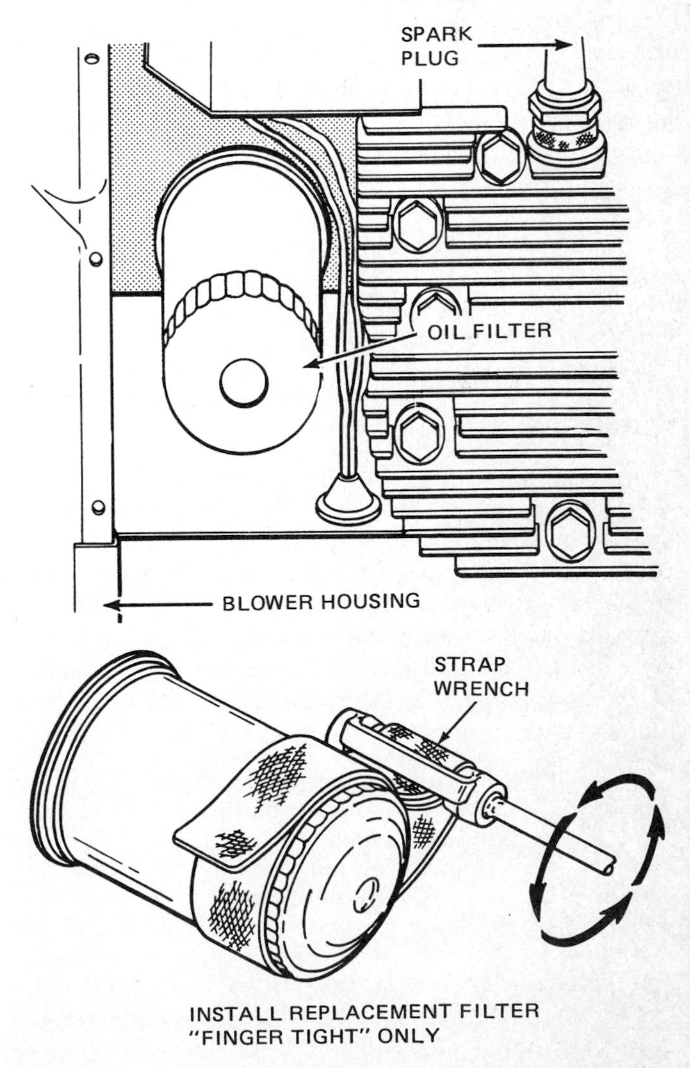

FIG. 18-10 Changing a spin-on type of oil filter. *(Kohler Company)*

information are often printed on the box in which the new filter is packaged.

To change a cartridge filter element, remove the top cover (Fig. 6-22) and pull out the old element. Have a container or rags handy to catch the oil that will drip from the element. Thoroughly clean the inside of the housing and the top. Replace the cover gasket, if needed. Then clean and install the drain plug in the housing. Install the new element. Place the top on the filter, and tighten the top to specifications.

After the new filter element is installed, add the amount of fresh oil to make up for the oil that was in the old element. One recommendation is to add 1 pint [0.47 L] when changing a spin-on filter and 1 quart [0.946 L] when changing a cartridge element. Start the engine. Then check for and correct any oil leaks.

Δ18-6 CLEANING THE ENGINE

One of the most important (and most neglected) maintenance services is to clean the engine and its components. Important parts to clean include the shrouds and fins on the cylinder and head. Wipe off the engine periodically to remove oil, grass clippings, dust, and mud. Foreign matter that collects around the engine acts as a blanket and causes overheating. On mowers, clean any accumulations of grass clippings from the inside of the housing.

CAUTION **Before working on the drive end of the engine and around the mower blade, disconnect the wire from the spark plug. Then connect a jumper wire from the terminal on the end of the spark-plug wire to a good ground on the engine. There must be no chance of the engine starting or the blade moving unexpectedly while you are working around it.**

One way to clean the engine is to brush it using a stiff bristle brush dipped in a strong mixture of household soap or detergent and water. The brush will clean away most of the grass clippings and trash. A degreasing compound can also be used. Carefully follow the safety cautions and instructions listed on the container. Never clean a hot engine. Allow it to cool first. Cold water or other liquid on a hot engine could crack the cylinder or head.

CAREFUL Never operate the engine with any part of the shroud or baffles removed. The engine will overheat. In addition, engines with an air-vane governor can overspeed and possibly self-destruct.

Δ18-7 CLEANING MUFFLER AND EXHAUST PORTS

In two-cycle engines, the exhaust ports tend to become clogged with carbon. Tecumseh recommends that the muffler and exhaust ports in two-cycle engines be cleaned after every 75 to 100 hours of engine operation.

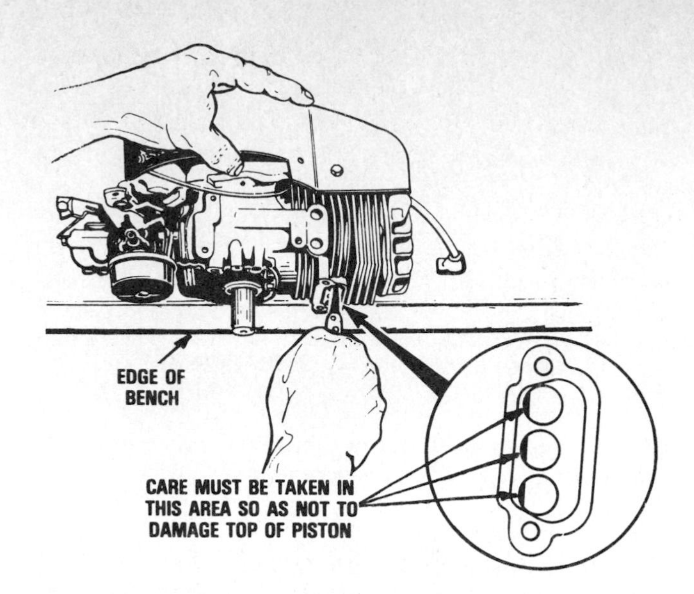

FIG. 18-11 Cleaning carbon deposits from the exhaust ports of a two-cycle engine. *(Tecumseh Products Company)*

If possible, remove the cylinder head. Then scrape the carbon from the cylinder head, the top of the piston, and the exhaust ports. Carbon can be removed without removing the head. Take off the muffler and spark plug. Position the piston at TDC. This prevents carbon from entering the cylinder since the piston is covering the port. Then place the engine on a workbench and scrape out the carbon using a knife or similar tool (Fig. 18-11). Some manufacturers recommend using a wooden stick to avoid scratching the metal surfaces. Be sure that no deposits are allowed to enter the cylinder or crankcase.

Clean the muffler in solvent. If the muffler remains obstructed, replace it. A new gasket may be required on installation. Then tighten the attaching screws securely.

Δ18-8 EQUIPMENT MAINTENANCE

The equipment or machine operated by a small engine also requires maintenance. Manufacturers' operator's manuals and service manuals include these instructions also. Listed below are three of the most common steps.

Tightening Fasteners Small engines operate with noticeable vibration. This can cause threaded fasteners such as nuts and bolts to loosen in service. Check the tightness of all bolts and nuts periodically. If loose fasteners are not retightened, parts may become damaged or lost.

Lubricating Lubricate all bearings outside the engine, such as the wheel bearings on a power mower. For example, make sure oil reservoirs are filled on the chainsaw lubricator.

Maintaining Blades and Saw Teeth Make sure that the blades or saw teeth are sharp and that the rest of the unit is in good condition.

Δ18-9 STORING GASOLINE Never store gasoline in a closed room, where gasoline vapors can accumulate. Explosions have resulted when gasoline vapors were ignited by the spark from a light switch being turned on.

Stored gasoline deteriorates with age. Old gasoline, or "stale gasoline," forms gum and other deposits on internal engine parts. These deposits, such as in carburetor jets, may cause poor engine performance, rapid wear, and even engine failure. A complete carburetor overhaul might be required to restore proper operation.

Before storing an engine, drain the fuel tank and carburetor.

Δ18-10 WINTER STORAGE Storing an engine for 30 days or less is called *short-term storage.* For a longer time, such as winter storage, drain the fuel tank. Then run the engine to use up the fuel in the carburetor. Remove the spark plug and squirt about a tablespoon [15 cc] of engine oil into the combustion chamber. Crank the engine slowly a few times to distribute the oil over the engine parts. Install the spark plug. Cover the engine with plastic or canvas. Then store the engine in a warm, dry place.

Δ18-11 ENGINE TUNEUP An engine tuneup restores power and performance that have been lost through wear, corrosion, and deterioration of engine parts. These changes take place gradually in many parts during normal operation. The procedure that follows includes the basic checks and adjustments recommended by several engine manufacturers.

1. Remove the spark plug. Note its condition. Check the cylinder compression by rocking the flywheel (with the plug installed) or using a compression gauge. Record the reading. If the compression reading is low, squirt about a tablespoon [15 cc] of engine oil through the spark-plug hole. Recheck the compression. If the compression rises significantly, the rings are probably worn. If the compression is unchanged, the valves are probably leaking. Tell the owner the engine cannot be tuned without repair or rebuilding.
2. Remove the air cleaner and check its condition. Note whether it has been serviced regularly.
3. Check the oil level and the condition of the oil in the crankcase. Then drain the oil and remove the oil filter.
4. Remove the cap from the fuel tank, and check for dirt and rust in the fuel tank. Drain the fuel tank, and clean it if necessary. Clean or replace the fuel filter. Inspect the fuel lines, and if they are in good condition, clean them.
5. Remove the blower housing and clean it. Check the starter or starting motor and drive. Examine the condition of the rope, and check the rewind assembly.
6. Clean the cooling fins on the cylinder and head. Then clean the entire engine.
7. Remove the carburetor. Disassemble it, and inspect it for worn or damaged parts. Wash the parts in solvent. Replace any worn or damaged parts, and then reassemble the carburetor. Follow the manufacturer's specifications and make the initial adjustments of the carburetor.
8. Inspect the intake manifold, crossover tube, and intake elbow. Tighten all nuts and bolts. Replace any gaskets that are leaking.
9. Check for governor blade, linkage, and spring for damage and wear. Check the adjustment of mechanical governors.
10. Remove the engine flywheel, and inspect it for cracks. Check the crankshaft-oil seals for leakage on the flywheel side and on the power-takeoff side of the engine. Check the flywheel key for wear and tightness in the grooves.
11. Remove the cover from the ignition breaker box, and check for proper sealing. Dirt inside the box indicates seal leakage.
12. Inspect the condition, alignment, and gap of the breaker points. Replace, or clean, and adjust as necessary. Check the condenser, and the cam or plunger that operates the breaker points.
13. Check the ignition coil for secure mounting, and check the wires connected to the coil for breaks and damaged insulation. Be sure none of the wires can touch the flywheel. Check the operation of the ignition stop switch, and its wiring.
14. Install the cover on the ignition breaker box. Use sealer to close the opening in the box through which the wires pass.
15. Install the flywheel. Time the engine, if necessary. Set the magneto armature air gap. Check the spark plug for sparks.
16. Remove the cylinder head. Check the gasket for signs of leakage. Clean carbon from the cylinder head, from the top of the piston, and from around the valves (four-cycle engines). Inspect the valves for proper seating.
17. Install the cylinder head. Torque it to specifications. Set the spark-plug gap, and install the spark plug. Discard worn or defective spark plugs. Gap all plugs, old and new, before installing.
18. Remove the muffler, and check it for restrictions and damage. Clean or replace it, as necessary. Clean any carbon deposits from the exhaust ports.

CHAPTER 18: REVIEW QUESTIONS

Select the *one* correct, best, or most probable answer to each question. You can find the answer in the section indicated at the right of each question.

1. Never pull the rope to start the engine in a rotary lawn mower until (Δ*18-1*)
   a. after the fuel tank is filled
   b. the tire pressures have been checked
   c. the choke is opened
   d. the mower is on level ground
2. Overloading a small engine can cause (Δ*18-2*)
   a. overheating
   b. rapid wear
   c. short engine life
   d. all of the above
3. If a small spark-ignition engine continues to run after you turn the ignition switch off, stop the engine by
   a. closing the choke (Δ*18-3*)
   b. opening wide the throttle
   c. both a and b
   d. neither a nor b
4. In a four-cycle engine, the crankcase-oil level should be checked (Δ*18-4*)
   a. after the engine has cooled
   b. immediately after stopping the engine
   c. while the engine is running
   d. with the engine under load
5. After changing the oil and filling the crankcase to the proper level on the dipstick, about half a quart is left over. Mechanic A says to add this oil to the crankcase now. Mechanic B says never overfill the engine crankcase. Who is right? (Δ*18-5*)
   a. mechanic A
   b. mechanic B
   c. both A and B
   d. neither A nor B
6. Dirt covering the fins of an air-cooled engine
   a. will damage the finish (Δ*18-6*)
   b. can cause the fins to rust
   c. will increase engine power
   d. will cause the engine to overheat
7. If the exhaust ports of a two-cycle engine become clogged with carbon (Δ*18-7*)
   a. use a wood stick to scrape out the carbon
   b. replace the ports
   c. clean the ports with a port reamer
   d. disassemble the engine to replace parts
8. Vibration from the operation of a small engine may cause (Δ*18-8*)
   a. oil leaks
   b. loose fasteners
   c. hard starting
   d. excessive fuel consumption
9. To prepare an engine for winter storage (Δ*18-10*)
   a. drain the fuel tank
   b. run the engine to use up all fuel in the carburetor
   c. squirt a tablespoon [15 cc] of oil into the combustion chamber
   d. all of the above
10. Mechanic A says the reason you should never operate an engine with the shroud and baffles removed is that the engine will overheat. Mechanic B says the reason is that the air-vane governor will not work. Who is right? (Δ*18-6*

CHAPTER

19

Spark-Ignition-Engine Trouble Diagnosis

After studying this chapter, you should be able to:

1. ***List five causes of small engine failure.***
2. ***Describe the checks and corrections for at least six spark-ignition engine troubles.***
3. ***Make quick checks of the fuel and ignition systems.***
4. ***Make a cylinder-compression test and interpret the results.***
5. ***Diagnose troubles in small spark-ignition engines.***

Δ19-1 CAUSES OF SMALL-ENGINE FAILURE Most small engines have relatively large crankshaft and connecting-rod bearings, considering the power these engines produce. Therefore, most engine failures are caused by neglect and lack of maintenance, and not design or manufacturing flaws.

For example, a minimum requirement to meet government specifications is that a small engine must be able to operate at full load and top speed for 1000 hours. In a lawn mower used 4 hours per week for 6 months, the engine would run 100 hours per year. At this rate, the engine should last about 10 years. Whether it lasts this long, or longer, depends primarily on how well the engine is maintained.

Listed below are some of the abuses that shorten engine life and cause (or contribute to) engine failure.

1. *Allowing Dirt to Get into the Engine* This will result from inadequate servicing of the air cleaner and fuel strainer, from improperly replacing spark plugs, and from contamination of the fuel.
2. *Failure to Check the Crankcase-Oil Level on Four-Cycle Engines* This failure can allow the oil to drop too low. The result is inadequate lubrication of the engine, which causes rapid engine wear and early engine failure.
3. *Failure to Feed the Two-Cycle Engine a Fuel-Oil Mixture* Without oil, the engine is inadequately lubricated, resulting in rapid wear and early failure. Improper lubrication means either that the proper amount of oil is not put into the gasoline for engines using the fuel-oil mix or that straight gasoline is used by mistake. On engines with an oil-injection system, failure of the system to deliver oil to the intake port or to the bearings will cause rapid engine wear and early engine failure.
4. *Overloading the Engine* Trying to make a small engine do a big engine's job will shorten the life of a small engine. If you change the governor setting on a lawn mower so the engine will run faster and handle heavier loads, the engine will wear out rapidly. Overspeeding and overloading an engine are two ways to shorten engine life.
5. *Failure to Store the Engine Properly* Many engines power machines that are in use only part of the year. When they are not to be used for several weeks or months, engines should be prepared for the idle period. Failure to do this can lead to early engine failure. Winter storage is described in **Δ18-10**.

Δ19-2 SMALL-ENGINE TROUBLE DIAGNOSIS Most engine complaints can be grouped under the following basic headings: engine will not crank, engine cranks but will not start, engine runs but misses, engine lacks power, engine overheats, engine has excessive fuel or oil consumption, and engine is noisy.

The chart that follows lists the engine troubles and the checks or corrections to be made for each condition. Some causes of trouble are in the engine itself. Later chapters cover engine-service operations that correct these troubles. For more information on ignition troubles, refer to the item indicated in the battery-ignition-system trouble-diagnosis chart (**Δ15-1**).

ENGINE TROUBLE-DIAGNOSIS CHART

Complaint	Possible cause	Check or correction
1. Engine will not crank	a. Run-down battery	Recharge or replace battery; start engine with jumper battery and cables
	b. Starting circuit open	Find and eliminate the open circuit; check for dirty or loose cables
	c. Starter drive jammed	Remove starter and free drive
	d. Starter jammed	Remove starter for disassembly and correction
	e. Engine jammed	Check engine to find trouble
	f. Transmission not in NEUTRAL, or safety-interlock switch out of adjustment	Check and adjust neutral switch if necessary
	g. See also causes listed under item 3; operator may have run battery down trying to start	
2. Engine cranks slowly but does not start	a. Run-down battery	Recharge or replace battery; start engine with jumper battery and cables
	b. Defective starter	Repair or replace
	c. Bad connections in starting circuit	Check for loose or dirty cables; clean and tighten
	d. See also causes listed under item 3; operator may have run battery down trying to start	
3. Engine cranks at normal speed but does not start	a. Defective ignition system	Make spark test; check timing, ignition system (see item 1, Δ15-1)
	b. Defective fuel pump, no fuel, or overchoking	Prime engine; check accelerator-pump discharge, fuel pump, fuel line, choke, carburetor
	c. Air leaks in intake manifold or carburetor	Tighten mounting; replace gaskets as needed
	d. Defect in engine	Check compression or leakage, valve action, timing
	e. Ignition coil or resistor burned out	Replace
	f. Plugged fuel filter	Clean or replace
	g. Plugged or collapsed muffler or exhaust system	Clean or replace parts
4. Engine runs but misses—one cylinder	a. Defective spark plug	Clean or replace
	b. Defective distributor cap or spark-plug cable	Replace
	c. Valve stuck open	Free valve; service valve guide
	d. Broken valve spring	Replace
	e. Burned valve	Replace
	f. Bent pushrod	Replace
	g. Flat cam lobe	Replace camshaft
	h. Defective piston or rings	Replace; service cylinder wall as necessary
	i. Defective head gasket	Replace
	j. Intake-manifold leak	Replace gasket; tighten manifold bolts
5. Engine runs but misses—different cylinders	a. Defective distributor advance, coil, condenser	Check distributor, coil, condenser (see item 3, Δ15-1)
	b. Defective fuel system	Check fuel pump, flex line, carburetor
	c. Cross-firing plug wires	Replace or relocate
	d. Loss of compression	Check compression or leakage
	e. Defective valve action	Check compression, leakage, vacuum
	f. Worn pistons and rings	Check compression, leakage, vacuum
	g. Overheated engine	Check cooling system
	h. Restricted exhaust	Check exhaust, ports, muffler; eliminate restriction

ENGINE TROUBLE-DIAGNOSIS CHART *(CONTINUED)*

Complaint	Possible cause	Check or correction
6. Engine lacks power—hot or cold	**a.** Defective ignition	Check timing, distributor, wiring, condenser, coil, and plugs
	b. Defective fuel system	Check carburetor, choke, filter, air cleaner, and fuel pump
7. Engine lacks power—hot only	**a.** Engine overheats	Check cooling system
	b. Choke stuck partly closed	Repair or replace
	c. Vapor lock	Use different fuel or shield fuel line
8. Engine lacks power—cold only	**a.** Choke stuck open	Repair or replace
	b. Cooling-system thermostat stuck open	Repair or replace
	c. Engine valve stuck open	Free valve; service valve stem and guide as needed
9. Engine overheats	**a.** Lack of coolant	Add coolant; look for leak
	b. Ignition timing late	Adjust timing
	c. Loose or broken fan belt	Tighten or replace
	d. Thermostat stuck closed	Replace
	e. Clogged water jackets	Clean
	f. Defective radiator hose	Replace
	g. Defective water pump	Repair or replace
	h. Insufficient oil	Add oil
	i. High-altitude, hot-climate operation	Adjust carburetor, ignition timing; keep radiator filled
	j. Valve timing late; slack timing chain has allowed chain to jump a tooth	Retime, adjust, or replace
10. Rough idle	**a.** Incorect carburetor idle adjustment	Readjust idle mixture and speed
	b. See also other causes listed under items 6 to 8	
11. Engine stalls cold or as it warms up	**a.** Choke valve stuck closed or will not close	Open choke valve; free or repair choke
	b. Fuel not getting to or through carburetor	Check fuel pump, lines, filter, float, and idle systems
	c. Idling speed set too low	Increase idling speed to specifications
12. Engine stalls after idling or slow-speed operation	**a.** Defective fuel pump	Repair or replace fuel pump
	b. Overheating	See item 9
	c. High carburetor float level	Adjust
	d. Incorrect idling adjustment	Adjust
13. Engine stalls after high-speed operation	**a.** Vapor lock	Use different fuel or shield fuel line
	b. Carburetor venting or idle-compensator valve defective	Check and repair
	c. Engine overheats	See item 9
	d. Fuel-tank vent plugged	Clear vent
14. Engine backfires	**a.** Ignition timing off	Adjust timing
	b. Spark plugs of wrong heat range	Install correct plugs
	c. Excessively rich or lean mixture	Repair or readjust fuel pump or carburetor
	d. Engine overheats	See item 9
	e. Carbon in engine	Clean
	f. Valves hot or stuck	Adjust, free, clean. Replace if bad
	g. Cracked distributor cap	Replace
	h. Cross-firing plug wires	Replace
15. Engine run-on, or dieseling	**a.** Incorrect idle adjustment	Adjust
	b. Engine overheats	See item 9
	c. Hot spots in cylinders	Check plugs, pistons, cylinders for carbon
	d. Timing advanced	Adjust

Complaint	Possible cause	Check or correction
16. Too much HC and CO in exhaust gas	a. Ignition miss	Check spark plug, wiring, cap, coil, etc.
	b. Incorrect ignition timing	Check choke, float level, idle-mixture screw, etc., as listed in item 20
17. Smoky exhaust		
a. Blue smoke	Excessive oil consumption	See item 18
b. Black smoke	Excessively rich mixture	See item 20
c. White smoke	Steam in exhaust	Replace cylinder-head gasket; tighten cylinder-head bolts to eliminate coolant leakage into combustion chamber
18. Excessive oil consumption	a. External leaks	Correct seals; replace gaskets
	b. Burning oil in combustion chamber	Check valve-stem clearance, piston rings, cylinder walls, rod bearings
	c. High-speed operation	Operate engine slower
19. Low oil pressure	a. Worn engine bearings	Replace
	b. Engine overheating	See item 9
	c. Oil dilution or foaming	Replace oil
	d. Lubricating-system defects	Check oil lines, oil pump, relief valve
20. Excessive fuel consumption	a. High-speed operation	Operate engine slower
	b. Excessive fuel-pump pressure or fuel-pump leakage	Reduce pressure; repair pump
	c. Choke partly closed after warm-up	Open; repair or replace choke
	d. Clogged air cleaner	Clean or replace
	e. High carburetor float level	Adjust
	f. Stuck or dirty float needle valve	Free and clean
	g. Worn carburetor jets	Replace
	h. Carburetor leaks	Replace gaskets; tighten screws
	i. Cylinder not firing	Check coil, condenser, timing, spark plug, contact points, wiring
	j. Loss of engine compression	Check compression or leakage
	k. Defective valve action (worn camshaft, chain slack, or jumped tooth)	Check with compression, leakage, or vacuum tester
	l. Excessive resistance from connected equipment	Correct defects causing resistance
	m. Clutch slippage	Adjust or repair
21. Engine noises		
a. Regular clicking	Valve and tappet	Readjust valve clearance, replace noisy valve
b. Ping, or spark knock, on load or acceleration	Detonation due to low-octane fuel, carbon, advanced ignition timing, or causes listed under item 14	Use higher-octane fuel; remove carbon; adjust ignition timing
c. Light knock or pound with engine floating	Worn connecting-rod bearings or crankpin, misaligned rod, lack of oil	Replace bearings, service crankpins, replace rod; add oil
d. Light, metallic double knock, usually most audible during idle	Worn or loose pin or lack of oil	Service pin and bushing; add oil
e. Chattering or rattling during acceleration	Worn rings, cylinder walls, low ring tension, or broken rings	Service cylinder walls; replace piston rings
f. Hollow, muffled bell-like sound (engine cold)	Piston slap due to worn pistons or walls, collapsed piston skirts, excessive clearance, misaligned connecting rods, or lack of oil	Replace or resize pistons; service cylinder walls; replace connecting rods; add oil
g. Dull, heavy metallic knock under load or acceleration, especially when cold	Regular noise; worn main bearings; irregular noise; worn thrust bearing knock on clutch engagement or on hard acceleration	Replace or service bearings and crankshaft
h. Miscellaneous noises (rattles, etc.)	Loosely mounted accessories: alternator, starter, water pump, etc.	Tighten mounting

Δ19-3 ENGINE WILL NOT START

Failure of the engine to start could be due to lack of fuel, failure of the fuel to reach the carburetor, failure of the carburetor to deliver fuel to the air passing through, clogged air filter, clogged exhaust ports or muffler, defective ignition system, or internal engine damage.

To locate the trouble, first make sure there is clean gasoline in the fuel tank. Then check that the fuel-tank vent (usually in the cap) is open. If the vent is clogged, the engine may start but soon stop. A clogged vent does not permit fuel to flow from the tank to the carburetor fast enough to keep the engine running.

Δ19-4 QUICK CHECK OF CYLINDER COMPRESSION

The cylinder-compression test (Δ19-5) measures the ability of the cylinder to hold compression pressure. On some engines, you can make a quick check of cylinder compression by slowly pulling the engine through the compression stroke with the starter (Fig. 19-1). Be sure the ignition switch is off.

If the starter is a rope-wind or rope-rewind type, you can judge the compression by feel. If the engine spins very easily, it has little compression pressure. This could be due to a loose cylinder head, leaking head gasket, loose spark plug, cracked head or cylinder, broken piston rings, or broken piston. On four-cycle engines, loss of compression may be due to the intake or exhaust valve sticking open.

On kick starters, you can judge compression by the ease with which you can kick over the engine. If it spins too easily, the engine has lost compression. If the engine uses a windup or electric starting motor, you can judge compresion by the way the engine sounds as it is cranked. With the windup starter, if release of the spring cranks the engine unusually fast or for a long time, the engine probably has low compression. With an electric starting motor, an engine that has low compression will spin abnormally fast.

If the engine has normal compression, it will resist the kick starter or the pull of the rope—or act normally when starting is attempted with the windup starter or electric starting motor. Another sign of good compression on small two-cycle engines is a sucking sound when the engine is spun fast. This should be followed by a sort of cough as the engine stops after the spin. The sound indicates that the engine is taking in air normally. If the engine has "easy-spin" starting, or a compression release, turn the crankshaft backward to check compression.

When checking compression, listen for unusual squeaks, squeals, scraping, or knocking sounds. Any of these could mean worn bearings, scored cylinder walls or pistons, or broken rings or other parts. If you hear such noise, do not try to start the engine. It must be disassembled for inspection.

Δ19-5 MAKING A CYLINDER-COMPRESSION TEST

An accurate check of cylinder-compression pressure can be made with a compression gauge (Fig. 19-2). Remove the spark plug and install the gauge in the spark-plug hole. Some gauges are threaded into the spark-plug hole. Others must be held there.

With the gauge in place, and the needle indicating zero, crank the engine for at least six revolutions, or until the needle stops rising. Compare the reading on the compression gauge with the engine manufacturer's specifications.

If you cannot locate the engine manufacturer's specifications, some manufacturers provide general rules to follow. Clinton suggests that on small general-purpose two-cycle engines, compression pressure should be above 60 psi [413 kPa]. For high-output two-cycle engines, such as chain saws and outboard engines, McCulloch recommends that an engine have at least 90 psi [621 kPa] of compression pressure. For comparison, a new Yamaha 125-cc two-cycle motorcycle engine has a compression pressure of about 120 psi [8274 kPa].

On engines with two or more cylinders, the compression pressure in the lowest cylinder should be within 10 to 15 percent of the highest cylinder.

Fig. 19-1 Check engine compression by slowly pulling the engine through the compression stroke with the starter rope. *(Lawn Boy Division of Outboard Marine Corporation)*

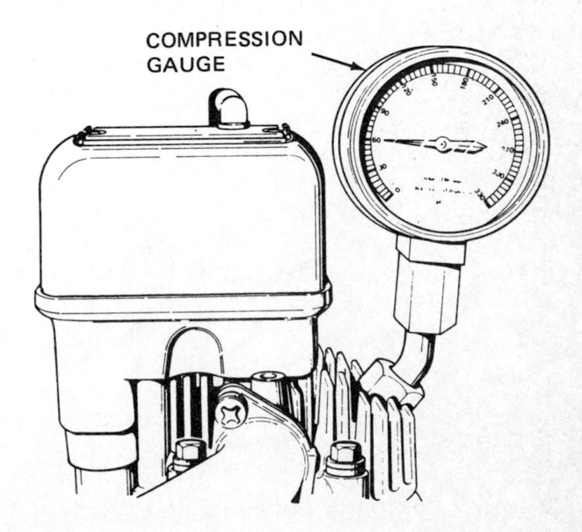

Fig. 19-2 Checking cylinder compression pressure with a compression gauge. *(Tecumseh Products Company)*

Δ19-6 QUICK CHECK OF IGNITION SYSTEM

If an engine cranks normally but will not start, check that there is fresh fuel in the tank, the choke is closed, and the ignition switch is turned on. Then crank the engine again. If the engine has normal compression but still will not start, then either the ignition system or the carburetor is probably at fault.

Make a spark test (**Δ**15-2). If no spark occurs, then the ignition system is probably at fault and should be checked (Chap. 15). Causes of trouble could be dirty or worn breaker points or points out of adjustment. Defects in the capacitor, high-voltage lead, ignition switch, or magneto coil could also cause trouble.

If a normal spark occurs during the spark test, remove the spark plug. Check for any condition that could prevent delivery of the spark to the combustion chamber (Fig. 15-5). A plug that is wet with gasoline must be dried or replaced (**Δ**15-4).

Δ19-7 QUICK CHECK OF FUEL SYSTEM

When the spark test indicates the problem is not in the ignition system, then the fuel system must be checked. After cranking the engine, remove the spark plug. If the end is wet with gasoline, fuel is getting into the cylinder. However, if the plug is not wet and you suspect that fuel is not reaching the cylinder, make a "thumb test."

Connect a jumper wire between the terminal on the end of the spark-plug wire and a good ground on the engine. Place you thumb over (but *not* in) the spark-plug hole (Fig. 19-3). Then crank the engine. Inspect your thumb. If fuel is being delivered, your thumb will be slightly damp and have a strong smell of gasoline.

If your thumb is dry, the carburetor is not delivering fuel. The cause could be a clogged fuel filter, strainer, or fuel line. Troubles in the carburetor, such as a clogged nozzle, incorrect adjustment, or defective float, could also cause the trouble.

Δ19-8 CHECKING THE CARBURETOR

If the carburetor is indicated as the cause of an engine failing to start, adjusting the carburetor may correct the problem. If not, then the carburetor must be removed for disassembly and repair.

Fig. 19-3 Using your thumb to find if fuel is entering the cylinder.

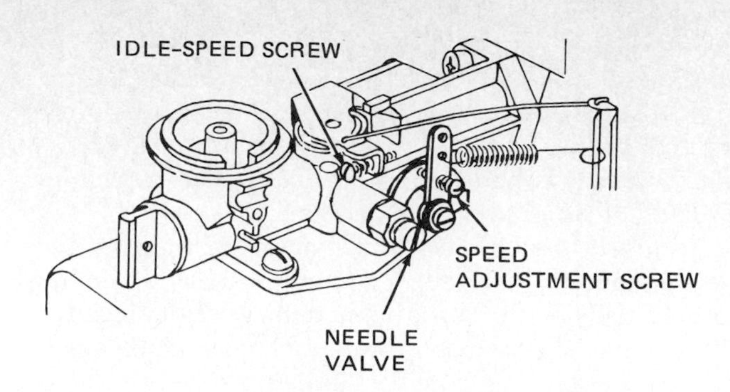

Fig. 19-4 Adjustment screws on a small-engine carburetor. *(Briggs & Stratton Corporation)*

To adjust a typical small-engine carburetor, turn the needle valve clockwise to seat the main-nozzle needle valve (Fig. 19-4). Then back off the needle 1½ turns. Close the choke and crank the engine while making a thumb test (**Δ**19-7). If gasoline now appears, install the spark plug and try to start the engine.

If the engine starts, open the choke as the engine warms up. If the engine runs roughly, it may be getting too much gasoline. Turn the needle valve in to produce a leaner mixture.

After the engine is warmed up, turn the needle valve in until the engine begins to die from an excessively lean mixture. Then back out the needle valve about a quarter turn. This should be the best adjustment for full-load operation. Adjustment procedures for various types of carburetors are given in Chap. 8.

Δ19-9 ENGINE STARTS BUT LACKS POWER

A common cause of this trouble in two-cycle engines is a clogged muffler and exhaust ports (**Δ**18-7). Remove the muffler and scrape out the carbon as shown in Fig. 18-11. If this is not the cause, then adjust the carburetor (**Δ**8-7). It may be supplying an excessively rich or lean mixture.

Δ19-10 ENGINE DEFECTS

If no other cause of lack of power is found, the trouble is probably in the engine itself. The trouble could be due to a worn piston and cylinder or to worn or broken piston rings. Another possible cause in two-cycle engines is a defective reed valve in the crankcase (**Δ**19-11).

Δ19-11 DEFECTIVE REED VALVE

If the reed valve is broken or not seating properly, it may not hold compression pressure in the crankcase (**Δ**3-7). Then not enough air-fuel mixture reaches the cylinder for the engine to develop full power.

Clean any oil or other deposits from the reeds, stops, and adapters. Serviceable reed plates will have locating smudge marks on the smooth side of the reed. Check the reeds for proper sealing against the adapter plate (Fig. 19-5). The reeds shown in Fig. 19-5 should not bend away from the plate more than 0.010 inch [0.25 mm]. Replace

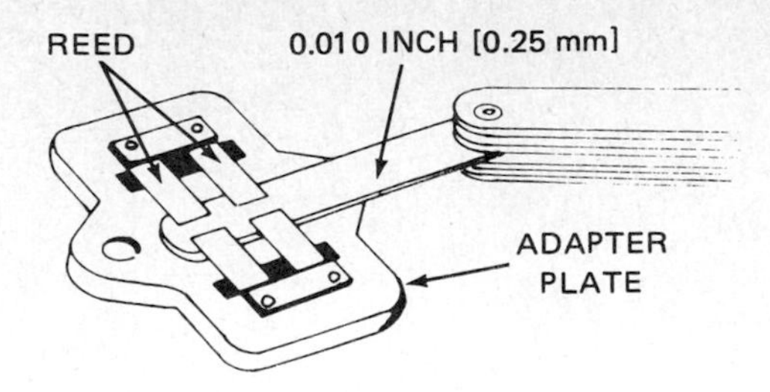

Fig. 19-5 Measuring how far the reeds bend away from the adapter plate. *(Tecumseh Products Company)*

the reed valve if it is broken, warped, or bent so it does not lie flat against the inlet holes.

Δ19-12 ENGINE SURGES If the engine surges (repeatedly speeds up and slows down), the trouble probably is in either the carburetor or the governor. Try adjusting the carburetor (Δ19-8). If this does not correct the condition, then check the governor. Look for binding linkage between the governor and the throttle valve, weak or damaged governor spring, and worn or binding governor parts. Governor service is covered in Δ8-13.

Δ19-13 ENGINE LOSES POWER If the engine loses power as it warms up, the trouble is probably in the fuel system. Possible causes are a clogged fuel-tank vent and a stuck needle valve in the float bowl. These allow too little gasoline to reach the carburetor. As a result, the engine slows because it is fuel-starved.

Lack of lubrication in the engine may also cause loss of power as the engine warms up. For example, failure to add oil to the gasoline of an engine that requires premix will cause slowing as scoring occurs on the piston and cylinder wall (Fig. 19-6). Then the engine will stop as the scoring becomes more severe and the bearings seize.

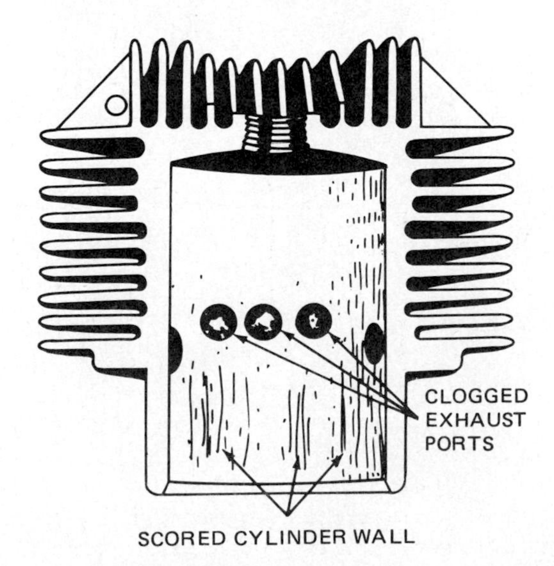

Fig. 19-6 Cutaway cylinder and head showing scored cylinder walls and clogged exhaust ports, caused by carbon deposits in the exhaust ports.

Δ19-14 IRREGULAR FIRING If the engine fires irregularly, the trouble could be a weak spark or improper carburetion. Check the spark by making a spark test (Δ15-2). Replace the coil or capacitor, clean and adjust or replace the breaker points (Δ15-7), and replace wires as necessary. Check and adjust the carburetor as described above and in Δ8-7.

A two-cylinder opposed piston engine may run fairly well when firing on only one cylinder if no load or a minimum load is applied. But the engine often loses power quickly when a full load is placed on the engine.

If this occurs, first check that fuel is getting into the cylinder (Δ19-7). Then check the intake-manifold gaskets and connections for air leaks. If no problem is found, check the ignition system for the cylinder that is misfiring. If no problem is found and the engine continues to misfire, make a compression test (Δ19-5).

However, a lean air-fuel mixture is the most common cause of a two-cylinder opposed engine running on only one cylinder. This condition is usually corrected by adjusting the carburetor (Δ8-7).

CHAPTER 19: REVIEW QUESTIONS

Select the *one* correct, best, or most probable answer to each question. You can find the answer in the section indicated at the right of each question.

1. Most engine failures are caused by (***Δ19-1***)
   a. design and manufacturing flaws
   b. neglect and lack of maintenance
   c. part-throttle operation
   d. using the wrong fuel
2. The cylinder-compression test measures (***Δ19-4***)
   a. compression pressure during cranking
   b. combustion pressure while running
   c. fuel-injection pressure for starting
   d. none of the above
3. To make a quick check of the ignition system, you should (***Δ19-6***)
   a. check for dirty or burned breaker points
   b. check for arcing of the points as they open
   c. remove the spark plug and make a thumb check
   d. make a spark test
4. A quick check of the fuel system can be made by
   a. removing the spark plug (***Δ19-7***)
   b. placing your thumb over the spark-plug hole
   c. cranking the engine
   d. all of the above
5. When a two-cycle engine lacks power, check for
   a. an empty fuel tank (***Δ19-9***)
   b. a weak spark
   c. clogged muffler and exhaust ports
   d. inadequate lubrication

CHAPTER
20

Two-Cycle-Engine Service

After studying this chapter, you should be able to:

1. ***Perform a top-end overhaul on a two-cycle engine.***
2. ***Perform a bottom-end overhaul on a two-cycle engine.***
3. ***Check for wear in the connecting-rod big-end bearing.***
4. ***Determine piston clearance.***
5. ***Measure bearing clearance.***
6. ***Service needle and roller bearings.***
7. ***Replace crankcase-oil seals.***
8. ***Perform a complete overhaul on a two-cycle engine.***

TOP-END OVERHAUL

Δ20-1 PREPARING FOR ENGINE DISASSEMBLY Before you begin work on an engine, be sure you have a clean place to work, trays for small parts you will remove, and the tools you will need. Then check that there is adequate working room around the engine. If it is installed in a lawn mower, for example, you may be able to remove the head and cylinder without removing the engine itself. However, many mechanics prefer to always remove the engine. Then it can be placed in a stand, or on the bench, for easier access.

After the engine is removed (or before starting any disassembly work), the engine must be cleaned. Cleaning the engine is described in **Δ**18-6.

Δ20-2 DISASSEMBLING THE TOP END Remove the deck, shroud, fuel tank, starter, or other parts that block access to the cylinder and head (Fig. 20-1). Next, disconnect the throttle and choke linkage, and remove the carburetor. Take off the exhaust pipe and muffler. Unscrew the spark plug.

Remove the screws or nuts holding the cylinder head and cylinder to the crankcase. Remove any other small parts as necessary. Pull the cylinder away from the crankcase about 2 inches [51 mm]. Put a clean cloth or shop towel around the connecting rod and over the crankcase opening (Fig. 20-2). This will help prevent any

FIG. 20-1 Removing the engine shroud, fuel tank, and starter as an assembly. *(Lawn Boy Division of Outboard Marine Corporation)*

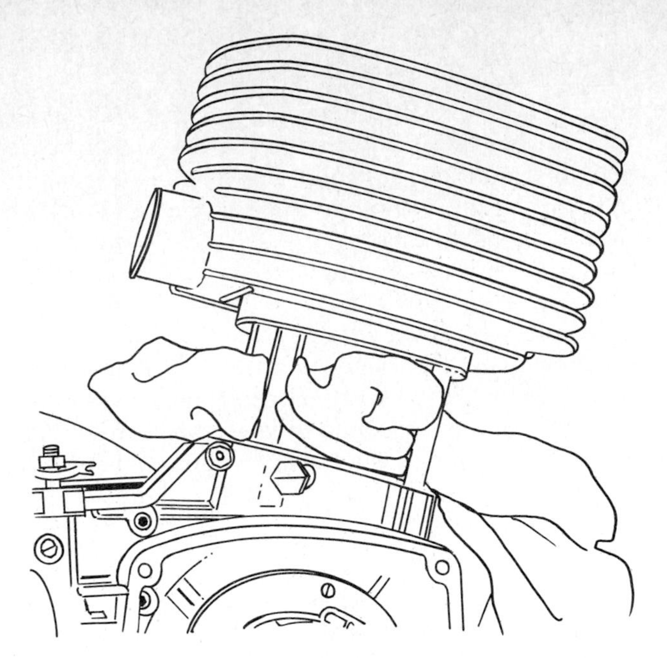

FIG. 20-2 Place a clean cloth around the connecting rod and crankcase opening to prevent dirt from entering the crankcase.

dirt from entering the crankcase as you remove the cylinder.

Now take the cylinder off the crankcase. Set the cylinder aside with the cylinder-head end down. Place it on a clean piece of cardboard. This protects the cylinder from damage to the crankcase sleeve or threads.

On some two-cycle engines, the head and cylinder are one piece (Fig. 20-3). To remove this type of cylinder

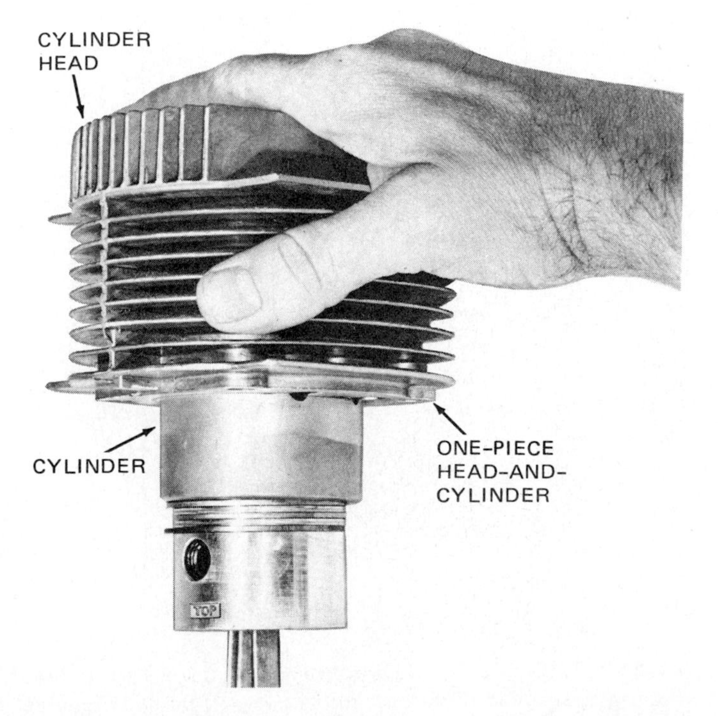

FIG. 20-3 Removing a one-piece type of cylinder-and-head assembly from the piston. *(Lawn Boy Division of Outboard Marine Corporation)*

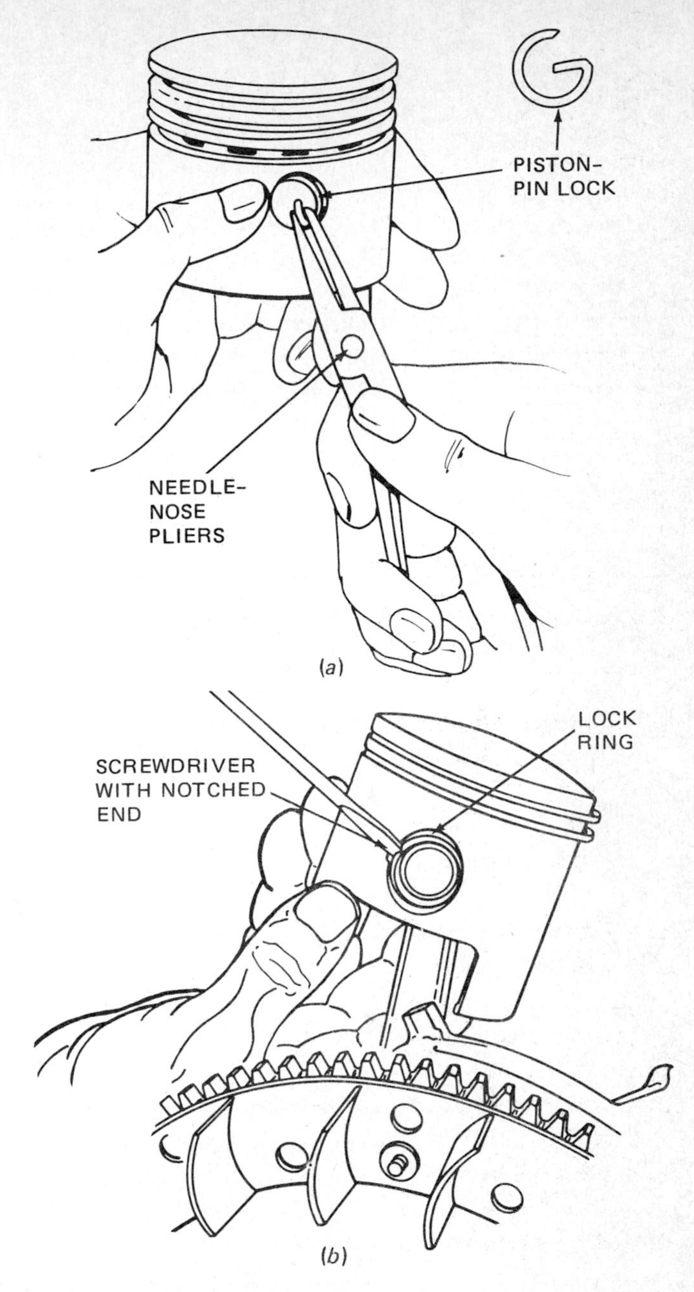

FIG. 20-4 Removing the piston-pin lock rings. *(Briggs & Stratton Corporation; Kohler Company)*

assembly, unbolt it from the crankcase. Then pull the cylinder from the crankcase and off the piston.

Complete the disassembly of the top end by detaching the piston from the connecting rod. If the piston pin is a press fit in the rod, the pin must be forced out in a press or driven out with a special punch. If the piston pin is held in place with retaining rings (*lock rings* or *Circlips*), remove the rings with needle-nose pliers or a screwdriver (Fig. 20-4). Then the pin can be pushed out and the piston removed from the connecting rod.

Δ20-3 CLEANING ENGINE PARTS After the top end of the engine is disassembled, the parts must be cleaned. Then they can be checked, inspected, and measured. The job of cleaning top-end parts is often called *decarboning* the engine.

Remove the carbon deposits from the transfer ports, exhaust ports and passages, piston head, and combustion-chamber surfaces. Cleaning the muffler and exhaust ports is described in Δ18-7.

Remove all the old gasket material from the cylinder head and the cylinder-to-crankcase surface. Wash all parts in clean solvent. Dry the parts with air or with a clean cloth. Inspect the threads in the spark-plug hole. If you can see carbon deposits in the threads, clean them with a thread chaser or a tap of the proper size.

Δ20-4 INSPECTING THE CYLINDER

Examine the cylinder for cracks, stripped threads in bolt holes, broken fins, and scores or other damage in the cylinder bore. Any of these require replacement of the cylinder. Sometimes stripped threads can be repaired with a thread insert.

If the cylinder appears to be in good condition, use an inside micrometer or a cylinder-bore gauge to check for wear, taper, and out-of-round (Fig. 20-5). A telescoping gauge and an outside micrometer can also be used to measure the cylinder bore (Fig. 20-6). Take the measurements at several places to check for taper and out-of-round (Fig. 20-6). The difference between wear, taper, and out-of-round is shown in Fig. 20-7.

Some small engines have cylinders of aluminum with cast-iron liners (Fig. 20-8). Check this type of cylinder for a loose liner. If there are blue spots on the bore (from the cast iron overheating) or if the liner is loose, discard the cylinder.

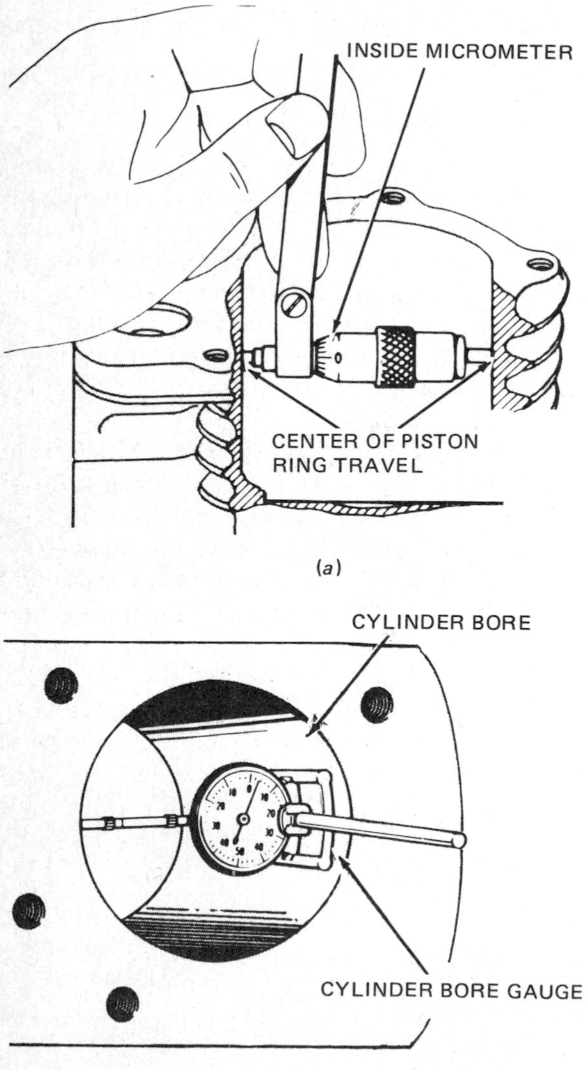

FIG. 20-5 Methods of measuring the cylinder bore. *(a)* Using an inside micrometer. *(b)* Using a cylinder-bore gauge.

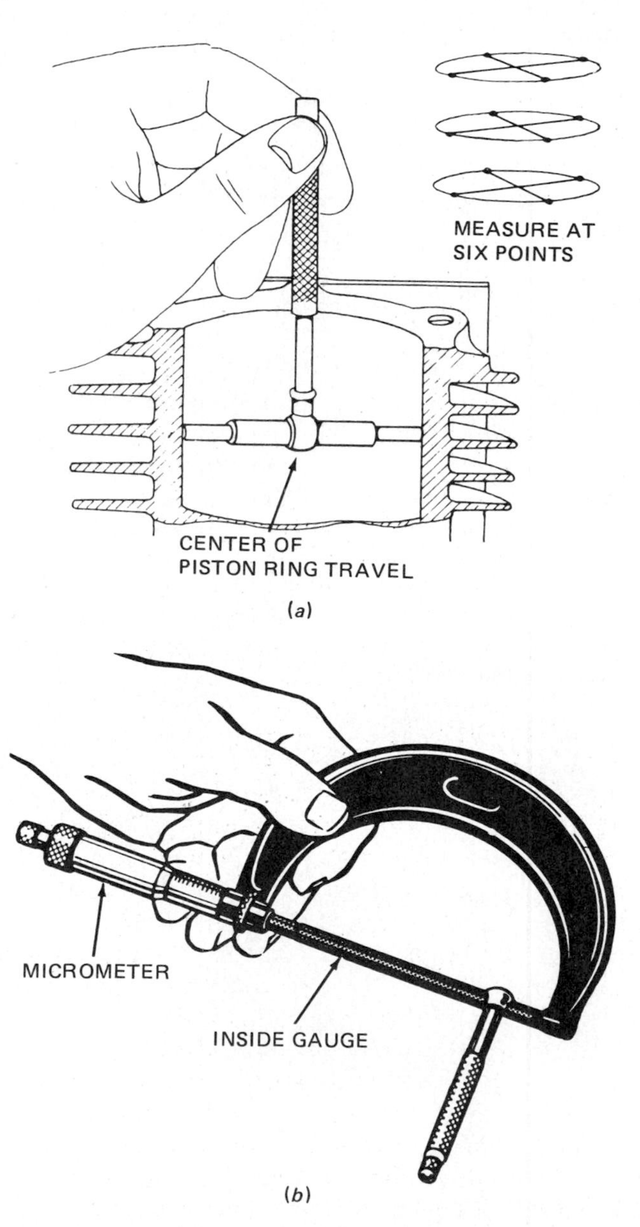

FIG. 20-6 Taking cylinder-bore measurements with a telescoping gauge and an outside micrometer. *(a)* Set the gauge to the bore diameter. *(b)* Then measure the gauge setting with the micrometer.

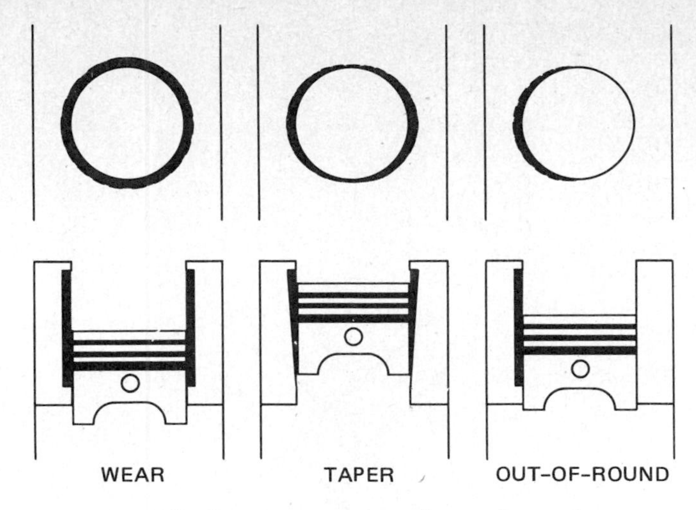

FIG. 20-7 Cylinder wear, taper, and out-of-round.

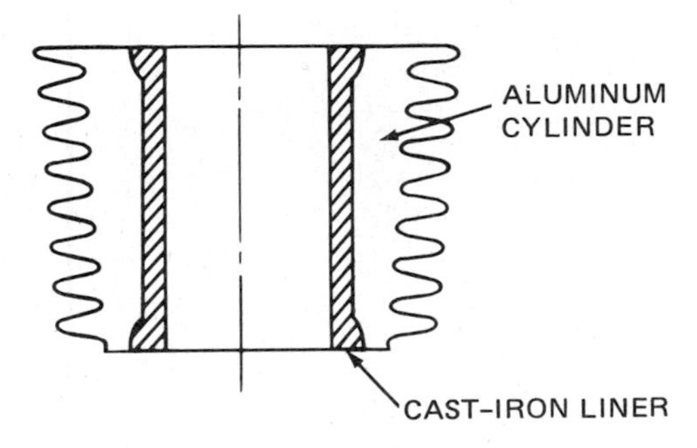

FIG. 20-8 Aluminum cylinder with a cast-iron liner.

Other small engines have chrome-plated cylinder bores. You can identify chrome-plated cylinders by their shiny appearance. These cylinders cannot be refinished. If they show signs of wear or scoring, replace the cylinder.

Δ20-5 REFINISHING THE CYLINDER If the cylinder is scored, worn, tapered, or out-of-round, it must be bored or honed to a larger size so that a larger piston and rings can be installed. Pistons are supplied in standard oversizes, such as 0.010, 0.020, and 0.030 inch [0.25, 0.50, and 0.76 mm]. The cylinder, unless it is a chrome-plated cylinder, must be refinished to take one of these *standard oversize pistons.* Chrome-plated cylinders cannot be refinished.

Figure 20-9 shows the honing procedure for a cylinder of a one-cylinder two-cycle engine with a detachable cylinder head. Some manufacturers recommend honing from the crankcase end of the cylinder. Honing also is done in a drill press and in special shop hones that have fixtures for holding the cylinder. Follow the instructions supplied by the hone manufacturer for installation of the hone in the cylinder, operating speed, and lubrication. Remove the hone and *measure the cylinder periodically so you do not remove too much metal.*

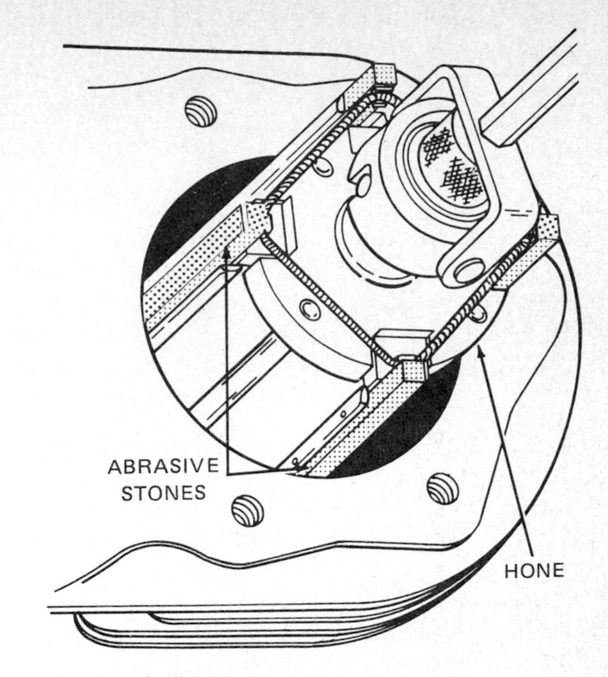

FIG. 20-9 Honing the cylinder. *(Onan Corporation)*

When the cylinder is approximately 0.002 inch [0.05 mm] within the desired size, change to fine stones to finish the honing operation. Usually, rough honing is done with 60-grit stones. Finish honing is done with 220-grit stones. When the honing job is finished, the cylinder wall should have a *crosshatch pattern* (Fig. 20-10). This finish requires that the hone be moved up and down at the right speed while the hone is rotating at the proper speed.

A one-piece-cylinder block-and-head assembly should be machine-bored (*not* honed) when cylinder-bore oversizing is necessary. If boring is not necessary, roughen the cylinder wall slightly by running the hone through the cylinder several times. This is called *breaking the glaze.* It helps new piston rings seat faster in a rebuilt engine.

After boring and honing, the cylinder wall must be thoroughly cleaned. This is to remove all particles of grit and metal that may have become embedded in the cylinder wall. To clean it, use soap and water and clean rags or a mop. Wash the cylinder wall until you can rub a clean cloth on it without getting the cloth dirty. Then dry the cylinder wall and coat it with engine oil.

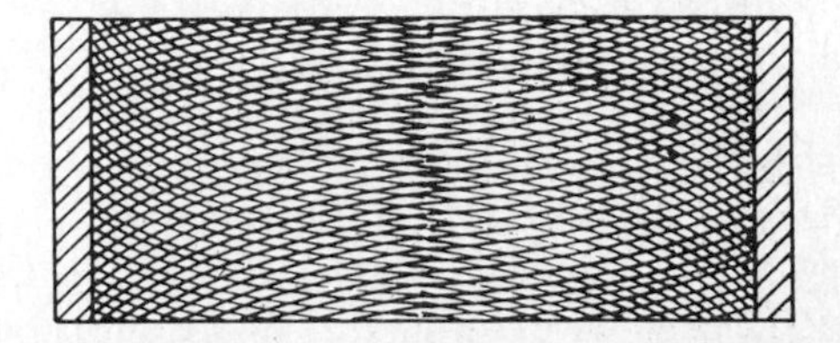
FIG. 20-10 Crosshatch appearance of a properly honed cylinder bore. *(Briggs & Stratton Corporation)*

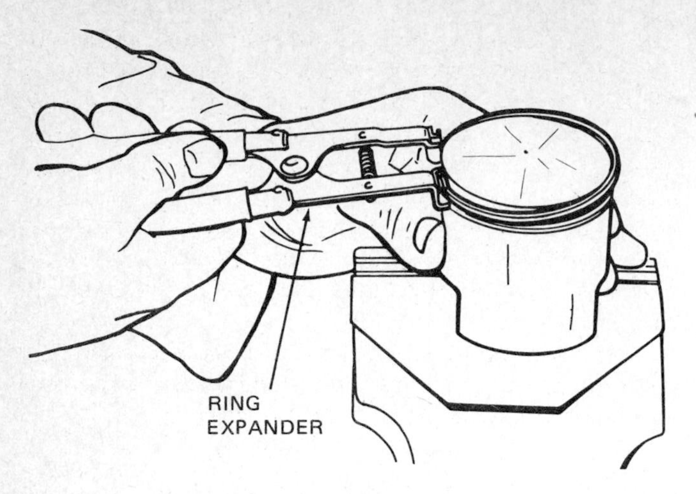

FIG. 20-11 Using a ring expander to remove the piston rings from the piston. *(Kohler Company)*

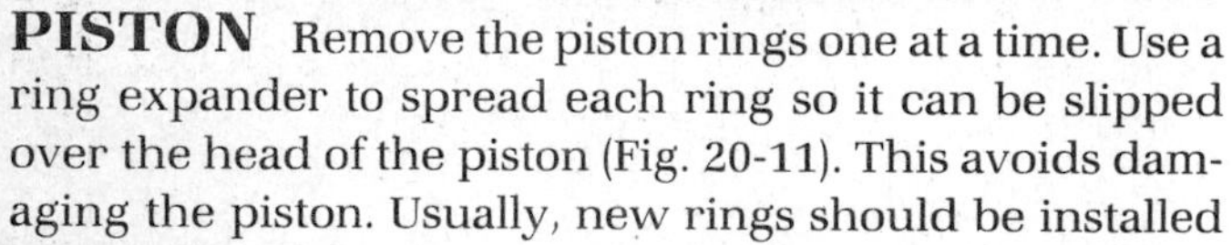

Δ20-6 CLEANING AND SERVICING THE PISTON Remove the piston rings one at a time. Use a ring expander to spread each ring so it can be slipped over the head of the piston (Fig. 20-11). This avoids damaging the piston. Usually, new rings should be installed every time the engine is overhauled. The exception is if the rings have been used only a short time. Then the rings can be cleaned and inspected. If they are in good condition, they can be reinstalled.

If the cylinder wall is worn so that the cylinder requires refinishing, then both new rings and a new piston will be required, as explained previously.

Inspect the piston for scuffing and scoring, cracks in the head or skirt, and damaged or broken ring lands. A ring land is the metal ring between the ring grooves. Cracks or similar damage means the piston should be discarded. However, some pistons that have rough or lightly scored piston skirts can be repaired and reused. One repair method is to polish the piston with a soap-filled steel-wool pad. Then rinse the piston in running water. When all soap and metal particles have been washed off, dry the piston and coat it with light engine oil.

Another way to smooth a rough piston skirt is to use an oil stone. Hold the piston in your hand and work the oil stone over the scuffed area. Follow the pattern shown in Fig. 20-12. If the damage does not clean up, discard the piston.

Check the piston dimensions with a micrometer (Fig. 20-13). Compare these dimensions with the cylinder dimensions previously checked (Figs. 20-5 and 20-6). The difference is the *piston clearance.*

Some engine manufacturers recommend checking the fit of the piston in the cylinder bore with a thickness gauge. The piston is inserted in the bore with the thickness gauge along its side (Fig. 20-14). If the piston moves too freely, the clearance is excessive. Too much clearance requires reboring the cylinder to a larger size and installing an oversize piston, as previously explained.

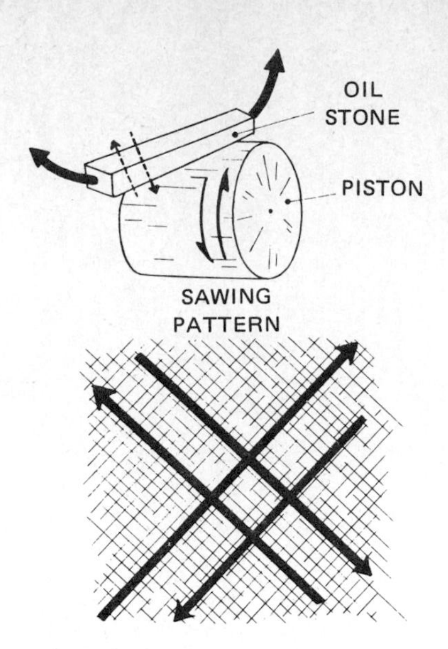

FIG. 20-12 Using an oilstone to clean up a scuffed piston. *(Yamaha Motor Company, Ltd.)*

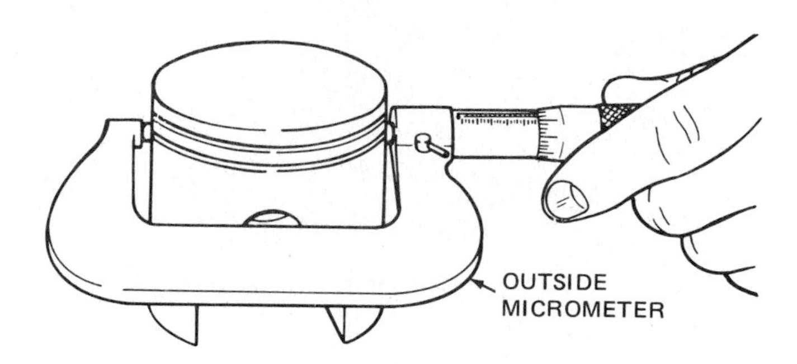

FIG. 20-13 Measuring the piston skirt just below the bottom ring and at right angles to the pin. *(Kohler Company)*

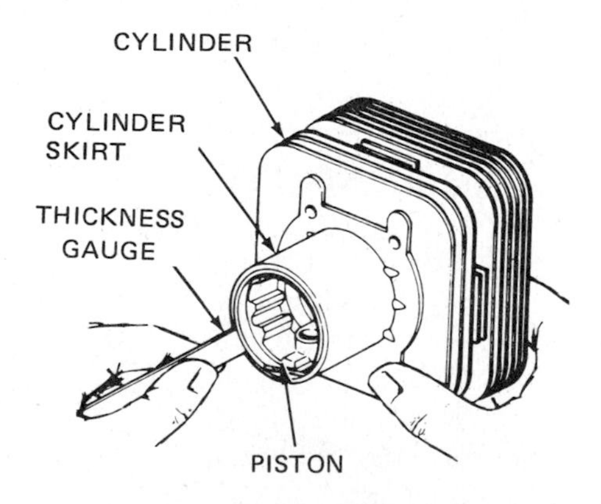

FIG. 20-14 Measuring piston clearance with a thickness gauge. *(Outboard Marine Corporation)*

If the piston checks out properly so far, inspect the ring grooves. Clean out the carbon with a groove cleaner (Fig. 20-15) or with a piece of a broken piston ring. Do not remove metal; remove carbon only. Another method is to soak the piston in old carburetor cleaner or a liquid chemical cleaner that is recommended for aluminum. Then scrape away any remaining carbon from the grooves.

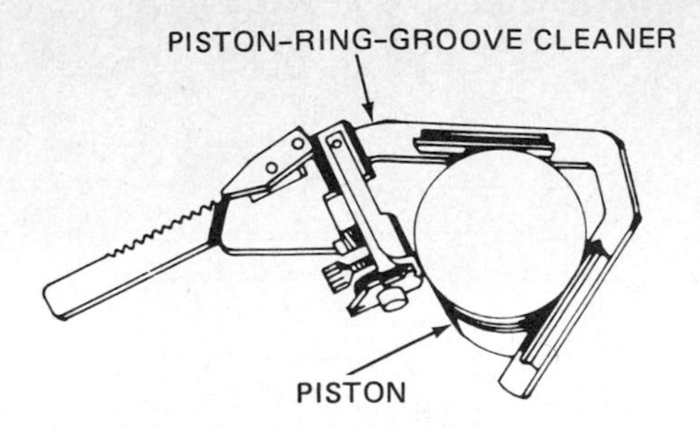

FIG. 20-15 Piston ring-groove cleaner.

Never use a wire brush on a piston. The wire bristles will scratch the piston and round off the outside edges of the piston ring lands.

Next, check for ring-groove wear. Use the rings you are to install on the piston to make this check. First, roll the ring around the groove (Fig. 20-16). The ring should roll freely in the groove all the way around. If the ring binds in any place, there probably is still some carbon in that spot, or there is a metal burr on the side of the piston-ring

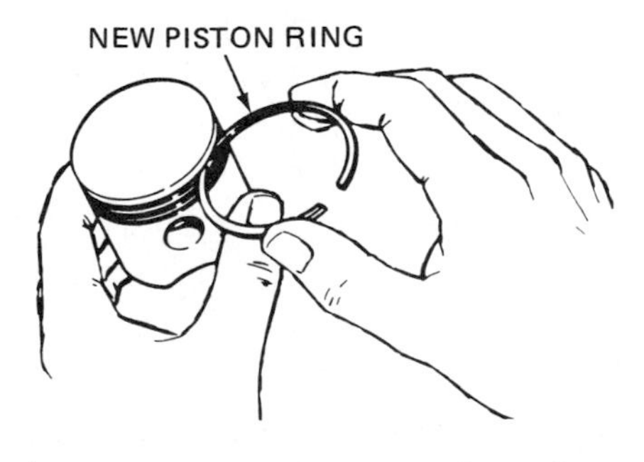

FIG. 20-16 Checking the piston ring for tightness or binding by rolling it around the groove. *(Outboard Marine Corporation)*

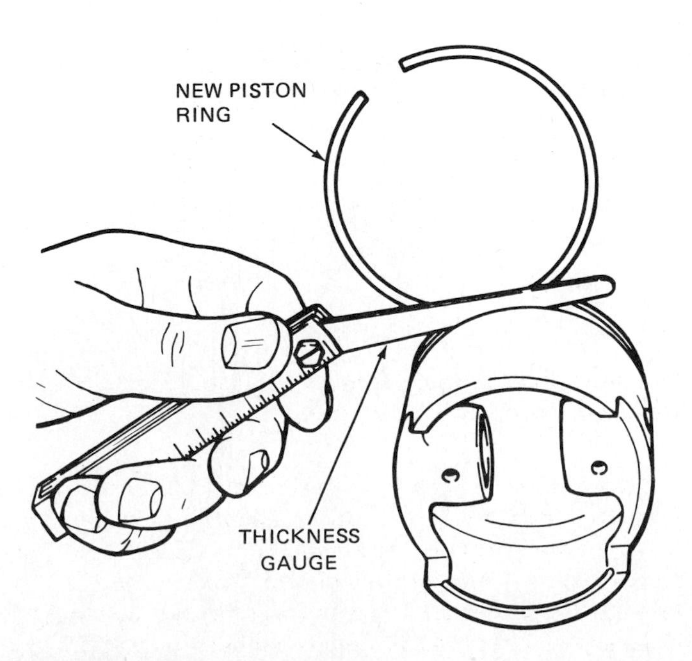

FIG. 20-17 Checking the side clearance of the piston ring in its groove. *(Kohler Company)*

land. This must be removed. Another way to check for ring-groove wear is to use a thickness gauge to measure the clearance between the ring and ring land (Fig. 20-17).

If the ring grooves appear satisfactory, check the ring gaps with the rings in the cylinder (Δ20-7). Then use the ring expander to install the rings on the piston (Fig. 20-11). In most two-cycle engines, the piston rings are pinned so they cannot move around (Fig. 20-18). The pin determines the type of ring and its proper position in the ring groove.

After the rings are in place on the piston, attach the piston to the connecting rod. Then install the cylinder and head as explained in Δ20-10.

Δ20-7 PISTON-RING SERVICE Before a piston ring is installed on a piston, the ring gap — the gap between the ends of the ring — must be checked. This is done by pushing the ring down in the cylinder with the piston turned upside down (Fig. 20-19).

Measure the gap with a thickness gauge. If the gap is excessive, do not use the ring. If there is no gap, you probably have the wrong ring for the job.

Δ20-8 CHECKING CONNECTING-ROD BEARINGS The piston is attached to the connecting rod in a variety of ways. In some, the piston pin is a press fit in the connecting rod. In others, the pin is supported in the rod with a sleeve or roller bearing (Fig. 20-20). The fit of the pin to the bearing should be checked, and if it is not correct, a new bearing, or a new connecting rod, will be required. In some engines using sleeve bearings in the rod small end, it is possible to ream the bearing to a larger size and use a larger-size pin. If this is done, the bearing surfaces in the piston must also be reamed to a larger size.

Before installing the cylinder or attaching the piston to the connecting rod, check the rod big-end bearing. Move the rod from one side to the other to see how much it will wobble (Fig. 20-21). If the rod small end can be moved more than a certain maximum, the rod big-end bearing is worn excessively. One manufacturer specifies that if the small end moves 0.080 inch [2 mm] or more, the big-end bearing requires service.

Δ20-9 INSTALLING THE PISTON After the rings are installed on the piston, attach the piston to the connecting rod with the piston pin. Coat the piston pin with oil before installing it. Some piston pins can be pushed in with a thumb. Others require a press or special tool (Figs. 20-22 and 20-23). If lock rings are used, install them.

Coat the cylinder and the piston and rings with oil, and install the cylinder over the piston and rings as explained in Δ20-10. A ring compressor is required to compress the rings into their grooves (Fig. 20-24). Then the cylinder can be slipped down over the piston.

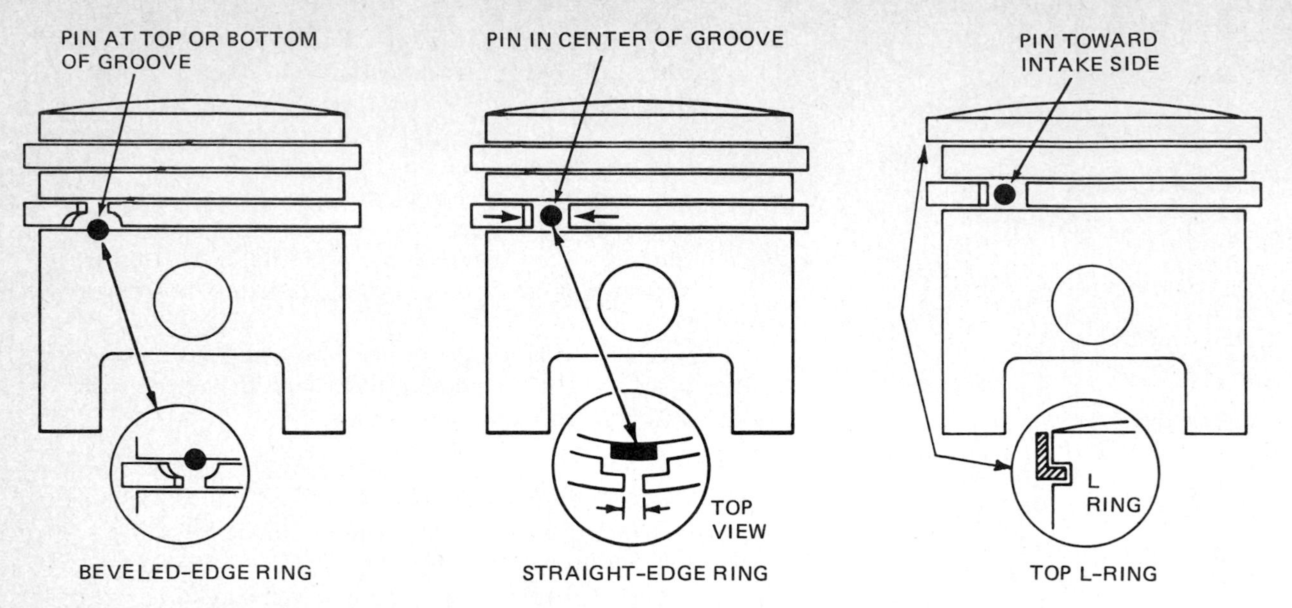

FIG. 20-18 Various pin-type piston rings used in small two-cycle engines. *(Kohler Company)*

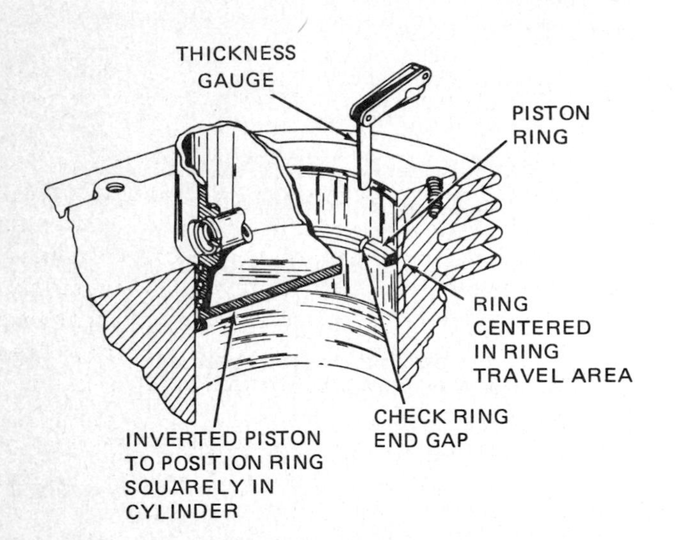

FIG. 20-19 Squaring ring in cylinder with a piston to measure the ring gap. *(Tecumseh Products Company)*

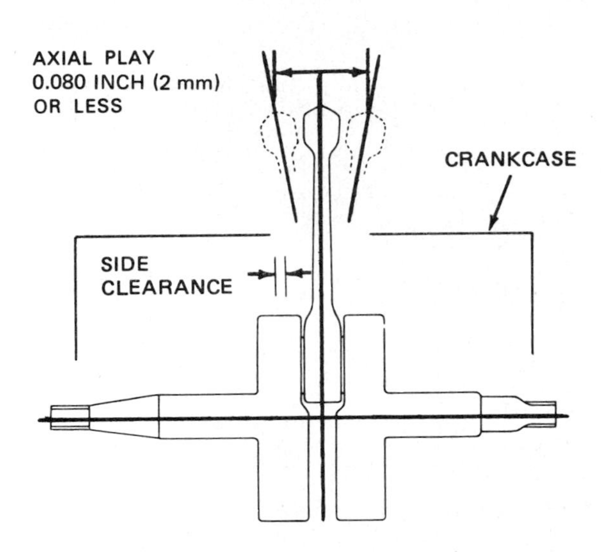

FIG. 20-21 Checking the condition of the big-end bearing by measuring the axial play in the connecting rod. *(Outboard Marine Corporation)*

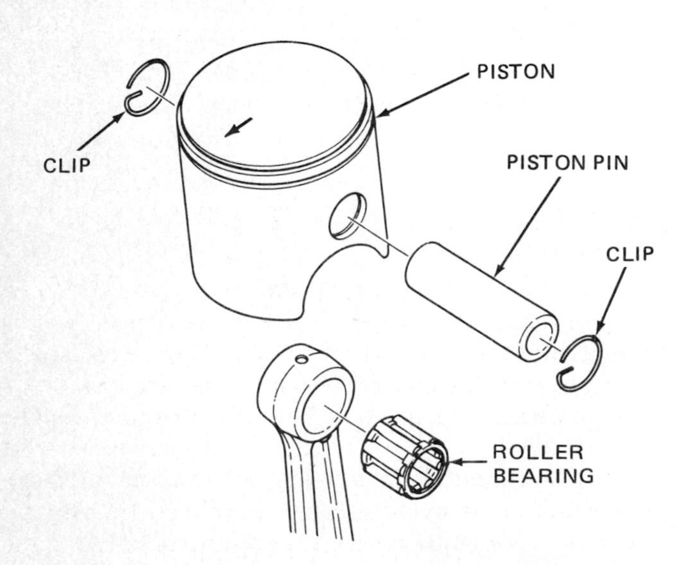

FIG. 20-20 Piston and connecting-rod assembly that uses a roller bearing in the small end of the connecting rod. *(Yamaha Motor Company)*

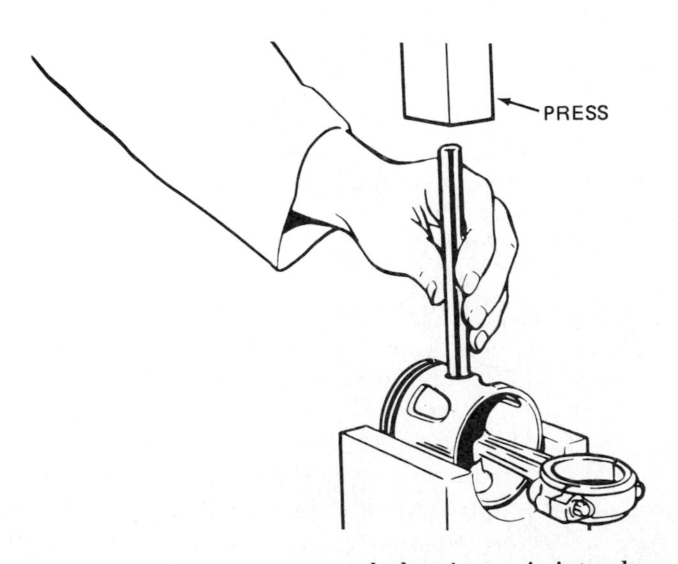

FIG. 20-22 Using a press to push the piston pin into place.

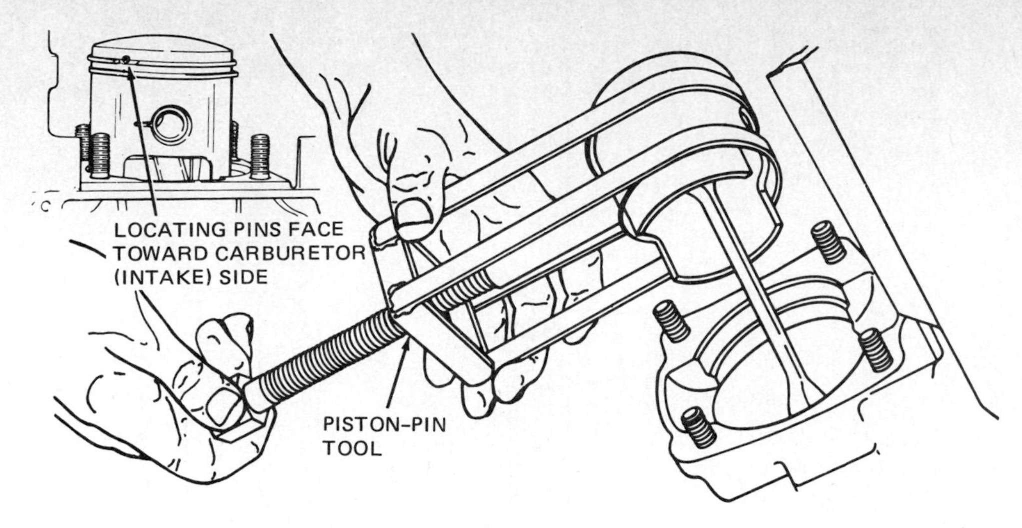

FIG. 20-23 **Using a special piston-pin tool to install a piston pin.** *(Kohler Company)*

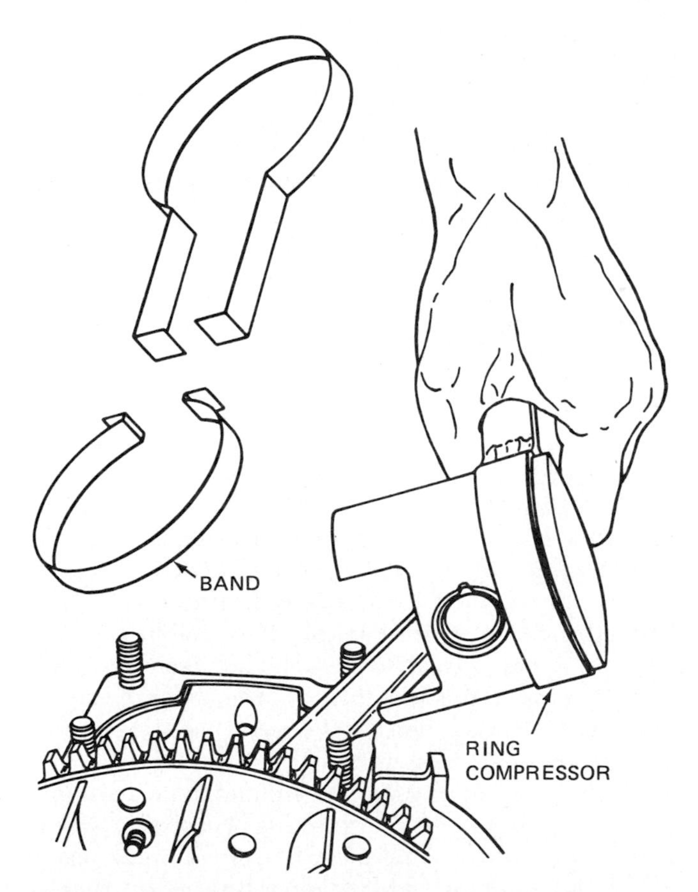

FIG. 20-24 A two-piece band-type ring compressor can be used to install the piston in many two-cycle engines. *(Kohler Company)*

Δ20-10 INSTALLING THE CYLINDER

Coat the cylinder bore with oil. Remove the cloth that you placed around the connecting rod to prevent dirt from entering the crankcase (Fig. 20-2). Install a new cylinder base gasket on the through bolts, and slide the gasket into place on the crankcase. Place the cylinder over the through bolts. Slide the skirt of the cylinder over the head of the piston. Use your fingers or a hinged ring compressor to compress each ring into the cylinder. When all rings are in the cylinder, slide the cylinder into place in the crankcase.

If a separate cylinder head is used, place a new head gasket on the cylinder. Then properly position the head and set it in place. Reinstall the washers and nuts on the through-bolt threads. Turn the nuts until they are finger-tight. Then use a torque wrench to torque the nuts to the manufacturer's specifications. Be sure to follow the torquing sequence recommended by the manufacturer.

Install the carburetor, being careful not to overtighten the nuts that hold it in place. Overtightening might cause the base of the carburetor to break off. Reinstall the shrouding and all other parts removed when you began the top-end overhaul.

Check that the engine has a supply of oil and gasoline mixed to the proper ratio in the fuel tank, or that the oil-injection oil tank is full. This assures ample lubrication during start-up. Then start the engine (Δ18-1).

Some manufacturers do not recommend any special break-in procedure. In general, operate the engine for the first few times at light load to give the piston rings and other parts a chance to seat. After 10 hours of operation, no further break-in procedures are required.

BOTTOM-END OVERHAUL

Δ20-11 CONNECTING-ROD SERVICE

The rod big end can be checked for looseness while it is still attached to the crankshaft (Δ20-8). The rod small end may have a needle bearing or a bushing. If a bushing or sleeve bearing is worn, it can be replaced in many rods. In others, the bushing can be reamed to a larger size and an oversize pin fitted. In some engines, the bushing is not serviceable. If the bushing is worn, replace the rod.

Some small engines have a built-up crankshaft (Δ3-12). In these engines, the crankshaft, connecting rod, and rod big-end bearing are replaced as an assembly. The big-end bearing usually is a needle bearing running in a one-

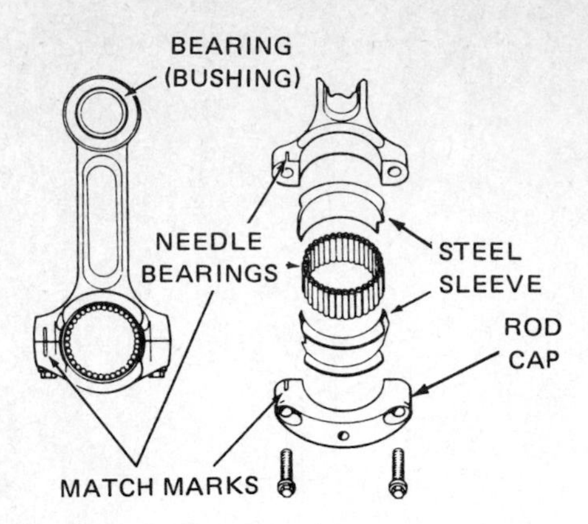

FIG. 20-25 When needle bearings are used in aluminum rods, steel liners or sleeves are placed between the bearings and the rod bore. *(Tecumseh Products Company)*

piece rod (Fig. 3-22). In most small engines, a cap can be removed from the big end of the rod (Fig. 20-25). Then the rod can be taken off the crankshaft and the needle bearing serviced.

Two-cycle engines use connecting rods made of either steel or aluminum. When needle bearings are used in aluminum rods, steel liners or sleeves are placed between the bearings and the rod bore (Fig. 20-25).

Δ20-12 CONNECTING-ROD NEEDLE-BEARING SERVICE

Before installing new needle bearings, check the crankshaft and crankpin. They must be clean and in good condition. Crankshaft-bearing service is described in Δ20-16.

Figure 20-26 shows two types of needle bearings: single-row and split-row. To install a new set of needles, lay the strip on your finger. Then carefully pull off the backing paper. Curl your finger, with the needles, around the crankpin (Fig. 20-27). The grease on the needles will hold them in place.

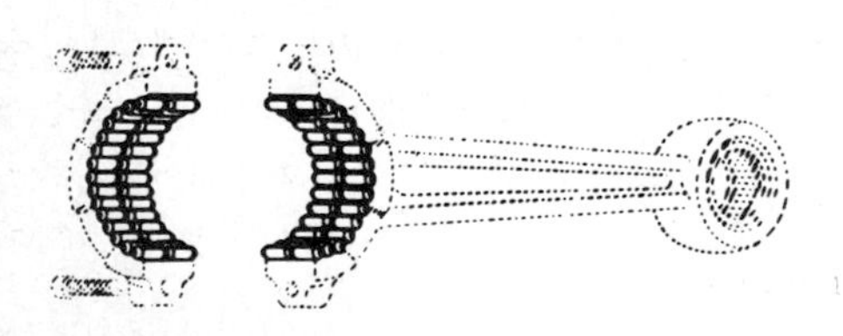

(*a*) SPLIT ROWS OF NEEDLE BEARINGS

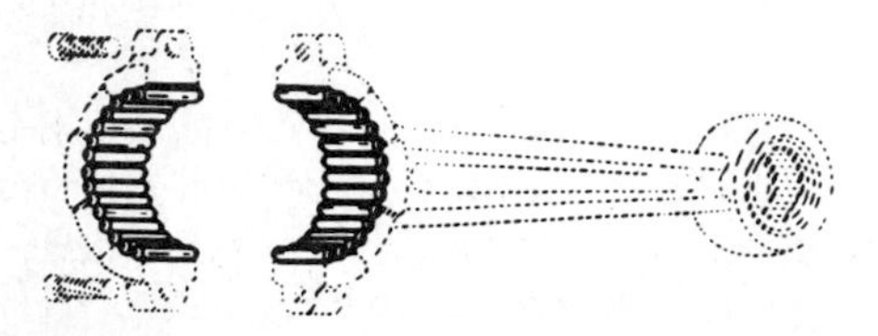

(*b*) SINGLE ROW OF NEEDLE BEARINGS

FIG. 20-26 Single-row and split-row needle bearings for connecting rods. *(Tecumseh Products Company)*

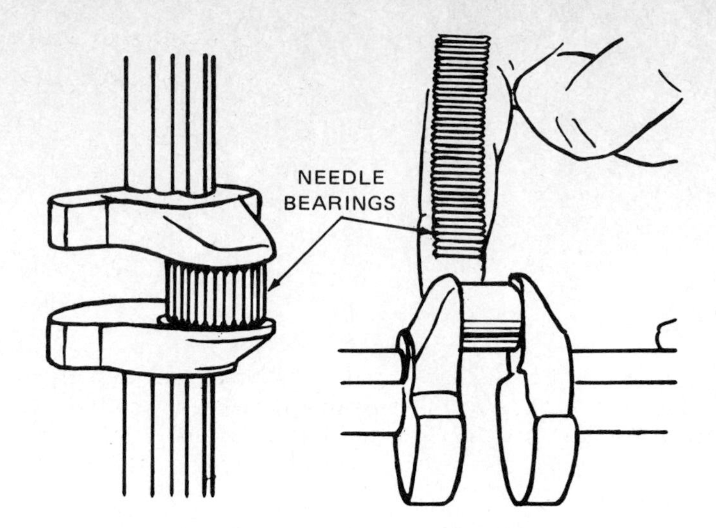

FIG. 20-27 Left, needles in place around the crankpin. Right, hold needles on finger to apply them to the crankpin. *(Lawn Boy Division of Outboard Marine Corporation)*

Δ20-13 CHECKING SLEEVE-BEARING CLEARANCE

There are three ways of checking the clearance between sleeve bearings (Δ2-17) and the crankshaft. These are with shim stock, Plastigage, and micrometer and telescope gauge. But before checking the clearance, inspect the crankpin for wear and roughness. If the crankpin is worn, it will require service or the crankshaft will require replacement (Δ20-20).

Shim Stock To check bearing clearance with shim stock, lubricate a strip of 0.001-inch [0.025-mm] shim stock that is about ¼ inch [6 mm] wide. Lay it lengthwise in the center of the bearing cap. Install the cap, and tighten the rod nuts to the specified torque. Try to move the rod endwise on the crankpin. If the rod has tightened up on the crankpin, the clearance between the bearing and the crankpin is less than the thickness of the shim stock. If the rod is still loose, the clearance is greater than the thickness of the shim stock.

Lay another shim on top of the first, and tighten the rod nuts again. Repeat the checking procedure. If the rod is still loose, add another piece of shim stock. Repeat the procedure until the rod locks up. The bearing clearance is the thickness of the shims required to lock up the rod. Compare this thickness with the manufacturer's specifications. Excessive clearance means a new bearing is required.

Plastigage Platigage is a plastic wire that flattens when pressure is applied to it. To use Plastigage, first make sure the bearing and crankpin are wiped clean of dirt and oil. Lay a strip of the Plastigage on the bearing in the rod cap (Fig. 20-28). Then install the cap and rod on the crankpin, and tighten the nuts to the specified torque. Do not move the crankshaft while the Plastigage is in place.

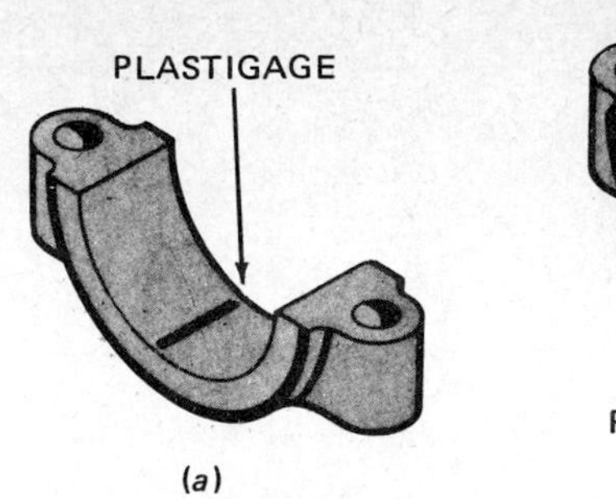

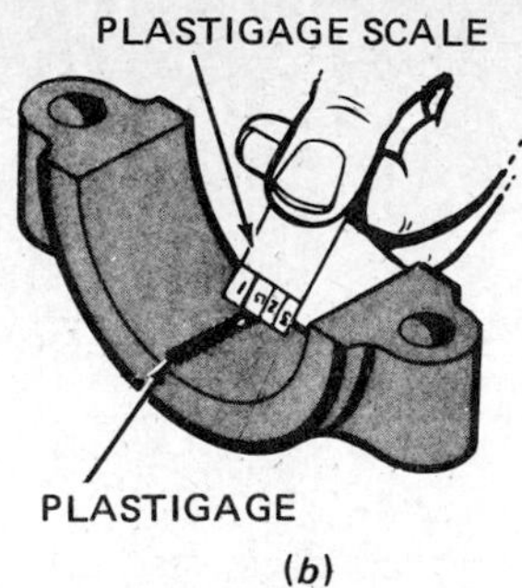

FIG. 20-28 Plastigage being used to check bearing clearance. *(a)* Lay a strip of Plastigage on the bearing in the rod cap. *(b)* After installing the cap and then removing it, measure the amount that the Plastigage has flattened to determine the clearance.

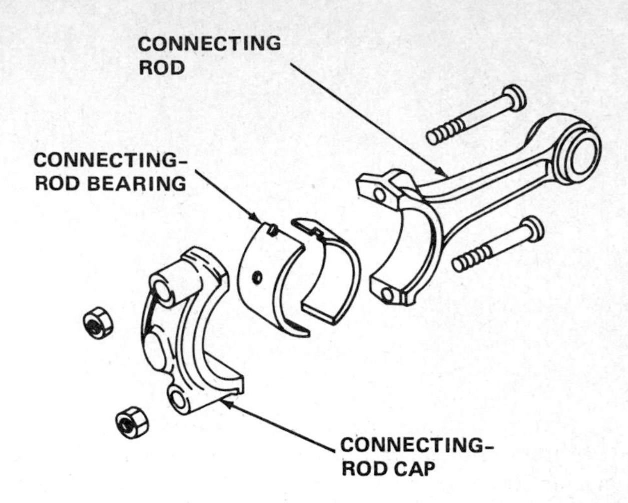

FIG. 20-29 Connecting-rod-and-bearing assembly. *(Federal-Mogul Corporation)*

Remove the cap and measure the amount that the Plastigage has flattened (Fig. 20-28). Use the scale that is printed on the Plastigage package. If the clearance is small, the Plastigage will have flattened considerably. If the clearance is relatively large, the Plastigage will not have flattened as much.

Micrometer and Telescope Gauge When a micrometer and telescope gauge are used, the diameter of the bearing is measured with the telescope gauge. Then the crankshaft is measured with a micrometer. The clearance is found by subtracting the crankshaft diameter from the bearing diameter.

Δ20-14 INSTALLING NEW SLEEVE BEARINGS

New connecting-rod bearings are required if the old ones are defective or have worn so much that the clearances are excessive. They also are required if the crankpins have become out-of-round or tapered. Then a new or reground crankshaft with new bearings must be installed (Δ20-20).

Keep the new bearings wrapped until you are ready to install them. Handle them carefully. Wipe each bearing with a clean cloth just before installing it. Be sure that the bore in the cap and rod is clean and not out-of-round.

Some manufacturers recommend a check of bore roundness with the bearing shells removed. The cap is attached and the nuts drawn up to specifications. Then a telescope gauge and micrometer are used to check the bore. If it is excessively out-of-round, a new rod should be installed. If the bore is satisfactory, install the bearing (Fig. 20-29). If the bearing halves or shells have locking tangs, be sure they enter the notches provided in the rod and cap.

Δ20-15 BONDED CONNECTING-ROD BEARING

On some connecting rods, the bearing is permanently bonded to the rod and cap. If this bearing is worn excessively, the complete rod must be replaced. However, some adjustment can be made to compensate for wear. This is done by removing shims from between the cap and rod. Clearance can be measured as for the replaceable sleeve bearings with shims, Plastigage, or telescope gauge and micrometer (Δ20-13). If clearance is excessive, shims can be removed from between the cap and rod to reduce it.

Δ20-16 CRANKSHAFT-BEARING SERVICE

The crankshaft of a single-cylinder engine is held at its ends by bearings (Δ2-17). A variety of bearings have been used. These include needle, ball, tapered roller, and sleeve.

Split-sleeve bearings are checked and serviced as described in Δ20-13 and Δ20-14. Other crankshaft sleeve bearings are complete bushings pressed into the crankcase end plates. The following sections describe the servicing of the bushing-type sleeve bearing and the needle, ball, and tapered roller bearings.

Δ20-17 CRANKSHAFT SLEEVE-BEARING SERVICE

Several different sleeve-bearing-and-crankcase combinations have been used on small engines. This is because of the various crankcase materials and designs. In some engines with aluminum crankcases, no separate bearing is used. Instead, the crankshaft is supported by, and turns in holes bored in, the aluminum itself. If these holes wear, bearings can be installed as explained later, or a new crankcase or end plates are required.

To check the bushing type of sleeve bearing, wipe the bearing clean. Then inspect it for wear, scoring, and other damage. One manufacturer supplies *reject* gauges for checking the bearings in its engine. If the gauge can enter the bearing (Fig. 20-30), the bearing is worn and should be replaced ("rejected"). If a reject gauge is not available, the bearing can be checked with a small-hole gauge or a telescope gauge and micrometer.

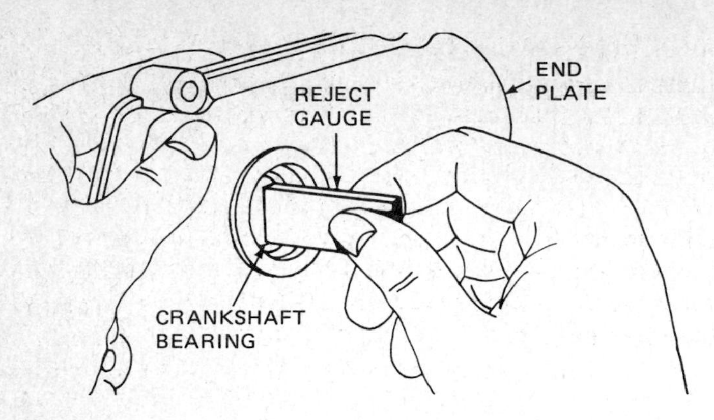

FIG. 20-30 Use a reject gauge to check the wear of a sleeve-type crankshaft bearing.

In single-cylinder engines, the exposed end of the crankshaft is called the *driven end,* or the *power-takeoff* (PTO) *end.* For example, the end of the crankshaft of a power mower on which the cutting blade mounts is the PTO end. The other end of the crankshaft is the magneto end. The two bearings that support the crankshaft often are identified as the PTO bearing and the magneto bearing.

If new sleeve bearings are required, reinstall the end plate. Remove the oil seal (Fig. 20-31). Then remove, install, and ream one bearing before removing the other bearing. In this way, the original bearing in the opposite end of the crankcase serves as a guide for reaming the new bearing. Then the newly installed and reamed bearing is used as a pilot for reaming the second bearing after it is installed.

To replace a sleeve bearing in the crankcase, first remove the PTO-end bearing. Use an arbor press or a bearing driver, and drive the old bearing toward the inside of the crankcase (Fig. 20-32). Support the crankcase or end plate around the bearing area while the old bearing is being removed and the new bearing installed. This prevents the bearing mounting area from distorting and the casting from cracking or breaking.

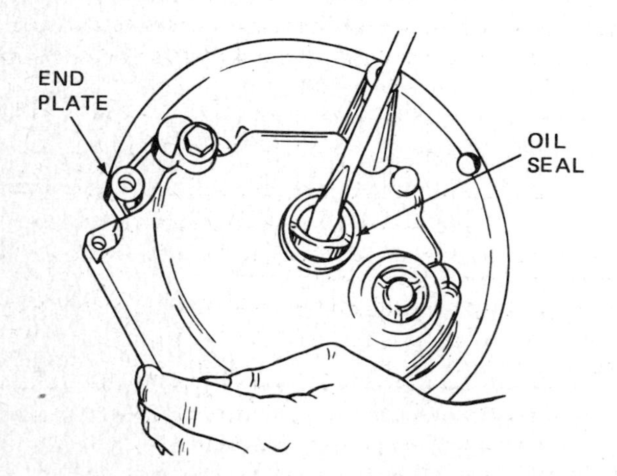

FIG. 20-31 Removing oil seal from the crankcase end plate. *(Tecumseh Products Company)*

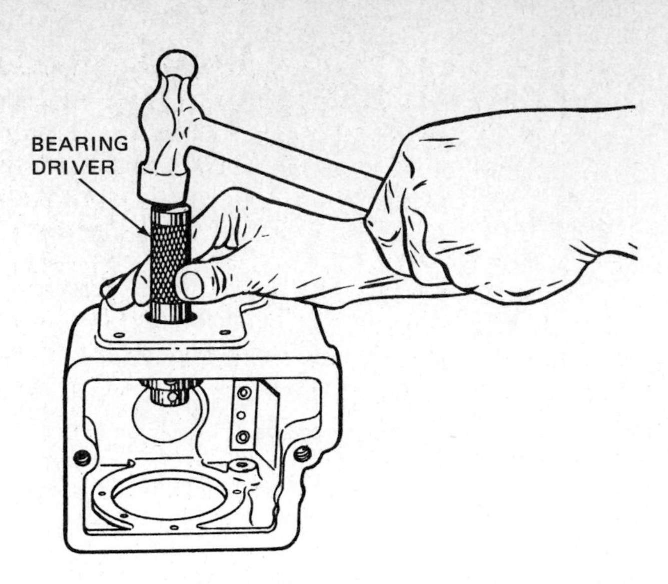

FIG. 20-32 Using a bearing driver to replace a sleeve bearing. *(Clinton Engines Corporation)*

Look for an oil hole in the crankcase and in the new bearing. If the holes are present, the new bearing must be installed so that the oil hole in the bearing aligns with the oil hole in the crankcase. Then press or drive the new bearing into place from the outside of the crankcase toward the inside. Drive the bearing into the crankcase to the proper depth, which is usually about 1/16 inch [1.6 mm]. This allows room for installation of the oil seal later.

When the sleeve bearing is in position, ream it to the correct size. Each engine manufacturer makes available the proper size of reamers.

Two different types of reamers are used. The difference is in the pilot used for the reamer. One type of reamer uses a guide bushing placed in the opposite bearing as the pilot (Fig. 20-33). With the guide bushing in place, reassemble the crankcase and perform the reaming operation.

Turn the reamer clockwise, slowly and steadily, until it is completely through the bearing. One manufacturer

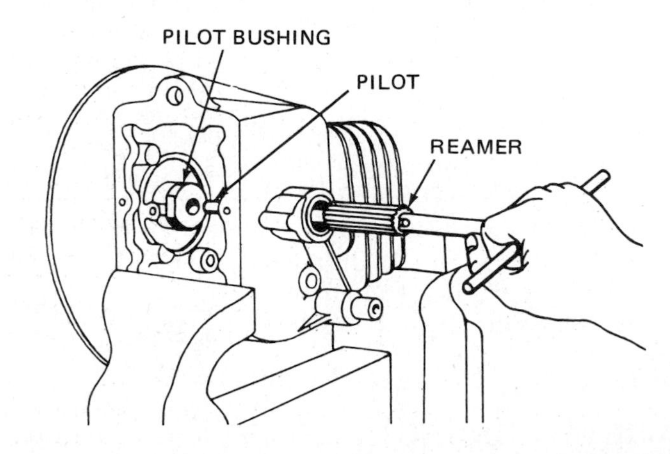

FIG. 20-33 Using a pilot and guide bushing in the original bearing to ream a newly installed sleeve bearing. *(Briggs & Stratton Corporation)*

recommends that the bearing be reamed dry, without oil. However, if the reamer cuts slightly large, then use oil. The oil causes the reamer to cut slightly smaller. When the reamer is completely through the bearing, remove the end plate and take out the reamer. *Do not back the reamer out of the bearing.* This will gouge the bearing surface and damage the reamer. Check the bearing diameter for correct size. Then carefully clean out all chips and metal particles.

Aluminum engines without removable bearings are reamed as outlined above to take a new replaceable bearing. However, a different size reamer is used. After the new bearing bore is reamed, *the bearing must be staked in place.* Make a notch in the bore with a chisel. The notch should be in the outer edge, opposite to where the split in the bearing will be after installation. Then install the bearing. With the chisel, notch the bearing above the notch in the bore. This will drive part of the bearing material into the outer notch and help prevent the bearing from turning. With the bearing staked in place, finish reaming it as outlined above.

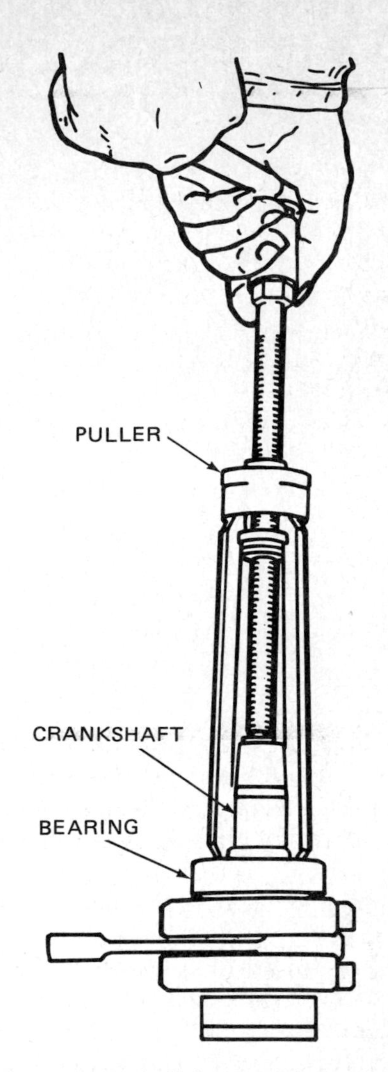

FIG. 20-34 Removing a ball bearing from the crankshaft with a puller. *(Kohler Company)*

△20-18 SERVICING CRANKSHAFT NEEDLE, ROLLER, AND BALL BEARINGS To service ball, needle, or tapered roller bearings, first determine if the bearings are damaged or worn. After the crankshaft is removed from the engine, wash the bearing and then dry it. Do not spin-dry the bearing with compressed air. Depending on the engine, the bearing will remain in the crankcase or on the crankshaft. *Do not remove the bearing until you have decided that it must be replaced.*

After the bearing is clean and dry, give it a thorough visual inspection for pits and discoloration. If the bearing appears to be in good condition, coat it with oil. Then rotate the inner race of the bearing so you can determine by *feel* whether the bearing is tight or has rough spots.

If the bearings are worn or damaged, they must be replaced. If the bearings are on the crankshaft, they should be pulled with a puller (Fig. 20-34), or pressed off with an arbor press. Then a new bearing can be pressed on.

If the bearings are in the crankcase, they may be pressed out or knocked out with a bearing driver. Another removal method is to put the crankcase half on a hot plate (Fig. 20-35). As the crankcase half reaches a temperature of about 400°F [204°C], the ball bearing should drop out. Tap the case lightly with a soft hammer to help loosen the bearing. The new bearing can be dropped into the case. Make sure the bearing seats all the way into the recess for it. Wear heavy gloves when you must handle the hot case.

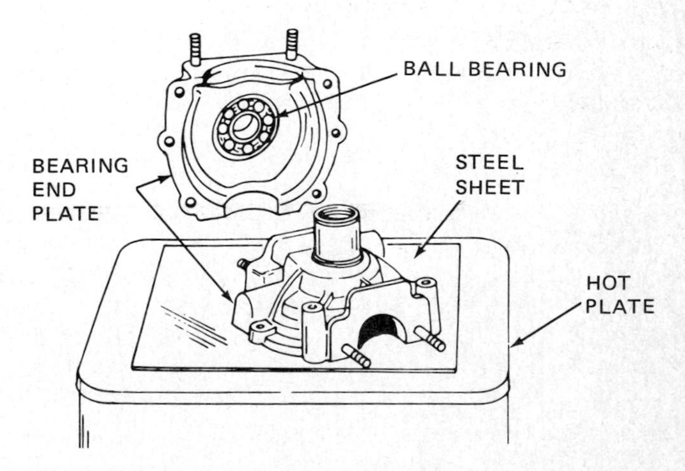

FIG. 20-35 Using a hot plate to heat the crankcase half for easier removal of the ball bearing. *(Tecumseh Products Company)*

When the crankshaft is mounted on roller bearings (Fig. 20-36), you can replace the outer race in the housing by pulling the race. Another way is to heat the housing until the race drops out. Then install a new race. The inner race on the crankshaft must be pulled out with a puller and a new race pressed on.

Crankshaft needle bearings should be cleaned, dried, and coated with oil. Then check the needle bearings and

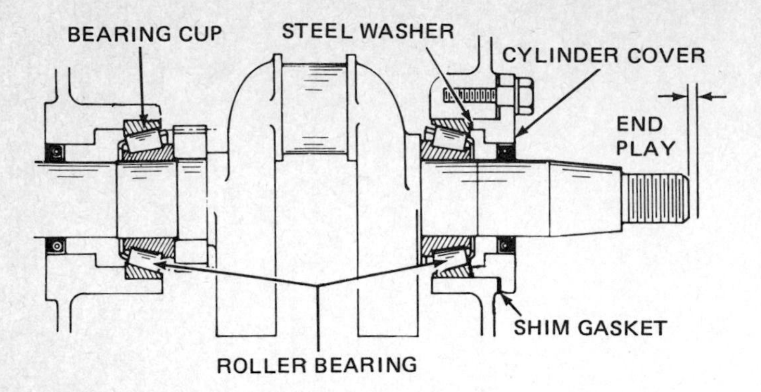

FIG. 20-36 Crankshaft mounted on tapered roller bearings. *(Tecumseh Products Company)*

their cage for wear. Replace the needle bearing when the needles are very loose or fall out of their cage. To install a new needle bearing, always drive the bearing on the end with the identification marks. But be careful! A damaged needle bearing will quickly damage the crankshaft.

△20-19 CRANKCASE-OIL-SEAL SERVICE

Crankcase-oil seals should be discarded and new seals installed every time the engine is given a complete overhaul. Usually, oil seals are damaged during removal. One oil-seal arrangement is shown in Fig. 20-37. The seal is held in place by a retainer and snap ring. To remove the snap ring and seal, use a pointed tool to pry the snap ring out of the spiral groove. This permits removal of the spring, retainer, and seal.

Be careful not to use too much force while removing the spring. Excessive force on the pointed tool may damage the thin crankcase or scratch the seal surface on the crankshaft.

When installing a new seal, it must be inserted squarely. The lip must not be deformed or torn during installation. Many technicians coat the outside of the seal case with Permatex No. 3 or other liquid gasket sealer. Then, while the seal is being driven into place, the liquid sealer will fill in any slight out-of-round condition of the seal or the seal bore. Some seals are made with a neoprene (plastic) coating for the same purpose. These can be installed without the use of liquid gasket sealer.

If you use a liquid sealer, be sure that it is applied to the seal and not to the bore in the block. Do not coat the bore in the block and then drive the seal in. Doing so allows the gasket sealer to get into the lip and may cause it to leak. Also, the sealer could possibly cause a leak by running into and blocking off the oil drain hole.

△20-20 CRANKSHAFT SERVICE

The crankshaft should be inspected for wear of the journals and for distortion. The crankshafts of power mowers, for example, can be bent if the cutting blade should strike a solid object a glancing blow. This could put a stress on the crankshaft that would bend it. A quick check for a bent crankshaft can be made with the crankshaft still in the engine. Remove the spark plug and crank the engine. Watch the end of the crankshaft for wobble. If there is wobble, the crankshaft is bent and it should be replaced.

With the crankshaft out of the engine, it can be inspected for roughness, discoloration, cracks and breaks, and stripped threads on the ends. Check the keyway for any enlarging or other damage that might have resulted from a loose flywheel, adapter, or pulley. Check the taper on the crankshaft where the flywheel mounts for wear or damage. Any damage, except for battered threads, means the crankshaft must be discarded. Bat-

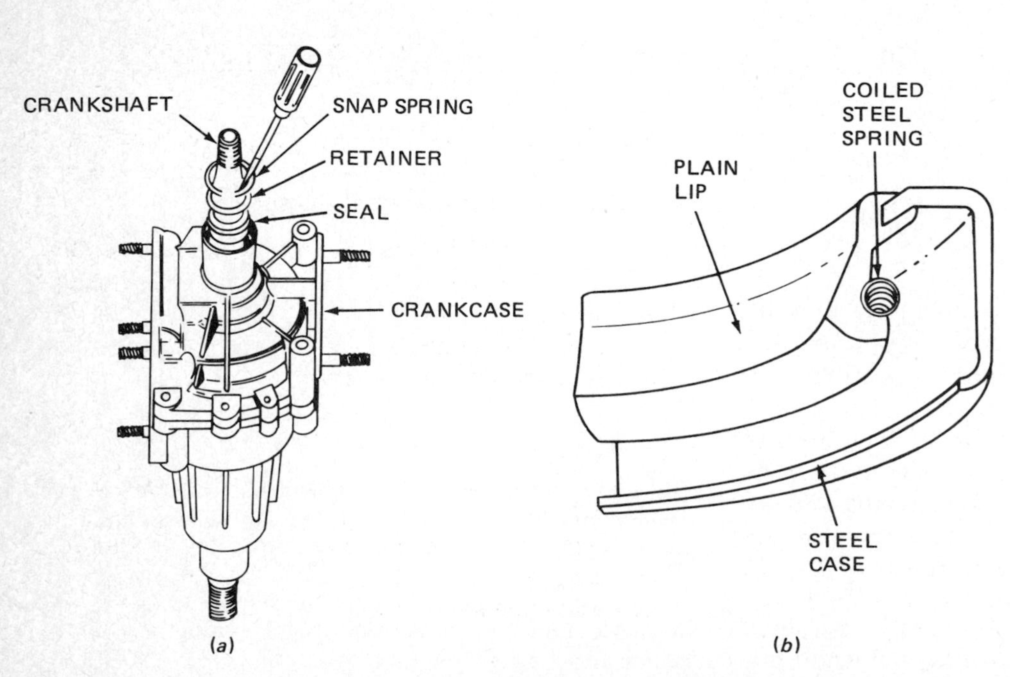

FIG. 20-37 *(a)* Removing the snap ring to permit removal of the oil seal. *(b)* A cutaway oil seal. *(Selastomer Division of Microdot, Inc.)*

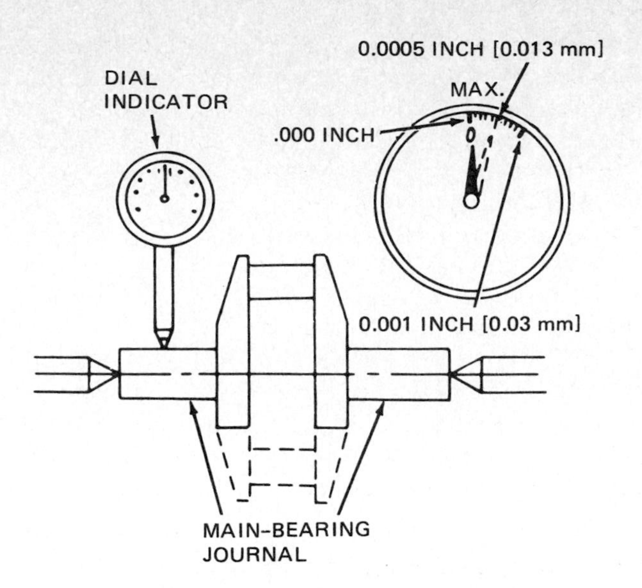

FIG. 20-38 Using a dial indicator to check main-bearing journals. *(Tecumseh Products Company)*

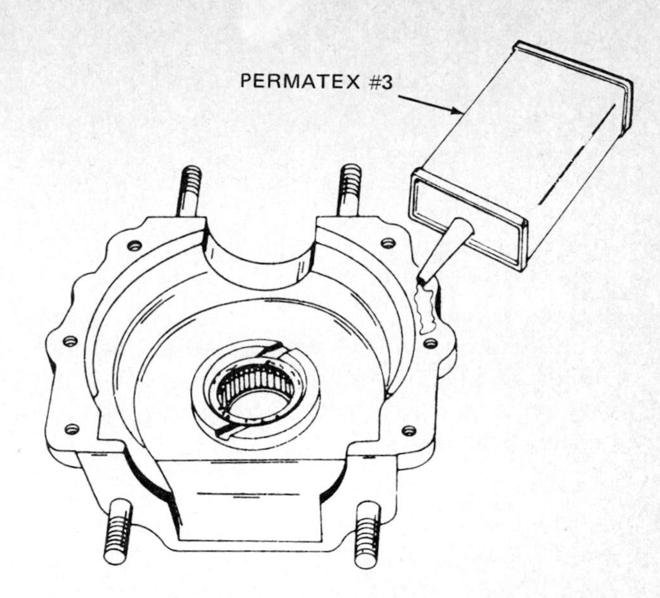

FIG. 20-39 Applying sealant to the contact face of one-half of the split crankcase. *(Tecumseh Products Company)*

tered threads can often be cleaned with a thread chaser.

Figure 20-38 shows various checks to be made on a crankshaft. After a thorough visual inspection, use a micrometer or dial indicator to check the crankpin and main journals. Mount the crankshaft on centers so it can be rotated. As the crankshaft rotates, any irregularity will cause the dial-indicator needle to move. If journals are rough, out-of-round, or tapered, the crankshaft should be discarded.

Δ20-21 REED-VALVE SERVICE When overhauling a two-cycle engine, always clean all dirt and oil from the reeds and the adapter or reed-valve plate. Do this carefully to avoid damaging reeds. If reeds are bent, damaged, or broken, replace the reed-valve assembly. Check also, on engines using a reed-valve stop, to make sure the stop is not bent or broken. Use a thickness gauge to check how much the reeds bend away from the base plate, or adapter (Fig. 19-5). One manufacturer specifies a maximum of 0.010 inch [0.25 mm]. If the reeds bend more than this, or are otherwise damaged, replace them.

Do not attempt to check reed-valve action with compressed air. This can damage the reeds. Reed valves are checked by visual inspection only.

Δ20-22 CRANKCASE GASKETS Always use new gaskets on engine reassembly. Old gaskets are probably hard and will not provide a good seal. In addition, they may have been damaged or destroyed during engine disassembly. Make sure that the sealing surfaces on the engine are clean, but do not scrape them. Instead, use lacquer thinner on a clean cloth to wipe traces of sealer or gasket material from the surfaces.

On the split-crankcase engine, the two halves of the crankcase are sealed by a bead of liquid gasket sealer. Apply the sealer to the contact face of one of the halves (Fig. 20-39).

Δ20-23 OVERHAULED-ENGINE BREAK-IN Allow a new or overhauled engine to work up to full power gradually. On two-cycle engines, adjust the carburetor for a fairly rich mixture for the first 10 hours. Follow the instructions on the nameplate attached to the engine or equipment. See also **Δ**18-2 for further information on operating a new or rebuilt engine.

CHAPTER 20: REVIEW QUESTIONS

Select the *one* correct, best, or most probable answer to each question. You can find the answer in the section indicated at the right of each question.

1. On some two-cycle engines, the head and cylinder (**Δ***20-2*)
   a. are one piece
   b. cannot be removed
   c. are high-strength steel
   d. are part of the piston-and-rod assembly
2. Worn or tapered cylinders should be (**Δ***20-5*)
   a. discarded
   b. honed or bored undersize and smaller pistons installed
   c. honed or bored oversize and larger pistons installed
   d. steam-cleaned

3. Piston clearance is (Δ*20-6*)
   a. the piston diameter plus ring diameter
   b. the difference between the piston diameter and ring diameter
   c. the difference between the piston diameter and cylinder diameter
   d. the distance between the piston head at TDC and the cylinder head
4. Plastigage is used to (Δ*20-13*)
   a. check connecting-rod sleeve-bearing clearance
   b. adjust connecting-rod bearing clearance
   c. reduce connecting-rod bearing clearance
   d. none of the above
5. Mechanic A says that a new seal must be inserted squarely. Mechanic B says the seal lip must not be deformed during installation. Who is right?
   a. mechanic A (Δ*20-19*)
   b. mechanic B
   c. both A and B
   d. neither A nor B

CHAPTER

21

Four-Cycle-Engine Service

After studying this chapter, you should be able to:

1. ***Adjust valve clearance on various types of small engines.***
2. ***Check the valve springs and tappets.***
3. ***Service valve guides.***
4. ***Reface valves.***
5. ***Refinish valve seats.***
6. ***Install a valve-seat insert.***
7. ***Inspect and check a camshaft.***
8. ***Perform a complete valve-service job.***

Δ21-1 CAMSHAFT AND VALVE-TRAIN SERVICE Many servicing procedures are the same for two-cycle and four-cycle engines. These include most operations on the cylinder, piston, piston rings, connecting rod, crankshaft, bearings, and seals. Chapter 20 covered these operations in detail.

However, the four-cycle engine requires additional service on the camshaft and valve train. These operations include adjusting valve clearance, reconditioning valves and valve seats, and servicing the camshaft and its bearings. The following sections describe how to perform these services.

Δ21-2 VALVE TROUBLE DIAGNOSIS The valves in a four-cycle engine must open and close with definite timing in relation to the piston position (Δ4-9). They must seal tightly against the valve seats and open and close promptly. The clearance between the valve stems and valve guides must be correct. Failure of the valves to meet any of these requirements means valve trouble.

Valve troubles include valve sticking, valve burning, valve and seat breakage, valve-face and seat wear, valve-seat recession, and valve deposits. The chart on the next page lists the valve troubles and the checks or corrections to be made for each condition.

ADJUSTING VALVE CLEARANCE

Δ21-3 VALVE ADJUSTMENTS Most small four-cycle engines must have the valves adjusted periodically. This adjustment sets the proper amount of clearance in the valve train (Fig. 21-1). There must be some clearance to assure complete closing of the valves. The adjustment procedure varies with the type and design of engine. The procedure has several names, such as *adjusting valve-lifter clearance, adjusting valve-tappet clearance,* and *adjusting valve lash.* However, all refer to the same basic adjustment.

In many engines, an adjustment screw is turned to change the valve clearance (Δ21-4 and 21-5). When no adjustment screw is provided, the clearance is adjusted by removing the valve and grinding off the tip end of the valve stem (Δ21-11).

Some engines using hydraulic valve lifters normally never require a valve-clearance adjustment. Others require checking and adjusting whenever valve-service work is performed. The sections that follow provide typical valve-adjustment procedures.

Δ21-4 L-HEAD ENGINE WITH MECHANICAL VALVE LIFTERS Some engines should be cold when the clearance is checked. Other engines

VALVE TROUBLE-DIAGNOSIS CHART

Complaint	Possible cause	Check or correction
1. Valve sticking	a. Deposits on valve stem	See item 6
	b. Worn valve guide	Replace guide
	c. Warped valve stem	Replace valve
	d. Insufficient oil	Service lubricating system; add oil
	e. Cold-engine operation	Valves become free as engine warms up
	f. Overheating valves	See item 2
2. Valve burning	a. Valve sticking	See item 1
	b. Distorted valve seat	Check cooling system; tighten cylinder-head bolts
	c. Valve-tappet clearance too small	Readjust
	d. Valve spring cocked or worn	Replace
	e. Overheated engine	Check cooling system
	f. Lean air-fuel mixture	Service fuel system
	g. Preignition	Clean carbon from engine; use cooler plugs
	h. Detonation	Adjust ignition timing; use higher-octane fuel
	i. Valve-seat leakage	Use an interference angle
	j. Overloaded engine	Reduce load or try heavy-duty valves
	k. Valve-stem stretching from strong spring or overheated engine	Use weaker spring; eliminate overheating
3. Valve breakage	a. Valve overheating	See item 2
	b. Detonation	Adjust ignition timing; use higher-octane fuel; clean carbon from engine
	c. Excessive tappet clearance	Readjust
	d. Seat eccentric to stem	Service
	e. Cocked spring or retainer	Service
	f. Scratches on stem from improper cleaning	Replace valve; avoid scratching stem when cleaning valves
4. Valve-face wear	a. Excessive tappet clearance	Readjust
	b. Dirt on valve face	Check air cleaner
	c. See also causes listed under item 2	
5. Valve-seat recession	a. Valve face cuts valve seat away	Use coated valves and valve-seat inserts
6. Valve deposits	a. Gum in fuel (intake valve)	Use proper fuel
	b. Carbon from rich mixture (intake valve)	Service fuel system
	c. Worn valve guides	Replace
	d. Carbon from poor combustion (exhaust valve)	Service fuel, ignition system, or engine as necessary
	e. Dirty or wrong oil	Service lubricating system; replace oil

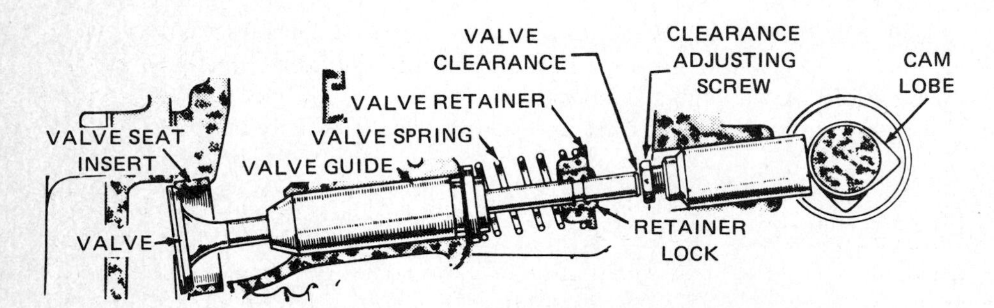

FIG. 21-1 Valve clearance in the valve train of an L-head engine. *(Onan Corporation)*

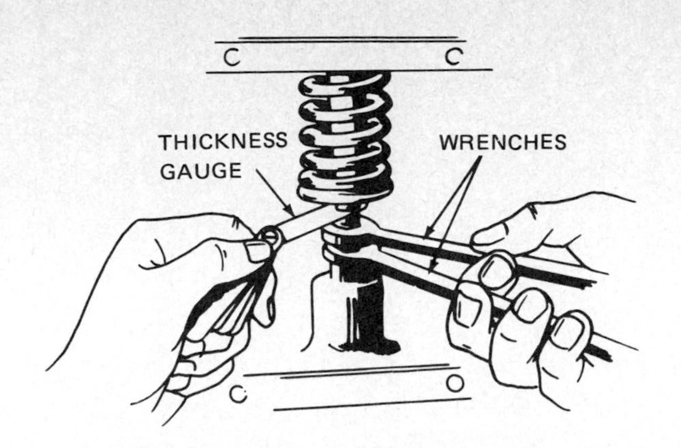

FIG. 21-2 Adjusting valve clearance on an L-head engine.

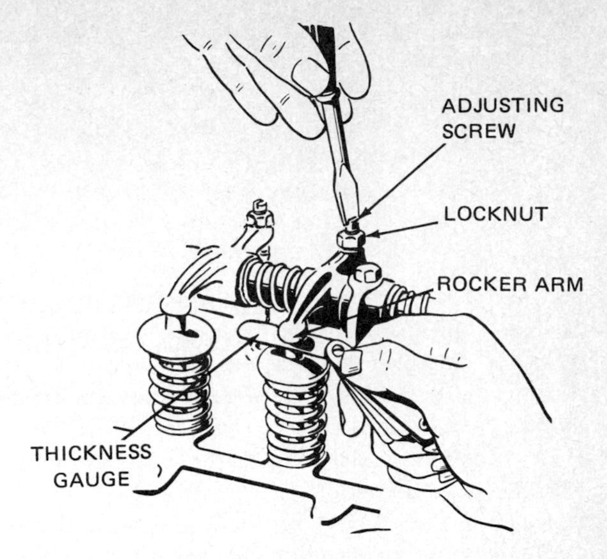

FIG. 21-3 To adjust valve clearance on an overhead-valve engine, loosen the locknut on the rocker arm. Then turn the adjusting screw until the thickness gauge is a slip-fit between the rocker arm and the valve stem.

should be warmed up and idling. Follow the manufacturer's recommendation for the engine you are working on.

To adjust the valves, remove the valve-cover plates. Then use a thickness gauge to check the clearance between the valve stem and the adjusting screw in the valve lifter (Fig. 21-2). If the clearance is not correct, turn the adjusting screw until the proper clearance is obtained. Adjustment is correct when the thickness gauge can be moved between the screw and the valve stem with a slight drag.

When a lock nut is used on the adjusting screw, two wrenches are required (Fig. 21-2). The lock nut should be tightened after the adjustment is made. Then the clearance should be checked again.

After the adjustment is complete, install the valve-cover plates. Use new valve-cover gaskets.

Δ21-5 OHV ENGINE WITH MECHANICAL VALVE LIFTERS Most manufacturers recommend checking valve clearance with the engine cold and not running. Remove the valve cover. Then measure the clearance between the valve stem and the tip end of the rocker arm (Fig. 21-3). The clearance is measured with the valve lifter on the base circle of the cam. Turn the crankshaft with the starter until the base circle of the cam is under the valve lifter.

On ball-stud-mounted rocker arms (Fig. 21-4), turn the self-locking nut to make the adjustment. Turning the nut down reduces the valve clearance.

Δ21-6 ENGINES WITH HYDRAULIC VALVE LIFTERS On some engines with hydraulic valve lifters, no adjustment is provided in the valve train. In normal service, no adjustment is necessary. The hydraulic valve lifter takes care of any small changes in the valve-train length. However, adjustment may be needed if valves and valve seats are ground. Unusual and severe wear of the pushrod ends, rocker arm, or valve stem may also require adjustment. Then some correction

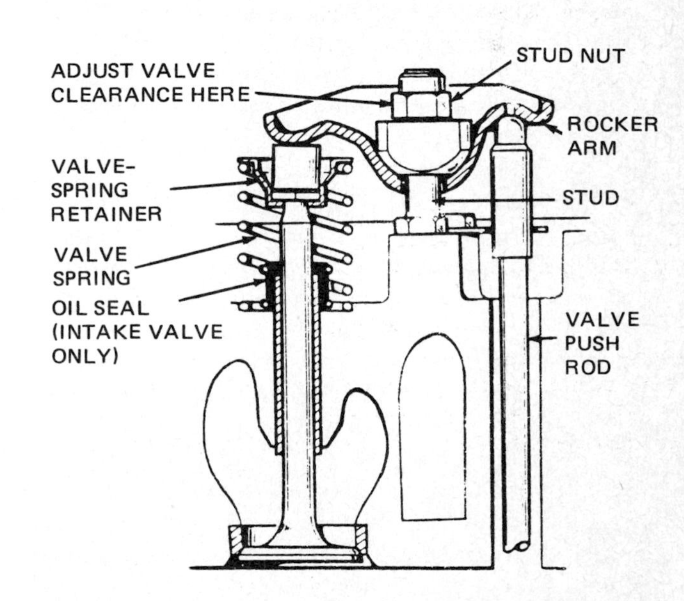

FIG. 21-4 Adjusting valve clearance on an overhead-valve engine that has the rocker arms mounted on studs in the cylinder head. Backing out the stud nut increases the clearance. *(Onan Corporation)*

may be required to reestablish the correct valve-train length. The procedures vary, so follow the steps in the manufacturer's service manual.

Δ21-7 STEPS IN THE COMPLETE VALVE JOB In addition to adjusting valves, other jobs relating to the valves may be required on four-cycle engines. These include removing the cylinder head, removing and servicing the valves, servicing the valve seats and valve guides, and installing new valve-seat inserts.

The procedures that follow apply generally to single-cylinder L-head and overhead-valve engines. They may also apply to other engines.

The details of valve and valve-seat service are described in the following sections. Listed below are the steps for performing a complete valve job. Some of the steps may not apply to a specific engine that you are servicing.

1. Remove the air cleaner and disconnect the throttle linkage, fuel line, and any air and vacuum hoses from the carburetor.
2. Remove or set aside the necessary lines and hoses to get at the cylinder head.
3. Disconnect the spark-plug wire and temperature-sending unit, if used.
4. Remove the crankcase ventilating system, if used.
5. On overhead-valve engines, remove the carburetor and intake manifold. On many L-head engines, it is not necessary to remove the carburetor.
6. Remove the rocker-arm cover or covers.
7. On engines with rocker arms supported on shafts, remove the shaft assembly or assemblies. Then remove the pushrods in order.
8. Remove the head bolts. Take the head off the engine.
9. Remove the valves and springs from the head or block. To avoid damaging or breaking the valve guide, do not pull or drive out a valve with a mushroomed stem end. The mushroom must be removed by grinding it off with a grinding stone. Keep the removed valves and springs in proper order so that they can be put back into the positions from which they were removed.
10. Check valves and valve seats. Clean the valve heads and stems on a wire wheel. Grind the valve seats and reface valves as necessary. Check valve seating. Reface and chamfer valve-stem ends if necessary. If you are installing new coated valves, do not reface them. Refacing or lapping coated valves removes the protective coating which shortens valve and seat life.
11. Check rocker arms for wear. Service or replace them as necessary.
12. Check valve guides for wear. Clean, replace, or knurl and ream for same-size valve stem if necessary. Or ream for a larger-diameter valve stem.
13. Reinstall valves and springs.
14. Install head, pushrods, rocker arms, rocker-arm cover, and other parts removed during head removal.
15. Check and adjust valve-stem clearance as necessary.

Δ21-8 VALVE REMOVAL Three types of valve-spring retainers are shown in Fig. 4-9. To remove the pin or split type, use a spring compressor (Figs. 21-5 and 21-6). Adjust the jaws of the compressor until they just touch the top and bottom of the valve chamber. This will keep the upper jaw from slipping into the coil of the spring.

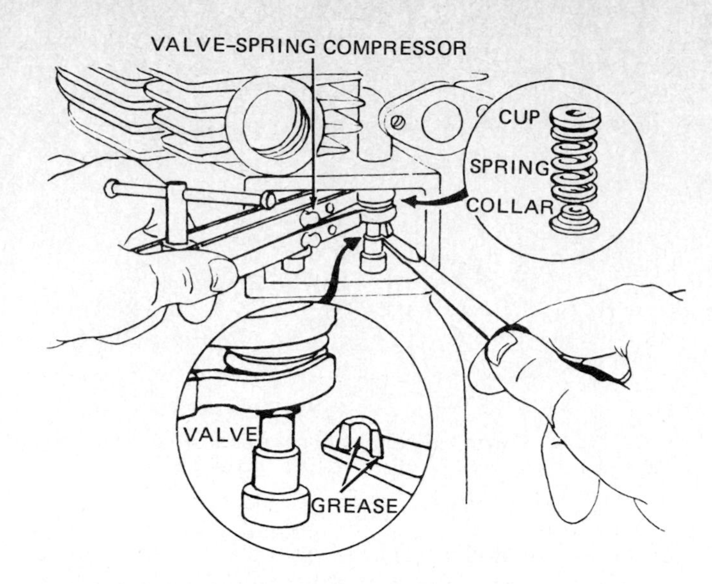

FIG. 21-5 Removing the valve spring on an engine using a split-collar retainer. *(Briggs & Stratton Corporation)*

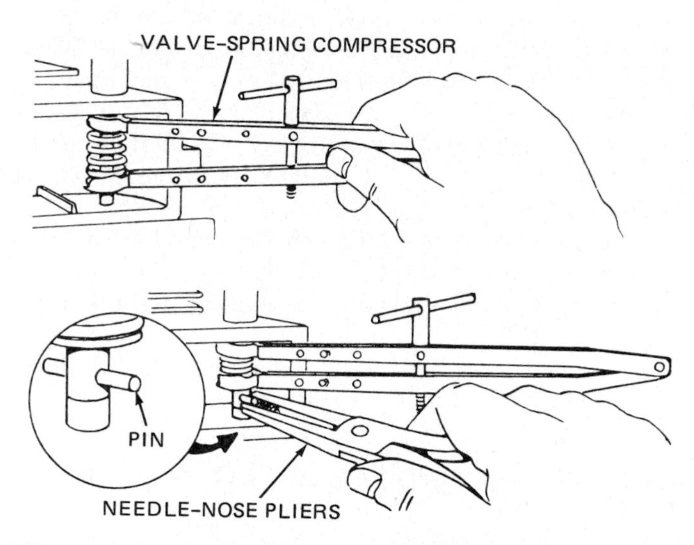

FIG. 21-6 Removing the valve spring on an engine using a pin retainer. *(Briggs & Stratton Corporation)*

Push the compressor in until the upper jaw slips over the upper end of the spring. Then compress the spring by tightening the jaws.

On the split-collar type, put a little grease on a screwdriver to remove the retainer (Fig. 21-5). On the pin type, use needle-nose pliers to pull out the pin (Fig. 21-6).

To remove the one-piece retainer (Fig. 21-7), move the retainer around so the larger part of the opening clears the undercut in the valve. Figure 21-8 shows a C-type spring compressor being used to compress the spring.

After the retainer is removed, lift the valve out and remove the compressor. The spring may come out in the jaws of the spring compressor (Fig. 21-6).

Do not interchange parts between valves or cylinders. Use a rack to keep valves and valve-train parts in proper order.

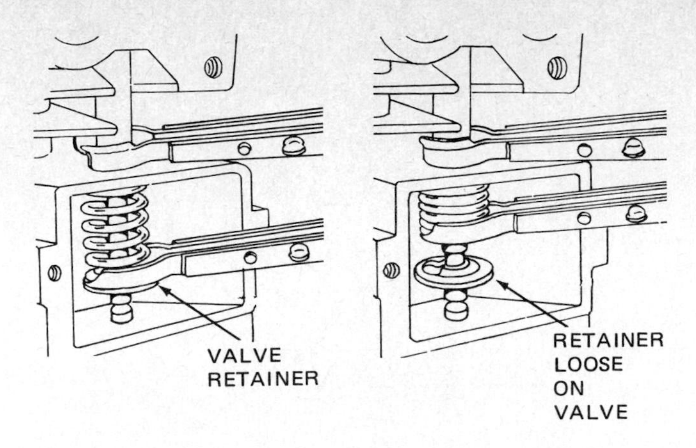

FIG. 21-7 Removing the valve spring on an engine using a one-piece retainer. *(Briggs & Stratton Corporation)*

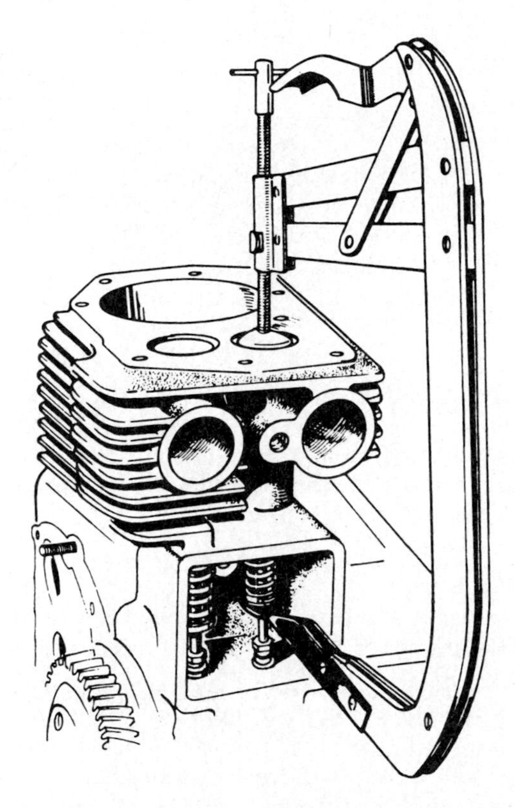

FIG. 21-8 Using a C-type spring compressor to compress the spring. *(Onan Corporation)*

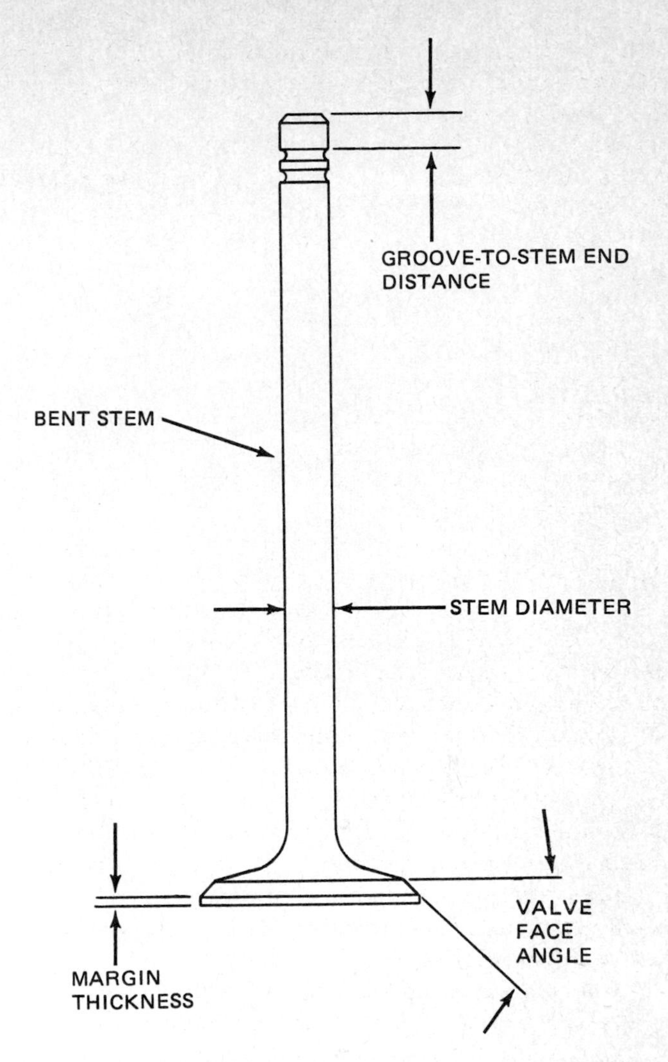

FIG. 21-9 Parts of the valve to be checked. For the dimensions, refer to the manufacturer's specifications.

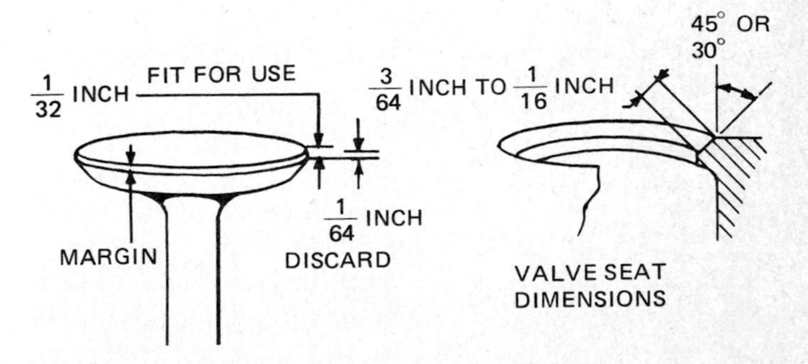

FIG. 21-10 Typical valve and valve-seat dimensions. *(Briggs & Stratton Corporation)*

Δ21-9 SERVICING VALVES

Valves should be inspected for wear, burned spots, pits, cracks, and other damage. If the valve face appears to be in good condition but worn, it can be refaced on a valve-refacing machine or special valve lathe. Figure 21-9 shows the valve parts to be checked. Figure 21-10 shows typical valve and seat dimensions recommended by Briggs & Stratton.

If any part of the valve is badly worn, bent, distorted, or damaged, replace the valve. After refacing (Δ21-10), any valve that has a margin of less than 1/32 inch [0.8 mm] should also be replaced.

Clean the carbon off the valves with a wire wheel. Always wear goggles to protect your eyes from flying particles of metal and dirt. Polish the stems, if necessary, with fine emery cloth. Do not take off more than the dirty coating on the surface. Never scratch the stem or remove any metal from it.

Δ21-10 REFACING VALVES

The next step in reconditioning a valve is to reface it. This usually requires a valve-refacing machine (Fig. 21-11).

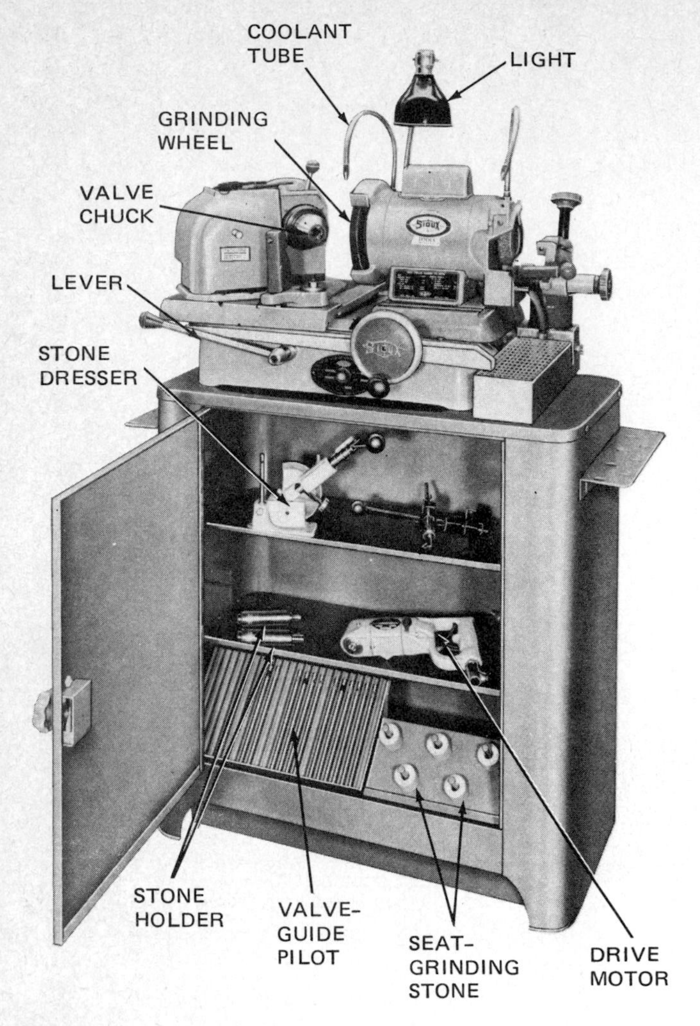

FIG. 21-11 A valve-refacing machine. The valve-seat grinding set is in the cabinet under the valve refacer. *(Sioux Tools, Inc.)*

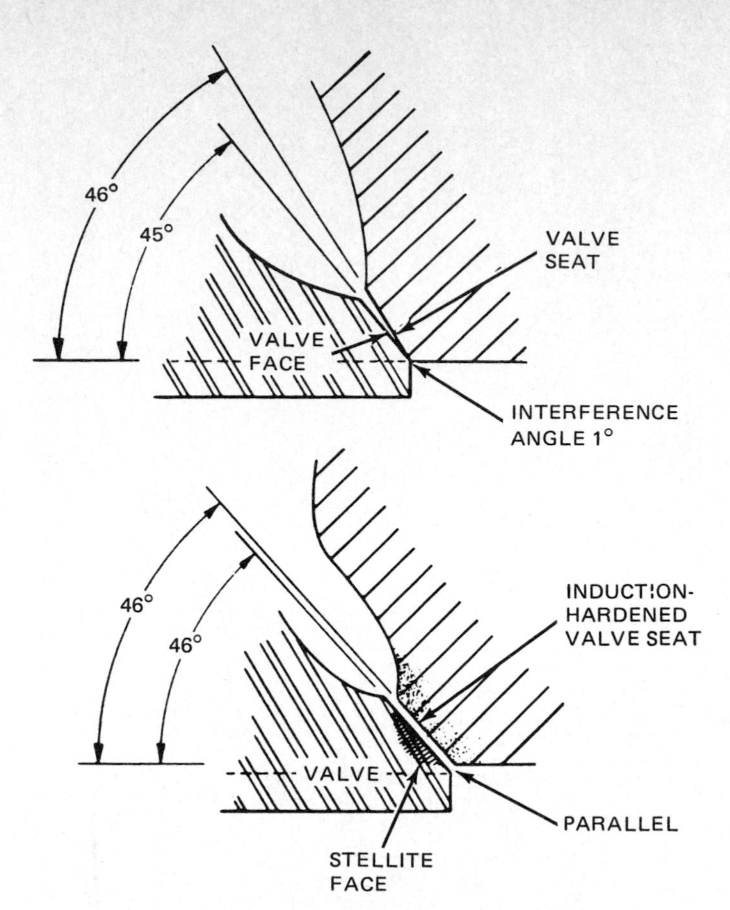

FIG. 21-12 Valves and valve-seat angles. Top, interference angle. Bottom, an induction-hardened valve seat. The valve shown is faced with stellite, which is extremely resistant to heat and wear. The angles on the valve face and seat (at bottom) are parallel.

The valve refacer has a grinding wheel, a coolant-delivery system, and a chuck which holds the valve for grinding. Set the chuck to grind the valve face at the specified angle. This angle must just match the valve-seat angle, or make an interference angle of ¼ to 1° (Fig. 21-12). Then put the valve into the chuck and tighten the chuck. The valve should be placed in the chuck so that the part of the stem that runs in the valve guide is gripped by the chuck.

To start the operation, align the coolant feed so that it sprays coolant on the rotating valve face. Then start the machine. Move the lever to carry the valve face across the grinding wheel. The first cut should be a light one. If this cut removes metal from only one-half or one-third of the face, the valve may not be centered in the chuck. Or the valve stem is bent, and the valve should be discarded. Cuts after the first should remove only enough metal to true the surface and remove pits. Do not take heavy cuts. If so much metal must be removed that the margin is lost, discard the valve. Loss of the margin causes the valve to run hot. Then it will soon fail.

If new valves are required, reface them lightly, provided they are not of the coated type. *Never reface or lap coated valves!*

Follow the operating instructions of the valve-refacer manufacturer. Dress the grinding wheel as necessary with the diamond-tipped dressing tool (Fig. 21-13). As the tool is moved across the rotating face of the grinding wheel, the diamond cleans and aligns the grinding face.

▵21-11 REFACING VALVE-STEM TIPS

If the tip of a valve stem is rough or worn unevenly, it can be ground lightly. Use the special attachment furnished with the valve-refacing machine (Fig. 21-14). The attachment allows you to swing the valve slightly and rotate it. In this way, the tip can be ground to produce a slightly crowned, or rounded, end. One recommendation is to grind off as much from the stem as you ground off the valve face. In that way, you make up for the amount the valve sinks into the seat as a result of face grinding.

The ends of some valve stems are hardened. These should have no more than 0.010 inch [0.25 mm] ground off them. Excessive grinding exposes soft metal. This causes the stem to wear rapidly.

Δ21-12 CHECKING VALVE SPRINGS AND TAPPETS

Valve springs should be tested for proper tension and for squareness. A valve-spring tester is required to check spring tension (Fig. 21-15). The force required to compress the spring to the proper length should be measured. Then the spring should be checked for squareness (Fig. 21-16).

Stand the spring on a surface plate and hold a steel square next to it. Rotate the spring slowly against the square and see whether the top coil moves away from the square. If the spring is excessively out-of-square or lacks sufficient tension, discard it.

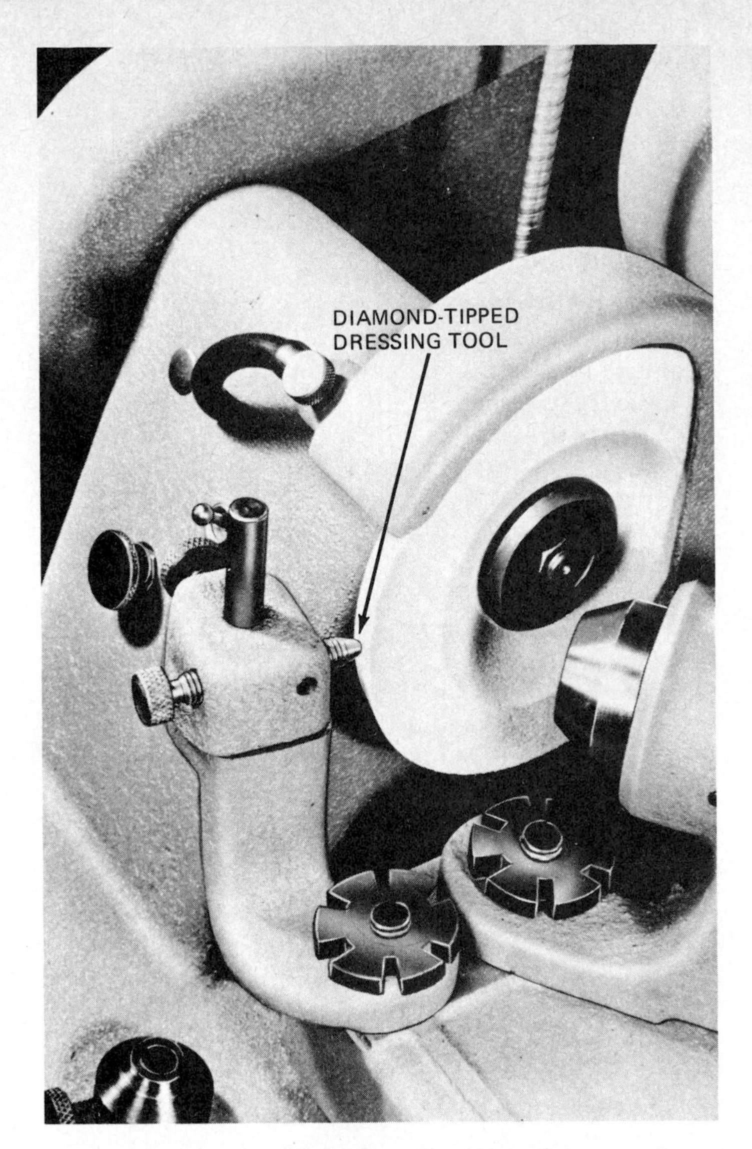

FIG. 21-13 Using a diamond-tipped dressing tool to dress the valve-grinding wheel. *(Snap-on Tools Corporation)*

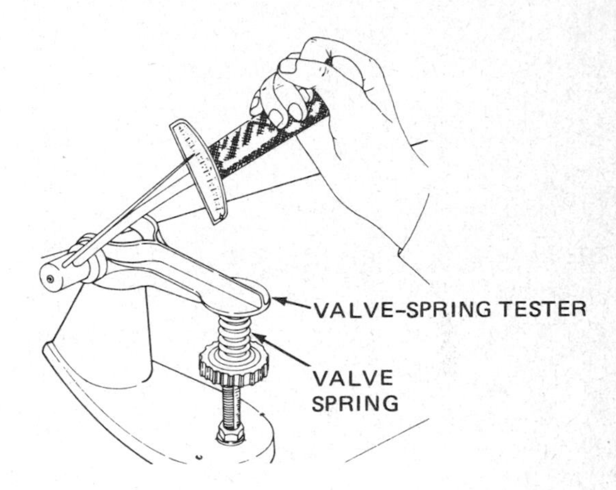

FIG. 21-15 Checking valve-spring tension. *(Outboard Marine Corporation)*

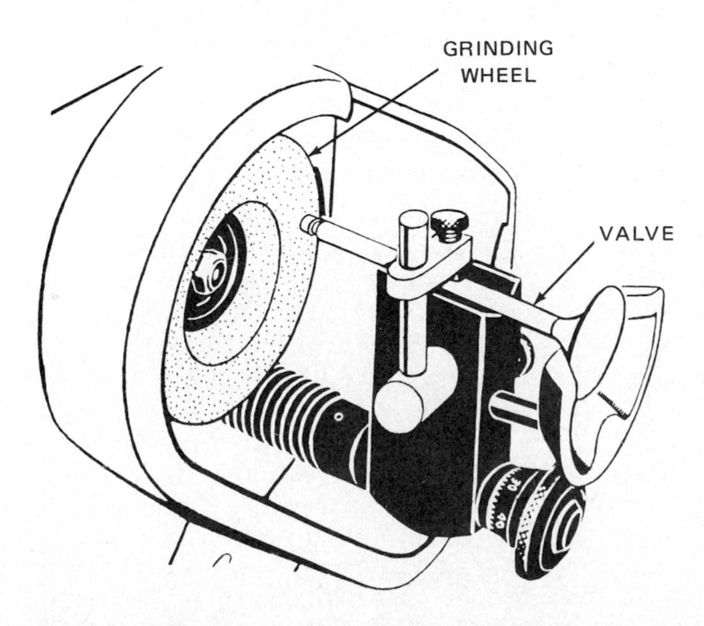

FIG. 21-14 Grinding the tip end of the valve stem. *(Snap-on Tools Corporation)*

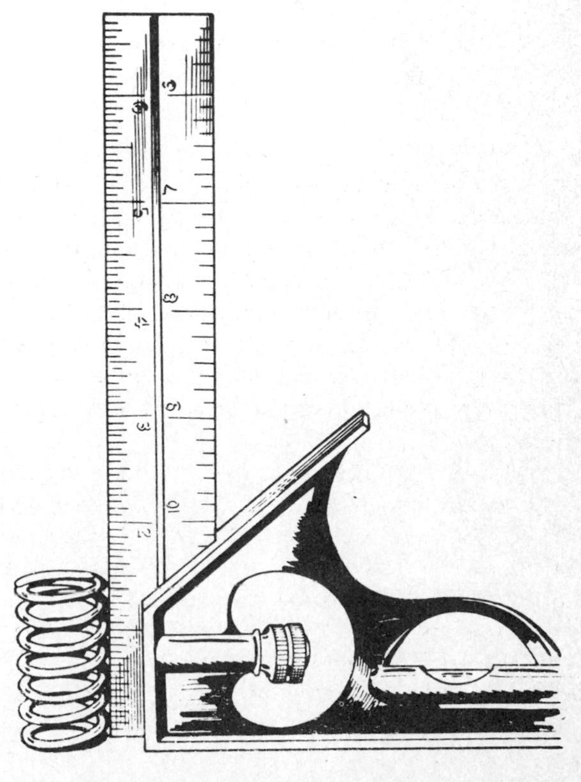

FIG. 21-16 Checking valve-spring squareness. *(Onan Corporation)*

Check the valve-tappet faces which ride on the cams for roughness or wear. Check for wear on the adjusting-screw head which is in contact with the valve stem. If the tappets do not have adjusting screws, check the stem end of the tappets. If you find wear or roughness, new tappets will be required. The camshaft must be removed to remove the tappets (Δ21-17).

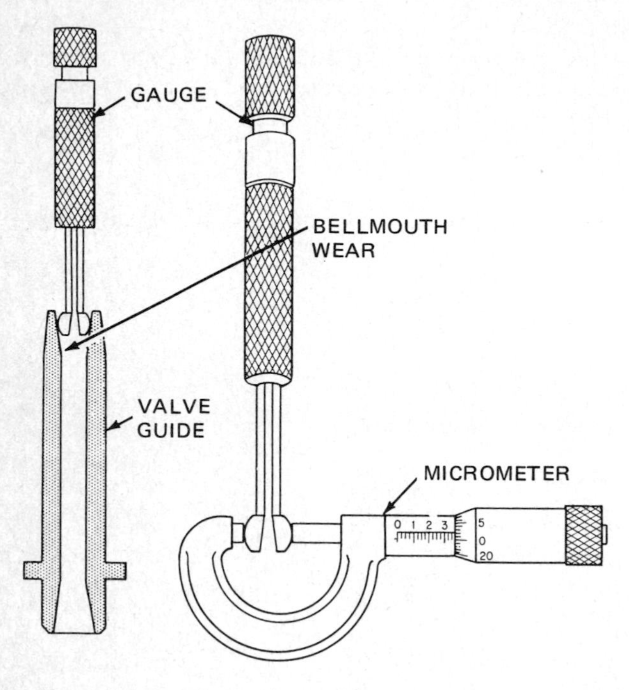

FIG. 21-17 Measuring valve-guide wear with a small-hole gauge. Left, the gauge is adjusted so that the split ball is a drag fit in the guide. Right, then the split ball is measured with a micrometer.

Δ21-13 SERVICING VALVE GUIDES

The valve guide must be clean and in good condition for normal valve seating. It must be serviced before the valve seats are reconditioned. Clean the valve guide with a wire brush or adjustable-blade cleaner. Then check the guide for wear.

A worn valve guide requires service. The type of service depends on whether the guide is replaceable or integral. If it is replaceable, the old guide should be pressed out. Then a new guide should be installed and reamed to size. If the guide is integral, you can service it in either of two ways:

1. By reaming the guide to a larger size and installing a valve with an oversize stem
2. By knurling and reaming the guide

The valve guide may wear bell-mouthed (Fig. 21-17) or oval-shaped. This is because the valve tends to wobble as it opens and closes. A small-hole gauge can be used to measure oval or bell-mouth wear. The split ball is adjusted until it is a light drag fit at the point being checked. Then the split ball is measured with a micrometer. By checking the guide at various points, any eccentricity will be detected.

If the guide is worn, it should be rebushed (Fig. 21-18). First, a bushing guide is used to center the reamer. Then the reamer is turned to ream out the worn guide. Ream to only 1/16 inch [1.6 mm] deeper than the valve-guide bushing. Use a soft-metal driver of brass or copper to avoid peening over the top end of the bushing. Finally, finish reaming the new bushing to the proper size so that a standard valve can be used.

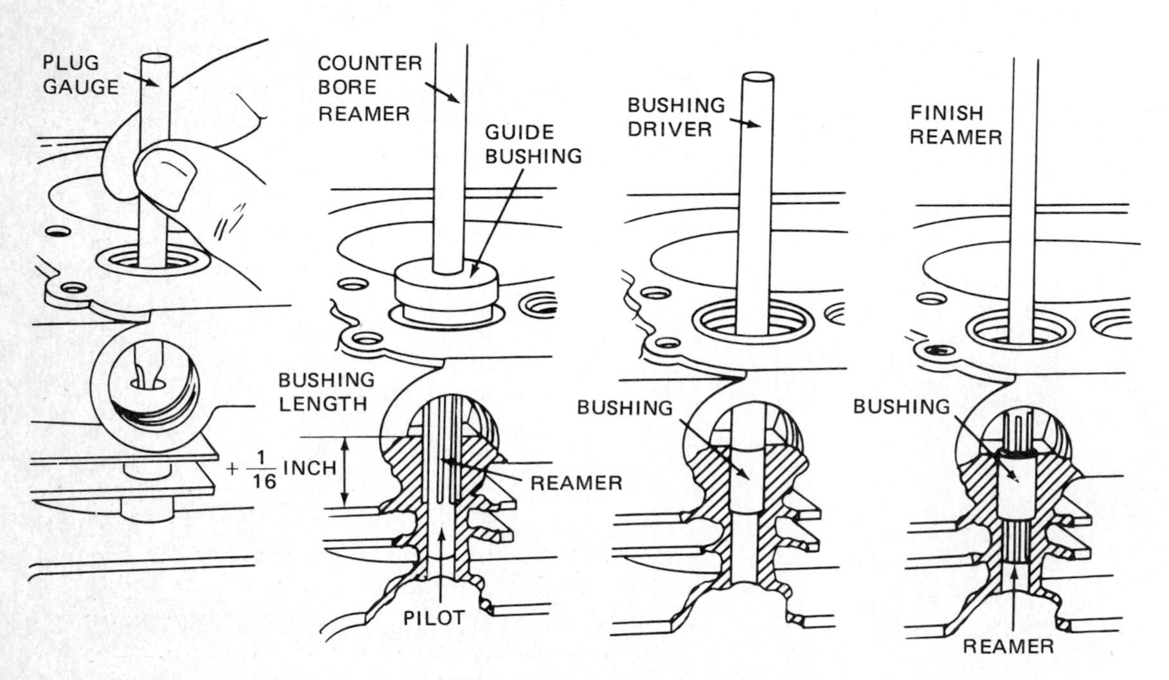

FIG. 21-18 Checking, installing, and reaming valve guides. *(Briggs & Stratton Corporation)*

To remove the replaceable type of valve guide on some L-head engines, drive the guide down into the valve-spring compartment.

New valve guides can be installed with a special driver or arbor press. Valve guides must be installed to the proper depth in the block or head. Then they must be reamed to size. This is usually done in two steps: a rough ream and then a second (or final) finishing ream. Figure 21-18 shows both the depth of installation of the new valve guide and the reaming dimensions for one model of small engine.

Δ21-14 RECONDITIONING VALVE SEATS

Valve seats are of two types. The integral type is actually the cylinder block or head. The insert type is a ring of special metal set into the block or head. Reconditioning valve seats is described below. Replacing valve-seat inserts is described in Δ21-16.

If the valve seat is pitted, scored, or otherwise damaged, it should be trued with a valve-seat grinder or a valve-seat cutter (Fig. 21-19). If the seat is so badly worn or burned that it will not clean up, it is possible to counterbore the seat and install a seat insert.

As a first step in grinding a valve seat, make sure that the valve guide is in good condition. This step is necessary because the seat-grinding stone is centered in the valve guide. How to check and service valve guides was described in Δ21-13.

The valve-seat grinder rotates a grinding stone of the proper shape on the valve seat (Fig. 21-20). The stone is kept concentric with the valve seat by a pilot installed in the valve guide (Fig. 21-21). This means that the valve guide must be cleaned and serviced before the valve seat is ground. In the grinding set shown in Fig. 21-20 and 21-21, the stone is automatically lifted about once a revolution. This permits the stone to clear itself of grit and dust by centrifugal force.

Figure 21-21 shows how the seat-grinding stone is centered by means of a pilot installed temporarily in the valve guide. In operation, an electric or air-powered motor rotates the grinding stone and smooths the seat. The angle of the stone determines the angle to which the seat will be ground. This angle matches the angle to which the valve face is ground in the valve-refacing machine (Δ21-10). To maintain the angle, the grinding stone must be dressed frequently with the diamond-tipped dressing tool. [Not having to do this is one advantage to using a valve-seat cutter (Fig. 21-19).]

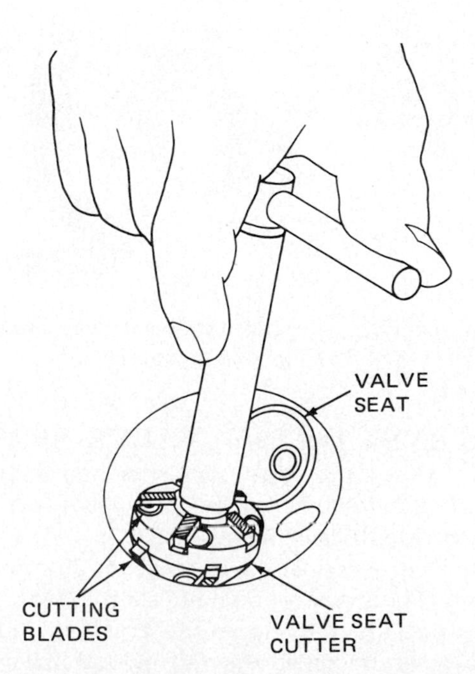

FIG. 21-19 Using a valve-seat cutter with carbide blades to recondition a valve seat. *(Neway Manufacturing, Inc.)*

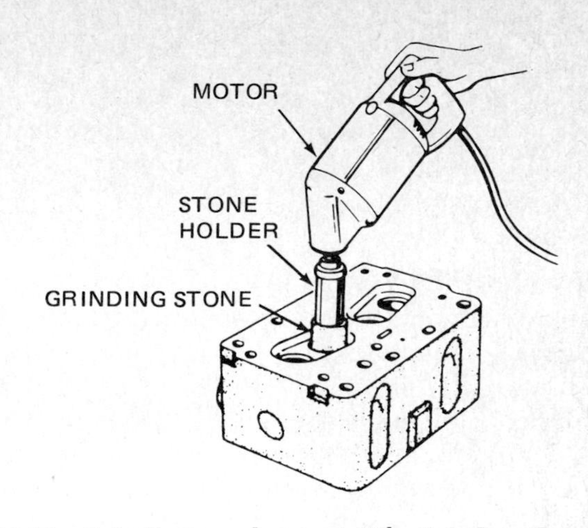

FIG. 21-20 Grinding a valve seat. *(The J. I. Case Company)*

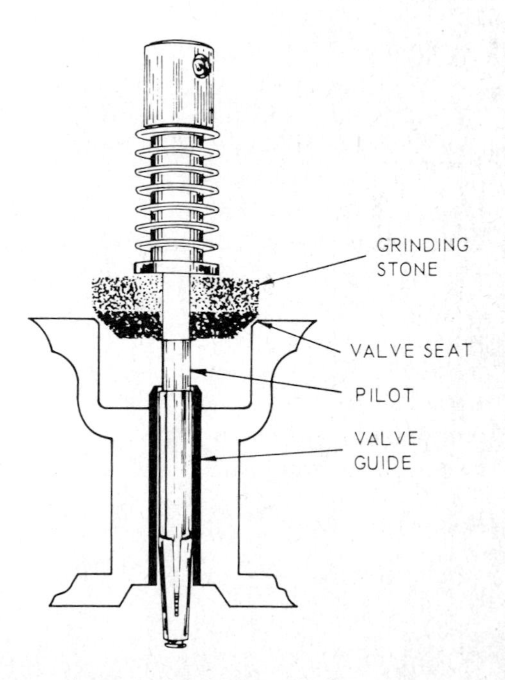

FIG. 21-21 Pilot on which the grinding stone rotates. The pilot keeps the stone concentric with the valve seat. *(Black and Decker Manufacturing Company)*

After the valve seat is ground, it may be too wide. If so, the seat must be narrowed. Use 15 and 75° grinding stones to grind away the upper and lower edges of the seat. A steel scale can be used to measure valve-seat width.

Δ21-15 LAPPING THE VALVES

Some small-engine manufacturers recommend lapping the valves. This is done when new or reground valves are installed and after the valve seat is refaced. Then the valve face and valve seat should be rubbed together with lapping compound to check and perfect the fit.

Lapping compound is an abrasive paste available in a tube or small can. To use it, place a small amount on your finger. Then apply the lapping compound to the valve face as shown in Fig. 21-22. When there is a light coat of lapping compound around the entire valve face, place the valve in its proper guide.

To lap the valve, a lapping tool is used. It is a stick with a small rubber suction cup on one end which holds to the valve head. The lapping tool is rotated back and forth between your hands several times (Fig. 21-23). The motion also causes the valve to bounce lightly up and down.

Remove the valve from the guide, and wipe the valve face clean with a cloth. Figure 21-24 shows the proper valve face-valve seat contact area that you should see on every valve face. After the lapping operation is completed, the valve seat and valve must be cleaned to remove all traces of lapping compound.

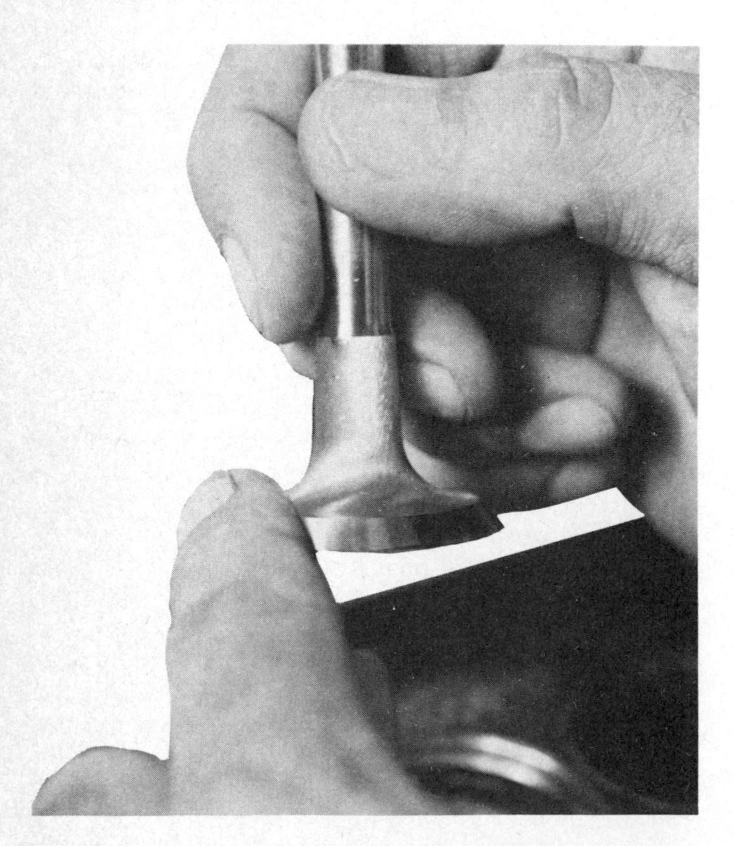

FIG. 21-22 Applying lapping compound to the valve face. *(TRW, Inc.)*

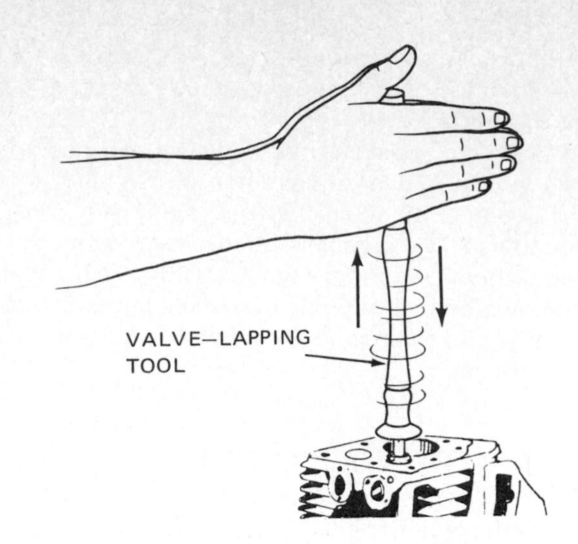

FIG. 21-23 Using a valve-lapping tool to check and improve the fit of the valve face to the valve seat. *(ATW)*

FIG. 21-24 Lapping compound shows the valve has the correct valve-seat to valve-face contact area.

Δ21-16 INSTALLING VALVE-SEAT INSERTS

When the valve port is located in an aluminum-alloy cylinder block or head, a *valve-seat insert* is used to provide the seating surface (Fig. 4-8). This is because the steel valve face will wear an aluminum seat excessively. Some cast-iron engines also have valve-seat inserts, especially for the exhaust valves. When the valve-seat insert becomes worn so much that it cannot be refaced again with a valve-seat grinder or cutter (**Δ21-14**), the insert can be replaced. In addition, inserts can be installed in engines that did not originally have them.

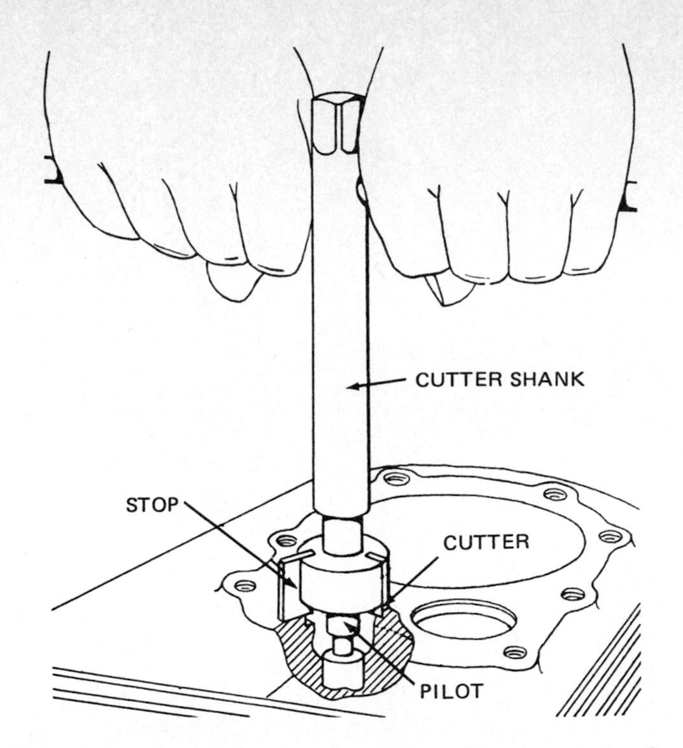

FIG. 21-25 Counterboring the seat to the correct size for the installation of the valve-seat insert. *(Briggs & Stratton Corporation)*

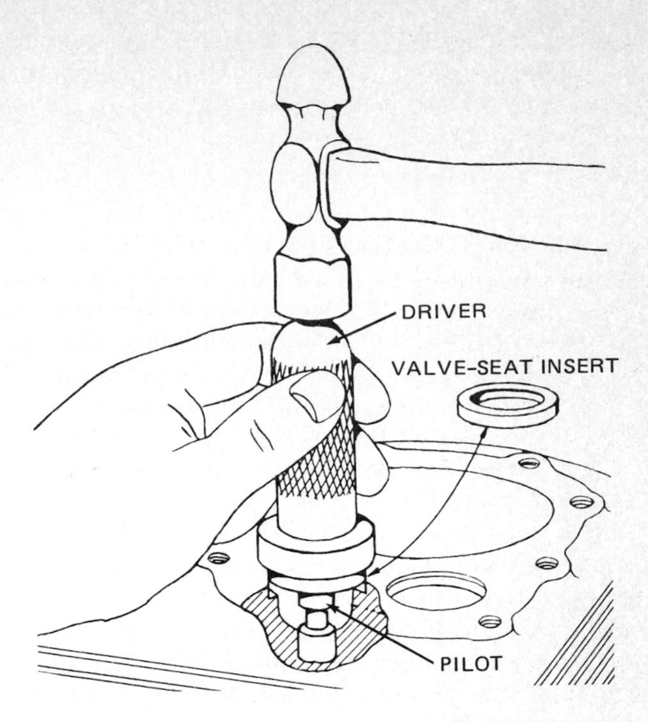

FIG. 21-27 Driving in a new valve-seat insert. *(Briggs & Stratton Corporation)*

The first step in installing an insert in an engine that has an integral seat is to cut away the old seat by counterboring it (Fig. 21-25). To do this, insert the pilot for the cutter in the valve guide. Then insert the cutter shank in the cutter, and place the assembly over the pilot. Turning the handle will counterbore the seat to the correct size for the installation of the insert.

If a worn valve-seat insert must be replaced, the old insert must be pulled with a *valve-seat puller* (Fig. 21-26). Install the puller. Then turn the bolt on the puller to remove the old insert.

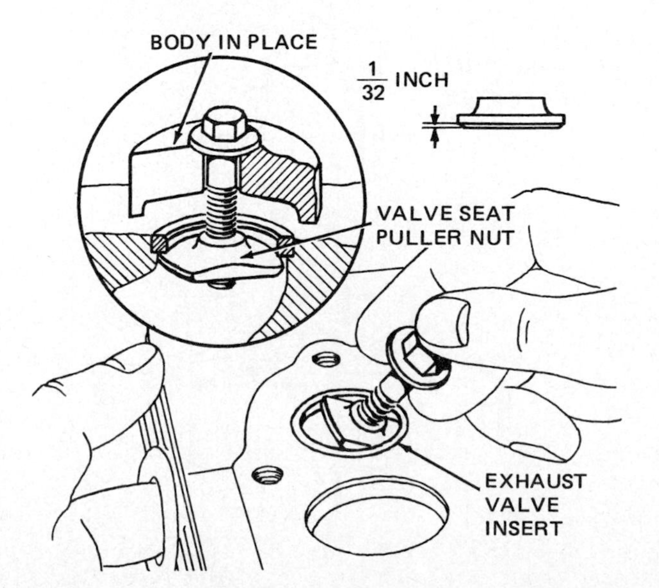

FIG. 21-26 Removing the worn valve-seat insert with a valve-seat-insert puller. *(Briggs & Stratton Corporation)*

To install the valve-seat insert, the proper pilot must be used with the correct driver. Insert the pilot into the valve guide. Then drive the insert into place with the driver (Fig. 21-27). After installation, grind the new valve-seat insert lightly. Then lap the valve into the seat with lapping compound (Δ21-15).

After installing the valve-seat insert in aluminum-alloy cylinder blocks, peen around the insert (Fig. 21-28). This

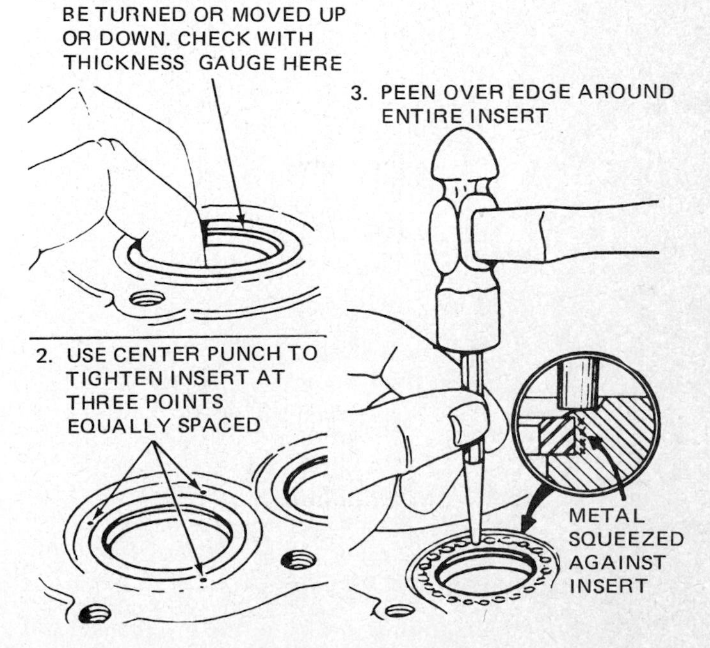

FIG. 21-28 Peening around the edge of the valve-seat insert to hold it in position. *(Briggs & Stratton Corporation)*

helps lock it in place. If a 0.005-inch [0.13-mm] thickness gauge will fit between the insert and the counterbore in the block, the block must be replaced.

∆21-17 CAMSHAFT SERVICE

If the camshaft or its bearings appear worn, remove the camshaft for inspection. Inspection includes checking the camshaft for straightness and checking the condition of the bearing journals and cam lobes. It also includes examining the teeth on the camshaft gear for chips, nicks, and wear pattern.

The procedure for removing the camshaft varies according to the engine design. The first step is to be sure that you can see the timing marks on the camshaft and crankshaft gears (Fig. 21-29). Some engines have a chamfer on the end of one tooth on the crankshaft gear.

If timing marks cannot be located, make them with dabs of metal paint or with a center punch or chisel. These marks *must* be aligned when the camshaft is installed to assure proper valve timing.

On some engines, the crankshaft must be removed before the camshaft can be removed. On other engines, the camshaft can be lifted out after the baseplate or gear cover is removed (Fig. 21-29). Some engines also have a thrust bearing that must be removed as the camshaft is removed.

Before attempting to remove the camshaft, turn it until the timing marks align on the compression stroke. In this position, both valves are closed and the valve tappets are free of the camshaft. Then the camshaft can be slipped out. Turn the engine on its side so the tappets will not fall out. Some engines using ball bearings require removal of the camshaft and crankshaft together. Also, some engines have the camshaft supported on a pin which must be driven out before the camshaft can be removed.

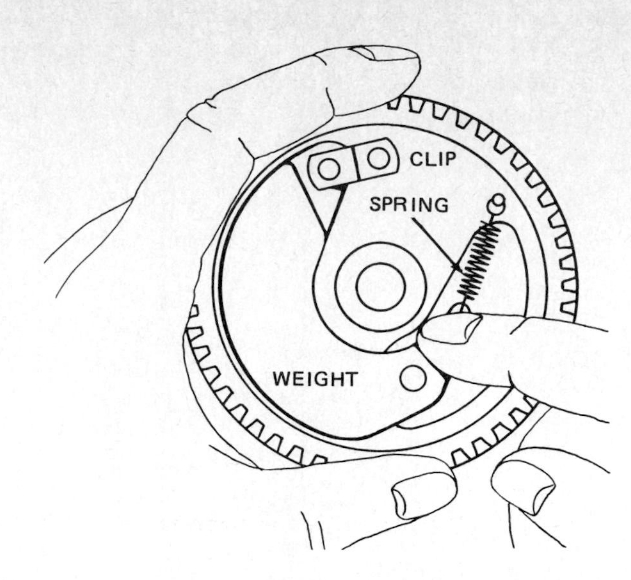

FIG. 21-30 Checking automatic spark advance on the camshaft gear for freeness of action. *(Briggs & Stratton Corporation)*

If the camshaft has an end-play washer, carefully note its position. Then remove it. During engine assembly, the washer must be reinstalled in its original position.

Wash the camshaft with solvent to clean off dirt and oil. Check the gears for wear or nicks. Check the automatic compression release or ignition-advance mechanism (Fig. 21-30) for freedom of action. If the camshaft has oil holes, blow them out with compressed air.

Figure 21-31 shows the various points at which the camshaft should be inspected. The camshaft dimensions should be checked with a micrometer. Compare the measurements with the manufacturer's specifications. Discard the camshaft if the wear is excessive.

Normal cam wear is close to the center of the cam. The reason for this is that the cam is slightly tapered in many

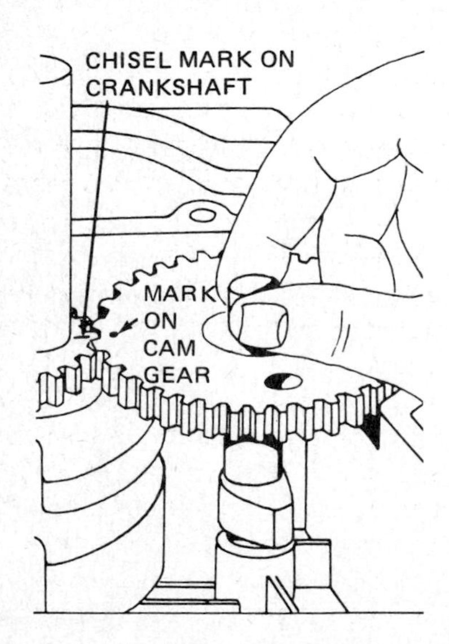

FIG. 21-29 Look for timing marks on the crankshaft and camshaft gears. Then lift the camshaft out of the engine. *(Briggs & Stratton Corporation)*

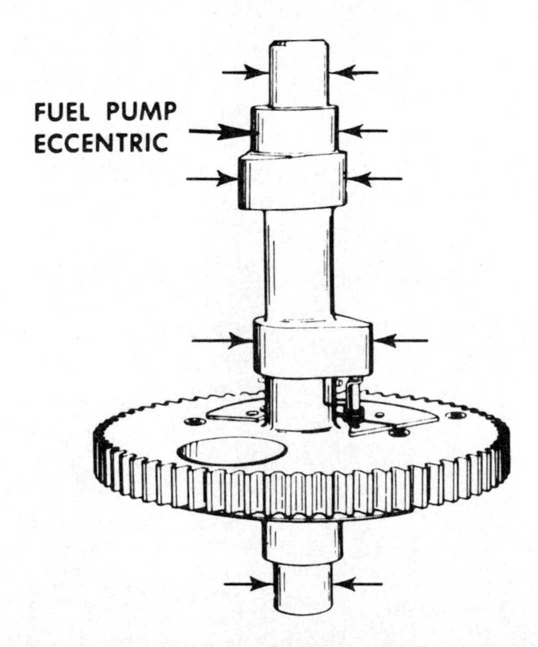

FIG. 21-31 Points on the camshaft which should be inspected. *(Tecumseh Products Company)*

engines. Also, the tappet foot may be slightly spherical or crowned in shape. Therefore, the contact pattern should normally appear as a narrow band around the cam lobe. If wear shows across the full width of the cam, a new camshaft is required.

The tappets should also be checked. Remove the tappets (Fig. 21-32) and examine them for wear. An excessively worn tappet must be replaced. If it is only slightly worn or pitted, it can often be reground on a valve-refacing machine and reused. A slight crown on the foot of the tappet can be produced by rocking and rotating the tappet during the finish grind.

Cam-lobe lift can be checked with the camshaft in or out of the engine. A dial indicator is needed to make the check with the camshaft in the engine. To measure lobe lift with the camshaft out of the engine, use a micrometer. Measure from the nose of the cam to the back of the cam. Then make another measurement at a right angle to the first measurement. Cam-lobe lift is the difference between these two measurements.

If the camshaft rides in sleeve bearings, or bushings, check them for wear and replace them if necessary. Replacement requires special drivers to force the old bushings out and new bushings in. Then the new bushings must be reamed to size. If the camshaft is supported by ball or roller bearings, check them as described in Chap. 20.

To install the camshaft, first coat all parts with oil. Then install the tappets in the same holes from which they were removed (Fig. 21-32). If the tappets are reversed during installation, they may not fit properly. On some engines, the two tappets are different lengths.

Push the tappets up out of the way, and then install the camshaft (Fig. 21-33). If the camshaft is supported by a camshaft pin, install the camshaft and crankshaft first. Then align the timing marks before positioning the camshaft and driving the camshaft pin into place. If the camshaft has additional parts attached, such as the oil pump, governor drive gear, ignition centrifugal-advance mechanism, or automatic compression release, make sure they are properly aligned.

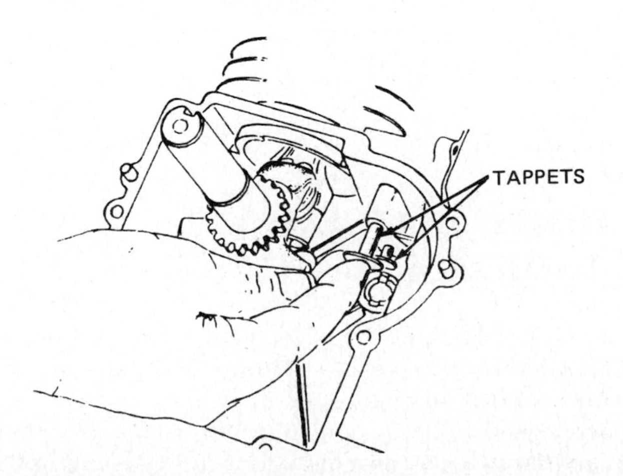

FIG. 21-32 With the camshaft removed, the valve tappets can be removed or installed. *(Clinton Engines Corporation)*

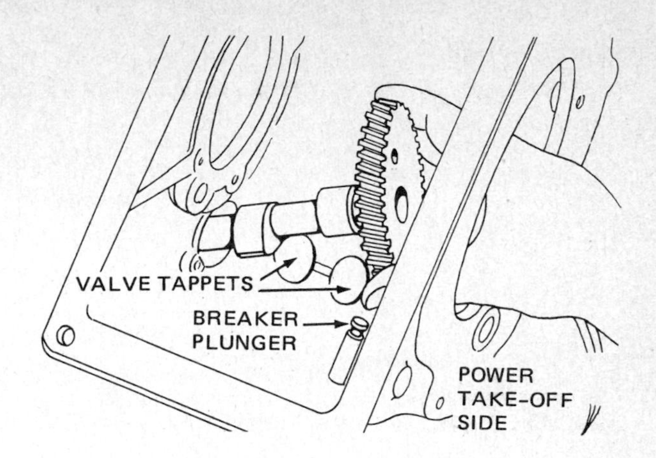

FIG. 21-33 Installing the camshaft. *(Briggs & Stratton Corporation)*

Δ21-18 INSTALLING VALVES

On engines with solid or nonadjustable valve tappets, install the valves in their proper positions in the cylinder block. Turn the crankshaft until one of the valves is in its highest position. Then turn the crankshaft one more complete revolution.

Check the valve clearance with a thickness gauge (Δ21-3). Repeat for the other valve. If the clearance is too small, as it may be if valves and seats have been ground, grind off the end of the valve stem the amount necessary to obtain the proper clearance.

If the valve tappets have an adjusting screw, turn the screw in or out to obtain the proper clearance (Δ21-4).

Check the valve springs carefully (Δ21-12). If a spring is held in place by a pin or a split collar (Fig. 4-9), put the spring with a retainer in the valve-spring compressor (Fig. 21-34). Insert the compressor with spring and retainer into position in the cylinder block, and then drop the valve into place. Install the retainer pin or collar, and release the spring compressor.

When installing the one-piece retainer (Fig. 4-9), move it around when dropping the valve into place. The valve stem must enter the larger part of the opening in the retainer. Lift the retainer up, and center it in the undercut on the valve stem. Then release the valve-spring compressor.

CYLINDER-HEAD SERVICE

Δ21-19 INSPECTING THE CYLINDER HEAD

If the cylinder head appears in good condition, clean off the carbon deposits (Fig. 21-35). Use a wood or plastic scraper on an aluminum head to avoid nicking or scratching it. Then wash the head in solvent and blow it dry with compressed air.

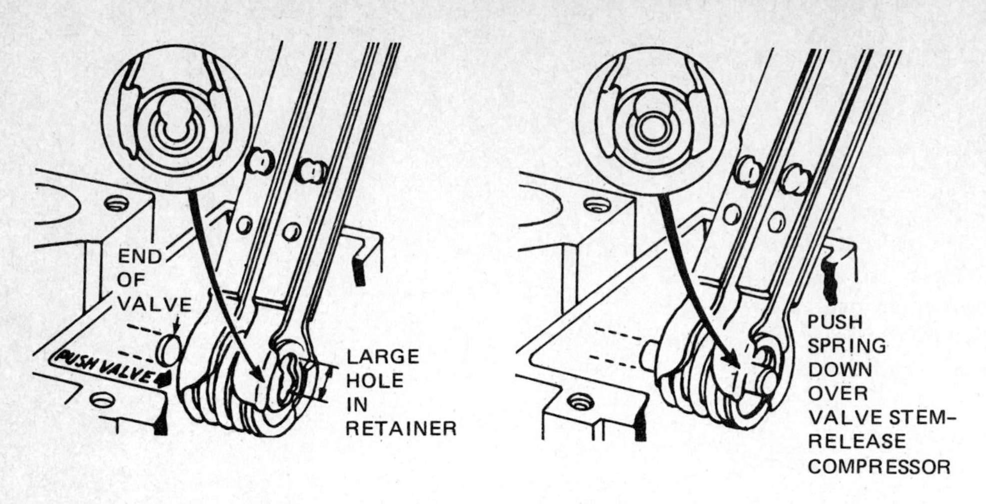

FIG. 21-34 **Compressing the valve spring in preparation for valve installation.** *(Briggs & Stratton Corporation)*

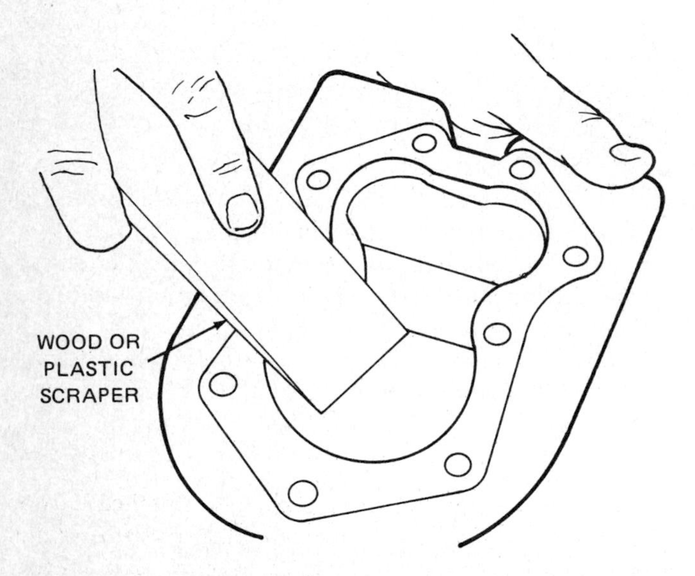

FIG. 21-35 **Scraping carbon deposits from the cylinder head.** *(Kohler Company)*

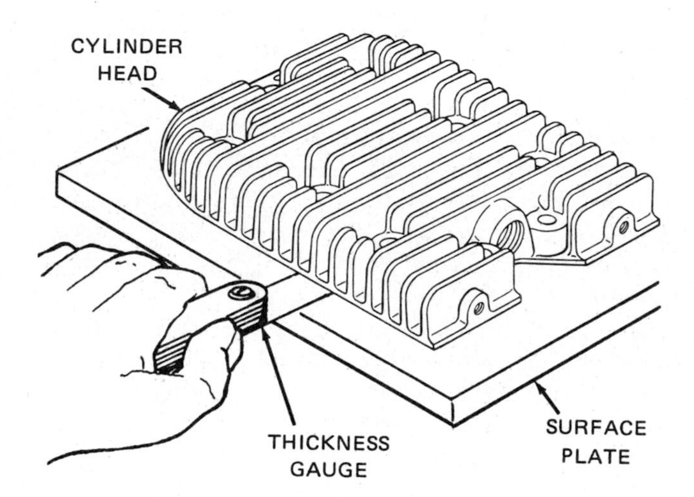

FIG. 21-36 **Checking for a warped cylinder head.** *(Kohler Company)*

Check the cylinder head for flatness (Fig. 21-36). Place the head on a surface plate or a piece of plate glass. Then measure between each cap-screw hole with a thickness gauge. A typical recommendation is that cylinder-head flatness should not vary more than 0.003 inch [0.08 mm]. If it does, replace the head.

If the head is warped or burned away, install a new head. Then install new screws in the head. The old screws could have been overheated, which would allow them to stretch when tightened.

Δ21-20 INSTALLING THE CYLINDER HEAD When installing the cylinder head, use a new head gasket. Do not use any sealer on the gasket. Use graphite grease on screws that go into aluminum cylinders. Tighten the screws down evenly by hand, and then use a torque wrench to finish the job. Tighten the screws in the sequence and to the torque shown in the manufacturer's specifications.

Do not tighten the screws down to the total torque the first time you put the wrench on them. Instead, tighten each screw in sequence only a little. Go around again and a third or fourth time, tightening the screws a little more each time until finally all are at the proper tightness. This assures even tension on all screws and guards against a warped cylinder head.

SHORT BLOCKS AND MINIBLOCKS

Δ21-21 SHORT BLOCKS Small engines are relatively inexpensive. Repair time and parts are relatively costly. If an engine is damaged or worn so badly that a new piston, rings, connecting rod, and bearings are required, it may cost more to rebuild the engine than to buy a new one. However, it may not be necessary to buy a complete engine.

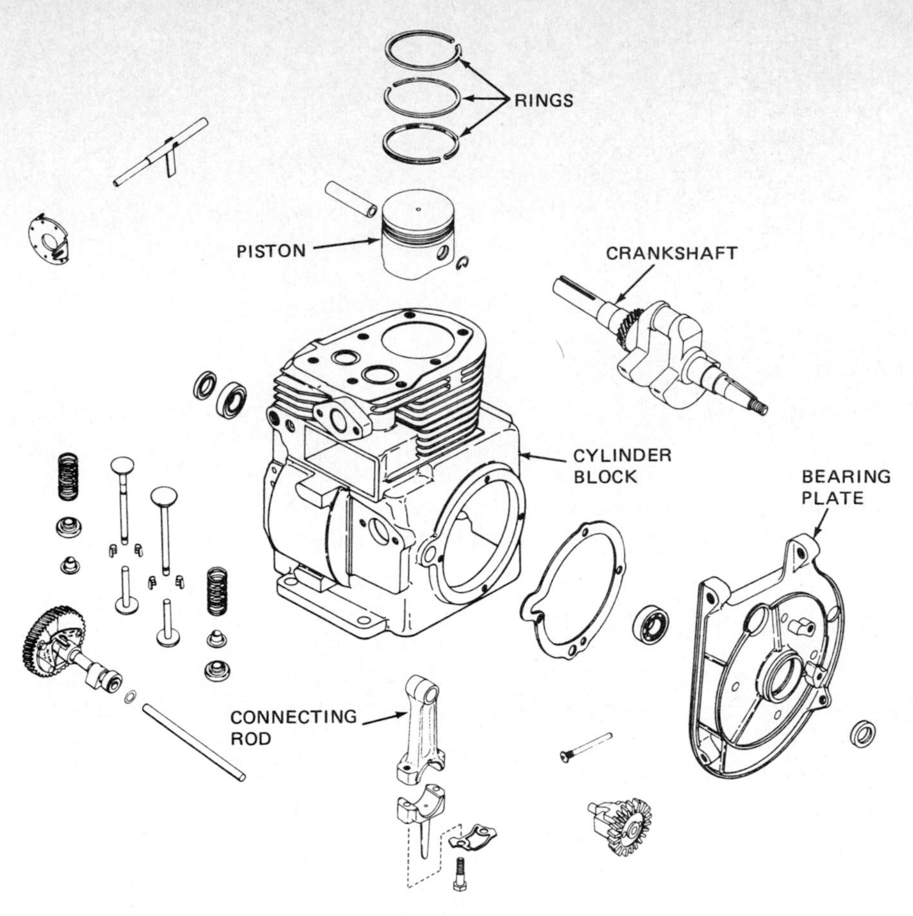

FIG. 21-37 Disassembled short-block assembly for a one-cylinder four-cycle engine. *(Kohler Company)*

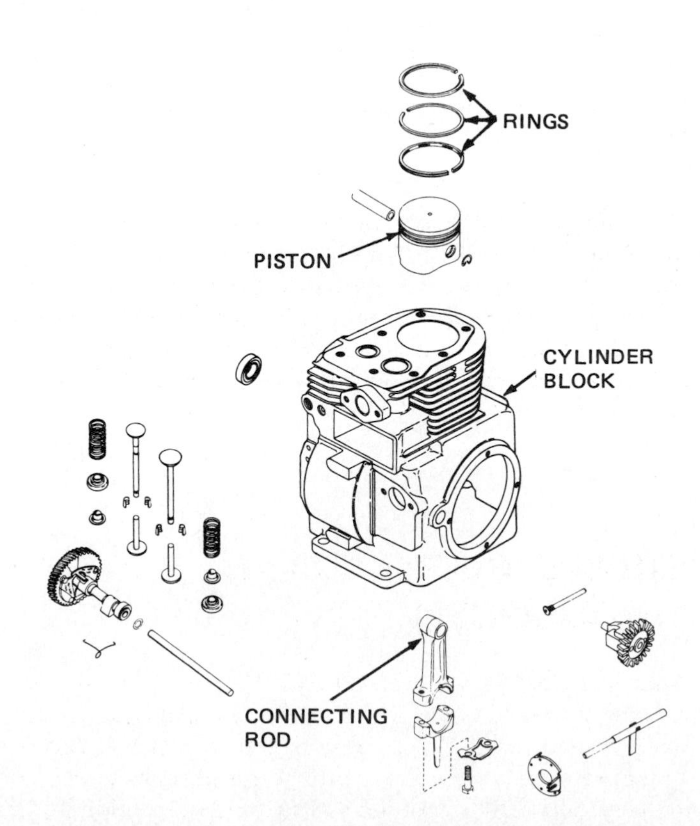

FIG. 21-38 Disassembled miniblock assembly for a one-cylinder four-cycle engine. *(Kohler Company)*

Some engine manufacturers supply *short-block* engine assemblies (Fig. 21-37). These are assembled engines minus magneto, cylinder head, carburetor, and starter. When these parts are still good, the short block may be the answer to a service problem. The short block includes the cylinder block, piston rings, connecting rod, crankshaft, and, on four-cycle engines, the valves and camshaft.

For engines that can be repaired with still fewer parts, a *miniblock* is available from some manufacturers (Fig. 21-38).

CHAPTER 21: REVIEW QUESTIONS

Select the *one* correct, best, or most probable answer to each question. You can find the answer in the section indicated at the right of each question.

1. Services that are similar for both two-cycle and four-cycle engines include (**Δ***21-1*)
   a. honing the cylinder
   b. installing piston rings
   c. replacing bearings
   d. all of the above

2. An L-head engine does not have an adjustment screw in the tappet. Mechanic A says the engine must have hydraulic valve lifters. Mechanic B says valve clearance is adjusted by grinding off the end of the valve stem. Who is right? (**Δ*21-3***)
   a. mechanic A
   b. mechanic B
   c. both A and B
   d. neither A nor B
3. Valve clearance is measured with (**Δ*21-5***)
   a. the valve lifter on the base circle of the cam
   b. the engine running at maximum speed
   c. the valve wide open
   d. the camshaft removed from the engine
4. If the end of the valve stem has mushroomed,
   a. twist it carefully as you pull it out (**Δ*21-7***)
   b. tap the stem from the end to move it through the valve guide
   c. remove the mushroom with a grinding stone before removing the valve
   d. saw off the end of the valve before removing it
5. Any valve that has a margin of less than 1/32 inch [0.8 mm] should be (**Δ*21-9***)
   a. refaced
   b. resurfaced
   c. reinstalled
   d. replaced
6. The amount ground off the valve-stem tip should never exceed (**Δ*21-11***)
   a. 0.010 inch [0.25 mm]
   b. 0.001 inch [0.03 mm]
   c. 0.100 inch [2.54 mm]
   d. 0.025 inch [0.63 mm]
7. Valve springs should be tested for (**Δ*21-11***)
   a. color and height
   b. coils and weight
   c. tension and squareness
   d. torque and twist
8. Integral valve guides can be serviced by (**Δ*21-13***)
   a. reaming the guide to a larger size and installing a valve with an oversize stem
   b. knurling and reaming the guide
   c. both a and b
   d. neither a nor b
9. To narrow a valve seat, you should use (**Δ*21-14***)
   a. 15 and 75° stones
   b. a coarse-cut file and then a fine-cut file
   c. 30 and 45° stones
   d. sandpaper and emery cloth
10. When the camshaft is removed, it should be inspected. Mechanic A says to check the condition of the bearing journals and cam lobes. Mechanic B says to inspect the teeth on the camshaft gear for chips, nicks, and wear. Who is right? (**Δ*21-17***)
    a. mechanic A
    b. mechanic B
    c. both A and B
    d. neither A nor B

CHAPTER
22

Small-Diesel-Engine Service

After studying this chapter, you should be able to:

1. ***Start and stop a small diesel engine.***
2. ***Diagnose troubles in the diesel fuel-injection system.***
3. ***Service diesel fuel filters.***
4. ***Prime (bleed) air from the fuel system.***
5. ***Test and service injection nozzles.***

STOPPING THE DIESEL ENGINE

Δ22-1 DIESEL-ENGINE SHUTDOWN PROCEDURE As long as fuel and air can get into the cylinder of a diesel engine, it will continue running. To stop the engine, the delivery of either the fuel or the air must be shut off. The normal method of stopping the engine is to stop the fuel delivery. Many diesel engines have a fuel-shutoff valve ("stopping control" in Fig. 22-1) for this purpose.

The proper shutdown procedure for a typical diesel engine is to reduce the engine speed to idle. Then let the engine idle for a few minutes. This lowers the temperature of the engine, oil, and coolant. Then close the fuel-shutoff valve to stop the engine.

Δ22-2 EMERGENCY STOPPING PROCEDURE Under certain conditions, a diesel engine may continue running even after the fuel-shutoff valve is closed. For example, this can occur if lubricating oil from the crankcase is getting into the cylinder.

Should a diesel engine fail to stop by the normal method, the engine can be stopped *in an emergency* by closing off the air supply. Some engines have an emergency *air-shutoff valve* for this purpose. On other engines, the intake air can be shut off by placing a piece of board or a metal plate over the air-intake pipe. However,

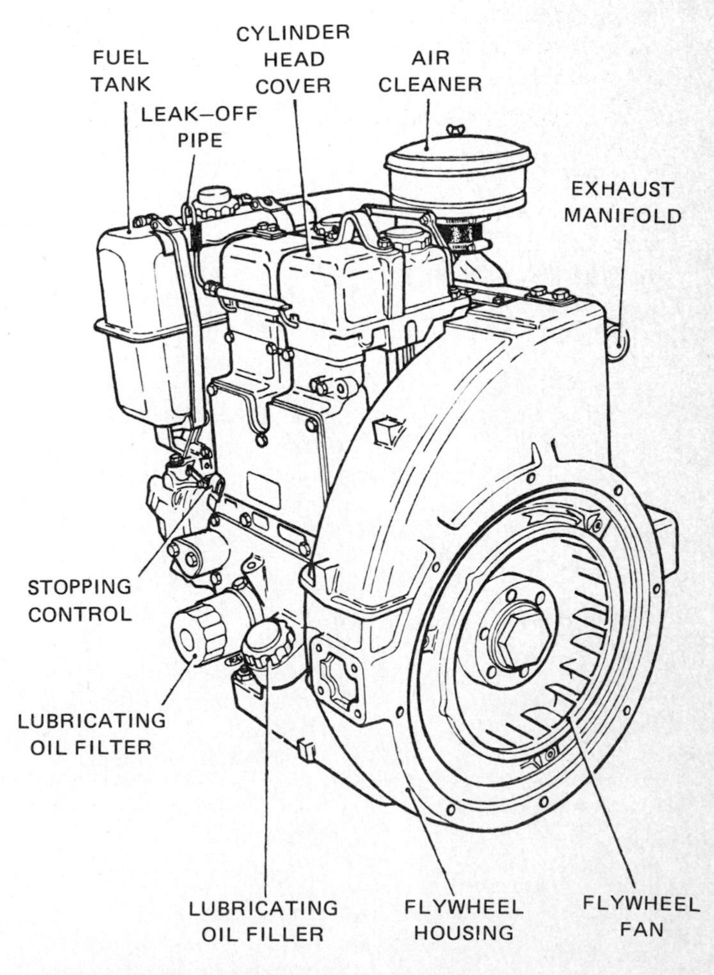

FIG. 22-1 Always know how to stop a diesel engine before starting it. *(Lister Diesels Division of Hawker Siddeley, Inc.)*

never shut down an engine using the emergency procedure *except* in an emergency.

CAUTION **Never try to stop a diesel engine by placing your hand or any part of your body or clothing over the air intake. Always use a board or a piece of metal.**

DIESEL-ENGINE TROUBLE DIAGNOSIS

Δ22-3 DIESEL FUEL-INJECTION-SYSTEM TROUBLE DIAGNOSIS Many diesel-engine troubles can be traced to the fuel system. The chart that follows lists various diesel engine troubles that could be caused by the fuel system and the checks or corrections to be made for each condition.

Δ22-4 ENGINE CRANKS NORMALLY BUT WILL NOT START If the fuel is dirty, it may have clogged the system. This requires cleaning of the pumps, fuel lines, and nozzles. The wrong fuel can also cause failure to start. Use the recommended fuel.

To check fuel nozzles for fuel flow, loosen the injection line at a nozzle. Do not disconnect it. Wipe the connection dry. Crank for 5 seconds. Fuel should flow from the loose connection. If it does not, check the fuel solenoid (if so equipped) and the fuel-supply line to the injection pump. Disconnect the line at the fuel inlet at the injection pump. Connect a hose from this line to a metal container. Crank the engine. If no fuel flows, the trouble is in the fuel-supply line or the fuel-supply pump. If fuel flows, the trouble is in the injection-pump fuel filter or the pump itself. Replace the filter first. If this does not cure the problem, replace the injection pump.

To check for a plugged fuel-return line, disconnect the line at the injection pump and connect this line to a metal

DIESEL FUEL-INJECTION-SYSTEM TROUBLE-DIAGNOSIS CHART*

Complaint	Possible cause	Check or correction
1. Engine cranks normally but will not start (Δ22-4)	a. Incorrect or dirty fuel	Flush system—use correct fuel
	b. No fuel to nozzles or injection pump	Check for fuel to nozzles
	c. Plugged fuel return	Check return, clean
	d. Pump timing off	Retime
	e. Inoperative glow plugs, incorrect starting procedure, or internal engine problems	
2. Engine starts but stalls on idle (Δ22-5)	a. Fuel low in tank	Fill tank
	b. Incorrect or dirty fuel	Flush system—use correct fuel
	c. Limited fuel to nozzles or injection pump	Check for fuel to nozzles and to pump
	d. Restricted fuel return	Check return, clean
	e. Idle incorrectly set	Reset idle
	f. Pump timing off	Retime
	g. Injection-pump trouble	Install new pump
	h. Internal engine problems	
3. Rough idle, no abnormal noise or smoke (Δ22-6)	a. Low idle incorrect	Adjust
	b. Injection line leaks	Fix leaks
	c. Restricted fuel return	Clear
	d. Nozzle trouble	Check, repair or replace
	e. Fuel-supply-pump problem	Check, replace if necessary
	f. Uneven fuel distribution to nozzles	Selectively replace nozzles until condition clears up
	g. Incorrect or dirty fuel	Flush system—use correct fuel
4. Rough idle with abnormal noise and smoke (Δ22-7)	a. Injection-pump timing off	Retime
	b. Nozzle trouble	Check cylinders in sequence to find defective nozzle
5. Idle okay but misfires as throttle opens (Δ22-8)	a. Plugged fuel filter	Replace filter
	b. Injection-pump timing off	Retime
	c. Incorrect or dirty fuel	Flush system—use correct fuel

* See Δ22-4 to 22-11 for explanations of the trouble causes and corrections listed.

Complaint	Possible cause	Check or correction
6. Loss of power (Δ22-9)	a. Incorrect or dirty fuel	Flush system—use correct fuel
	b. Restricted fuel return	Clear
	c. Plugged fuel-tank vent	Clean
	d. Restricted fuel supply	Check fuel lines, fuel-supply pump, injection pump
	e. Plugged fuel filter	Replace filter
	f. Plugged nozzles	Selectively test nozzles, replace as necessary
	g. Internal engine problems, loss of compression, compression leaks	
7. Noise—"rap" from one or more cylinders (Δ22-10)	a. Air in fuel system	Check for cause, correct
	b. Gasoline in fuel system	Replace fuel
	c. Air in high-pressure line	Bleed system
	d. Nozzle sticking open or with low opening pressure	Replace defective nozzle
	e. Engine problems	
8. Combustion noise with excessive black smoke (Δ22-11)	a. Timing off	Reset
	b. Injection-pump trouble	Replace pump
	c. Nozzle sticking open	Clean or replace
	d. Internal engine problems	

* See Δ22-4 to 22-11 for explanations of the trouble causes and corrections listed.

container. Connect a hose from the injection-pump connection to the metal container. Crank the engine. If it starts and runs, the trouble is in the fuel-return line.

Δ22-5 ENGINE STARTS BUT STALLS ON IDLE This could be caused by any of the troubles discussed in Δ22-4. These include incorrect or dirty fuel, limited fuel to nozzles or injection pump, a restricted fuel return, incorrect pump timing, or defects in the injection pump. Also, stalling on idle after starting could be caused by an incorrectly set idle or by low fuel in the tank.

Δ22-6 ROUGH IDLE, NO ABNORMAL NOISE OR SMOKE First check if the low-idle speed is correctly set. Then look for injection-line leaks. Wipe off the lines and run the engine. If there are leaks, tighten connections or replace lines as necessary to eliminate them. Next, check for a restricted fuel-return system (Δ22-4). Disconnect the fuel-return line to see if the engine runs normally.

To check for a defective nozzle, start the engine and then loosen the injection-line fitting at each nozzle in turn. This relieves the pressure and prevents normal injection-nozzle action. Be careful to avoid spraying of fuel onto hot engine parts. When a nozzle that is good is prevented from operating in this way, the rhythm of the engine will change. The engine will run more roughly. If you find a nozzle that does not change the idle when partly disconnected, then that nozzle was not working and should be replaced.

To check for internal fuel leaks at the fuel nozzle, disconnect the fuel-return system from the nozzles on one bank at a time. With the engine running, note the fuel seepage at the nozzles. Replace any nozzle with excessive fuel leakage.

A rough idle can also be caused by a fuel-supply problem. The fuel-supply pump, line, and fuel filter should be checked. In addition, rough idle can also be caused by dirty fuel or the wrong fuel.

Δ22-7 ROUGH IDLE WITH ABNORMAL NOISE AND SMOKE First check the injection-pump timing. Then disable the nozzles, one at a time, to check their operation. Do this by loosening the injection-line connection at the nozzle (Δ22-6). When you find a nozzle that does not change the noise or smoke when it is disabled, that is the bad nozzle. It should be replaced.

Δ22-8 IDLES OK BUT MISFIRES AS THROTTLE OPENS This can be caused by a plugged fuel filter, incorrect injection-pump timing, or incorrect or dirty fuel. A plugged fuel filter should be replaced. If the condition is caused by incorrect injection-pump timing, reset the timing. Incorrect or dirty fuel should be flushed out. Then the tank should be filled with the correct fuel.

Δ22-9 LOSS OF POWER This is a general complaint that could result from many conditions in the engine systems or engine as well as outside the engine.

For example, dragging brakes, excessive resistance in the power train, or underinflated tires can produce an impression of low power. In the fuel system, loss of power could result from incorrect or dirty fuel, restricted fuel return, a restricted fuel supply, or plugged nozzles. Previous sections describe how to check for these conditions. Another possible cause is a plugged fuel-tank vent. This would prevent normal fuel flow to the injection pump and engine so that the engine could not produce full power. Also, if the engine overheats, it will lose power.

Δ22-10 NOISE (RAP) FROM ONE OR MORE CYLINDERS One possible cause of this condition could be air in the fuel system. Air could cause a very uneven flow of fuel to the nozzles. This air expands and contracts with changing pressure, causing too much or too little fuel to feed. To correct the problem, loosen the injection line at each nozzle to allow the air to bleed from the system. Another cause of the noise or rap could be a nozzle sticking open or having a very low opening pressure. Loosening the injection lines at each nozzle in turn will locate a defective nozzle.

Δ22-11 COMBUSTION NOISE WITH EXCESSIVE BLACK SMOKE This could be caused by incorrect injection-pump timing or by injection-pump troubles. The cylinders receive too much fuel or receive fuel at the wrong time. It could also be caused by an injection nozzle sticking open and by internal engine problems.

FUEL-SYSTEM SERVICE

Δ22-12 SERVICING DIESEL FUEL FILTERS Diesel engines always have one or more fuel filters (Δ5-18). In addition, a fuel strainer is often used to trap larger particles in the fuel before they reach the filters. A typical recommendation for small diesel engines is to clean the strainer and change the filter at least once a year, or after every 100 hours of engine operation.

The typical fuel filter has a top cover that mounts to the engine and has connections for the fuel lines. A bottom cover or bowl holds the filter element in place. Most diesel fuel filters have replaceable filter elements (Fig. 22-2).

To replace the fuel-filter elements, stop the engine. Then clean all dirt from the fuel-line connections and from around the filter. Unscrew the center bolt in the top of the filter and remove the bottom cover and filter element. Remove the old filter element and seals. Using diesel fuel as a solvent, clean the top and bottom cover. Then air-dry the parts thoroughly.

NOTE Never use a cloth to clean fuel-filter parts. Lint from the cloth could get into the fuel system and damage the injection pump or an injection nozzle.

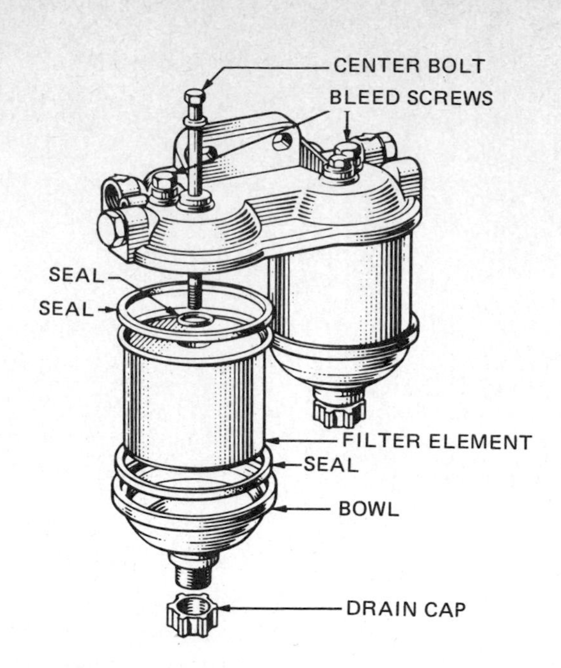

FIG. 22-2 Changing the filter element in a diesel-engine fuel filter.

Position the new seals in the bottom cover and on the top of the filter element. Be sure they lie flat. Assemble the bottom bowl to the element and install them in the top cover. Install the center bolt and tighten it. Then bleed all trapped air from the fuel system (Δ22-13).

Δ22-13 PRIMING THE FUEL SYSTEM On any diesel engine, whenever fuel lines are removed, the filter element is replaced, or the fuel tank runs dry, air will get into the fuel system. Since the fuel system cannot handle air (only liquid), the air must be removed before the engine will run. This procedure is called *priming* or *bleeding*.

To bleed air from the fuel system, loosen the bleed screw in the cover of the fuel filter (Fig. 22-2). If there is no bleed screw in the fuel filter, loosen the fuel outlet connection at the filter cover. Then operate the priming lever on the fuel transfer pump (Fig. 22-3) until a solid stream of air-free fuel flows from the bleed screw or connection.

Retighten the screw or connection while continuing to operate the priming lever. This eliminates any air in the fuel system between the fuel tank and the fuel filter.

Next, loosen the bleed screw in the top of the fuel-injection pump (Fig. 22-4). Again, operate the priming lever on the transfer pump until a solid stream of fuel flows from around the loosened plug. Retighten the bleed screw while continuing to operate the priming lever. After the bleed screw is tight, operate the priming lever for at least 12 more strokes. Then loosen the fuel-supply line at the injection nozzle (Fig. 5-5).

Set the throttle fully open and the stop control in the RUN position. Now crank the engine until a steady stream of fuel flows from around the loosened connection at the

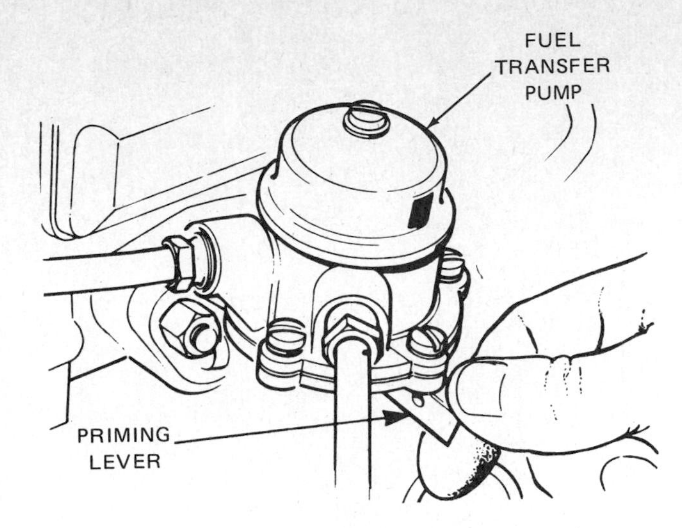

FIG. 22-3 Priming lever on the fuel transfer pump.

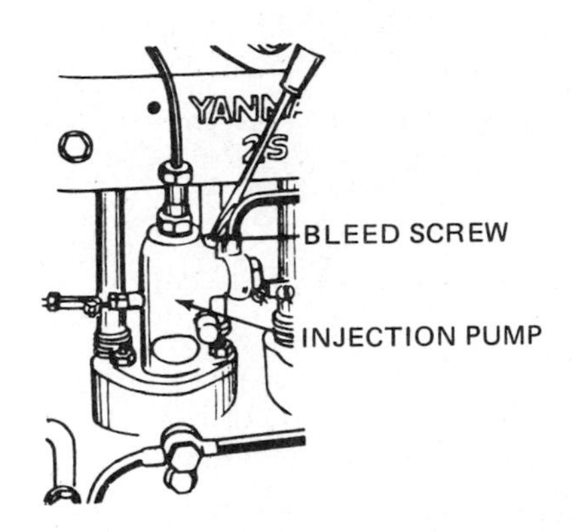

FIG. 22-4 To bleed air from the fuel system, loosen the bleed screw at the fuel-injection pump. Then operate the priming lever until a solid stream of fuel emerges from around the screw. *(Yanmar Diesel Engine Company, Limited)*

nozzle. Retighten the connection while continuing to operate the priming lever.

Start the engine and let it run at idle. Check all bleed screws and fuel-line connections for leaks.

▵22-14 SERVICING INJECTION NOZZLES

Most fuel-injection nozzles can be disassembled, cleaned, and sometimes, adjusted. However, all injection nozzles must be inspected periodically. One reason is that diesel engines often run at idle for long periods of time. Prolonged idling may cause carbon deposits on the spray tip. The deposits may block holes in the tip or interfere with the normal spray pattern (Fig. 22-5).

The more severe the engine operating conditions, the more frequently the nozzles should be inspected. One manufacturer recommends that injection nozzles should be inspected at least once a year, or after every 100 hours of engine operation. In addition, the nozzle opening pressure, spray pattern, and leakage should be checked every third year. This requires removing the nozzle from the engine and testing the nozzle on a *nozzle tester* (Fig. 22-6).

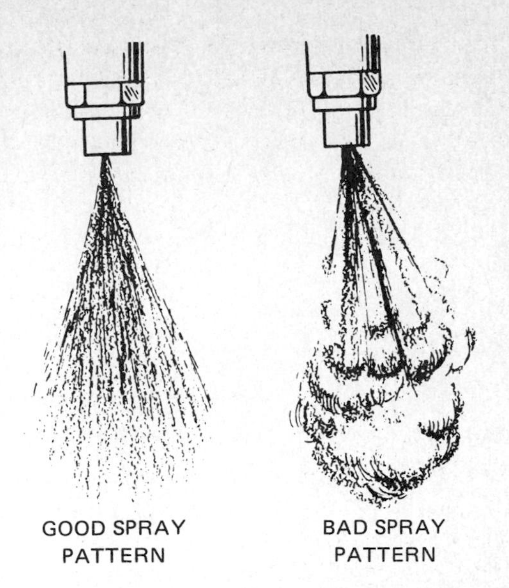

FIG. 22-5 Nozzle spray patterns. *(Onan Corporation)*

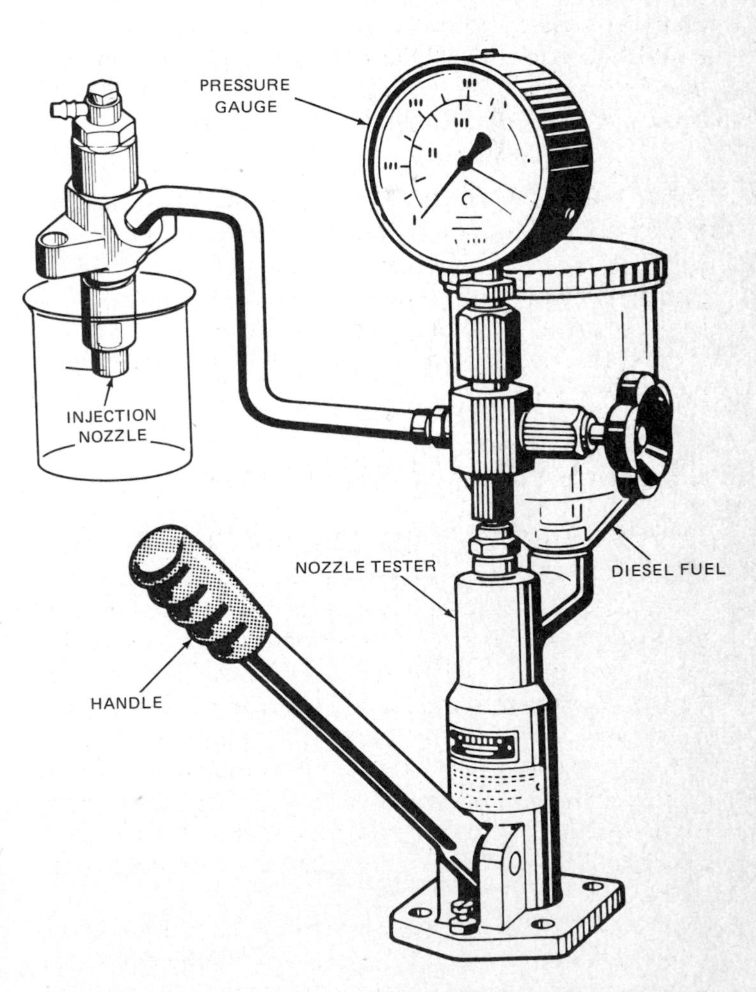

FIG. 22-6 Test the injection nozzle by attaching it to a nozzle tester.

Some manufacturers recommend replacing defective injection nozzles with new or rebuilt units. Other manufacturers provide instructions on how to disassemble and clean nozzles. However, a nozzle tester (Fig. 22-6) is

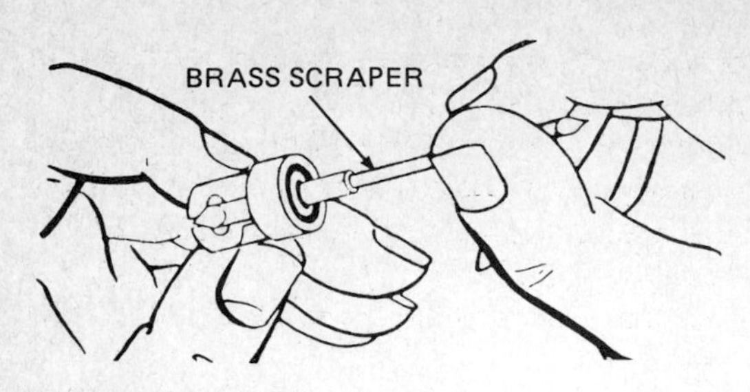

(*a*) REMOVE CARBON FROM VALVE SEAT

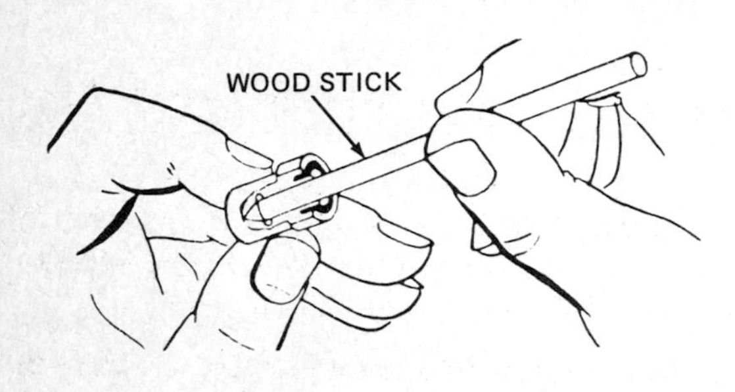

(*b*) POLISH VALVE SEAT WITH TALLOW

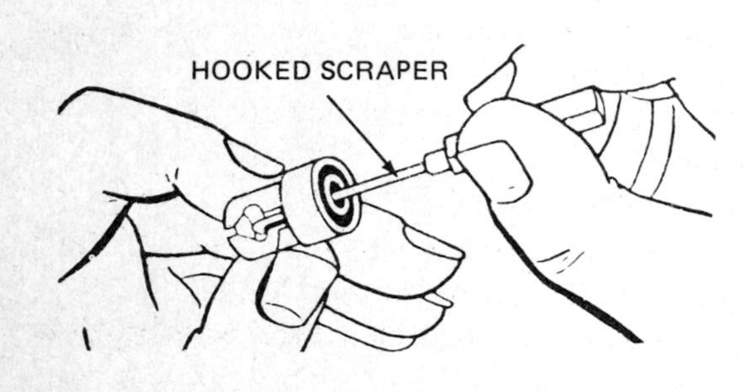

(*c*) CLEAN PRESSURE–CHAMBER GALLERY

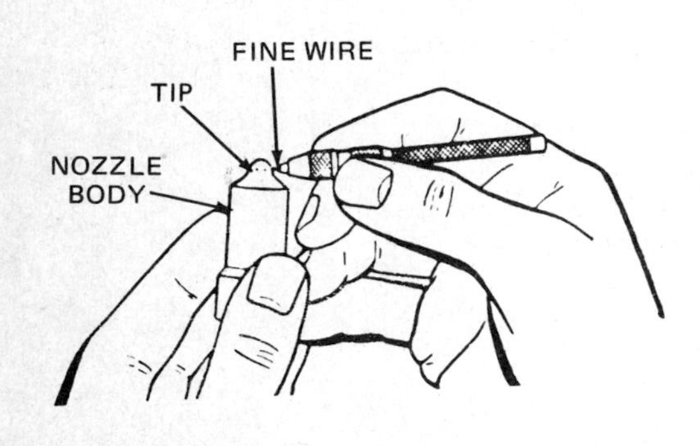

(*d*) CLEAN HOLES IN TIP

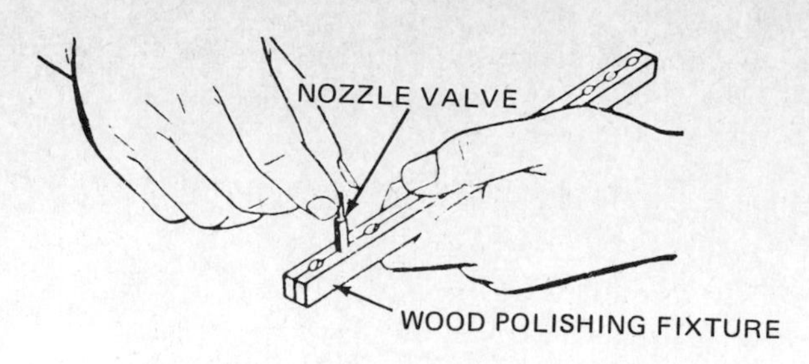

(*e*) POLISH NOZZLE VALVE WITH TALLOW

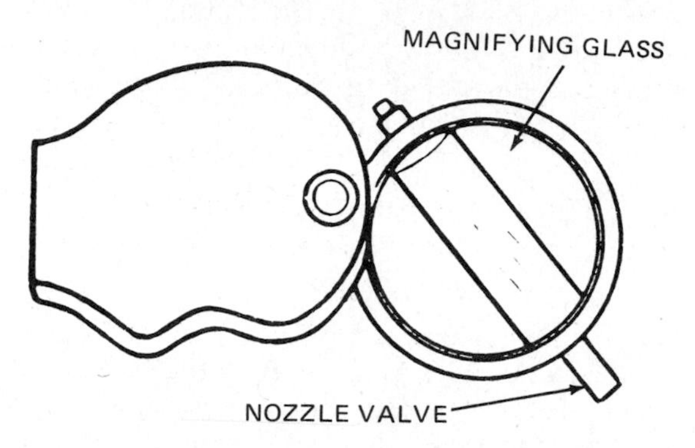

(*f*) INSPECT NOZZLE VALVE FOR SCORES AND EROSION

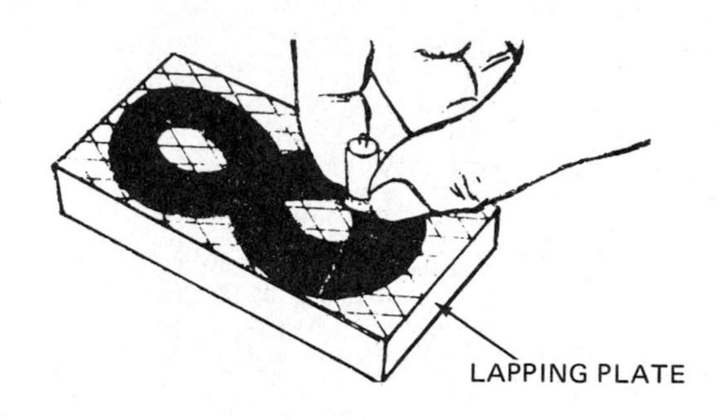

(*g*) LAP NOZZLE PARTS USING A FIGURE 8 MOTION

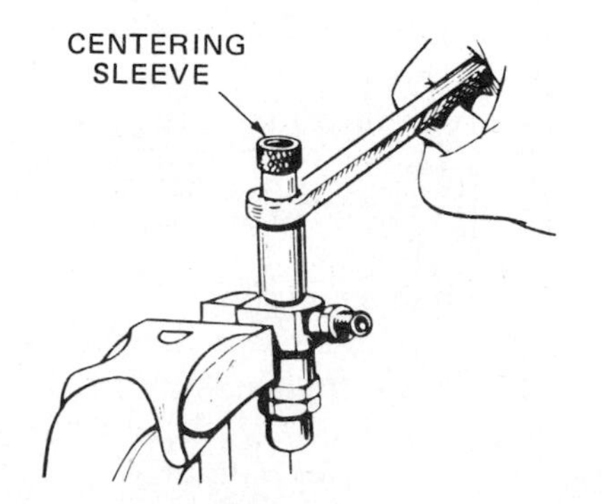

(*h*) USE CENTERING SLEEVE TO CENTER NOZZLE BODY IN CAP NUT DURING REASSEMBLY

FIG. 22-7 **Steps in cleaning one type of pintle nozzle.** *(Onan Corporation)*

needed to adjust the opening (injection) pressure on certain types of nozzles.

Δ22-15 TESTING INJECTION NOZZLES

If you think an injection nozzle is operating improperly, it can be checked on a nozzle tester (Fig. 22-6). This tester is basically a hand-operated fuel-injection pump with a pressure gauge to which the nozzle is connected. Then the pump handle is moved up and down.

As the pressure increases, it finally forces the valve off its seat, and fuel sprays from the tip end of the nozzle. This allows you to observe the opening pressure, the spray pattern (Fig. 22-5), and any leakage or dribble that occurs from the nozzle before and after injection. If faulty fuel delivery is caused by deposits in the nozzle, some manufacturers recommend disassembling and cleaning it (Δ22-16). If a problem still exists, replace the nozzle.

CAUTION **Never allow your hand, skin, or any part of your body to get in the way of fuel spraying from a fuel-injection nozzle. The fuel is under high pressure and can penetrate your skin. This may cause serious injury and blood poisoning.**

Δ22-16 CLEANING INJECTION NOZZLES

Nozzle service must be done in a clean, dust-free area. After removing the nozzle from the engine, cover all openings and fuel lines to prevent dirt and other contaminants from entering. Then use diesel fuel and a brass-wire brush to remove any dirt, paint, or carbon from the mating surfaces of the nozzle and the engine.

Disassemble the nozzle and soak the parts in diesel fuel or solvent to loosen the carbon. After the carbon is removed, thoroughly examine the spray tip to see if it has turned blue. This indicates overheating or other damage.

The steps in cleaning a typical pintle nozzle are shown in Fig. 22-7. When all parts are clean, reassemble the nozzle. Then install it on the nozzle tester. Adjust the opening pressure, and check the operation and spray pattern. After cleaning, a nozzle that cannot be adjusted to within specifications, fails to operate properly, or has a bad spray pattern must be replaced.

Δ22-17 INJECTION-PUMP SERVICE

No periodic service is required on most fuel-injection pumps. Many manufacturers recommend replacing a defective injection pump with a new or rebuilt unit. However, some injection pumps can be disassembled for cleaning and inspection (Fig. 5-7).

If the injection pump is removed and disassembled, the pump must be reassembled and installed properly to maintain the correct injection timing. Otherwise, the engine may not deliver maximum speed and power. The injection pump on each engine must be timed to that particular engine.

In some engines, shims of various thicknesses are installed or removed from between the engine and the injection-pump body. This varies the distance between the lobe on the engine camshaft and the pump plunger. As a result, the injection timing is varied.

CHAPTER 22: REVIEW QUESTIONS

Select the *one* correct, best, or most probable answer to each question. You can find the answer in the section indicated at the right of each question.

1. To stop a diesel engine in an emergency (Δ22-2)
   a. remove the air cleaner and place a wood board over the air-intake pipe
   b. disconnect the battery
   c. cover the air cleaner with cardboard
   d. close the choke
2. In cold weather, if the wire to the glow plug is left disconnected, the engine will (Δ22-3)
   a. start but stall on idle
   b. idle OK but misfire as the throttle opens
   c. have loss of power
   d. crank normally but will not start
3. Excessive black exhaust indicates (Δ22-11)
   a. the cylinder is receiving too much fuel
   b. the cylinder is receiving the fuel at the wrong time
   c. both a and b
   d. neither a nor b
4. Priming the fuel system is required to (Δ22-13)
   a. remove any water from the fuel system
   b. remove any air from the fuel system
   c. start an engine if the filter is clogged
   d. operate the engine if the transfer pump has failed
5. After cleaning, the injection nozzle is usually checked for (Δ22-15)
   a. opening vacuum
   b. spray odor
   c. volume of fuel injected
   d. none of the above

Glossary

abrasive A substance used for cutting, grinding, lapping, or polishing metal.

AC or ac Abbreviation for alternating current.

additive A substance added to gasoline or oil which improves some property of the gasoline or oil.

adjust To bring the parts of a component or system to a specified relationship, dimension, or pressure.

adjustments Necessary or desired changes in clearances, fit, or settings.

aerobic gasket material A type of formed-in-place gasket, also known as *self-curing, room-temperature-vulcanizing* (RTV), and *silicone rubber.* A material that cures only in the presence of air, normally used on surfaces that flex or vibrate.

air cleaner A device, mounted on or connected to the engine air intake, for filtering dirt and dust out of the air being drawn into the engine.

air-cooled engine An engine that is cooled by the passage of air around the cylinders, not by the passage of a liquid through water jackets.

air filter A filter that removes dirt and dust particles from air passing through it.

air-fuel mixture The air and fuel traveling to the combustion chamber after being mixed by the carburetor.

air-fuel ratio The proportions of air and fuel (by weight) supplied to the engine cylinders for combustion.

air gap A small space between parts that are related magnetically, as in an alternator, or electrically, as the electrodes of a spark plug.

air pressure Atmospheric pressure; also the pressure produced by an air compressor, or by compression of air in a cylinder.

alignment The act of lining up, or the state of being in a true line.

alternating current An electric current that flows first in one direction and then in the other.

alternator The device in the electric system that converts mechanical energy into electric energy for charging the battery and operating electric accessories. Also known as an *ac generator.*

aluminum cylinder block An engine cylinder block cast from aluminum alloy and usually provided with cast-iron sleeves for use as the cylinder bore.

ammeter A meter for measuring the amount of current (in amperes) flowing through an electric circuit.

anaerobic sealant A material that cures only when subjected to pressure and the absence of air; the material hardens when squeezed tightly between two surfaces.

antifreeze A chemical, usually ethylene glycol, that is added to the engine coolant to raise its boiling point and lower its freezing point.

antifriction bearing Name given to almost any type of ball, roller, or tapered roller bearing.

antiseize compound A coating placed on some threaded fasteners to prevent corrosion, seizing, galling, and pitting.

arc Name given to the spark that jumps an air gap between two electric conductors; for example, the arc between the ignition breaker points.

armature A part moved by magnetism, or a part moved through a magnetic field to produce an electric current.

ATDC Abbreviation for after top dead center; any position of the piston between top dead center and bottom dead center, on the downward stroke.

atmospheric pressure The weight of the atmosphere per unit area. Atmospheric pressure at sea level is 14.7 psi absolute [101 kPa]; it decreases as altitude increases.

automatic choke A device that positions the choke valve automatically in accordance with engine temperature or time.

backfire Noise made by the explosion of the air-fuel mixture in the intake or exhaust system, usually during cranking or deceleration.

backlash In gearing, the clearance between the meshing teeth of two gears.

ball bearing An antifriction bearing with an inner race and an outer race and one or more rows of balls between them.

battery An electrochemical device for storing energy in chemical form so that it can be released as electricity; a group of electric cells connected together.

battery acid The electrolyte used in a battery; a mixture of sulfuric acid and water.

BDC Bottom dead center.

bearing A part that transmits a load to a support and in so doing absorbs most of the friction and wear of the moving parts. A bearing is usually replaceable.

bearing caps In the engine, caps held in place by bolts or nuts, which, in turn, hold bearing halves in place.

bearing clearance The space between the bearing and the shaft rotating within it.

bearing spin A type of bearing failure in which a lack of lubrication causes the bearing to overheat until it seizes on the shaft, shears its locking lip, and rotates in the housing or bore.

belt tension The tightness of a drive belt.

bleeding A process by which air is removed from a hydraulic system (such as a diesel fuel-injection system) by draining part of the fluid or by operating the system to work out the air.

block See *cylinder block.*

blowby Leakage of compressed air-fuel mixture and burned gases (from combustion) past the piston rings into the crankcase.

bore An engine cylinder, or any cylindrical hole; also used to described the process of enlarging or accurately refinishing a hole, as "to bore an engine cylinder." The bore size is the diameter of the hole.

bottom dead center (BDC) The piston position at the lower limit of its travel in the cylinder, when the cylinder volume is at its maximum.

brake horsepower Power available from the engine crankshaft to do work; bhp = torque × rpm/5252.

breaker points In the breaker-point ignition system, the stationary and movable points which open and close the ignition primary circuit.

breather The valve or filter over the crankcase opening in four-cycle engines that allows fresh air to circulate through the crankcase, thereby providing crankcase ventilation.

brush A block of conducting substance, such as carbon, which rests against a rotating ring or commutator to form a continuous electric circuit.

BTDC Abbreviation for before top dead center; any position of the piston between bottom dead center and top dead center, on the upward stroke.

bushing A one-piece sleeve placed in a bore to serve as a bearing surface.

cables Stranded conductors, usually covered with insulating material, used for connections between electric devices.

calibrate To check or correct the initial setting of a test instrument.

cam A rotating lobe or eccentric that can be used with a cam follower to change rotary motion to reciprocating motion.

cam follower See *valve lifter.*

camshaft The shaft in the four-cycle engine which has a series of cams for operating the valve mechanisms. It is driven by gears, or by sprockets and a toothed belt or chain from the crankshaft.

camshaft bearings Full-round bearings that support the camshaft.

capacitor See *condenser.*

capacitor-discharge ignition An ignition system in which the primary energy is stored in a large capacitor that discharges when an electronic switch or transistor closes the primary circuit.

capacity The ability to perform or to hold.

carbon A black deposit left on engine parts such as pistons, rings, and valves by the combustion of fuel, and which inhibits their action.

carbon dioxide A colorless, odorless gas that results from complete combustion.

carbon monoxide A colorless, odorless, tasteless, poisonous gas that results from incomplete combustion.

carburetion The actions that take place in the carburetor; converting liquid gasoline to vapor and mixing it with air to form a combustible mixture.

carburetor The device in an engine fuel system which mixes gasoline with air and supplies the combustible mixture to the intake manifold.

CDI See *capacitor-discharge ignition.*

celsius A thermometer scale on which water boils at 100° and freezes at 0°. The formula $C = \frac{5}{9}(F - 32)$ converts Fahrenheit readings to Celsius readings.

centimeter (cm) A unit of linear measure in the metric system; equal to approximately 0.390 inch.

centrifugal advance A rotating-weight mechanism in the distributor that advances and retards ignition timing through the centrifugal force resulting from changes in the rotational speed of the distributor shaft.

cetane number A measure of the ignition quality of diesel fuel, or how high a temperature is required to ignite it. The lower the cetane number, the higher the temperature required to ignite a diesel fuel.

charging rate The amperage flowing from the alternator or generator into the battery.

check To verify that a component, system, or measurement complies with specifications.

check valve A valve that opens to permit the passage of air or other fluid in one direction only, or operates to prevent (check) some undesirable action.

chemical reaction The formation of one or more new substances when two or more substances are brought together.

chip One or more integrated circuits manufactured as a very small package capable of performing many functions.

choke In the carburetor, a device used when starting a cold engine. It "chokes off" the airflow through the air horn, producing a partial vacuum in the air horn for greater fuel delivery and a richer mixture.

circuit The complete path of an electric current, including the current source. When the path is continuous, the circuit is closed and current flows. When the path is broken, the circuit is open and no current flows. Also used to refer to fluid paths, as in hydraulic systems.

circuit breaker A resettable protective device that opens an electric circuit to prevent damage when the circuit is overheated by excess current flow.

clearance The space between two moving parts, or between a moving and a stationary part, such as a journal and a bearing. The bearing clearance is filled with lubricating oil when the mechanism is running.

closed crankcase ventilation system A system in which the crankcase vapors (blowby gases) are discharged into the engine intake system and pass through to the engine cylinders rather than being discharged into the air.

clutch A coupling that connects and disconnects a shaft from its drive while the drive mechanism is running.

CO See *carbon monoxide.*

CO_2 See *carbon dioxide.*

coil In the ignition system, a transformer used to step up the primary voltage (by induction) to the high voltage required to fire the spark plug.

combustion Burning; fire produced by the proper combination of fuel, heat, and oxygen. In the engine, the rapid burning of air and fuel in the combustion chamber.

combustion chamber The space between the top of the piston and the cylinder head, in which the fuel is burned.

commutator A series of copper bars at one end of a generator or starting-motor armature, electrically insulated from the armature shaft and insulated from one another by mica. The brushes rub against the bars of the commutator, which form a rotating connector between the armature windings and brushes.

compression Reduction in the volume of a gas by squeezing it into a smaller space. Increasing the pressure reduces the volume and increases the density and temperature of the gas.

compression-ignition engine An engine operating on the diesel cycle, in which the fuel is injected into the cylinders, where the heat of compression ignites it.

compression pressure The pressure in the combustion chamber at the end of the compression stroke.

compression ratio The volume of the cylinder and combustion chamber when the piston is at BDC, divided by the volume when the piston is at TDC.

compression ring The upper ring or rings on a piston, designed to hold the compression and combustion pressure in the combustion chamber, thereby preventing blowby.

compression stroke The piston movement from BDC to TDC immediately following the intake stroke, during which both the intake and exhaust valves are closed while the air or air-fuel mixture in the cylinder is compressed.

compression tester An instrument for testing the amount of pressure, or compression, developed in an engine cylinder during cranking.

condenser In the inductive-ignition system, a capacitor connected across the breaker points to reduce arcing by providing a storage place for electricity (electrons) as the breaker points open.

conductor Any material or substance that allows electric current or heat to flow easily.

connecting rod In the engine, the rod that connects the crank on the crankshaft with the piston.

connecting-rod bearing See *rod bearing.*

connecting rod cap The part of the connecting-rod assembly that attaches the rod to the crankpin.

connecting-rod journal See *crankpin.*

contact points See *breaker points.*

coolant In the liquid-cooling system, the liquid mixture of about 50 percent antifreeze and 50 percent water used to carry heat out of the engine.

cooling system In the engine, the system that removes heat by the circulation of liquid coolant or of air to prevent engine overheating.

corrosion Chemical action, usually by an acid, that eats away, or decomposes, a metal.

crank A device used for converting reciprocating motion into rotary motion.

crankcase The part of the engine in which the crankshaft rotates.

crankcase dilution Dilution of the lubricating oil in the crankcase by liquid fuel seeping down the cylinder wall.

crankcase emission Pollutants emitted into the atmosphere from any portion of the engine crankcase ventilating or lubricating systems.

crankcase ventilation The circulation of air through the crankcase of a running engine to remove water, blowby, and other vapors; prevents oil dilution, contamination, sludge formation, and pressure buildup.

cranking motor See *starting motor.*

crankpin The part of a crankshaft to which the connecting rod is attached.

crankshaft The main rotating member, or shaft, of the engine with the cranks to which the connecting rods are attached.

crankshaft gear A gear, or sprocket, mounted on the crankshaft; used to drive the camshaft gear, chain, or toothed belt.

cross-firing In a multicylinder engine, jumping of the high-voltage surge in the ignition secondary circuit to the wrong high-voltage lead, so that the wrong spark plug fires. Usually caused by improper routing of the spark-plug cables, faulty insulation, or a defective distributor cap or rotor.

cubic centimeter (cc) A unit of volume in the metric system; equal to approximately 0.061 cubic inch.

cubic inch displacement (CID) The cylinder volume swept out by the pistons of an engine as they move from BDC to TDC, measured in cubic inches.

current A flow of electrons, measured in amperes.

cycle A series of events that continuously repeat themselves. In an engine, the four piston strokes (or two piston strokes) that together produce the power.

cylinder A round hole or tubular-shaped structure in a block or casting in which a piston reciprocates. In an engine, the circular bore in the block in which the piston moves up and down.

cylinder block The basic framework of the engine, in and on which the other engine parts are attached. It includes the engine cylinders and the upper part of the crankcase.

cylinder-compression tester See *compression tester.*

cylinder head The part of the engine that covers and encloses the cylinders. It contains cooling fins or water jackets and may include the valves.

cylinder liner See *cylinder sleeve.*

cylinder sleeve A replaceable sleeve, or liner, set into the cylinder block to form the cylinder bore.

DC or dc Abbreviation for direct current.

detonation Commonly referred to as spark knock or ping. In the combustion chamber of a spark-ignition engine, an uncontrolled second explosion (after the spark occurs at the spark plug) with spontaneous combustion of the remaining compressed air-fuel mixture, resulting in a pinging sound.

device A mechanism, tool, or other piece of equipment designed to serve a special purpose or perform a special function.

diagnosis A procedure followed in locating the cause of a malfunction; answers the question "What is wrong?"

diaphragm A thin dividing sheet or partition that separates an area into compartments; used in fuel pumps, vacuum-advance units, and other control devices.

diesel cycle An engine operating cycle in which air is compressed and then fuel oil is injected into the compressed air at the end of the compression stroke. The heat produced by the compression ignites the fuel oil, eliminating the need for an electric-ignition system.

diesel engine An engine operating on the diesel cycle and burning diesel fuel oil instead of gasoline.

diesel fuel A light oil sprayed into the cylinders of a diesel engine near the end of the compression stroke.

dieseling A condition in which a spark-ignition engine continues to run after the ignition is turned off; caused by carbon deposits or hot spots in the combustion chamber glowing sufficiently to furnish heat for combustion.

diode A solid-state electronic device that allows the passage of an electric current in one direction only. Used in the alternator to change alternating current to direct current for charging the battery.

dipstick See *oil-level indicator.*

direct current Electric current that flows in one direction only.

disassemble To take apart.

displacement In an engine, the total volume of air or air-fuel mixture an engine is theoretically capable of drawing into all cylinders during one operating cycle. Also, the volume swept out by the piston in moving from one end of a stroke to the other.

distributor Any device that distributes. In the ignition system of a multicylinder engine, the rotary switch that directs high-voltage surges to the engine cylinders in the proper sequence. See *ignition distributor.*

distributor advance See *centrifugal advance, ignition advance,* and *vacuum advance.*

distributor cam The cam on the top end of the distributor shaft which rotates to open and close the breaker points.

distributor timing See *ignition timing.*

DOHC See *double-overhead-camshaft engine.*

double-overhead-camshaft engine An engine with two camshafts in each cylinder head; one camshaft operates the intake valves, and the other operates the exhaust valves.

dwell In a breaker-point distributor, the number of degrees of distributor-cam rotation that the points stay closed before they open again.

dwell meter A test instrument used to measure dwell.

dynamometer A device for measuring the power available from an engine to do work.

eccentric A disk or offset section (of a shaft, for example) used to convert rotary motion to reciprocating motion. Sometimes called a *cam.*

efficiency The ratio between the effect produced and the power expended to produce the effect; the ratio between the actual and the theoretical.

electric current A movement of electrons through a conductor such as a copper wire; measured in amperes.

electric system The system that electrically cranks the engine for starting, furnishes high-voltage sparks to the engine cylinders to fire the compressed air-fuel charges, and lights the lights and powers the other electric accessories. May include the starting motor, wiring, battery, generator or alternator, rectifier and regulator, ignition coil, and ignition distributor.

electrode An electric conductor (usually solid) through which current enters or leaves. In a spark plug, the spark jumps between two electrodes. The wire passing through the insulator is the center electrode. The small piece of metal welded to the spark-plug shell (and to which the spark jumps) is the side, or ground, electrode.

electrolyte The mixture of sulfuric acid and water used in lead-acid storage batteries (about 60 percent water and 40 percent sulfuric acid in a fully charged battery). The acid enters into chemical reaction with active material in the plates to produce voltage and current.

electromagnetic induction The characteristic of a magnetic field that causes an electric current to be created in a conductor if it passes through the field, or if the field builds up and collapses around the conductor.

electron A negatively charged particle that circles the nucleus of an atom. The movement of electrons is an electric current.

electronic-ignition system An ignition system that uses transistors and other semiconductor devices (instead of breaker points) as an electric switch to turn the primary current on and off.

electronics Electric assemblies, circuits, and systems that use electronic devices such as transistors and diodes.

emission control Any device or modification added onto or designed into a motor vehicle for the purpose of reducing air-polluting emissions.

emission standards Allowable emission levels, set by local, state, and federal legislation.

end play As applied to the crankshaft, the distance that it can move forward and back in the cylinder block or crankcase.

energy The capacity or ability to do work. The most common forms are heat, mechanical, electric, and chemical. Usually measured in work units of pound-feet [kilogram-meters], but also expressed in heat-energy units (Btus [Joules]).

engine A machine that converts heat energy into mechanical energy. A device that burns fuel to produce mechanical power; sometimes referred to as a *power plant.*

engine efficiency The ratio of the power actually delivered to the power that could be delivered if the engine operated without any power loss.

engine fan See *fan.*

engine mounts Flexible rubber insulators through which the engine is bolted to the frame.

engine power The power available from the crankshaft to do work. Usually measured in units of horsepower and kilowatts.

engine tuneup A procedure for inspecting, testing, and adjusting an engine, and replacing any worn parts, to restore the engine to its best performance.

ethylene glycol Chemical name of a widely used type of permanent antifreeze.

exhaust emissions Pollutants emitted into the atmosphere through any opening downstream of the exhaust ports of an engine.

exhaust gas The burned and unburned gases that remain (from the air-fuel mixture) after combustion.

exhaust manifold A device with several passages through which exhaust gases leave the engine combustion chambers and enter the exhaust piping system.

exhaust pipe The pipe connecting the exhaust manifold to the next component in the exhaust system.

exhaust stroke The piston stroke (from BDC to TDC) immediately following the power stroke, during which the exhaust valve is open so that the exhaust gases can escape from the cylinder.

exhaust system The system that collects the exhaust gases and discharges them into the air. May include the exhaust manifold, exhaust pipe, muffler, tail pipe, and spark arrestor.

exhaust valve The valve that is open during the exhaust stroke, allowing burned gases to flow from the cylinder.

expansion plug A slightly dished plug that is used to seal core passages in the cylinder block and cylinder head of a liquid-cooled engine. When driven into place, the expansion plug is flattened and expanded to fit tightly.

expansion tank A tank connected by a hose to the filler neck of the radiator on a liquid-cooled engine; the tank provides room for heated coolant to expand and to give off any air that may be trapped in the coolant.

fan The bladed device that rotates in back of the radiator to draw cooling air through the radiator (or around the engine cylinders); an air blower.

fastener Any device that holds or joins other parts together.

fatigue failure A type of metal failure resulting from repeated stress which finally alters the character of the metal so that it cracks. In engine bearings, frequently caused by excessive idling or slow engine idle speed.

field coil A coil, or winding, in a generator or starting motor which produces a magnetic field as current passes through it.

filter A device through which air, gases, or liquids are passed to remove impurities.

fins On a radiator or heat exchanger, thin metal projections over which cooling air flows to remove heat from hot liquid flowing through internal passages. On an air-cooled engine, thin metal projections on the cylinder and head which greatly increase the area of the heat-dissipating surfaces and help cool the engine.

firing order The order in which the engine cylinders fire, or deliver their power strokes, beginning with the number 1 cylinder.

flat-head engine See *L-head engine.*

flat rate Method of paying mechanics and technicians by use of a manual which indicates the time normally required to do each service job.

float bowl In a carburetor, the reservoir from which fuel is metered into the passing air.

float level The float position at which the needle valve closes the fuel inlet to the carburetor, to prevent further delivery of fuel.

float system In the carburetor, the system that controls the entry of fuel and the fuel level in the float bowl.

flooded Term used to indicate that the engine cylinders received "raw" or liquid gasoline, or an air-fuel mixture too rich to burn.

fluid Any liquid or gas.

flywheel The rotating metal wheel attached to the crankshaft which helps even out the power surges from the power strokes and many also serve as part of the clutch and the engine cranking system.

flywheel ring gear A gear, fitted around the flywheel, that is engaged by teeth on the starting-motor drive to crank the engine.

force Any push or pull exerted on an object; measured in pounds and ounces in the USCS or in newtons in the metric system.

formed-in-place gasket A gasket formed by a bead of plastic gasket material applied by a machine or squeezed from a tube.

four-cycle See *four-stroke cycle.*

four-stroke cycle The four piston strokes—intake, compression, power, and exhaust—that make up the complete cycle of events in the four-stroke-cycle engine. Also called *four-cycle* and *four-stroke.*

franchise A locally owned business that carries a nationally known name. The local business has been granted authority or license to market a product, service, or method, often exclusively within a certain area.

friction The resistance to motion between two bodies in contact with each other.

friction bearing Bearing in which there is sliding contact between the moving surfaces. Sleeve bearings, such as those used in connecting rods, are friction bearings.

friction horsepower The power that an engine uses to overcome its own internal friction.

fuel Any combustible substance. In a spark-ignition engine, the fuel (usually gasoline) is burned and the heat of combustion expands the resulting gases, which force the piston downward rotating the crankshaft.

fuel filter A device located in the fuel line that removes dirt and other contaminants and keeps them from passing through.

fuel gauge A gauge that indicates the amount of fuel in the fuel tank.

fuel-injection system A system that delivers fuel under pressure into the combustion chamber or into the intake airflow.

fuel line The pipe or tubes through which fuel flows from the fuel tank to the carburetor or fuel-injection system.

fuel nozzle The tube in the carburetor through which gasoline feeds from the float bowl into the passing air.

fuel pump The electric or mechanical device in the fuel system which forces fuel from the fuel tank to the carburetor or fuel-injection system.

fuel system The system that delivers the fuel and air to the engine cylinders. May include the fuel tank and lines, gauge, fuel pump, carburetor or fuel-injection system, and intake manifold.

fuel tank The storage tank for engine fuel.

full throttle Wide-open throttle position.

fuse A device designed to open an electric circuit when the current is excessive, to protect wiring and equipment in the circuit.

fusible link A type of circuit protector in which a special wire melts to open the circuit when excessive current flows.

gap The air space between two electrodes, as the spark-plug gap or the breaker-point gap.

gas The state of matter in which the matter has neither a definite shape nor a definite volume; air is a mixture of several gases. In the engine, the discharge from the muffler or tailpipe is called the *exhaust gas.* Also, *gas* is a slang expression for the liquid fuel gasoline.

gas engine An engine that uses a gas (*not* a liquid) for fuel.

gasket A thin layer of soft material, such as paper, cork, rubber, or copper, placed between two flat surfaces to make a tight seal.

gasket cement A liquid adhesive material, or sealer, used to install gaskets; in some applications, a layer of gasket cement is used as the gasket.

gasohol An engine fuel made by mixing 10 percent ethyl alcohol with 90 percent unleaded gasoline.

gasoline A liquid blend of hydrocarbons, obtained from crude oil; used as the fuel for most spark-ignition engines.

gassing Hydrogen gas escaping from a battery during charging.

gear ratio The number of revolutions of a driving gear required to turn a driven gear through one complete revolution. For a pair of gears, the ratio is found by dividing the number of teeth on the driven gear by the number of teeth on the driving gear.

gears Mechanical devices that transmit power or turning force from one shaft to another; gears contain teeth that mesh as the gears turn.

gear-type pump A pump using a pair of matching gears that rotate; meshing of the gears forces oil (or other liquid) from between the teeth through the pump outlet.

generator A device that converts mechanical energy into electric energy; can produce either ac or dc electricity.

glaze The very smooth, mirrorlike finish that develops on engine cylinder walls.

glow plug A small electric heater installed in the precombustion chamber of a diesel engine to preheat the chamber for easier starting in cold weather.

governor A device that controls, or governs, another device, usually on the basis of speed or load.

grease Lubricating oil to which thickening agents have been added.

ground The return path for current in an electric circuit.

ground-return system System of wiring in which the engine and frame are used as part of the electric return circuit to the battery or alternator; also known as the single-wire system.

HC See *hydrocarbon.*

head See *cylinder head.*

heat A form of energy released by the burning of fuel.

heat of compression Increase of temperature brought about by the compression of air or air-fuel mixture; the source of ignition in a diesel engine.

Heli-Coil See *thread insert.*

hone A tool with abrasive stones that is rotated in a bore or bushing to remove metal.

horsepower A measure of mechanical power, or the rate at which work is done. One horsepower equals 33,000 ft-lb (foot-pounds) of work per minute, the power necessary to raise 33,000 pounds a distance of 1 foot in 1 minute.

hydraulic pressure Pressure exerted through the medium of a liquid.

hydraulics The use of a liquid under pressure to transfer force or motion, or to increase an applied force.

hydraulic valve lifter A valve lifter that uses oil pressure from the engine lubricating system to keep the lifter in constant contact with the cam lobe and with the valve stem, pushrod, or rocker arm.

hydrocarbon (HC) A compound containing only carbon and hydrogen atoms, usually derived from fossil fuels such as petroleum, natural gas, and coal; an agent in the formation of photochemical smog. Gasoline is a blend of liquid hydrocarbons refined from crude oil.

hydrogen (H) A colorless, odorless, highly flammable gas whose combustion produces water; the simplest and lightest element.

hydrometer A device used to measure specific gravity. In engine servicing, a device used to measure the specific gravity of battery electrolyte to determine the battery's state of charge; also, a device used to measure the specific gravity of coolant to determine its freezing temperature.

idle mixture The air-fuel mixture supplied to the engine during idling.

idle-mixture screw The adjustment screw that can be turned in or out to lean or enrich the idle mixture.

idle speed The speed, or rpm, at which the engine runs without load when the throttle is closed.

idle system In the carburetor, the passages through which fuel is delivered while the engine is idling.

ignition The action of the spark in starting the burning of the compressed air-fuel mixture in the combustion chamber of a spark-ignition engine. In a diesel engine, the start of the burning of fuel after its temperature has been raised by the heat of compression.

ignition advance The moving forward, in time, of the ignition spark relative to the piston position. TDC or 1° ATDC is considered advanced as compared with 2° ATDC.

ignition coil The ignition-system component that acts as a transformer to step up (increase) the primary voltage to many thousands of volts. The high-voltage surge from the coil is transmitted to the spark plug to ignite the compressed air-fuel mixture.

ignition distributor The unit in the ignition system of a multicylinder engine which usually contains the mechanical or electronic switch that closes and opens the primary circuit to the coil at the proper time and then distributes the resulting high-voltage surges to the spark plugs.

ignition lag In a diesel engine, the delay in time between the injection of fuel and the start of combustion.

ignition switch The switch in the ignition system (usually operated with a key) that opens and closes the ignition-coil primary circuit. May also be used to open and close other electric circuits.

ignition system The system that furnishes high-voltage sparks to the engine cylinders to fire the compressed air-fuel mixture. May include the battery, ignition coil, ignition distributor, ignition switch, wiring, and spark plug.

ignition timing The delivery of the spark from the coil to the spark plug at the proper time for the power stroke, relative to the piston position.

I-head engine An overhead-valve engine; an engine with the valves in the cylinder head.

indicator A device using a light or a dial and pointer to make some condition known; for example, the temperature indicator or oil-pressure indicator.

induction The action of producing a voltage in a conductor or coil by moving the conductor or coil through a magnetic field, or by moving the field past the conductor or coil.

inertia The property of an object that causes it to resist any change in its speed or direction of travel.

in-line engine A multicylinder engine in which all the cylinders are located in a single row or line.

inspect To examine a part or system for condition or function.

install To set up for use on a vehicle any part, accessory, option, or kit.

insulation Material that stops the travel of electricity (electrical insulation) or heat (heat insulation).

insulator A poor conductor of electricity or heat.

intake manifold A casting or assembly of several passages through which air or air-fuel mixture flows from the air intake or carburetor to the ports in the cylinder head or cylinder block.

intake stroke The piston stroke from TDC to BDC immediately following the exhaust stroke, during which the intake valve is open and the cylinder fills with air or air-fuel mixture.

integral Built into, as part of the whole.

integrated circuit Many very small solid-state devices capable of performing as a complete electronic circuit, with one or more integrated circuits manufactured as a "chip."

internal-combustion (IC) engine An engine in which the fuel is burned inside the engine itself, rather than in a separate device (as in a steam engine).

jet A calibrated passage in the carburetor through which fuel flows.

journal The part of a rotating shaft which turns in a bearing.

jump-starting Staring an engine that has a dead battery by connecting a charged battery to the starting system.

key A wedgelike metal piece, usually rectangular or semicircular, inserted in grooves to hold two parts in relative position and which may be used to transmit torque. The small strip of metal with coded peaks and grooves used to operate a lock, such as in the ignition switch.

kilogram (kg) In the metric system, a unit of weight and mass, approximately equal to 2.2 pounds.

kilometer (km) In the metric system, a unit of linear measure, equal to 0.621 mile.

kilowatt (kW) A unit of power, equal to about 1.34 hp.

kinetic energy The energy of motion; the energy stored in a moving body through its momentum; for example, the kinetic energy stored in a rotating flywheel.

knock A heavy metallic engine sound that varies with engine speed; usually caused by a loose or worn bearing; name also used for *detonation, pinging,* and *spark knock.* See *detonation.*

kW See *kilowatt.*

laminated Made up of several thin sheets or layers.

lapping Coating the valve face with lapping compound and then turning the valve back and forth on the seat to check the fit.

lash The amount of free motion in a gear train, between gears, or in a mechanical assembly, such as the lash in a valve train.

lead (pronounced "leed") A cable or conductor that carries electric current.

lead (pronounced "led") A heavy metal; used in lead-acid storage batteries.

leaded gasoline Gasoline to which small amounts of tetraethyl lead are added to improve engine performance and reduce detonation.

lean mixture An air-fuel mixture that has a relatively high proportion of air and a relatively low proportion of fuel. An air-fuel ratio of 16 : 1 is a lean mixture, compared with an air-fuel mixture of 13 : 1.

L-head engine A type of engine in which the valves are located in the cylinder block.

lifter See *valve lifter.*

lines of force See *magnetic lines of force.*

linkage A hydraulic system, or an assembly of rods and links, used to transmit motion.

liquefied petroleum gas (LPG) An engine fuel obtained from petroleum and natural gas; stored as a liquid under pressure, it vaporizes at atmospheric pressure. Butane and propane are liquefied gases used as engine fuels.

liquid-cooled engine An engine that is cooled by the circulation of liquid coolant around the cylinders.

liter (L) In the metric system, a measure of volume; approximately equal to 0.26 gallon (U.S.), or about 61 cubic inches. Used as a metric measure of engine-cylinder displacement.

lobe Projecting part; for example, the rotor lobe or the cam lobe.

LPG See *liquefied petroleum gas.*

lubricating system The system in the engine that supplies engine parts with lubricating oil to prevent contact between any two moving metal surfaces.

lugging Low-speed, full-throttle engine operation in which the engine is heavily loaded and overworked.

machining The process of using a machine to remove metal from a metal part.

magnetic Having the ability to attract iron. This ability may be permanent, or it may depend on a current flow, as in an electromagnet.

magnetic field The space around a magnet which is filled by invisible lines of force.

magnetic lines of force The imaginary lines by which a magnetic field may be visualized.

magnetic switch A switch with a winding which, when energized by connection to a battery, generator, or alternator, causes the switch to open or close a circuit.

magnetism The ability, either natural or produced by a flow of electric current, to attract iron.

magneto An engine-driven device that generates its own primary current, transforms that current into high-voltage surges, and delivers them to the proper spark plugs.

main bearings In the engine, the bearings that support the crankshaft.

main jet The fuel nozzle, or jet, in the carburetor that supplies the fuel when the throttle is partially to fully open.

maintenance-free battery A battery without removable vent plugs so that water cannot be added.

malfunction Improper or incorrect operation.

manifold A device with several inlet or outlet passageways through which a gas or liquid is gathered or distributed. See *exhaust manifold, intake manifold.*

manifold vacuum The vacuum in the intake manifold that develops as a result of the vacuum in the cylinders on their intake strokes.

matter Anything that has weight and occupies space.

measuring The act of determining the size, capacity or quantity of an object.

mechanical efficiency In an engine, the ratio between brake horsepower and indicated horsepower.

mechanism A system of interrelated parts that make up a working assembly.

member Any essential part of a machine or assembly.

meshing The mating, or engaging, of the teeth of two gears.

meter (m) A unit of linear measure in the metric system, equal to 39.37 inches. Also, the name given to any test instrument that measures a property of a substance passing through it, as an ammeter measures electric current. Also, any device that measures and controls the flow of a substance passing through it, as a carburetor jet meters fuel flow.

micrometer A precision measuring device that measures small distances, such as crankshaft or cylinder-bore diameter or thickness of an object. Also called a *mike.*

millimeter (mm) In the metric system, a unit of linear measure approximately equal to 0.039 inch.

misfire In the engine, a failure to ignite the air-fuel mixture in one or more cylinders. This condition may be intermittent or continuous.

miss See *misfire.*

modification An alteration; a change from the original.

motor A device that converts electric energy into mechanical energy; for example, a starting motor.

muffler In the engine exhaust system, a device through which the exhaust gases must pass and which reduces the exhaust noise.

multiple-viscosity oil An engine oil that has a low viscosity when cold (for easier cranking) and a higher viscosity when hot (to provide adequate engine lubrication).

mutual induction The condition in which a voltage is induced in one coil by a changing magnetic field caused by a changing current in another coil. The magnitude of the induced voltage depends on the number of turns in the two coils.

needle bearing An antifriction bearing of the roller type, in which the rollers are very small (needle size) in diameter.

needle valve A small, tapered, needle-pointed valve that can move into or out of a seat to close or open the passage through it. Used to control the fuel level in the carburetor float bowl.

negative One of the two poles of a magnet, or one of the two terminals of an electric device.

negative terminal The terminal from which electrons flow in a complete circuit. On a battery, the negative terminal can be identified as the battery post with the smaller diameter; the minus sign (−) is often also used to identify the negative terminal.

neutral-start switch A switch wired into the ignition system to prevent engine cranking unless the transmission selector lever is in NEUTRAL.

nitrogen (N) A colorless, tasteless, odorless gas that constitutes 78 percent of the atmosphere by volume and is part of all living tissues.

nitrogen oxides (NO_x) Any chemical compound of nitrogen and oxygen; a basic air pollutant.

nonconductor Same as *insulator.*

north pole The pole from which the lines of force leave a magnet.

NO_x See *nitrogen oxides.*

nozzle The opening, or jet, through which fuel or air passes as it is discharged.

octane rating A measure of the antiknock properties of gasoline. The higher the octane rating, the more resistant the gasoline is to spark knock or detonation.

OHC See *overhead-camshaft engine.*

ohm The unit of electrical resistance.

ohmmeter An instrument used to measure electrical resistance.

OHV See *overhead-valve engine.*

oil A liquid lubricant usually made from crude oil and used for lubrication between moving parts. In a diesel engine, oil is used for fuel.

oil clearance The space between the bearing and the shaft rotating within it.

oil cooler A small radiator that lowers the temperature of oil flowing through it.

oil dilution Thinning of oil in the crankcase; caused by liquid fuel from the combustion chamber leaking past the piston rings.

oil filter A filter that removes impurities from the engine oil passing through it.

oil-level indicator The dipstick that is removed and inspected to check the level of oil in the crankcase of an engine or compressor.

oil pan On some four-cycle engines, the detachable lower part of the engine which encloses the crankcase and acts as an oil reservoir.

oil-pressure indicator A gauge that indicates the oil pressure in the engine lubricating system, or a light that comes on if the oil pressure drops too low.

oil pump In the lubricating system, the device that forces oil from the crankcase to the moving engine parts.

oil pumping Leakage of oil past the piston rings and into the combustion chamber, usually as a result of defective rings or worn cylinder walls.

oil ring The lower ring or rings on a piston; designed to prevent excessive oil from working up the cylinder walls and into the combustion chamber. Also called an *oil-control ring.*

oil seal A seal placed around a rotating shaft or other moving part to prevent leakage of oil.

oil strainer A wire-mesh screen placed at the inlet end of the oil-pump pickup tube to prevent dirt and other large particles from entering the oil pump.

oil viscosity A rating of an oil's resistance to flow. The higher the number, the higher the viscosity.

one-wire system Use of the engine and frame as a path for the ground side of the electric circuits; eliminates the need for a second wire as a return path to the battery, generator, or alternator.

open circuit In an electric circuit, a break or opening which prevents the passage of current.

orifice A small opening, or hole, into a cavity.

O-ring A type of sealing ring made of a special rubberlike material; in use, the O-ring is compressed into a groove to provide the sealing action.

Otto cycle The cycle of events in a four-stroke-cycle engine. Named for the inventor, Dr. Nikolaus Otto.

overcharging Continued charging of a battery after it has reached the charged condition. This action damages the battery and shortens its life.

overhaul To disassemble a unit completely, clean and inspect all parts, reassemble it with the original or new parts, and make all adjustments necessary for proper operation.

overhead-camshaft (OHC) engine An engine in which the camshaft is mounted over the cylinder head, instead of inside the cylinder block.

overhead-valve (OHV) engine An engine in which the valves are mounted in the cylinder head above the combustion chamber, instead of in the cylinder block. In this type of engine, the camshaft is mounted in the cylinder block, and the valves are actuated by pushrods.

overrunning clutch A drive unit that transmits rotary motion in one direction only. In the other direction, the driving member overruns and does not pass the motion to the other member. Widely used as the drive mechanism for starting motors.

oxides of nitrogen See *nitrogen oxides.*

oxygen (O) A colorless, tasteless, odorless, gaseous element that makes up about 21 percent of air. Capable of combining rapidly with all elements except the inert gases in the oxidation process that is called *burning*. Combines very slowly with many metals in the oxidation process that is sometimes called *rusting*.

pan See *oil pan*.

pancake engine An engine with two rows of cylinders which are opposed and on the same plane, usually set horizontally.

parallel The quality of objects being the same distance from each other at all points; usually applied to lines and, in engine work, to machined surfaces.

parallel circuit The electric circuit formed when two or more electric devices have their terminals connected, positive to positive and negative to negative, so that each may operate independently of the others from the same power source.

particle A very small piece of metal, dirt, or other impurity which may be contained in the air, fuel, or lubricating oil used in an engine.

passage A small hole or gallery in an assembly or casting, through which air, coolant, fuel, or oil flows.

petroleum The crude oil from which gasoline, lubricating oil, and other such products are refined.

ping Characteristic sound of engine spark knock or detonation. Caused by excessive advance of ignition timing or low-octane fuel.

pinion gear The smaller of two meshing gears.

pin press A small press used to force the piston pin in and out of the piston.

piston A movable part, fitted to a cylinder, which can receive or transmit motion as a result of pressure changes in a fluid. In the engine, the round plug that slides up and down in the cylinder and which, through the connecting rod, forces the crankshaft to rotate.

piston clearance The space between the piston and the cylinder wall.

piston displacement The cylinder volume displaced by the piston as it moves from the bottom to the top of the cylinder during one complete stroke.

piston pin The cylindrical or tubular metal piece that attaches the piston to the connecting rod. Also called the *wrist pin*.

piston rings Rings fitted into grooves in the piston. There are two types: compression rings, for sealing the compression pressure in the combustion chamber, and oil rings, for scraping excessive oil off the cylinder wall. See *compression ring* and *oil ring*.

piston skirt The lower part of the piston, below the piston-pin hole.

piston slap Hollow, muffled, bell-like sound made by an excessively loose piston slapping the cylinder wall.

pivot A pin or shaft upon which another part rests or turns.

Plastigage A plastic material available in strips of various sizes; used to measure crankshaft main-bearing and connecting-rod-bearing clearances.

plate In a battery, a rectangular sheet of sponge lead. Sulfuric acid in the electrolyte chemically reacts with the lead to produce an electric current.

polarity The quality of an electric component or circuit that determines the direction of current flow.

pollutant Any substance that adds to the pollution of the atmosphere; any such substance that is in the exhaust gas from the engine or that evaporates from the fuel tank or carburetor.

pollution Any gas or substance in the air which makes it less fit to breathe. Also, *noise pollution* is the name applied to excessive noise from machinery or engines.

poppet valve A mushroom-shaped valve, used in four-cycle engines.

port In the engine, the passage to the cylinder opened and closed by a valve, and through which gases flow to enter and leave the cylinder.

positive crankcase ventilation (PCV) A crankcase ventilation system that uses an intake-manifold vacuum to return the crankcase vapors and blowby gases from the crankcase to the intake manifold to be burned, thereby preventing their escape into the atmosphere.

positive terminal The terminal to which electrons flow in a complete electric circuit. On a battery, the positive terminal can be identified as the battery post with the larger diameter; the plus sign (+) is also used to identify the positive terminal.

post A point at which a cable is connected to the battery.

power The rate at which work is done. A common power unit is the horsepower, which is equal to 33,000 ft-lb/min (foot-pounds per minute).

power plant The engine, or power-producing mechanism.

power stroke The piston stroke from TDC to BDC immediately following the compression stroke, during which both valves are closed and the fuel burns, expanding the compressed air, thereby forcing the piston down to transmit power to the crankshaft.

precombustion chamber In some engines, a separate small combustion chamber where combustion begins.

preignition Ignition of the air-fuel mixture in the combustion chamber by some unwanted means, before the ignition spark occurs at the spark plug.

preload In bearings, the amount of load placed on a bearing before actual operating loads are imposed. Proper preloading requires bearing adjustment and ensures alignment and minimum looseness in the system.

press fit A fit so tight that the pin must be pressed into place, usually with a shop press.

pressure Force per unit area, or force divided by area. Usually measured in pounds per square inch (psi) and kilopascals (kPa).

pressure cap A radiator cap with valves which cause the liquid-cooling system to operate under pressure at a higher and more efficient temperature.

pressure-feed oil system A type of engine lubricating system that uses an oil pump to force oil through tubes and passages to the various engine parts requiring lubrication.

pressure regulator A device that operates to prevent excessive pressure from developing. In hydraulic systems, a valve that opens to release oil from a line when the oil pressure reaches a specified maximum.

pressure-relief valve A valve in the oil line that opens to relieve excessive pressure.

pressure tester An instrument that clamps in the radiator filler neck; used to pressure-test the cooling system for leaks.

pressurize To apply more than atmospheric pressure to a gas or liquid.

prevailing torque fasteners Nuts and bolts designed to have a continuous resistance to turning.

primary The low-voltage circuit of the ignition system.

primary winding The outer winding of relatively heavy wire in an ignition coil.

psi Abbreviation for pounds per square inch, a measure of fluid pressure.

pulley A small wheel with a V-shaped groove around the rim; drives, or is driven by, a belt.

pump A device that transfers gas or liquid from one place to another.

pushrod In the overhead-valve engine, the rod between the valve lifter and the rocker arm; transmits cam-lobe lift.

quench The space in some combustion chambers which absorbs enough heat to quench, or extinguish, the combustion flame front as it approaches a relatively cold cylinder wall. This prevents detonation of the end gas, but results in hydrocarbon emissions.

quick charger A battery charger that produces a high charging current which charges, or boosts, a battery in a short time.

races The metal rings on which ball or roller bearings rotate.

radiator In the liquid-cooling system, the heat exchanger that removes heat from coolant passing through it; takes hot coolant from the engine and returns the coolant to the engine at a lower temperature.

radiator pressure cap See *pressure cap.*

ratio Proportion; the relative amounts of two or more substances in a mixture. Usually expressed as a numerical relationship, as in 2:1.

readout The visual delivery or display of information from an electronic device, circuit, or system.

reassembly Putting back together the parts of an assembly, device, or unit.

rebore To increase the diameter of a cylinder.

recharging The action of forcing electric current into a battery in the direction opposite to that in which current normally flows during use. Reverses the chemical reaction between the plates and electrolyte.

reciprocating motion Motion of an object between two limiting positions; motion in a straight line, back and forth or up and down.

rectifier A device that changes alternating current to direct current; diodes are used as rectifiers in ac charging systems.

reed valve A type of valve used in the crankcase of some two-cycle engines. Air-fuel mixture enters the crankcase through the reed valve, which then closes as pressure builds up in the crankcase.

regulator In the charging system, a device that controls generator or alternator output to prevent excessive voltage.

relay An electric device that opens or closes a circuit or circuits in response to a voltage signal.

relief valve A valve that opens when a preset pressure is reached. This relieves or prevents excessive pressures.

remove and reinstall (R and R) To perform a series of servicing procedures on an original part or assembly; includes removal, inspection, lubrication, all necessary adjustments, and reinstallation.

replace To remove a used part or assembly and install a new part or assembly in its place; includes cleaning, lubricating, and adjusting as required.

resistance The opposition to a flow of current through a circuit or electric device; measured in ohms. A voltage of one volt will cause one ampere to flow through a resistance of one ohm. This is known as *Ohm's law,* which can be written in three ways: amperes = volts/ohms; ohms = volts/amperes; and volts = amperes × ohms.

retard To delay the occurrence of the spark in the combustion chamber. Usually associated with the engine spark-timing mechanisms; the opposite of spark advance.

return spring A pull-back spring.

rich mixture An air-fuel mixture that has a relatively high proportion of fuel and a relatively low proportion of air. An air-fuel ratio of 13:1 indicates a rich mixture, compared with an air-fuel ratio of 16:1.

ring See *compression ring* and *oil ring.*

ring gap The gap between the ends of the piston ring when it is in place in the cylinder.

ring grooves Grooves cut in a piston, into which the piston rings are assembled.

ring ridge The ridge formed at the top of a cylinder as the cylinder wall below is worn away by piston-ring movement.

rocker arm In some engines with the valves in the cylinder head, a part in the valve train that the cam lobe causes to rock or pivot, thereby opening the valve.

rod bearing In an engine, the split insert-type bearing in the connecting rod in which a crankpin of the crankshaft rotates. Also called a *connecting-rod bearing.*

roller tappet A valve lifter with a hardened steel roller on the end riding against the camshaft.

room temperature 68 to 72°F [20 to 22°C].

room-temperature-vulcanizing sealant See *aerobic gasket material.*

rotary The motion of a part that continually rotates or turns.

rotor A revolving part of a machine, such as an alternator rotor or distributor rotor.

rotor oil pump A type of oil pump using a pair of rotors, one inside the other, to produce the oil pressure required to circulate oil to engine parts.

rpm Abbreviation for revolutions per minute, a measure of rotational speed.

RTV sealant See *aerobic gasket material.*

run-on See *dieseling.*

runout Wobble.

SAE Abbreviation for Society of Automotive Engineers. Used to indicate a grade or weight of oil measured according to Society of Automotive Engineers standards.

schematic A pictorial representation, most often in the form of a line drawing. A systematic positioning of components and their relationship to one another or to the overall function.

scored Scratched or grooved, as a cylinder may be scored by abrasive particles moved up and down by the piston rings.

screen A fine-mesh screen in the fuel and lubricating systems that prevents large particles from entering the system.

scuffing A type of wear of moving parts, characterized by transfer of metal from one to the other part and pits or grooves in the mating surfaces.

seal A material, shaped around a shaft, used to close off the operating compartment of the shaft, preventing fluid leakage.

sealer A thick, tacky compound, usually spread with a brush, which may be used as a gasket, or sealant, to seal small openings or surface irregularities.

seat The surface upon which another part rests, as a valve seat. Also, to wear into a good fit.

secondary circuit The high-voltage circuit of the ignition system; may include the coil secondary winding, rotor, distributor cap, spark-plug cables, and spark plugs.

segments The copper bars of a commutator.

self-discharge Chemical activity in the battery which causes the battery to discharge even though it is not furnishing current to a load.

self-induction The inducing of a voltage in a current-carrying coil of wire because the current in that wire is changing.

semiconductor A material that acts as an insulator under some conditions and as a conductor under other conditions.

sensor Any device that receives and reacts to a signal, such as a change in voltage, temperature, or pressure.

series circuit An electric circuit in which the devices are connected end to end, positive terminal to negative terminal. The same current flows through all the devices in the circuit.

service manual The book published by each manufacturer listing specifications and service procedures for each make and model of engine built. Also called *shop manual.*

service rating A designation that indicates the type of service for which an engine lubricating oil is best suited.

shim A slotted strip of metal used as a spacer.

short circuit A defect in an electric circuit which permits current to take a short path, or circuit, instead of following the desired path.

shrink fit A tight fit of one part in another achieved by heating or cooling one part and then assembling it with the other part. If heated, the part then shrinks on cooling to provide the fit. If cooled, the parts expands on warming to provide the fit.

shroud A hood or cover placed around an engine fan to improve and direct air flow.

side clearance The clearance between the sides of moving parts when the sides do not serve as load-carrying surfaces.

silicone rubber See *aerobic gasket material.*

sludge An accumulation of water, dirt, and oil in the crankcase; sludge is very viscous and tends to reduce lubrication.

smog A term coined from the words *smoke* and *fog.* First applied to the foglike layer that hangs in the air under certain atmospheric conditions; now often used to describe any condition of dirty air and/or fumes or smoke.

smoke Small gas-borne or airborne particles, exclusive of water vapor, that result from combustion; such particles emitted by an engine into the atmosphere in sufficient quantities to be visible.

solenoid An electromechanical device which, when connected to an electric source such as a battery, produces a mechanical movement. This movement can be used to control a valve or to produce other movements.

solenoid switch A switch that is opened and closed magnetically, by the movement of a solenoid core. Usually, the core also causes a mechanical action, such as the movement of the starting-motor drive pinion into mesh with flywheel teeth for cranking.

solid-state device A device that has no moving parts except electrons. Diodes and transistors are examples.

solvent A cold liquid cleaner used to wash away grease and dirt.

south pole The pole at which magnetic lines of force enter a magnet.

spark advance See *centrifugal advance* and *vacuum advance.*

spark-ignition (SI) engine An engine operating on the Otto cycle, in which the fuel is ignited by the heat from an electric spark as it jumps the gap at the end of the spark plug.

spark knock See *detonation.*

spark plug The assembly, which includes a pair of electrodes and an insulator, that provides a spark gap in the engine cylinder.

spark-plug heat range The distance heat must travel from the center electrode to reach the outer shell of the spark plug and enter the cylinder head.

spark test A quick check of the ignition system; made by holding the metal spark-plug end of a spark-plug cable about ¼-inch [6 mm] from the cylinder head, or block; cranking the engine; and checking for a spark.

specifications Information provided by the manufacturer for each engine and its components, operation, and clearances. Also, the service procedures that must be followed for a system to operate properly.

specific gravity The weight per unit volume of a substance as compared with the weight per unit volume of water.

specs Short for specifications.

splash-feed oil system A type of engine lubricating system that depends on splashing of the oil for lubricating moving engine parts.

splines Slots or grooves cut in a shaft or bore. Splines on a shaft are matched to splines in a bore, to ensure that the two parts turn together.

spring A device that changes shape under stress or force but returns to its original shape when the stress or force is removed.

spring retainer In the valve train, the piece of metal that holds the spring in place and is itself locked in place by the valve-spring retainer locks.

squish The action in some combustion chambers in which the last part of the compressed mixture is pushed, or squirts, out of a decreasing space between the piston and cylinder head.

starter See *starting motor.*

starting motor The electric motor that cranks the engine, or turns the crankshaft, for starting.

starting-motor drive The drive mechanism and gear on the end of the starting-motor armature shaft; used to couple the starting motor to, and disengage it from, the flywheel ring-gear teeth.

static balance The balance of an object while it is not moving.

stator The stationary member of a machine, such as an electric motor or generator, in or about which a rotor revolves.

stoichiometric ratio In a spark-ignition engine, the ideal air-fuel-mixture ratio of 14.7 : 1, which must be maintained on automotive engines with dual-bed and three-way catalytic converters.

storage battery A lead-acid battery that changes chemical energy into electric energy; that part of the electric system which acts as a reservoir for electric energy, storing it in chemical form.

stratified charge In a spark-ignition engine, an air-fuel charge with a small layer or pocket of rich air-fuel mixture. The rich mixture is ignited first; then ignition spreads to the leaner mixture filling the rest of the combustion chamber. The diesel engine is a stratified-charge engine.

stroke In an engine cylinder, the distance that the piston moves in traveling from BDC to TDC or from TDC to BDC.

sulfation The lead sulfate that forms on battery plates as a result of the battery action that produces electric current.

sulfuric acid See *electrolyte.*

swept volume See *piston displacement.*

switch A device that opens and closes an electric circuit.

synthetic oil An artificial oil that is manufactured; not a natural mineral oil made from petroleum.

tachometer A device for measuring engine speed, or rpm.

tank unit The part of the fuel-level-indicating system that is mounted in the fuel tank.

taper A gradual reduction in the width of a shaft or hole; in an engine cylinder, uneven wear, with more at the top than at the bottom.

tappet See *valve lifter.*

tappet noise A regular clicking noise in the valve train that increases with engine speed.

TDC Top dead center.

temperature The measure of heat intensity, in degrees. Temperature is *not* a measure of heat quantity.

temperature indicator A gauge that indicates the temperature of the engine coolant, or a light that comes on if the coolant gets too hot.

tetraethyl lead A chemical which, when added to gasoline, increases its octane rating, and reduces its tendency to detonate.

thermal Of or pertaining to heat.

thermal efficiency Relationship between the power output and the energy in the fuel burned to produce the output.

thermostat A device for the automatic regulation of temperature; usually contains a temperature-sensitive element that expands or contracts to open or close off the flow of air, a gas, or a liquid.

thermostatic gauge An indicating device (for fuel quantity, oil pressure, coolant temperature) that contains a thermostatic blade or blades.

thread insert A threaded coil that is used to restore the original thread size to a hole with damaged threads; the hole is drilled oversize and tapped; then the insert is threaded into the tapped hole.

throttle valve The round disk valve or plate in the carburetor throttle body that can be turned to admit more or less air, thereby controlling the speed of a spark-ignition engine.

thrust bearing In the engine, the main bearing that has thrust faces to prevent excessive end play, or forward and backward movement of the crankshaft.

timing In an engine, delivery of the ignition spark or operation of the valves (in relation to the piston position) for the power stroke. See *ignition timing* and *valve timing.*

timing belt A toothed belt that is driven by a sprocket on the crankshaft and that drives the sprocket on the camshaft.

timing chain A chain that is driven by a sprocket on the crankshaft and that drives the sprocket on the camshaft.

timing gear A gear on the crankshaft that drives the camshaft by meshing with a gear on it.

timing light A light that can be connected to the ignition system to flash each time the number 1 spark plug fires; used for adjusting the timing of the ignition spark.

timing marks Marks on gears, sprockets, and other parts that must be properly aligned when the engine is assembled. Lines or numbers on the crankshaft pulley or flywheel used to adjust ignition timing so that the spark plugs fire at the right time.

top dead center The piston position when the piston has reached the upper limit of its travel in the cylinder and the center line of the connecting rod is parallel to the cylinder walls.

torque Turning or twisting force, usually measured in pound-feet, kilogram-meters, or newton-meters.

transistor A solid-state electronic device that can be used as an electric switch or as an amplifier.

trouble diagnosis The detective work necessary to find the cause of a trouble.

tuneup A procedure for inspecting, testing, and adjusting an engine, and replacing any worn parts, to restore engine performance.

turbulence The state of being violently disturbed. In the engine, the rapid swirling motion imparted to the air-fuel mixture entering the cylinder.

two-stroke cycle The two piston strokes during which fuel and air intake, compression, combustion, and exhaust take place in a two-stroke-cycle engine.

unit An assembly or device that can perform its function only if it is not further divided into its components.

unleaded gasoline Gasoline to which no lead compounds have been intentionally added.

vacuum Negative gauge pressure, or a pressure less than atmospheric pressure. Vacuum can be measured in pounds per square inch (psi) but is usually measured in inches or millimeters of mercury (Hg); a reading of 30 inches [762 mm] Hg would indicate a perfect vacuum.

vacuum advance The advancing (or retarding) of ignition timing by changes in intake-manifold vacuum, reflecting throttle opening and engine load. Also, a mechanism on the ignition distributor that uses intake-manifold vacuum to advance the timing of the spark to the spark plugs.

vacuum gauge A device that measures intake-manifold vacuum and thereby indicates actions of engine components.

valve A device that can be opened or closed to allow or stop the flow of a liquid or gas.

valve clearance The clearance in the valve train when the valve is closed.

valve float A condition in which the engine valves do not close completely or fail to close at the proper time.

valve grinding Refacing a valve in a valve-refacing machine.

valve guide The cylindrical part in the cylinder block or head in which the valve is assembled and moves up and down.

valve lash See *valve clearance.*

valve lifter A cylindrical part of the engine which rests on a cam of the camshaft and is lifted, by cam action, so that the valve is opened. Also called a *lifter, tappet, valve tappet,* or *cam follower.*

valve overlap The number of degrees of crankshaft rotation during which the intake and exhaust valves are open at the same time.

valve rotator A device installed in place of the valve-spring retainer, which turns the valve slightly as it opens.

valve seat The surface against which the valve comes to rest to provide a seal against leakage.

valve-seat insert Metal ring installed in the cylinder head or block to act as a valve seat.

valve spring The spring in each valve assembly which has the job of closing the valve.

valve-spring retainer The device on the valve stem that holds the spring in place.

valve-spring-retainer lock The locking device on the valve stem that locks the spring retainer in place.

valve stem The long, thin section of the valve that fits in the valve guide.

valve-stem seal A device placed on or around the valve stem to reduce the amount of oil that can get on the stem and then work its way down into the combustion chamber.

valve tappet See *valve lifter*.

valve timing The timing of the opening and closing of the valves in relation to the piston position.

valve train The valve-operating mechanism of a four-cycle engine; includes all components from the camshaft to the valve.

vapor A gas; any substance in the gaseous state, as distinguished from the liquid or solid state.

vapor lock A condition in the fuel system in which gasoline vaporizes in the fuel line or fuel pump; bubbles of gasoline vapor restrict or prevent fuel delivery to the carburetor.

vent An opening through which air can leave an enclosed chamber.

venturi In the carburetor, a narrowed passageway or restriction that increases the velocity of air moving through it; produces the vacuum responsible for the discharge of fuel from the fuel nozzle.

vibration A rapid back-and-forth motion; an oscillation.

viscosity The resistance to flow exhibited by a liquid. A thick oil has greater viscosity than a thin oil.

viscosity rating An indicator of the viscosity of engine oil. There are separate ratings for winter use and for summer use. The winter grades are SAE5W, SAE10W, and SAE20W. The summer grades are SAE20, SAE30, SAE40, and SAE50. Many oils have multiple viscosity ratings, as, for example, SAE10W-30.

viscous Thick; tending to resist flowing.

viscous friction Friction between layers of a liquid.

volatility A measure of the ease with which a liquid vaporizes; has a direct relationship to the flammability of a fuel.

voltage The force which causes electrons to flow in a conductor. The difference in electric pressure (or potential) between two points in a circuit.

voltage regulator A device that prevents excessive generator or alternator voltage by alternately inserting and removing a resistance in the field circuit.

voltmeter A device for measuring the potential difference (voltage) between two points, such as the terminals of a battery or alternator, or two points in an electric circuit.

volumetric efficiency A measure of how completely the cylinder fills during the intake stroke of a spark-ignition engine.

V-type engine An engine with two banks or rows of cylinders, set at an angle to form a V.

warranty work Repair work that the manufacturer agrees to pay for—if it is required—while the engine (or a certain part) is new.

water jackets The spaces between the inner and outer shells of the cylinder block or head, through which coolant circulates.

water pump In the cooling system, the device that circulates coolant between the engine water jackets and the radiator.

wiring harness A group of individually insulated wires, wrapped together to form a neat, easily installed bundle.

work The moving of an object against an opposing force; measured in foot-pounds, meter-kilograms, or joules. The product of a force and the distance through which the force acts.

WOT Abbreviation for wide-open throttle.

wrist pin Piston pin.

zener diode A special type of diode that will conduct current in its normally blocked (or reverse) direction under certain conditions.

Index